Student Solutions Guide for

Brief Calculus with Applications

Second Edition

Larson and Hostetler

Dianna L. Zook

D.C. Heath and Company
Lexington, Massachusetts   Toronto

Copyright © 1987 by D. C. Heath and Company.
Previous edition copyright © 1983.

All rights reserved. No part of this publication may be
reproduced or transmitted in any form or by any means,
electronic or mechanical, including photocopy, recording,
or any information storage or retrieval system, without
permission in writing from the publisher.

Published simultaneously in Canada.

Printed in the United States of America.

International Standard Book Number: 0-669-12061-8

Preface

This student solutions guide is a supplement to BRIEF CALCULUS WITH APPLICATIONS, Second Edition, by Roland E. Larson and Robert P. Hostetler. All references to chapters, sections, theorems, and definitions apply to the text. The purpose of this supplement is to guide you through calculus by providing the basic solution steps for each of the odd-numbered exercises in the text. This solutions guide is not intended to be used as a substitute for working homework problems yourself. It is one thing to be able to read and understand a solution, but quite another thing to be able to derive the solution on your own. To use this solutions guide correctly, I suggest the following study pattern.

1. Attempt each problem before looking at its solution.

2. Even after you have attempted to solve a problem, don't be too anxious to look up the solution. Check your work or try to come up with an alternative approach. It is through trial and error that you will sharpen your mathematical skills. These skills, like muscles, will develop only with consistent exercise.

3. Once you have completed an exercise, compare your solution with the one given in this guide. Sometimes the given solution will be more efficient than your own and will provide you with a better way of doing that type of problem in the future. Other times your own solution may require fewer steps and you can claim a small victory for your ingenuity. Be careful, however, not to sacrifice accuracy for efficiency.

Good luck in your study of calculus. If you have any corrections or other suggestions for improving this guide, I would appreciate hearing from you.

I would like to thank several people who helped in the production of this guide: David E. Heyd, who assisted the authors of the text and double checked my solutions, Linda L. Matta, who was in charge of the production, Timothy R. Larson, who proof read my solutions and prepared the art, and Linda M. Bollinger who typed the solutions. I would also like to thank the students in my mathematics classes. Finally, I would like to thank my husband Edward L. Schlindwein for his support during the many months I have worked on this project.

Dianna L. Zook
The Pennsylvania State University
The Behrend College
Erie, Pennsylvania 16563

Contents

CHAPTER 0 A Precalculus Review
- 0.1 The real line and order ... 1
- 0.2 Absolute value and distance on the real line ... 5
- 0.3 Exponents and radicals ... 9
- 0.4 Factoring and polynomials ... 13
- 0.5 Fractions and rationalization ... 20
- Review Exercises for Chapter 0 ... 26
- Practice Test for Chapter 0 ... 32

CHAPTER 1 Functions, Graphs, and Limits
- 1.1 The Cartesian plane and the Distance Formula ... 34
- 1.2 Graphs of equations ... 38
- 1.3 Lines in the plane, slope ... 46
- 1.4 Functions ... 52
- 1.5 Limits ... 59
- 1.6 Continuity ... 65
- Review Exercises for Chapter 1 ... 69
- Practice Test for Chapter 1 ... 77

CHAPTER 2 Differentiation
- 2.1 The derivative and the slope of a curve ... 79
- 2.2 Some rules for differentiation ... 83
- 2.3 Rates of change: velocity and marginals ... 87
- 2.4 The Product and Quotient Rules ... 92
- 2.5 The Chain Rule ... 97
- 2.6 Higher-order derivatives ... 102
- 2.7 Implicit differentiation ... 106
- 2.8 Related rates ... 110
- Review Exercises for Chapter 2 ... 114
- Practice Test for Chapter 2 ... 120

CHAPTER 3 Applications of the Derivative
- 3.1 Increasing and Decreasing functions ... 122
- 3.2 Extrema and the First-Derivative Test ... 128
- 3.3 Concavity and the Second-Derivative Test ... 134
- 3.4 Optimization Problems ... 140
- 3.5 Business and economics applications ... 146
- 3.6 Asymptotes ... 152
- 3.7 Curve sketching: a summary ... 157
- 3.8 Differentials ... 162
- Review Exercises for Chapter 3 ... 165
- Practice Test for Chapter 3 ... 171

CHAPTER 4 Integration
- 4.1 Antiderivatives and the indefinite integral ... 172
- 4.2 The General Power Rule ... 177
- 4.3 Area and the Fundamental Theorem of Calculus ... 185
- 4.4 The area of a region between two curves ... 193
- 4.5 The definite integral as the limit of a sum ... 200
- 4.6 Volumes of solids of revolution ... 203
- Review Exercises for Chapter 4 ... 207
- Practice Test for Chapter 4 ... 215

CHAPTER 5 Exponential and Logarithmic Functions

- 5.1 Exponential functions — 217
- 5.2 Differentiation and integration of exponential functions — 222
- 5.3 The natural logarithmic function — 227
- 5.4 Logarithmic functions: differentiation and integration — 233
- 5.5 Exponential growth and decay — 239
- Review Exercises for Chapter 5 — 243
- Practice Test for Chapter 5 — 249

CHAPTER 6 Techniques of Integration

- 6.1 Integration by substitution — 251
- 6.2 Integration by parts — 260
- 6.3 Partial fractions — 267
- 6.4 Integration by tables — 275
- 6.5 Numerical integration — 280
- 6.6 Improper integrals — 287
- 6.7 Random variables and probability — 291
- 6.8 Expected value, standard deviation, and median — 295
- Review Exercises for Chapter 6 — 302
- Practice Test for Chapter 6 — 311

CHAPTER 7 Differential Equations

- 7.1 Solutions of differential equations — 313
- 7.2 Separation of variables — 318
- 7.3 First-order linear differential equations — 323
- 7.4 Applications of differential equations — 327
- Review Exercises for Chapter 7 — 333
- Practice Test for Chapter 7 — 339

CHAPTER 8 Functions of Several Variables

- 8.1 The three-dimensional coordinate system — 340
- 8.2 Surfaces in space — 343
- 8.3 Functions of several variables — 346
- 8.4 Partial derivatives — 350
- 8.5 Extrema of functions of two variables — 356
- 8.6 Lagrange multipliers and constrained optimization — 362
- 8.7 The method of least squares — 367
- 8.8 Double integrals and area in the plane — 371
- 8.9 Applications of double integrals — 376
- Review Exercises for Chapter 8 — 380
- Practice Test for Chapter 8 — 388

CHAPTER 9 Taylor Polynomials and Series

- 9.1 Sequences — 390
- 9.2 Series and convergence — 394
- 9.3 p-Series and the Ratio Test — 399
- 9.4 Power Series and Taylor's Theorem — 403
- 9.5 Taylor polynomials — 411
- 9.6 Newton's Method — 413
- Review Exercises for Chapter 9 — 418
- Practice Test for Chapter 9 — 425

CHAPTER 10 The Trigonometric Functions
    10.1 Radian measure of angles   427
    10.2 The trigonometric functions   430
    10.3 Graphs of trignometric functions   436
    10.4 Derivatives of trigonometric functions   440
    10.5 Integrals of trigonometric functions   446
         Review Exercises for Chapter 10   453
         Practice Test for Chapter 10   458

Answers to Practice Tests   460

Chapter 0   A Precalculus Review

Section 0.1   The Real Line and Order

1. Determine whether 0.7 is rational or irrational.
   Solution: Since $0.7 = 7/10$, it is rational.

3. Determine whether $3\pi/2$ is rational or irrational.
   Solution: $3\pi/2$ is irrational because π is irrational.

5. Determine whether $4.3\overline{451}$ is rational or irrational.
   Solution:

   $4.3\overline{451}$ is rational because it has a repeating decimal expansion.

7. Determine whether $\sqrt[3]{64}$ is rational or irrational.
   Solution: Since $\sqrt[3]{64} = 4$, it is rational.

9. Determine whether $\sqrt[3]{60}$ is rational or irrational.
   Solution: $\sqrt[3]{60}$ is irrational.

11. Determine whether the given value of x satisfies the inequality in $5x - 12 > 0$.
    Solution:
    (a) Yes, if $x = 3$, then $5(3) - 12 = 3 > 0$.
    (b) No, if $x = -3$, then $5(-3) - 12 = -27 < 0$.
    (c) Yes, if $x = 5/2$, then $5(5/2) - 12 = 1/2 > 0$.
    (d) No, if $x = 3/2$, then $5(3/2) - 12 = -9/2 < 0$.

13. Determine whether the given value of x satisfies the inequality

    $$0 < \frac{x-2}{4} < 2$$

    Solution:

    $$0 < \frac{x-2}{4} < 2$$
    $$0 < x - 2 < 8$$
    $$2 < x < 10$$

    (a) Yes, if $x = 4$, then $2 < x < 10$.
    (b) No, if $x = 10$, then x is not less than 10.
    (c) No, if $x = 0$, then x is not greater than 2.
    (d) Yes, if $x = 5$, then $2 < x < 10$.

15. Complete the given table by filling in the appropriate interval notation, inequality, and graph.
Solution:

Interval Notation	Inequality Notation	Graph
[−2, 0)	−2 ≤ x < 0	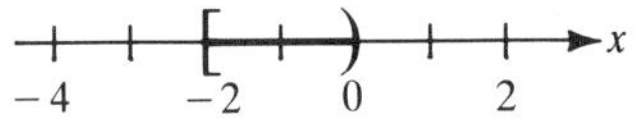 number line with [at −2 and) at 0
(−∞, −4]	x ≤ −4	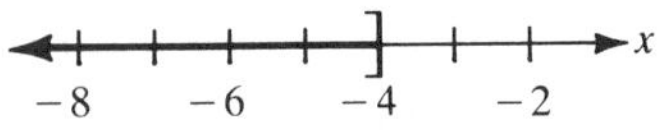 number line with arrow left and] at −4
[3, $\frac{11}{2}$]	3 ≤ x ≤ $\frac{11}{2}$	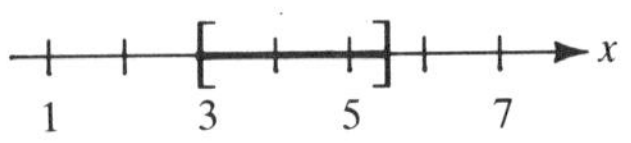 number line with [at 3 and] at 5
(−1, 7)	−1 < x < 7	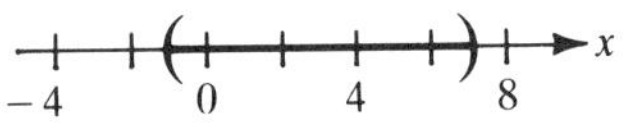 number line with (at −1 and) at 7

17. Solve the inequality x − 5 ≥ 7 and graph the solution on the real line.
Solution: 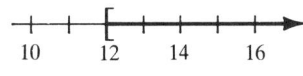
$$x - 5 \geq 7$$
$$x - 5 + 5 \geq 7 + 5$$
$$x \geq 12$$

19. Solve the inequality 4x + 1 < 2x and graph the solution on the real line.
Solution:
$$4x + 1 < 2x$$
$$2x < -1$$
$$x < -1/2$$

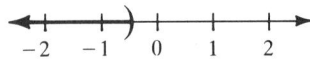

21. Solve the inequality 2x − 1 ≥ 0 and graph the solution on the real line. 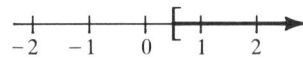
Solution:
$$2x - 1 \geq 0$$
$$2x \geq 1$$
$$x \geq 1/2$$

23. Solve the inequality 4 − 2x < 3x − 1 and graph the solution on the real line.
Solution:
$$4 - 2x < 3x - 1$$
$$4 - 5x < -1$$
$$-5x < -5$$
$$x > 1$$

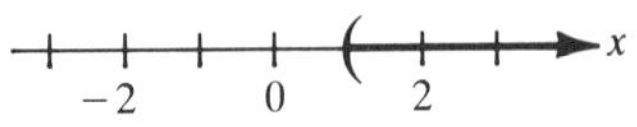

Section 0.1   3

25. Solve the inequality $-4 < 2x - 3 < 4$ and graph the solution on the real line.
Solution:
$$-4 < 2x - 3 < 4$$
$$-4 + 3 < 2x - 3 + 3 < 4 + 3$$
$$-1 < 2x < 7$$
$$-\frac{1}{2} < x < \frac{7}{2}$$

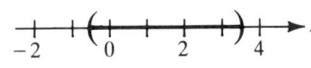

27. Solve the inequality $3/4 > x + 1 > 1/4$ and graph the solution on the real line. 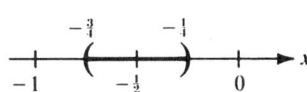
Solution:
$$\frac{3}{4} > x + 1 > \frac{1}{4}$$
$$-\frac{1}{4} > x > -\frac{3}{4}$$
$$-\frac{3}{4} < x < -\frac{1}{4}$$

29. Solve the inequality $(x/2) + (x/3) > 5$ and graph the solution on the real line. 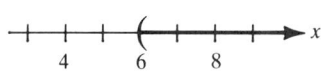
Solution:
$$\frac{x}{2} + \frac{x}{3} > 5$$
$$3x + 2x > 30$$
$$5x > 30$$
$$x > 6$$

31. Solve the inequality $x^2 \leq 3 - 2x$ and graph the solution on the real line. 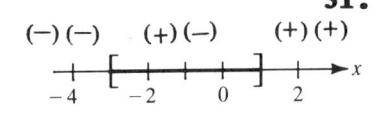
Solution:
$$x^2 \leq 3 - 2x$$
$$x^2 + 2x - 3 \leq 0$$
$$(x + 3)(x - 1) \leq 0$$

Therefore, the solution is $-3 \leq x \leq 1$.

33. Solve the inequality $x^2 + x - 1 \leq 5$ and graph the solution on the real line. 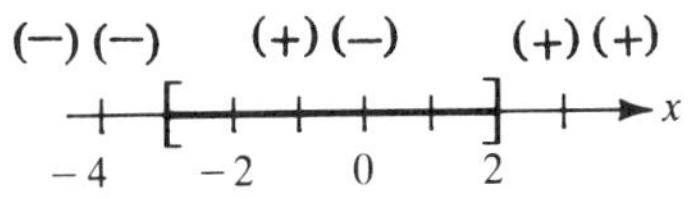
Solution:
$$x^2 + x - 1 \leq 5$$
$$x^2 + x - 6 \leq 0$$
$$(x + 3)(x - 2) \leq 0$$

Therefore, the solution is $-3 \leq x \leq 2$.

35. P dollars is invested at a (simple) interest rate of r. After t years the balance in the account is given by $A = P + Prt$. In order for an investment of $1000 to grow to more than $1250 in two years, what must the interest rate be?
Solution: Since $P = 1000$ and $t = 2$, we have $A > 1250$, and
$$P + Prt > 1250$$
$$1000 + 1000r(2) > 1250$$
$$2000r > 250$$
$$r > \frac{250}{2000} = 0.125 = 12.5\%$$

37. The revenue for selling x units of a product is $R = 115.95x$ and the cost of producing x units is $C = 95x + 750$. In order to obtain a profit, the revenue must be greater than the cost. For what values of x will this product return a profit?
Solution: Since $R = 115.95x$ and $C = 95x + 750$, we have
$$R > C$$
$$115.95x > 95x + 750$$
$$20.95x > 750$$
$$x > \frac{750}{20.95} = 35.7995...$$
Therefore, $x \geq 36$ units.

39. A square region is to have an area of at least 500 square meters. What must the length of the sides of the region be?
Solution: Let x = length of the side of the square. Then, the area of the square is x^2, and we have
$$x^2 \geq 500$$
$$x \geq \sqrt{500}$$
$$x \geq 10\sqrt{5}$$

41. Determine which real number is greater.

(a) π or $\frac{355}{113}$ (b) π or $\frac{22}{7}$

Solution:
(a) $\pi \approx 3.1415926536$

$\frac{355}{113} \approx 3.1415929204$

Therefore, $355/113 > \pi$.

(b) $\pi \approx 3.1415926536$

$\frac{22}{7} \approx 3.1428571429$

Therefore, $22/7 > \pi$.

43. Determine whether each statement is true or false, given a < b.
Solution:
(a) The statement −2a < −2b is false.
(b) The statement a + 2 < b + 2 is true.
(c) The statement 6a < 6b is true.
(d) The statement (1/a) < (1/b) is true if ab < 0 and false if ab > 0.

● Section 0.2 Absolute Value and Distance on the Real Line

1. Let a = −1 and b = 3. Find (a) the directed distance from a to b, (b) the directed distance from b to a, and (c) the distance between a and b.
Solution:
(a) The directed distance from a to b is 3 − (−1) = 4.
(b) The directed distance from b to a is −1 − 3 = −4.
(c) The distance between a and b is |3 − (−1)| = 4.

3. Let a = −5/2 and b = 13/4. Find (a) the directed distance from a to b, (b) the directed distance from b to a, and (c) the distance between a and b.
Solution:
(a) The directed distance from a to b is

$$\frac{13}{4} - \left(-\frac{5}{2}\right) = \frac{23}{4}$$

(b) The directed distance from b to a is

$$-\frac{5}{2} - \frac{13}{4} = -\frac{23}{4}$$

(c) The distance between a and b is

$$\left|\frac{13}{4} - \left(-\frac{5}{2}\right)\right| = \frac{23}{4}$$

5. Let a = 126 and b = 75. Find (a) the directed distance from a to b, (b) the directed distance from b to a, and (c) the distance between a and b.
Solution:
(a) The directed distance from a to b is

75 − 126 = −51

(b) The directed distance from b to a is

126 − 75 = 51

(c) The distance between a and b is

|75 − 126| = 51

Section 0.2

7. Let a = 9.34 and b = −5.65. Find (a) the directed distance from a to b, (b) the directed distance from b to a, and (c) the distance between a and b.
Solution:
(a) The directed distance from a to b is
$-5.65 - 9.34 = -14.99$
(b) The directed distance from b to a is
$9.34 - (-5.65) = 14.99$
(c) The distance between a and b is
$|-5.65 - 9.34| = 14.99$

9. Find the midpoint of the interval [−1, 3].
Solution:
$$\text{Midpoint} = \frac{-1 + 3}{2} = 1$$

11. Find the midpoint of the interval [−5, −3/2].
Solution:
$$\text{Midpoint} = \frac{-5 + (-3/2)}{2} = -\frac{13}{4}$$

13. Find the midpoint of the interval [7, 21].
Solution:
$$\text{Midpoint} = \frac{7 + 21}{2} = 14$$

15. Find the midpoint of the interval [−6.85, 9.35].
Solution:
$$\text{Midpoint} = \frac{-6.85 + 9.35}{2} = 1.25$$

17. Solve the inequality $|x| < 5$ and graph the solution on the real line.
Solution:
$-5 < x < 5$

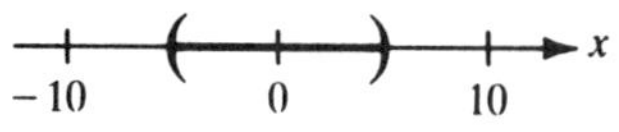

19. Solve the inequality $|x/2| > 3$ and graph the solution on the real line. 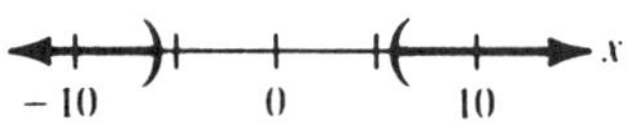
Solution:
$$\frac{x}{2} < -3 \quad \text{or} \quad \frac{x}{2} > 3$$
$x < -6 \quad \text{or} \quad x > 6$

21. Solve the inequality $|x + 2| < 5$ and graph the solution on the real line.
Solution:
$-5 < x + 2 < 5$
$-7 < x < 3$

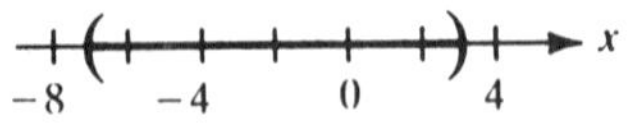

23. Solve the inequality $|(x - 3)/2| \geq 5$ and graph the solution on the real line.
Solution:
$$-\frac{x - 3}{2} \geq 5 \quad \text{or} \quad \frac{x - 3}{2} \geq 5$$
$$-\frac{x - 3}{2}(-2) \leq 5(-2) \quad \text{or} \quad \frac{x - 3}{2}(2) \geq 5(2)$$
$x - 3 \leq -10 \quad \text{or} \quad x - 3 \geq 10$
$x - 3 + 3 \leq -10 + 3 \quad \text{or} \quad x - 3 + 3 \geq 10 + 3$
$x \leq -7 \quad \text{or} \quad x \geq 13$

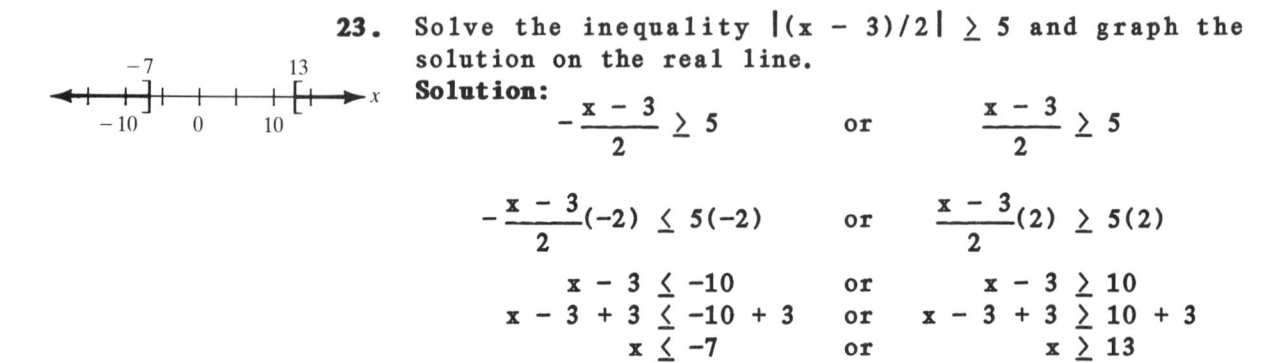

Section 0.2

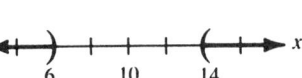

25. Solve the inequality $|10 - x| > 4$ and graph the solution on the real line.
    Solution:
    $$10 - x < -4 \quad \text{or} \quad 10 - x > 4$$
    $$-x < -14 \quad \text{or} \quad -x > -6$$
    $$x > 14 \quad \text{or} \quad x < 6$$

27. Solve the inequality $|9 - 2x| < 1$ and graph the solution on the real line.
    Solution:
    $$-1 < 9 - 2x < 1$$
    $$-10 < -2x < -8$$
    $$5 > x > 4$$

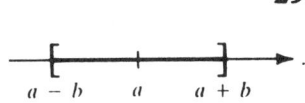

29. Solve the inequality $|x - a| \leq b$ and graph the solution on the real line.
    Solution:
    $$-b \leq x - a \leq b$$
    $$a - b \leq x \leq a + b$$

31. Use absolute values to describe the interval $[-2, 2]$.
    Solution: The midpoint of the interval is
    $$\text{Midpoint} = \frac{-2 + 2}{2} = 0$$
    Therefore, the distance between the midpoint and each endpoint is 2 and we have
    $$|x| \leq 2$$

33. Use absolute values to describe the union of the intervals $(-\infty, -2)$ and $(2, \infty)$.
    Solution: The midpoint between the two intervals is
    $$\text{Midpoint} = \frac{-2 + 2}{2} = 0$$
    Since the distance between the midpoint and ± 2 is 2, we have
    $$|x| > 2$$

35. Use absolute values to describe the interval $[2, 6]$.
    Solution: The midpoint of the interval is
    $$\text{Midpoint} = \frac{2 + 6}{2} = 4$$
    Therefore, the distance between the midpoint and each endpoint is 2 and we have
    $$|x - 4| \leq 2$$

37. Use absolute values to describe the union of the intervals $(-\infty, 0)$ and $(4, \infty)$.
    Solution: The midpoint between the two intervals is
    $$\text{Midpoint} = \frac{0 + 4}{2} = 2$$
    Therefore, the distance between the midpoint and each endpoint is 2 and we have
    $$|x - 2| > 2$$

39. Use absolute values to describe all numbers <u>less than</u> 2 units from 4 on the real line.
Solution:
$$|x - 4| < 2$$

41. Use absolute values to describe the interval where y is <u>at most</u> 2 units from a on the real line.
Solution:
$$|y - a| \leq 2$$

43. The heights, h, of two-thirds of the members of a certain population satisfy the inequality
$$\left|\frac{h - 68.5}{2.7}\right| \leq 1$$
where h is measured in inches. Determine the solution interval for these heights.
Solution:
$$\left|\frac{h - 68.5}{2.7}\right| \leq 1$$
$$-1 \leq \frac{h - 68.5}{2.7} \leq 1$$
$$-2.7 \leq h - 68.5 \leq 2.7$$
$$65.8 \leq h \leq 71.2$$

45. The estimated daily production, x, at a refinery is given by $|x - 2,250,000| < 125,000$, where x is measured in barrels of oil. Determine the high and low production levels.
Solution:
$$|x - 2,250,000| < 125,000$$
$$-125,000 < x - 2,250,000 < 125,000$$
$$2,125,000 < x < 2,375,000$$

47. Use the definition of absolute value to prove $|ab| = |a||b|$.
Solution:
Case 1: If $ab \geq 0$, then either

$a \geq 0$ and $b \geq 0$ or $a < 0$ and $b < 0$

If $a \geq 0$ and $b \geq 0$, then $|a||b| = ab = |ab|$. If $a < 0$ and $b < 0$, then $|a||b| = (-a)(-b) = ab = |ab|$.

Case 2: If $ab < 0$, then either

$a < 0$ and $b > 0$ or $a > 0$ and $b < 0$

If $a < 0$ and $b > 0$, then $|a||b| = (-a)(b) = -ab = |ab|$.
If $a > 0$ and $b < 0$, then $|a||b| = (a)(-b) = -ab = |ab|$.

Therefore, $|ab| = |a||b|$.

Section 0.3

49. Use the definition of absolute value to prove

$$\left|\frac{a}{b}\right| = \frac{|a|}{|b|}, \quad b \neq 0$$

Solution:
Case 1: If $a/b \geq 0$, then either

$$a \geq 0 \text{ and } b > 0 \quad \text{or} \quad a < 0 \text{ and } b < 0$$

If $a \geq 0$ and $b > 0$, then

$$\frac{|a|}{|b|} = \frac{a}{b} = \left|\frac{a}{b}\right|$$

If $a < 0$ and $b < 0$, then

$$\frac{|a|}{|b|} = \frac{-a}{-b} = \frac{a}{b} = \left|\frac{a}{b}\right|$$

Case 2: If $a/b < 0$, then either

$$a < 0 \text{ and } b > 0 \quad \text{or} \quad a > 0 \text{ and } b < 0$$

If $a < 0$ and $b > 0$, then

$$\frac{|a|}{|b|} = \frac{-a}{b} = -\frac{a}{b} = \left|\frac{a}{b}\right|$$

If $a > 0$ and $b < 0$, then

$$\frac{|a|}{|b|} = \frac{a}{-b} = -\frac{a}{b} = \left|\frac{a}{b}\right|$$

Therefore, we have proved that

$$\left|\frac{a}{b}\right| = \frac{|a|}{|b|}$$

● Section 0.3 Exponents and radicals

1. Evaluate $-3x^3$ when $x = 2$.
 Solution:
 $-3(2)^3 = -3(8) = -24$

3. Evaluate $4x^{-3}$ when $x = 2$.
 Solution:
 $4(2)^{-3} = 4(\frac{1}{8}) = \frac{1}{2}$

5. Evaluate $(1 + x^{-1})/x^{-1}$ when $x = 2$.
 Solution:
 $$\frac{1 + (2)^{-1}}{(2)^{-1}} = \frac{1 + (1/2)}{1/2} = 3$$

7. Evaluate $3x^2 - 4x^3$ when $x = -2$.
 Solution:
 $3(-2)^2 - 4(-2)^3 = 3(4) - 4(-8) = 12 + 32 = 44$

Section 0.3

9. Evaluate $6x^0 - (6x)^0$ when $x = 10$.
    Solution:
    $$6(10)^0 - (6(10))^0 = 6(1) - 1 = 5$$

11. Evaluate $\sqrt[3]{x^2}$ when $x = 27$.
    Solution:
    $$\sqrt[3]{27^2} = (\sqrt[3]{27})^2 = 3^2 = 9$$

13. Evaluate $x^{-1/2}$ when $x = 4$.
    Solution:
    $$4^{-1/2} = \frac{1}{\sqrt{4}} = \frac{1}{2}$$

15. Evaluate $x^{-2/5}$ when $x = -32$.
    Solution:
    $$(-32)^{-2/5} = \frac{1}{(\sqrt[5]{-32})^2} = \frac{1}{(-2)^2} = \frac{1}{4}$$

17. Evaluate $500x^{60}$ when $x = 1.01$.
    Solution:
    $$500(1.01)^{60} \approx 908.3483$$

19. Evaluate $\sqrt[3]{x}$ when $x = -154$.
    Solution:
    $$\sqrt[3]{-154} \approx -5.3601$$

21. Simplify the expression $5x^4(x^2)$.
    Solution:
    $$5x^4(x^2) = 5x^6$$

23. Simplify the expression $6y^2(2y^4)^2$.
    Solution:
    $$6y^2(2y^4)^2 = 6y^2(4y^8) = 24y^{10}$$

25. Simplify the expression $10(x^2)^2$.
    Solution:
    $$10(x^2)^2 = 10x^4$$

27. Simplify the expression $7x^2/x^{-3}$.
    Solution:
    $$\frac{7x^2}{x^{-3}} = 7x^5$$

29. Simplify the expression $12(x + y)^3/9(x + y)$.
    Solution:
    $$\frac{12(x + y)^3}{9(x + y)} = \frac{4}{3}(x + y)^2$$

31. Simplify the expression $3x\sqrt{x}/x^{1/2}$.
    Solution:
    $$\frac{3x\sqrt{x}}{x^{1/2}} = \frac{3x\sqrt{x}}{\sqrt{x}} = 3x$$

33. Simplify the expression $(\sqrt{2}\sqrt{x^3}/\sqrt{x})^4$.
    Solution:
    $$\left(\frac{\sqrt{2}\sqrt{x^3}}{\sqrt{x}}\right)^4 = \left(\frac{\sqrt{2}(x\sqrt{x})}{\sqrt{x}}\right)^4$$
    $$= (\sqrt{2}x)^4 = (\sqrt{2})^4 x^4 = 4x^4$$

Section 0.3

35. Simplify (a) $\sqrt{8}$ and (b) $\sqrt{18}$ by removing all possible factors from the radical.
    Solution:
    (a) $\sqrt{8} = \sqrt{4\cdot 2} = \sqrt{4}\sqrt{2} = 2\sqrt{2}$
    (b) $\sqrt{18} = \sqrt{9\cdot 2} = \sqrt{9}\sqrt{2} = 3\sqrt{2}$

37. Simplify (a) $\sqrt[3]{16x^5}$ and (b) $\sqrt[4]{32x^4z^5}$ by removing all possible factors from the radical.
    Solution:
    (a) $\sqrt[3]{16x^5} = \sqrt[3]{(8x^3)(2x^2)} = \sqrt[3]{8x^3}\sqrt[3]{2x^2} = 2x\sqrt[3]{2x^2}$
    (b) $\sqrt[4]{32x^4z^5} = \sqrt[4]{16x^4z^4 2z} = \sqrt[4]{16x^4z^4}\sqrt[4]{2z} = 2|x|z\sqrt[4]{2z}$

    Note: Since x^4 is under the radical, x could be positive or negative. For z^5 to be under the radical, z must be positive.

39. Simplify (a) $\sqrt{75x^2y^{-4}}$ and (b) $\sqrt{5(x-y)^3}$ by removing all possible factors from the radical.
    Solution:

    (a) $\sqrt{75x^2y^{-4}} = \sqrt{\dfrac{25x^2}{y^4}\cdot 3} = \dfrac{5\sqrt{3}|x|}{y^2}$

    (b) $\sqrt{5(x-y)^3} = \sqrt{(x-y)^2 5(x-y)}$
    $= (x-y)\sqrt{5(x-y)}$

41. Insert the required factor in the parentheses in the equation
    $$y^4 - 4y^2 = y^2(y+2)(\quad)$$
    Solution:
    $$y^4 - 4y^2 = y^2(y+2)(y-2)$$

43. Insert the required factor in the parentheses in the equation
    $$\frac{3}{4}x + \frac{1}{2} = \frac{1}{4}(\quad)$$
    Solution:
    $$\frac{3}{4}x + \frac{1}{2} = \frac{3}{4}x + \frac{2}{4} = \frac{1}{4}(3x+2)$$

45. Insert the required factor in the parentheses in the equation
    $$\sqrt{x} + x\sqrt{x} = \sqrt{x}(\quad)$$
    Solution:
    $$\sqrt{x} + x\sqrt{x} = \sqrt{x}(1+x)$$

47. Insert the required factor in the parentheses in the equation
    $$x^2(x^3-1)^4 = (\quad)(x^3-1)^4(3x^2)$$
    Solution:
    $$x^2(x^3-1)^4 = (\tfrac{1}{3})(x^3-1)^4(3x^2)$$

49. Insert the required factor in the parentheses in the equation
    $$5x\sqrt[3]{1+x^2} = (\quad)\sqrt[3]{1+x^2}(2x)$$
    Solution:
    $$5x\sqrt[3]{1+x^2} = (\tfrac{5}{2})\sqrt[3]{1+x^2}(2x)$$

51. Insert the required factor in the parentheses in the equation
$$3x^{1/2} + 4x^{3/2} = x^{1/2}(\quad)$$
Solution:
$$3x^{1/2} + 4x^{3/2} = x^{1/2}(3 + 4x)$$

53. Insert the required factor in the parentheses in the equation
$$3x^{-1/2} + 4x^{3/2} = x^{-1/2}(\quad)$$
Solution:
$$3x^{-1/2} + 4x^{3/2} = x^{-1/2}(3 + 4x^2)$$

55. Insert the required factor in the parentheses in the equation
$$\frac{1}{2}x(x+1)^{-1/2} + (x+1)^{1/2} = \frac{1}{2}(x+1)^{-1/2}(\quad)$$
Solution:
$$\frac{1}{2}x(x+1)^{-1/2} + (x+1)^{1/2}$$
$$= \frac{1}{2}(x+1)^{-1/2}[x + 2(x+1)^1]$$
$$= \frac{1}{2}(x+1)^{-1/2}(3x + 2)$$

57. Find the domain of $\sqrt{x-1}$.
Solution: $\sqrt{x-1}$ is defined when $x \geq 1$. Therefore, the domain is $[1, \infty)$.

59. Find the domain of $\sqrt{x^2+3}$.
Solution: $\sqrt{x^2+3}$ is defined for all real numbers. Therefore, the domain is $(-\infty, \infty)$.

61. Find the domain of $1/\sqrt[3]{x-1}$.
Solution: $1/\sqrt[3]{x-1}$ is defined for all real numbers except $x = 1$. Therefore, the domain is $(-\infty, 1)$ and $(1, \infty)$.

63. Find the domain of $1/\sqrt[4]{2x-6}$.
Solution: $1/\sqrt[4]{2x-6}$ is defined when $x > 3$. Therefore, the domain is $(3, \infty)$.

65. Find the domain of $\sqrt{x-1} + \sqrt{5-x}$.
Solution: $\sqrt{x-1}$ is defined when $x > 1$ and $\sqrt{5-x}$ is defined when $x < 5$. Therefore, the domain of $\sqrt{x-1} + \sqrt{5-x}$ is $1 \leq x \leq 5$.

Section 0.4

Section 0.4 Factoring Polynomials

Quadratic Formula: $ax^2 + bx + c = 0$ implies that
$$x = \frac{-b \pm \sqrt{b^2 - 4ac}}{2a}$$

1. Use the Quadratic Formula to find all real zeros of $6x^2 - x - 1$.
   Solution: Since $a = 6$, $b = -1$, and $c = -1$, we have
   $$x = \frac{1 \pm \sqrt{1 - (-24)}}{12} = \frac{1 \pm 5}{12}$$
   Thus,
   $$x = \frac{1 + 5}{12} = \frac{1}{2} \quad \text{or} \quad x = \frac{1 - 5}{12} = -\frac{1}{3}$$

3. Use the Quadratic Formula to find all real zeros of $4x^2 - 12x + 9$.
   Solution: Since $a = 4$, $b = -12$, and $c = 9$, we have
   $$x = \frac{12 \pm \sqrt{144 - 144}}{8} = \frac{12}{8} = \frac{3}{2}$$

5. Use the Quadratic Formula to find all real zeros of $y^2 + 4y + 1$.
   Solution: Since $a = 1$, $b = 4$, and $c = 1$, we have
   $$y = \frac{-4 \pm \sqrt{16 - 4}}{2} = \frac{-4 \pm 2\sqrt{3}}{2} = -2 \pm \sqrt{3}$$

7. Use the Quadratic Formula to find all real zeros of $3x^2 - 2x - 2$.
   Solution: Since $a = 3$, $b = -2$, and $c = -2$, we have
   $$x = \frac{2 \pm \sqrt{4 - (-24)}}{6} = \frac{2 \pm 2\sqrt{7}}{6} = \frac{1 \pm \sqrt{7}}{3}$$

9. Use the Quadratic Formula to find all real zeros of $2s^2 - 7s + 4$.
   Solution: Since $a = 2$, $b = -7$, and $c = 4$, we have
   $$s = \frac{7 \pm \sqrt{49 - 32}}{4} = \frac{7 \pm \sqrt{17}}{4}$$

11. Use the Quadratic Formula to find all real roots of $x^2 - 2x + 3 = 0$.
    Solution: Since $a = 1$, $b = -2$, and $c = 3$, we have
    $$x = \frac{2 \pm \sqrt{4 - 12}}{2} = \frac{2 \pm \sqrt{-8}}{2}$$
    Since $\sqrt{-8}$ is imaginary, there are no real roots.

13. Use the Quadratic Formula to find all real roots of $x + 1 = 3/x$.
    Solution: By writing $x + 1 = 3/x$ as $x^2 + x - 3 = 0$, we find that $a = 1$, $b = 1$, and $c = -3$. Therefore,
    $$x = \frac{-1 \pm \sqrt{1 - (-12)}}{2} = \frac{-1 \pm \sqrt{13}}{2}$$

15. Write $x^2 - 4x + 4$ as the product of two linear factors.
   Solution:
   $$x^2 - 4x + 4 = (x - 2)^2$$

17. Write $4x^2 + 4x + 1$ as the product of two linear factors.
    Solution:
    $$4x^2 + 4x + 1 = (2x + 1)^2$$

19. Write $x^2 + x - 2$ as the product of two linear factors.
    Solution:
    $$x^2 + x - 2 = (x + 2)(x - 1)$$

21. Write $3x^2 - 5x + 2$ as the product of two linear factors.
    Solution:
    $$3x^2 - 5x + 2 = (3x - 2)(x - 1)$$

23. Write $x^2 - 4xy + 4y^2$ as the product of two linear factors.
    Solution:
    $$x^2 - 4xy + 4y^2 = (x - 2y)^2$$

25. Write $2y^2 + 21yz - 36z^2$ as the product of two linear factors.
    Solution:
    $$2y^2 + 21yz - 36z^2 = (2y - 3z)(y + 12z)$$

27. Write $16y^2 - 9$ as the product of two linear factors.
    Solution:
    $$16y^2 - 9 = (4y + 3)(4y - 3)$$

29. Write $(x - 1)^2 - 4$ as the product of two linear factors.
    Solution:
    $$(x - 1)^2 - 4 = [(x - 1) + 2][(x - 1) - 2]$$
    $$= (x + 1)(x - 3)$$

31. Completely factor $81 - y^4$.
    Solution:
    $$81 - y^4 = (9 + y^2)(9 - y^2)$$
    $$= (9 + y^2)(3 + y)(3 - y)$$

33. Completely factor $x^3 - 8$.
    Solution:
    $$x^3 - 8 = x^3 - 2^3 = (x - 2)(x^2 + 2x + 4)$$

35. Completely factor $y^3 + 64$.
    Solution:
    $$y^3 + 64 = y^3 + 4^3 = (y + 4)(y^2 - 4y + 16)$$

37. Completely factor $x^3 - 27$.
    Solution:
    $$x^3 - 27 = x^3 - 3^3 = (x - 3)(x^2 + 3x + 9)$$

Section 0.4   15

39. Completely factor $x^3 - 4x^2 - x + 4$.
Solution:
$$x^3 - 4x^2 - x + 4 = x^2(x - 4) - (x - 4)$$
$$= (x - 4)(x^2 - 1)$$
$$= (x - 4)(x + 1)(x - 1)$$

41. Completely factor $2x^3 - 3x^2 + 4x - 6$.
Solution:
$$2x^3 - 3x^2 + 4x - 6 = x^2(2x - 3) + 2(2x - 3)$$
$$= (2x - 3)(x^2 + 2)$$

43. Completely factor $2x^3 - 4x^2 - x + 2$.
Solution:
$$2x^3 - 4x^2 - x + 2 = 2x^2(x - 2) - (x - 2)$$
$$= (x - 2)(2x^2 - 1)$$

45. Find all real roots of $x^2 - 5x = 0$.
Solution:
$$x^2 - 5x = 0$$
$$x(x - 5) = 0$$
$$x = 0, 5$$

47. Find all real roots of $x^2 - 9 = 0$.
Solution:
$$x^2 - 9 = 0$$
$$(x + 3)(x - 3) = 0$$
$$x = -3, 3$$

49. Find all real roots of $x^2 - 3 = 0$.
Solution:
$$x^2 - 3 = 0$$
$$(x + \sqrt{3})(x - \sqrt{3}) = 0$$
$$x = -\sqrt{3}, \sqrt{3}$$

51. Find all real roots of $(x - 3)^2 - 9 = 0$.
Solution:
$$(x - 3)^2 - 9 = 0$$
$$x^2 - 6x + 9 - 9 = 0$$
$$x(x - 6) = 0$$
$$x = 0, 6$$

53. Find all real roots of $x^2 + x - 2 = 0$.
Solution:
$$x^2 + x - 2 = 0$$
$$(x + 2)(x - 1) = 0$$
$$x = -2, 1$$

55. Find all real roots of $x^2 - 5x + 6 = 0$.
Solution:
$$x^2 - 5x + 6 = 0$$
$$(x - 2)(x - 3) = 0$$
$$x = 2, 3$$

Section 0.4

57. Find all real roots of $2x^2 - x - 1 = 0$.
Solution:
$$2x^2 - x - 1 = 0$$
$$(2x + 1)(x - 1) = 0$$
$$x = -\frac{1}{2}, 1$$

59. Find all real roots of $(x - 5)(x + 3) = 33$.
Solution:
$$(x - 5)(x + 3) = 33$$
$$x^2 - 2x - 15 = 33$$
$$x^2 - 2x - 48 = 0$$
$$(x + 6)(x - 8) = 0$$
$$x = -6, 8$$

61. Find all real roots of $x^3 + 64 = 0$.
Solution:
$$x^3 + 64 = 0$$
$$x^3 = -64$$
$$x = \sqrt[3]{-64} = -4$$

63. Find all real roots of $x^4 - 16 = 0$.
Solution:
$$x^4 - 16 = 0$$
$$x^4 = 16$$
$$x = \pm\sqrt[4]{16} = \pm 2$$

65. Find all real roots of $x^3 - x^2 - 4x + 4 = 0$.
Solution:
$$x^3 - x^2 - 4x + 4 = 0$$
$$x^2(x - 1) - 4(x - 1) = 0$$
$$(x - 1)(x^2 - 4) = 0$$
$$(x - 1)(x - 2)(x + 2) = 0$$
$$x = 1, 2, -2$$

67. Find all real roots of
$$(x + 2)^2(x - 1) + (x + 2)(x - 1)^2 = 0$$
Solution:
$$(x + 2)^2(x - 1) + (x + 2)(x - 1)^2 = 0$$
$$(x + 2)(x - 1)[(x + 2) + (x - 1)] = 0$$
$$(x + 2)(x - 1)(2x + 1) = 0$$
$$x = -2, 1, -\frac{1}{2}$$

69. Find the intervals on which $\sqrt{x^2 - 7x + 12}$ is defined.
Solution: Since
$$\sqrt{x^2 - 7x + 12} = \sqrt{(x - 3)(x - 4)}$$
the roots are $x = 3$ and $x = 4$. By testing points inside and outside the interval [3, 4], we find that the expression is defined when $x \leq 3$ or $x \geq 4$.

Section 0.4   17

71. Find the interval on which $\sqrt{4 - x^2}$ is defined.
    Solution: Since
    $$\sqrt{4 - x^2} = \sqrt{(2 + x)(2 - x)}$$
    the roots are $x = \pm 2$. By testing points inside and outside the interval $[-2, 2]$, we find that the expression is defined when $-2 \leq x \leq 2$.

73. Find the interval on which $\sqrt{12 - x - x^2}$ is defined.
    Solution: Since
    $$\sqrt{12 - x - x^2} = \sqrt{(4 + x)(3 - x)}$$
    the roots are $x = -4$ and $x = 3$. By testing points inside and outside the interval $[-4, 3]$, we find that the expression is defined when $-4 \leq x \leq 3$.

75. Find the interval on which $\sqrt{x^2 - 3x + 3}$ is defined.
    Solution: Since $x^2 - 3x + 3$ has no real roots and is positive when $x = 0$, we conclude that the expression is defined for all real numbers.

77. Use synthetic division to complete the indicated factorization: $x^3 + 8 = (x + 2)(\quad)$.
    Solution:
    ```
    -2 | 1    0    0    8
       |     -2    4   -8
       |_____
         1   -2    4    0
    ```
    Therefore, the factorization is
    $$x^3 + 8 = (x + 2)(x^2 - 2x + 4)$$

79. Use synthetic division to complete the indicated factorization: $2x^3 - x^2 - 2x + 1 = (x - 1)(\quad)$.
    Solution:
    ```
    1 | 2   -1   -2    1
      |      2    1   -1
      |_____
        2    1   -1    0
    ```
    Therefore, the factorization is
    $$2x^3 - x^2 - 2x + 1 = (x - 1)(2x^2 + x - 1)$$

81. Use synthetic division to complete the indicated factorization:
    $$x^4 + 2x^3 - 6x^2 - 18x - 27 = (x - 3)(\quad).$$
    Solution:
    ```
    3 | 1    2   -6   -18   -27
      |      3   15    27    27
      |_____
        1    5    9     9     0
    ```
    Therefore, the factorization is
    $$x^4 + 2x^3 - 6x^2 - 18x - 27 = (x - 3)(x^3 + 5x^2 + 9x + 9)$$

83. Use the Rational Zero Theorem as an aid in finding all real roots of $x^3 - x^2 - x + 1 = 0$.
Solution: Possible rational roots: ± 1
Using synthetic division for $x = 1$, we have

$$
\begin{array}{r|rrrr}
1 & 1 & -1 & -1 & 1 \\
  &   & 1 & 0 & -1 \\
\hline
  & 1 & 0 & -1 & 0
\end{array}
$$

Therefore, we have
$$x^3 - x^2 - x + 1 = 0$$
$$(x - 1)(x^2 - 1) = 0$$
$$(x - 1)(x - 1)(x + 1) = 0$$
$$x = 1, -1$$

85. Use the Rational Zero Theorem as an aid in finding all real roots of $x^3 - 6x^2 + 11x - 6 = 0$.
Solution: Possible rational roots: $\pm 1, \pm 2, \pm 3, \pm 6$
Using synthetic division for $x = 1$, we have

$$
\begin{array}{r|rrrr}
1 & 1 & -6 & 11 & -6 \\
  &   & 1 & -5 & 6 \\
\hline
  & 1 & -5 & 6 & 0
\end{array}
$$

Therefore, we have
$$x^3 - 6x^2 + 11x - 6 = 0$$
$$(x - 1)(x^2 - 5x + 6) = 0$$
$$(x - 1)(x - 2)(x - 3) = 0$$
$$x = 1, 2, 3$$

87. Use the Rational Zero Theorem as an aid in finding all real roots of $4x^3 - 4x^2 - x + 1 = 0$.
Solution:
Possible rational roots: $\pm 1, \pm \frac{1}{2}, \pm \frac{1}{4}$

Using synthetic division for $x = 1$, we have

$$
\begin{array}{r|rrrr}
1 & 4 & -4 & -1 & 1 \\
  &   & 4 & 0 & -1 \\
\hline
  & 4 & 0 & -1 & 0
\end{array}
$$

Therefore, we have
$$4x^3 - 4x^2 - x + 1 = 0$$
$$(x - 1)(4x^2 - 1) = 0$$
$$(x - 1)(2x + 1)(2x - 1) = 0$$
$$x = 1, -\frac{1}{2}, \frac{1}{2}$$

Section 0.4   19

89. Use the Rational Zero Theorem as an aid in finding all real roots of $x^3 - 3x^2 - 3x - 4 = 0$.
Solution: Possible rational roots: $\pm 1, \pm 2, \pm 4$
Using synthetic division for $x = 4$, we have

```
4 | 1   -3   -3   -4
  |      4    4    4
  |_____
    1    1    1    0
```

Therefore, we have
$$x^3 - 3x^2 - 3x - 4 = 0$$
$$(x - 4)(x^2 + x + 1) = 0$$

Since $x^2 + x + 1$ has no real solutions, $x = 4$ is the only real solution.

91. Use the Rational Zero Theorem as an aid in finding all real roots of $z^3 + 8z^2 + 11z - 2 = 0$.
Solution: Possible rational roots: $\pm 1, \pm 2$
Using synthetic division for $x = -2$ we have

```
-2 | 1    8   11   -2
   |     -2  -12    2
   |_____
     1    6   -1    0
```

Therefore, we have
$$z^3 + 8z^2 + 11z - 2 = 0$$
$$(z + 2)(z^2 + 6z - 1) = 0$$
$$z = -2, -3 \pm \sqrt{10}$$

Note: Use the Quadratic Formula on $z^2 + 6z - 1$.

93. Use the Rational Zero Theorem as an aid in finding all real roots of $x^4 - 13x^2 + 36 = 0$.
Solution: Possible rational roots: $\pm 1, \pm 2, \pm 3, \pm 4,$
$\pm 6, \pm 9, \pm 12, \pm 18, \pm 36$
Using synthetic division for $x = \pm 2$, we have

```
 2 | 1    0   -13    0    36
   |      2    4   -18   -36
   |_____
     1    2   -9   -18    0
-2 |     -2    0    18
   |_____
     1    0   -9     0
```

Therefore, we have
$$x^4 - 13x^2 + 36 = 0$$
$$(x - 2)(x + 2)(x^2 - 9) = 0$$
$$(x - 2)(x + 2)(x - 3)(x + 3) = 0$$
$$x = \pm 2, \pm 3$$

95. Use the Binomial Formula
$$(x + a)^5 = x^5 + 5x^4a + 10x^3a^2 + 10x^2a^3 + 5xa^4 + a^5$$
to factor $x^5 - 10x^4 + 40x^3 - 80x^2 + 80x - 32$.
Solution: Letting $a = -2$, we have
$$(x - 2)^5 = x^5 - 10x^4 + 40x^3 - 80x^2 + 80x - 32$$

Section 0.5 Fractions and rationalization

1. Perform the indicated operations and simplify your answer:
$$\frac{5}{x-1} + \frac{x}{x-1}$$
Solution:
$$\frac{5}{x-1} + \frac{x}{x-1} = \frac{5+x}{x-1} = \frac{x+5}{x-1}$$

3. Perform the indicated operations and simplify your answer:
$$\frac{2x}{x^2+2} - \frac{1-3x}{x^2+2}$$
Solution:
$$\frac{2x}{x^2+2} - \frac{1-3x}{x^2+2} = \frac{2x-(1-3x)}{x^2+2} = \frac{5x-1}{x^2+2}$$

5. Perform the indicated operations and simplify your answer:
$$\frac{4}{x} - \frac{3}{x^2}$$
Solution:
$$\frac{4}{x} - \frac{3}{x^2} = \frac{4x}{x^2} - \frac{3}{x^2} = \frac{4x-3}{x^2}$$

7. Perform the indicated operations and simplify your answer:
$$\frac{2}{x+2} - \frac{1}{x-2}$$
Solution:
$$\frac{2}{x+2} - \frac{1}{x-2} = \frac{2(x-2)-(x+2)}{(x+2)(x-2)}$$
$$= \frac{x-6}{x^2-4}$$

9. Perform the indicated operations and simplify your answer:
$$\frac{5}{x-3} + \frac{3}{3-x}$$
Solution:
$$\frac{5}{x-3} + \frac{3}{3-x} = \frac{5}{x-3} + \frac{-3}{x-3} = \frac{2}{x-3}$$

Section 0.5   21

11. Perform the indicated operations and simplify your answer:

$$\frac{1}{x^2 - x - 2} - \frac{x}{x^2 - 5x + 6}$$

Solution:

$$\frac{1}{x^2 - x - 2} - \frac{x}{x^2 - 5x + 6}$$

$$= \frac{1}{(x + 1)(x - 2)} - \frac{x}{(x - 2)(x - 3)}$$

$$= \frac{(x - 3) - x(x + 1)}{(x + 1)(x - 2)(x - 3)}$$

$$= \frac{-x^2 - 3}{(x + 1)(x - 2)(x - 3)}$$

$$= -\frac{x^2 + 3}{(x + 1)(x - 2)(x - 3)}$$

13. Perform the indicated operations and simplify your answer:

$$\frac{A}{x - 6} + \frac{B}{x + 3}$$

Solution:

$$\frac{A}{x - 6} + \frac{B}{x + 3} = \frac{A(x + 3) + B(x - 6)}{(x - 6)(x + 3)}$$

$$= \frac{Ax + 3A + Bx - 6B}{(x - 6)(x + 3)}$$

$$= \frac{(A + B)x + 3(A - 2B)}{(x - 6)(x + 3)}$$

15. Perform the indicated operations and simplify your answer:

$$\frac{A}{x - 5} + \frac{B}{x + 5} + \frac{C}{(x + 5)^2}$$

Solution:

$$\frac{A}{x - 5} + \frac{B}{x + 5} + \frac{C}{(x + 5)^2}$$

$$= \frac{A(x + 5)^2 + B(x - 5)(x + 5) + C(x - 5)}{(x - 5)(x + 5)^2}$$

$$= \frac{A(x^2 + 10x + 25) + B(x^2 - 25) + Cx - 5C}{(x - 5)(x + 5)^2}$$

$$= \frac{(A + B)x^2 + (10A + C)x + 5(5A - 5B - C)}{(x - 5)(x + 5)^2}$$

17. Perform the indicated operations and simplify your answer:

$$-\frac{1}{x} + \frac{2}{x^2 + 1}$$

Solution:

$$-\frac{1}{x} + \frac{2}{x^2 + 1} = \frac{-(x^2 + 1) + 2x}{x(x^2 + 1)}$$

$$= \frac{-x^2 + 2x - 1}{x(x^2 + 1)}$$

$$= \frac{-(x^2 - 2x + 1)}{x(x^2 + 1)} = \frac{-(x - 1)^2}{x(x^2 + 1)}$$

19. Perform the indicated operations and simplify your answer:

$$\frac{-x}{(x + 1)^{3/2}} + \frac{2}{(x + 1)^{1/2}}$$

Solution:

$$\frac{-x}{(x + 1)^{3/2}} + \frac{2}{(x + 1)^{1/2}} = \frac{-x + 2(x + 1)}{(x + 1)^{3/2}}$$

$$= \frac{x + 2}{(x + 1)^{3/2}}$$

21. Perform the indicated operations and simplify your answer:

$$\frac{2 - t}{2\sqrt{1 + t}} - \sqrt{1 + t}$$

Solution:

$$\frac{2 - t}{2\sqrt{1 + t}} - \sqrt{1 + t} = \frac{2 - t}{2\sqrt{1 + t}} - \frac{\sqrt{1 + t}}{1} \cdot \frac{2\sqrt{1 + t}}{2\sqrt{1 + t}}$$

$$= \frac{(2 - t) - 2(1 + t)}{2\sqrt{1 + t}}$$

$$= \frac{-3t}{2\sqrt{1 + t}}$$

23. Perform the indicated operations and simplify your answer:

$$\left(2x\sqrt{x^2 + 1} - \frac{x^3}{\sqrt{x^2 + 1}}\right) \div (x^2 + 1)$$

Solution:

$$\left(2x\sqrt{x^2 + 1} - \frac{x^3}{\sqrt{x^2 + 1}}\right) \div (x^2 + 1)$$

$$= \frac{2x(x^2 + 1) - x^3}{\sqrt{x^2 + 1}} \cdot \frac{1}{x^2 + 1}$$

$$= \frac{x^3 + 2x}{\sqrt{x^2 + 1}(x^2 + 1)} = \frac{x(x^2 + 2)}{(x^2 + 1)^{3/2}}$$

Section 0.5   23

25. Perform the indicated operations and simplify your answer:
$$\left[\frac{1}{(x+\Delta x)^2} - \frac{1}{x^2}\right] \div \Delta x$$

Solution:
$$\left[\frac{1}{(x+\Delta x)^2} - \frac{1}{x^2}\right] \div \Delta x$$
$$= \frac{x^2 - (x^2 + 2x\Delta x + (\Delta x)^2)}{x^2(x+\Delta x)^2} \cdot \frac{1}{\Delta x}$$
$$= \frac{-2x\Delta x - (\Delta x)^2}{x^2(x+\Delta x)^2} \cdot \frac{1}{\Delta x} = \frac{-2x - \Delta x}{x^2(x+\Delta x)^2}$$

27. Perform the indicated operations and simplify your answer:
$$\frac{(x^2+2)^{1/2} - x^2(x^2+2)^{-1/2}}{x^2}$$

Solution:
$$\frac{(x^2+2)^{1/2} - x^2(x^2+2)^{-1/2}}{x^2}$$
$$= \frac{(x^2+2)^{-1/2}[(x^2+2) - x^2]}{x^2}$$
$$= \frac{2}{x^2\sqrt{x^2+2}}$$

29. Perform the indicated operations and simplify your answer:
$$\frac{\frac{\sqrt{x+1}}{\sqrt{x}} - \frac{\sqrt{x}}{\sqrt{x+1}}}{2(x+1)}$$

Solution:
$$\frac{\frac{\sqrt{x+1}}{\sqrt{x}} - \frac{\sqrt{x}}{\sqrt{x+1}}}{2(x+1)} = \frac{(x+1) - x}{\sqrt{x}\sqrt{x+1}} \cdot \frac{1}{2(x+1)}$$
$$= \frac{1}{2\sqrt{x}(x+1)^{3/2}}$$

31. Rationalize the denominator and simplify: $3/\sqrt{27}$
Solution:
$$\frac{3}{\sqrt{27}} = \frac{3}{3\sqrt{3}} = \frac{1}{\sqrt{3}} \cdot \frac{\sqrt{3}}{\sqrt{3}} = \frac{\sqrt{3}}{3}$$

33. Rationalize the numerator and simplify: $\sqrt{2}/3$
Solution:
$$\frac{\sqrt{2}}{3} = \frac{\sqrt{2}}{3} \cdot \frac{\sqrt{2}}{\sqrt{2}} = \frac{2}{3\sqrt{2}}$$

35. Rationalize the denominator and simplify: $x/\sqrt{x-4}$
Solution:
$$\frac{x}{\sqrt{x-4}} = \frac{x}{\sqrt{x-4}} \cdot \frac{\sqrt{x-4}}{\sqrt{x-4}} = \frac{x\sqrt{x-4}}{x-4}$$

37. Rationalize the numerator and simplify: $\sqrt{y^3}/6y$
Solution:
$$\frac{\sqrt{y^3}}{6y} = \frac{y\sqrt{y}}{6y} = \frac{\sqrt{y}}{6} = \frac{y}{6\sqrt{y}}$$

39. Rationalize the denominator and simplify:
$$\frac{49(x-3)}{\sqrt{x^2-9}}$$
Solution:
$$\frac{49(x-3)}{\sqrt{x^2-9}} = \frac{49(x-3)}{\sqrt{x^2-9}} \cdot \frac{\sqrt{x^2-9}}{\sqrt{x^2-9}}$$
$$= \frac{49(x-3)\sqrt{x^2-9}}{(x+3)(x-3)} = \frac{49\sqrt{x^2-9}}{x+3}$$

41. Rationalize the denominator and simplify: $5/(\sqrt{14}-2)$
Solution:
$$\frac{5}{\sqrt{14}-2} = \frac{5}{\sqrt{14}-2} \cdot \frac{\sqrt{14}+2}{\sqrt{14}+2}$$
$$= \frac{5(\sqrt{14}+2)}{14-4}$$
$$= \frac{\sqrt{14}+2}{2}$$

43. Rationalize the denominator and simplify: $2x/(5-\sqrt{3})$
Solution:
$$\frac{2x}{5-\sqrt{3}} = \frac{2x}{5-\sqrt{3}} \cdot \frac{5+\sqrt{3}}{5+\sqrt{3}}$$
$$= \frac{2x(5+\sqrt{3})}{25-3}$$
$$= \frac{x(5+\sqrt{3})}{11}$$

45. Rationalize the denominator and simplify: $1/(\sqrt{6}+\sqrt{5})$
Solution:
$$\frac{1}{\sqrt{6}+\sqrt{5}} = \frac{1}{\sqrt{6}+\sqrt{5}} \cdot \frac{\sqrt{6}-\sqrt{5}}{\sqrt{6}-\sqrt{5}}$$
$$= \frac{\sqrt{6}-\sqrt{5}}{6-5}$$
$$= \sqrt{6}-\sqrt{5}$$

Section 0.5   25

47. Rationalize the numerator and simplify: $\dfrac{\sqrt{3} - \sqrt{2}}{x}$

Solution:

$$\dfrac{\sqrt{3} - \sqrt{2}}{x} = \dfrac{\sqrt{3} - \sqrt{2}}{x} \cdot \dfrac{\sqrt{3} + \sqrt{2}}{\sqrt{3} + \sqrt{2}}$$

$$= \dfrac{3 - 2}{x(\sqrt{3} + \sqrt{2})}$$

$$= \dfrac{1}{x(\sqrt{3} + \sqrt{2})}$$

49. Rationalize the numerator and simplify:

$$\dfrac{2x - \sqrt{4x - 1}}{2x - 1}$$

Solution:

$$\dfrac{2x - \sqrt{4x - 1}}{2x - 1} = \dfrac{2x - \sqrt{4x - 1}}{2x - 1} \cdot \dfrac{2x + \sqrt{4x - 1}}{2x + \sqrt{4x - 1}}$$

$$= \dfrac{4x^2 - (4x - 1)}{(2x - 1)(2x + \sqrt{4x - 1})}$$

$$= \dfrac{(2x - 1)^2}{(2x - 1)(2x + \sqrt{4x - 1})}$$

$$= \dfrac{2x - 1}{2x + \sqrt{4x - 1}}$$

51. Rationalize the denominator and simplify:

$$\dfrac{x + 1}{\sqrt{x^2 - 2} - \sqrt{x}}$$

Solution:

$$\dfrac{x + 1}{\sqrt{x^2 - 2} - \sqrt{x}} = \dfrac{x + 1}{\sqrt{x^2 - 2} - \sqrt{x}} \cdot \dfrac{\sqrt{x^2 - 2} + \sqrt{x}}{\sqrt{x^2 - 2} + \sqrt{x}}$$

$$= \dfrac{(x + 1)(\sqrt{x^2 - 2} + \sqrt{x})}{x^2 - 2 - x}$$

$$= \dfrac{(x + 1)(\sqrt{x^2 - 2} + \sqrt{x})}{(x + 1)(x - 2)}$$

$$= \dfrac{\sqrt{x^2 - 2} + \sqrt{x}}{x - 2}$$

53. Rationalize the denominator and simplify: $\dfrac{8x}{\sqrt{17x - 1}}$

Solution:

$$\dfrac{8x}{\sqrt{17x - 1}} = \dfrac{8x}{\sqrt{17x - 1}} \cdot \dfrac{\sqrt{17x - 1}}{\sqrt{17x - 1}}$$

$$= \dfrac{8x(\sqrt{17x - 1})}{17x - 1}$$

Review Exercises for Chapter 0

1. Place the appropriate inequality sign (< or >) between the two real numbers.
Solution:

(a) $\dfrac{3}{2} < 7$ 	(b) $\pi > -6$

3. Place the appropriate inequality sign (< or >) between the two real numbers.
Solution:

(a) $-\dfrac{3}{7} > -\dfrac{8}{7}$ 	(b) $\dfrac{3}{4} > \dfrac{11}{16}$

5. Sketch the solution set of $3 - 2x \leq 0$.
Solution:

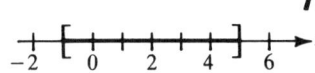

$$3 - 2x \leq 0$$
$$-2x \leq -3$$
$$x \geq 3/2$$

7. Sketch the solution set of $|x - 2| \leq 3$.
Solution:

$$-3 \leq x - 2 \leq 3$$
$$-1 \leq x \leq 5$$

9. Sketch the solution set of $4 < (x + 3)^2$.
Solution:

$$4 < (x + 3)^2$$
$$4 < x^2 + 6x + 9$$
$$0 < x^2 + 6x + 5$$
$$0 < (x + 1)(x + 5)$$

Therefore, the solution set is $x < -5$ or $x > -1$.

11. Find the midpoint of the interval [7/8, 5/2].
Solution:

$$\text{Midpoint} = \frac{(7/8) + (5/2)}{2} = \frac{27}{16}$$

13. Use inequality notation to describe the expression:
  x is nonnegative.
Solution:

$x \geq 0$

15. Use inequality notation to describe the expression:
  The area A is <u>no more than</u> 12 square meters.
Solution:

$0 \leq A \leq 12$

(Area is a nonnegative quantity.)

Review Exercises for Chapter 0

17. Use absolute value notation to describe the expression: The distance between x and 5 is <u>no more than</u> 3.
Solution:
$$|x - 5| \leq 3$$

19. Use absolute value notation to describe the expression: y is <u>at least</u> 2 units from a.
Solution:
$$|y - a| \geq 2$$

21. Simplify $(10xy)^2(3y^3)$.
Solution:
$$(10xy)^2(3y^3) = (100x^2y^2)(3y^3) = 300x^2y^5$$

23. Simplify $4x^4y^3/(3xy)^4$.
Solution:
$$\frac{4x^4y^3}{(3xy)^4} = \frac{4x^4y^3}{81x^4y^4} = \frac{4}{81y}$$

25. Simplify $(4a^{-2}b^3)^{-3}$.
Solution:
$$(4a^{-2}b^3)^{-3} = 4^{-3}a^6b^{-9}$$
$$= \frac{a^6}{4^3 b^9} = \frac{a^6}{64b^9}$$

27. Simplify $(\sqrt[3]{(x-y)^2})^6$.
Solution:
$$(\sqrt[3]{(x-y)^2})^6 = [(x-y)^{2/3}]^6 = (x-y)^4$$

29. Simplify $\sqrt[4]{(3x^2y^3)^2}$.
Solution:
$$\sqrt[4]{(3x^2y^3)^2} = (3x^2y^3)^{2/4}$$
$$= (3x^2|y^3|)^{1/2}$$
$$= \sqrt{3x^2|y^3|} = |xy|\sqrt{3|y|}$$

31. Completely factor the polynomial $2x^3 - 2x^2 - 4x$.
Solution:
$$2x^3 - 2x^2 - 4x = 2x(x^2 - x - 2)$$
$$= 2x(x+1)(x-2)$$

33. Completely factor the polynomial $63rs^2 - 7r^3$.
Solution:
$$63rs^2 - 7r^3 = 7r(9s^2 - r^2)$$
$$= 7r(3s + r)(3s - r)$$

35. Completely factor the polynomial $4x^3y - 4x^2y^2 + xy^3$.
   Solution:
   $$4x^3y - 4x^2y^2 + xy^3 = xy(4x^2 - 4xy + y^2)$$
   $$= xy(2x - y)^2$$

37. Completely factor the polynomial $6x^4 - 48xy^3$.
   Solution:
   $$6x^4 - 48xy^3 = 6x(x^3 - 8y^3)$$
   $$= 6x(x - 2y)(x^2 + 2xy + 4y^2)$$

39. Completely factor the polynomial $(x^2 + 2y^2)^2 - 9x^2y^2$.
   Solution:
   $$(x^2 + 2y^2)^2 - 9x^2y^2$$
   $$= [(x^2 + 2y^2) + 3xy][(x^2 + 2y^2) - 3xy]$$
   $$= (x^2 + 3xy + 2y^2)(x^2 - 3xy + 2y^2)$$
   $$= (x + y)(x + 2y)(x - y)(x - 2y)$$

41. Completely factor the polynomial
   $$4x^2(2x - 1) + 2x(2x - 1)^2$$
   Solution:
   $$4x^2(2x - 1) + 2x(2x - 1)^2 = 2x(2x - 1)[2x + (2x - 1)]$$
   $$= 2x(2x - 1)(4x - 1)$$

43. Completely factor the polynomial $xy + 2y^2 + 2xz + 4yz$.
   Solution:
   $$xy + 2y^2 + 2xz + 4yz = y(x + 2y) + 2z(x + 2y)$$
   $$= (x + 2y)(y + 2z)$$

45. Insert the missing factor in
   $$\frac{3}{4}x^2 - \frac{5}{6}x + 4 = \frac{1}{12}(\qquad)$$
   Solution:
   $$\frac{3}{4}x^2 - \frac{5}{6}x + 4 = \frac{9}{12}x^2 - \frac{10}{12}x + \frac{48}{12}$$
   $$= \frac{1}{12}(9x^2 - 10x + 48)$$

47. Insert the missing factor in
   $$x^3 - 1 = (x - 1)(\qquad)$$
   Solution:
   $$x^3 - 1 = x^3 - 1^3 = (x - 1)(x^2 + x + 1)$$

Review Exercises for Chapter 0

49. Insert the missing factor in
$$x^4 - 2x^2y^2 + y^4 = (x + y)^2(\quad)^2$$
Solution:
$$x^4 - 2x^2y^2 + y^4 = (x^2 - y^2)^2$$
$$= [(x + y)(x - y)]^2$$
$$= (x + y)^2(x - y)^2$$

51. Use the Quadratic Formula to find the real roots of $16x^2 + 8x - 3 = 0$.
Solution: Since $a = 16$, $b = 8$, and $c = -3$, we have
$$x = \frac{-8 \pm \sqrt{64 - 4(16)(-3)}}{2(16)}$$
$$= \frac{-8 \pm \sqrt{256}}{32} = \frac{-8 \pm 16}{32} = -\frac{3}{4}, \frac{1}{4}$$

53. Use the Quadratic Formula to find the real roots of $x^2 + 8x - 4 = 0$.
Solution: Since $a = 1$, $b = 8$, and $c = -4$, we have
$$x = \frac{-8 \pm \sqrt{64 - 4(1)(-4)}}{2(1)}$$
$$= \frac{-8 \pm \sqrt{80}}{2} = \frac{-8 \pm 4\sqrt{5}}{2} = -4 \pm 2\sqrt{5}$$

55. Use the Quadratic Formula to find the real roots of $16t^2 + 4t + 3 = 0$.
Solution: Since $a = 16$, $b = 4$, and $c = 3$, we have
$$x = \frac{-4 \pm \sqrt{16 - 4(16)(3)}}{2(16)}$$
$$= \frac{-4 \pm \sqrt{-176}}{32}$$
Since the discriminant is negative, there are no real roots.

57. Use the Quadratic Formula to find the real roots of $5.1x^2 - 1.7x - 3.2 = 0$.
Solution: Since $a = 5.1$, $b = -1.7$, and $c = -3.2$, we have
$$x = \frac{1.7 \pm \sqrt{(-1.7)^2 - 4(5.1)(-3.2)}}{2(5.1)}$$
$$= \frac{1.7 \pm \sqrt{2.89 + 65.28}}{10.2}$$
$$= \frac{1.7 \pm \sqrt{68.17}}{10.2} \approx -0.643, 0.976$$

59. Use synthetic division to simplify
$$\frac{3x^3 - 17x^2 + 15x - 25}{x - 5}$$
Solution:

```
5 | 3   -17    15   -25
  |      15   -10    25
  |_____
    3    -2     5     0
```

Therefore, we have
$$\frac{3x^3 - 17x^2 + 15x - 25}{x - 5} = 3x^2 - 2x + 5$$

61. Use synthetic division to simplify
$$\frac{5 - 3x + 2x^2 - x^3}{x + 1}$$
Solution:

```
-1 | -1    2    -3     5
   |       1    -3     6
   |_____
     -1    3    -6    11
```

Therefore, we have
$$\frac{5 - 3x + 2x^2 - x^3}{x + 1} = -x^2 + 3x - 6 + \frac{11}{x + 1}$$

63. Use the Rational Zero Theorem as an aid in finding all real zeros of $2x^3 - 9x^2 - 6x + 5$.
Solution:

Possible rational roots: $\pm 1, \pm 5, \pm\frac{1}{2}, \pm\frac{5}{2}$

Using synthetic division for $x = -1$, we have

```
-1 | 2    -9    -6     5
   |      -2    11    -5
   |_____
     2   -11     5     0
```

Therefore, we have
$$2x^3 - 9x^2 - 6x + 5 = 0$$
$$(x + 1)(2x^2 - 11x + 5) = 0$$
$$(x + 1)(2x - 1)(x - 5) = 0$$
$$x = -1, \frac{1}{2}, 5$$

65. Use the Rational Zero Theorem as an aid in finding all real zeros of $x^4 - 4x^3 + 3x^2 + 2x$.
Solution:

$$x^4 - 4x^3 + 3x^2 + 2x = x(x^3 - 4x^2 + 3x + 2)$$

Possible rational roots: $0, \pm 1, \pm 2$
Using synthetic division for $x = 2$, we have

```
2 | 1   -4    3    2
  |      2   -4   -2
  |_____
    1   -2   -1    0
```

Therefore, we have
$$x^4 - 4x^3 + 3x^2 + 2x = 0$$
$$x(x - 2)(x^2 - 2x - 1) = 0$$
$$x = 0, 2, 1 \pm \sqrt{2}$$

67. Perform the indicated operations and simplify

$$x - 1 + \frac{1}{x + 2} + \frac{1}{x - 1}$$

Solution:

$$x - 1 + \frac{1}{x + 2} + \frac{1}{x - 1}$$

$$= \frac{(x - 1)^2(x + 2) + (x - 1) + (x + 2)}{(x + 2)(x - 1)}$$

$$= \frac{x^3 - 3x + 2 + 2x + 1}{(x + 2)(x - 1)} = \frac{x^3 - x + 3}{(x + 2)(x - 1)}$$

69. Rationalize the numerator in $\frac{\sqrt{x + \Delta x} - \sqrt{x}}{\Delta x}$.

Solution:

$$\frac{\sqrt{x + \Delta x} - \sqrt{x}}{\Delta x} = \frac{\sqrt{x + \Delta x} - \sqrt{x}}{\Delta x} \cdot \frac{\sqrt{x + \Delta x} + \sqrt{x}}{\sqrt{x + \Delta x} + \sqrt{x}}$$

$$= \frac{(x + \Delta x) - x}{\Delta x(\sqrt{x + \Delta x} + \sqrt{x})}$$

$$= \frac{\Delta x}{\Delta x(\sqrt{x + \Delta x} + \sqrt{x})} = \frac{1}{\sqrt{x + \Delta x} + \sqrt{x}}$$

71. The daily cost of producing x units of a certain product is $C = 800 + 0.04x + 0.0002x^2$, $0 \leq x$. If the cost is $1680, how many units are produced?
Solution: When $C = 1680$, we have

$$800 + 0.04x + 0.0002x^2 = 1680$$
$$0.0002x^2 + 0.04x - 880 = 0$$
$$2x^2 + 400x - 8,800,000 = 0$$
$$2(x - 2000)(x + 2200) = 0$$

Since $x \geq 0$, $x = 2000$ units.

PRACTICE TEST FOR CHAPTER 0

1. Determine whether $\sqrt[4]{81}$ is rational or irrational.

2. Determine whether the given value of x satisfies the inequality $3x + 4 \leq x/2$.
   (a) $x = -2$ (b) $x = 0$
   (c) $x = -8/5$ (d) $x = -6$

3. Solve the inequality $3x + 4 \geq 13$.

4. Solve the inequality $x^2 < 6x + 7$.

5. Determine which of the two given real numbers is greater, $\sqrt{19}$ or $13/3$.

6. Given the interval $[-3, 7]$, find (a) the distance between -3 and 7 and (b) the midpoint of the interval.

7. Solve the inequality $|3x + 1| \leq 10$.

8. Solve the inequality $|4 - 5x| > 29$.

9. Solve the inequality $\left|3 - \dfrac{2x}{5}\right| < 8$.

10. Use absolute value to describe the interval $[-3, 5]$.

11. Simplify: $\dfrac{12x^3}{4x^{-2}}$

12. Simplify: $\left(\dfrac{\sqrt{3}\sqrt{x^3}}{x}\right)^0$, $x \neq 0$.

13. Remove all possible factors from the radical: $\sqrt[3]{32x^4y^3}$

14. Complete the factorization:
    $$\dfrac{3}{2}(x + 1)^{-1/3} + \dfrac{1}{4}(x + 1)^{2/3} = \dfrac{1}{4}(x + 1)^{-1/3}(\qquad)$$

15. Find the domain: $\dfrac{1}{\sqrt{5 - x}}$

16. Factor completely: $3x^2 - 19x - 14$

17. Factor completely: $25x^2 - 81$

18. Factor completely: $x^3 + 8$

19. Use the Quadratic Formula to find all real roots of $x^2 + 6x - 2 = 0$.

20. Use the Rational Zero Theorem to find all real roots of $x^3 - 4x^2 + x + 6 = 0$.

Practice Test for Chapter 0

21. Combine terms and simplify: $\dfrac{x}{x^2 + 2x - 3} - \dfrac{1}{x - 1}$

22. Combine terms and simplify: $\dfrac{3 - x}{2\sqrt{x + 3}} + \sqrt{x + 3}$

23. Combine terms and simplify: $\dfrac{\dfrac{\sqrt{x + 2}}{\sqrt{x}} - \dfrac{\sqrt{x}}{\sqrt{x + 2}}}{2(x + 2)}$

24. Rationalize the denominator: $\dfrac{3y}{\sqrt{y^2 + 9}}$

25. Rationalize the numerator: $\dfrac{\sqrt{x} + \sqrt{x + 7}}{14}$

Chapter 1 Functions, Graphs, and Limits

Section 1.1 The Cartesian Plane and the Distance Formula

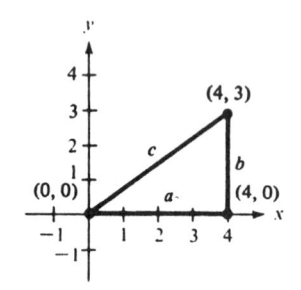

1. (a) Find the length of each side of the right triangle shown in the accompanying graph. (b) Show that these lengths satisfy the Pythagorean Theorem.
   Solution:
   (a) $a = 4$, $b = 3$, $c = \sqrt{(4-0)^2 + (3-0)^2} = 5$
   (b) $a^2 + b^2 = 16 + 9 = 25 = c^2$

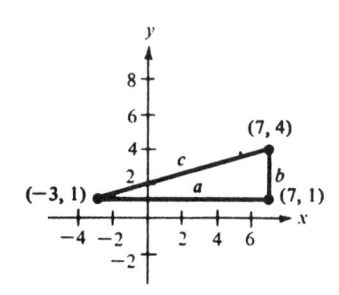

3. (a) Find the length of each side of the right triangle shown in the accompanying graph. (b) Show that these lengths satisfy the Pythagorean Theorem.
   Solution:
   (a) $a = 10$, $b = 3$, $c = \sqrt{(7+3)^2 + (4-1)^2} = \sqrt{109}$
   (b) $a^2 + b^2 = 100 + 9 = 109 = c^2$

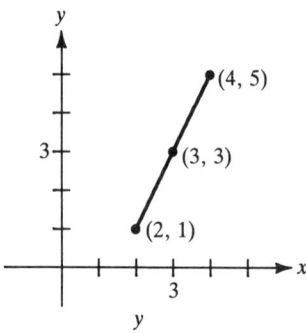

5. (a) Plot the points $(2, 1)$, $(4, 5)$, (b) find the distance between the points, and (c) find the midpoint of the line segment joining the points.
   Solution:
   (a) See graph.
   (b) $d = \sqrt{(4-2)^2 + (5-1)^2} = \sqrt{4 + 16} = \sqrt{20} = 2\sqrt{5}$
   (c) Midpoint $= \left(\dfrac{2+4}{2}, \dfrac{1+5}{2}\right) = (3, 3)$

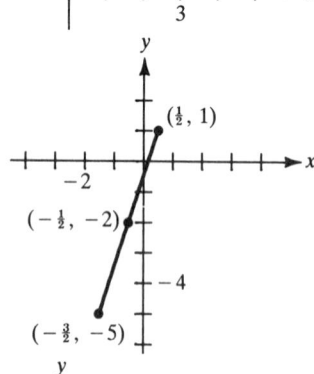

7. (a) Plot the points $(1/2, 1)$ and $(-3/2, -5)$, (b) find the distance between the points, and (c) find the midpoint of the line segment joining the points.
   Solution:
   (a) See graph.
   (b) $d = \sqrt{[-(3/2) - (1/2)]^2 + (-5 - 1)^2}$
       $= \sqrt{4 + 36} = \sqrt{4(1 + 9)} = 2\sqrt{10}$
   (c) Midpoint $= \left(\dfrac{(1/2) + (-3/2)}{2}, \dfrac{1 + (-5)}{2}\right) = \left(-\dfrac{1}{2}, -2\right)$

9. (a) Plot the points $(2, 2)$, $(4, 14)$, (b) find the distance between the points, and (c) find the midpoint of the line segment joining the points.
   Solution:
   (a) See graph.
   (b) $d = \sqrt{(4-2)^2 + (14-2)^2} = \sqrt{4 + 144}$
       $= \sqrt{4(1 + 36)} = 2\sqrt{37}$
   (c) Midpoint $= \left(\dfrac{2+4}{2}, \dfrac{2+14}{2}\right) = (3, 8)$

Section 1.1

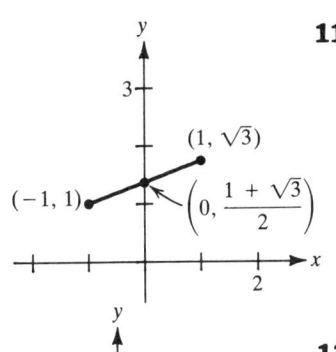

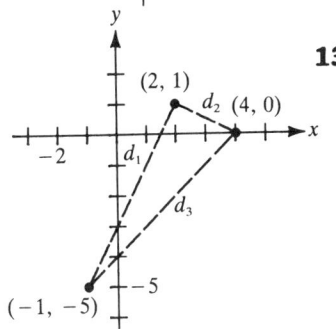

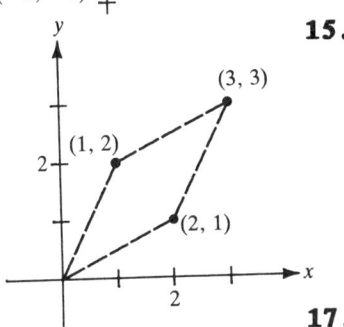

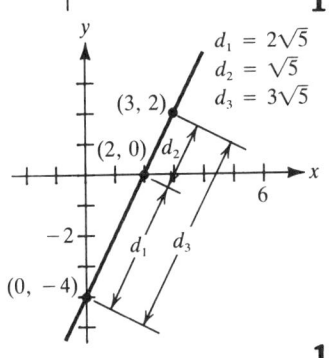

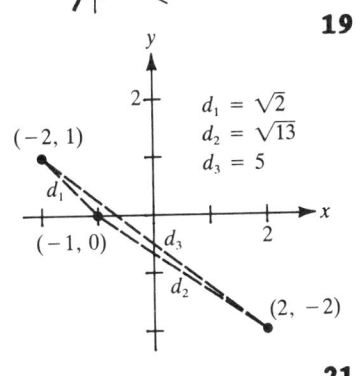

11. (a) Plot the points $(1, \sqrt{3})$, $(-1, 1)$, (b) find the distance between the points, and (c) find the midpoint of the line segment joining the points.
Solution:
(a) See graph.

(b) $d = \sqrt{(-1 - 1)^2 + (1 - \sqrt{3})^2} = \sqrt{8 - 2\sqrt{3}}$

(c) Midpoint $= (\dfrac{1 + (-1)}{2}, \dfrac{\sqrt{3} + 1}{2}) = (0, \dfrac{\sqrt{3} + 1}{2})$

13. Show that the points $(4, 0)$, $(2, 1)$, $(-1, -5)$ form the vertices of a right triangle.
Solution:
$d_1 = \sqrt{[2 - (-1)]^2 + [1 - (-5)]^2} = \sqrt{45}$
$d_2 = \sqrt{(4 - 2)^2 + (0 - 1)^2} = \sqrt{5}$
$d_3 = \sqrt{(-1 - 4)^2 + (-5 - 0)^2} = \sqrt{50}$

Since $d_1^2 + d_2^2 = d_3^2$, triangle is a right triangle.

15. Show that the points $(0, 0)$, $(1, 2)$, $(2, 1)$, $(3, 3)$ form the vertices of a rhombus. (A rhombus is a polygon whose sides are all of the same length.)
Solution:
$d_1 = \sqrt{(1 - 0)^2 + (2 - 0)^2} = \sqrt{5}$
$d_2 = \sqrt{(3 - 1)^2 + (3 - 2)^2} = \sqrt{5}$
$d_3 = \sqrt{(2 - 3)^2 + (1 - 3)^2} = \sqrt{5}$
$d_4 = \sqrt{(0 - 2)^2 + (0 - 1)^2} = \sqrt{5}$

17. Use the Distance Formula to determine whether the points $(0, -4)$, $(2, 0)$, $(3, 2)$ are collinear (lie on the same line). (Hint: Use the model shown in Example 4 and determine whether $d_1 + d_2 = d_3$.)
Solution:
$d_1 = \sqrt{(2 - 0)^2 + (0 + 4)^2} = \sqrt{20} = 2\sqrt{5}$
$d_2 = \sqrt{(3 - 2)^2 + (2 - 0)^2} = \sqrt{5}$
$d_3 = \sqrt{(3 - 0)^2 + (2 + 4)^2} = \sqrt{45} = 3\sqrt{5}$

Since $d_1 + d_2 = d_3$, the points are collinear.

19. Use the Distance Formula to determine whether the points $(-2, 1)$, $(-1, 0)$, $(2, -2)$ are collinear (lie on the same line). (Hint: Use the model shown in Example 4 and determine whether $d_1 + d_2 = d_3$.)
Solution:
$d_1 = \sqrt{(-2 + 1)^2 + (1 - 0)^2} = \sqrt{2}$
$d_2 = \sqrt{(-1 - 2)^2 + (0 + 2)^2} = \sqrt{13}$
$d_3 = \sqrt{(-2 - 2)^2 + (1 + 2)^2} = \sqrt{25} = 5$

Since $d_1 + d_2 \neq d_3$, the points are not collinear.

21. Find x so that the distance between $(0, 0)$ and $(x, -4)$ is 5.
Solution:
$d = \sqrt{(x - 0)^2 + (-4 - 0)^2} = 5$
$\sqrt{x^2 + 16} = 5$
$x^2 + 16 = 25$
$x^2 = 9, \quad x = \pm 3$

23. Find y so that the distance between (0, 0) and (3, y) is 8.
Solution:
$$d = \sqrt{(3-0)^2 + (y-0)^2} = 8$$
$$\sqrt{9 + y^2} = 8$$
$$9 + y^2 = 64$$
$$y^2 = 55, \quad y = \pm\sqrt{55}$$

25. Find the relationship between x and y so that (x, y) is equidistant from (4, -1) and (-2, 3).
Solution:
$$\sqrt{(4-x)^2 + (-1-y)^2} = \sqrt{(-2-x)^2 + (3-y)^2}$$
$$x^2 + y^2 - 8x + 2y + 17 = x^2 + y^2 + 4x - 6y + 13$$
$$8y = 12x - 4$$
$$2y = 3x - 1$$
$$3x - 2y - 1 = 0$$

27. Use the Midpoint Formula successively to find the three points that divide the line segment joining (x_1, y_1) and (x_2, y_2) into four equal parts.
Solution:
$$\text{Midpoint} = \left(\frac{x_1 + x_2}{2}, \frac{y_1 + y_2}{2}\right)$$

The point one-fourth of the way between (x_1, y_1) and (x_2, y_2) is the midpoint of the line segment from (x_1, y_1) to $([x_1 + x_2]/2, [y_1 + y_2]/2)$, which is

$$\left(\frac{x_1 + \frac{x_1 + x_2}{2}}{2}, \frac{y_1 + \frac{y_1 + y_2}{2}}{2}\right) = \left(\frac{3x_1 + x_2}{4}, \frac{3y_1 + y_2}{4}\right)$$

The point three-fourths of the way between (x_1, y_1) and (x_2, y_2) is the midpoint of the line segment from $([x_1 + x_2]/2, [y_1 + y_2]/2)$ to (x_2, y_2), which is

$$\left(\frac{\frac{x_1 + x_2}{2} + x_2}{2}, \frac{\frac{y_1 + y_2}{2} + y_2}{2}\right) = \left(\frac{x_1 + 3x_2}{4}, \frac{y_1 + 3y_2}{4}\right)$$

Thus, $([3x_1 + x_2]/4, [3y_1 + y_2]/4)$, $([x_1 + x_2]/2, [y_1 + y_2]/2)$, and $([x_1 + 3x_2]/4, [y_1 + 3y_2]/4)$ are the three points that divide the line segment joining (x_1, y_1) and (x_2, y_2) into four equal parts.

29. Use the result of Exercise 27 to find the points that divide the line segment joining the given points into four equal parts.
(a) (1, -2), (4, -1)   (b) (-2, -3), (0, 0)
Solution:

(a) $\left(\frac{3(1) + 4}{4}, \frac{3(-2) - 1}{4}\right) = \left(\frac{7}{4}, -\frac{7}{4}\right)$

$\left(\frac{1 + 4}{2}, \frac{-2 - 1}{2}\right) = \left(\frac{5}{2}, -\frac{3}{2}\right)$

$\left(\frac{1 + 3(4)}{4}, \frac{-2 + 3(-1)}{4}\right) = \left(\frac{13}{4}, -\frac{5}{4}\right)$

Section 1.1

(b) $\left(\dfrac{3(-2)+0}{4}, \dfrac{3(-3)+0}{4}\right) = \left(-\dfrac{3}{2}, -\dfrac{9}{4}\right)$

$\left(\dfrac{-2+0}{2}, \dfrac{-3+0}{2}\right) = \left(-1, -\dfrac{3}{2}\right)$

$\left(\dfrac{-2+3(0)}{4}, \dfrac{-3+3(0)}{4}\right) = \left(-\dfrac{1}{2}, -\dfrac{3}{4}\right)$

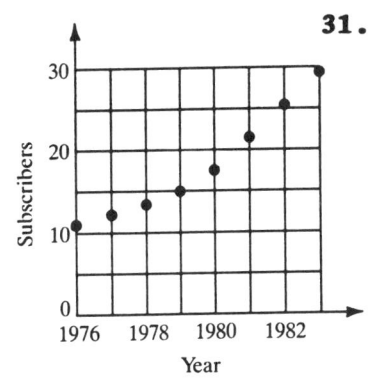

31. Plot the points given in the following table on a rectangular coordinate system. The number of subscribers to cable TV is given in millions. (Use Example 2 as a model.)

Year	1976	1977	1978	1979	1980	1981	1982	1983
Subscribers	11.0	12.2	13.4	15.0	17.5	21.5	25.4	29.4

Solution:
See accompanying graph.

33. The base and height of the trusses for the roof of a house are 32 feet and 5 feet, respectively. (a) Find the distance from the eaves to the peak of the roof. (b) Use the result of part (a) to find the number of square feet of roofing required for the house if the length of the house is 40 feet.
Solution:
(a) $x^2 = 16^2 + 5^2$, $x > 0$
$x^2 = 281$
$x = \sqrt{281} \approx 16.76$ feet
(b) $A = 2(40)(\sqrt{281}) = 80\sqrt{281} \approx 1341.04$ square feet

35. Use the Midpoint Formula to estimate the sales of a company for 1983, given the sales in 1980 and 1986. Assume that the annual increase in sales followed a linear pattern.

Year	1980	1986
Sales	$520,000	$740,000

Solution:
Midpoint $= \left(\dfrac{1980+1986}{2}, \dfrac{520,000+740,000}{2}\right)$
$= (1983, 630,000)$

Thus, the 1983 sales were $\approx \$630,000.00$.

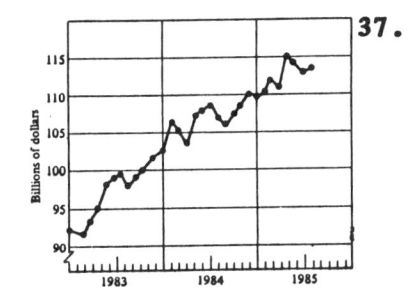

37. Use the figure showing retail sales in billions of dollars to approximate retail sales for
(a) June 1983   (b) January 1984
(c) January 1985   (d) June 1985
Solution:
(a) Sales were approximately $99 billion.
(b) Sales were approximately $103 billion.
(c) Sales were approximately $110 billion.
(d) Sales were approximately $113 billion.

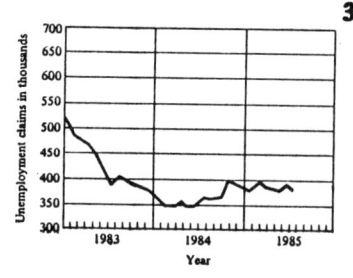

39. Use the figure showing unemployment claims in thousands to approximate the number of claims for
    (a) January 1983   (b) June 1983
    (c) June 1984   (d) June 1985
    Solution:
    (a) Claims were approximately 510,000.
    (b) Claims were approximately 410,000.
    (c) Claims were approximately 350,000.
    (d) Claims were approximately 380,000.

41. A line segment has (x_1, y_1) as one endpoint and (x_m, y_m) as its midpoint. Find (x_2, y_2), the other endpoint of the line segment.
    Solution: The midpoint is $([x_1 + x_2]/2, [y_1 + y_2]/2)$. Thus,
    $$x_m = \frac{x_1 + x_2}{2} \quad \text{and} \quad y_m = \frac{y_1 + y_2}{2}$$
    $$2x_m = x_1 + x_2 \qquad\qquad 2y_m = y_1 + y_2$$
    $$2x_m - x_1 = x_2 \qquad\qquad 2y_m - y_1 = y_2$$

    and the other endpoint is $(2x_m - x_1, 2y_m - y_1)$.

● Section 1.2 Graphs of Equations

1. Determine whether the points are solution points for the equation $2x - y - 3 = 0$.
   (a) $(1, 2)$ (b) $(1, -1)$ (c) $(4, 5)$
   Solution:
   (a) This is a not solution point since
       $2x - y - 3 = 2(1) - 2 - 3 = -3 \neq 0$

   (b) This is a solution point since
       $2x - y - 3 = 2(1) - (-1) - 3 = 0$

   (c) This is a solution point since
       $2x - y - 3 = 2(4) - 5 - 3 = 0$

3. Determine whether the points are solution points for the equation $x^2y - x^2 + 4y = 0$.
   (a) $(1, 1/5)$ (b) $(2, 1/2)$ (c) $(-1, -2)$
   Solution:
   (a) This is a solution point since
       $$x^2y - x^2 + 4y = (1)^2\left(\frac{1}{5}\right) - (1)^2 + 4\left(\frac{1}{5}\right) = 0$$

   (b) This is a solution point since
       $$x^2y - x^2 + 4y = (2)^2\left(\frac{1}{2}\right) - (2)^2 + 4\left(\frac{1}{2}\right) = 0$$

   (c) This is not a solution point since
       $$x^2y - x^2 + 4y = (-1)^2(-2) - (-1)^2 + 4(-2) = -11$$

Section 1.2

5. Find the intercepts of the graph of $2x - y - 3 = 0$.
   Solution: To find the y-intercept, let $x = 0$ to obtain
   $$2(0) - y - 3 = 0$$
   $$y = -3$$
   Thus, the y-intercept is $(0, -3)$. To find the x-intercept, let $y = 0$ to obtain
   $$2x - (0) - 3 = 0$$
   $$x = 3/2$$
   Thus, the x-intercept is $(3/2, 0)$.

7. Find the intercepts of the graph of $y = x^2 + x - 2$.
   Solution: The y-intercept occurs at $(0, -2)$. To find the x-intercepts, let $y = 0$ to obtain
   $$(x + 2)(x - 1) = 0$$
   $$x = -2, 1$$
   Thus, the x-intercepts are $(-2, 0)$ and $(1, 0)$.

9. Find the intercepts of the graph of $y = x^2\sqrt{9 - x^2}$.
   Solution: The y-intercept occurs at $(0, 0)$. To find the x-intercepts, let $y = 0$ to obtain
   $$x^2\sqrt{9 - x^2} = 0$$
   $$x = 0, -3, 3$$
   Thus, the x-intercepts are $(0, 0)$, $(-3, 0)$, and $(3, 0)$.

11. Find the intercepts of the graph of
    $$y = \frac{x - 1}{x - 2}$$
    Solution: The y-intercept occurs at $(0, 1/2)$, and the x-intercept occurs at $(1, 0)$.

13. Find the intercepts of the graph of $x^2y - x^2 + 4y = 0$.
    Solution: The x-intercept and y-intercept both occur at $(0, 0)$.

15. Match $y = x - 2$ with the correct graph.
    Solution: The graph of $y = x - 2$ is a straight line with y-intercept at $(0, -2)$. Thus, it matches (c).

17. Match $y = x^2 + 2x$ with the correct graph.
    Solution: The graph of $y = x^2 + 2x$ is a parabola opening up with vertex at $(-1, -1)$. Thus, it matches (b).

19. Match $y = |x| - 2$ with the correct graph.
    Solution: The graph of $y = |x| - 2$ matches (a).

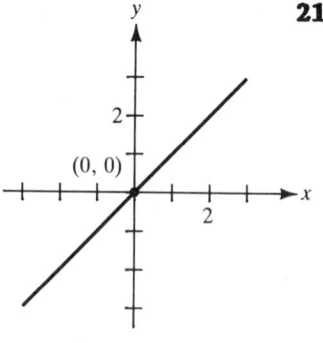

21. Sketch the graph of y = x and plot the intercepts.
Solution: The graph is a straight line with intercept at (0, 0).

x	0	1	2	3
y	0	1	2	3

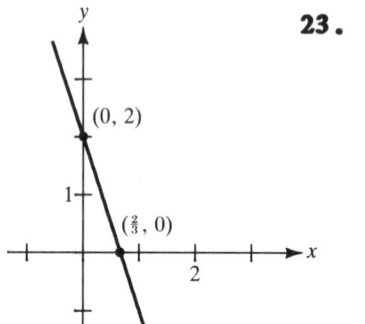

23. Sketch the graph of y = -3x + 2 and plot the intercepts.
Solution: The graph is a straight line with intercepts at (2/3, 0) and (0, 2).

x	0	2/3	1	2
y	2	0	-1	-4

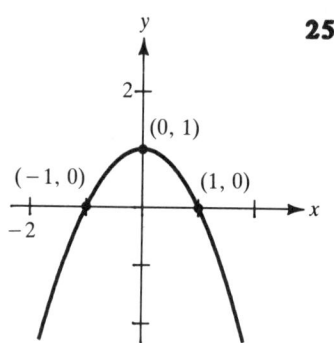

25. Sketch the graph of $y = 1 - x^2$ and plot the intercepts.
Solution: The graph is a parabola with vertex at (0, 1) and intercepts at (1, 0), (-1, 0), and (0, 1).

x	0	±1	±2
y	1	0	-3

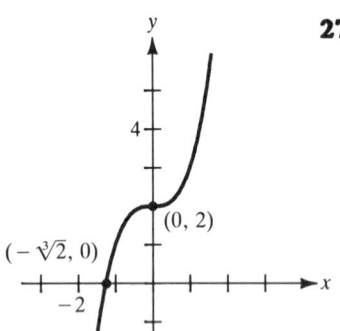

27. Sketch the graph of $y = x^3 + 2$ and plot the intercepts.
Solution:
Intercepts: $(-\sqrt[3]{2}, 0)$ and (0, 2)

x	-2	-1	0	1	2
y	-6	1	2	3	10

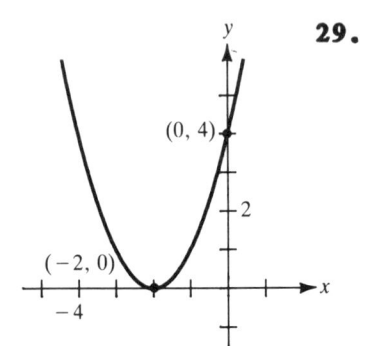

29. Sketch the graph of $y = (x + 2)^2$ and plot the intercepts.
Solution: The graph is a parabola with vertex at (-2, 0) and intercepts at (-2, 0) and (0, 4).

x	-3	-2	-1	0	1
y	1	0	1	4	9

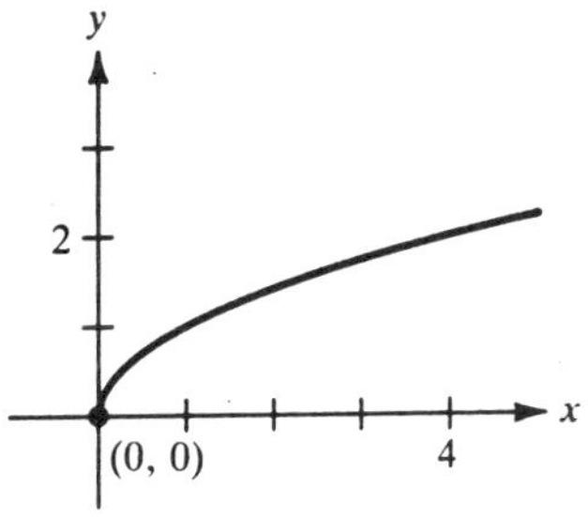

Section 1.2

31. Sketch the graph of $y = \sqrt{x}$ and plot the intercepts.
Solution:
$x \geq 0$, intercept: (0, 0)

x	0	1	4	9
y	0	1	2	3

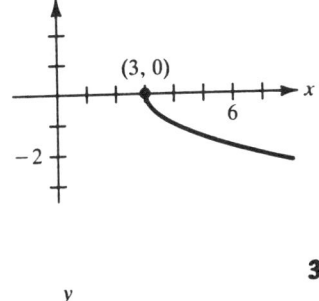

33. Sketch the graph of $y = -\sqrt{x - 3}$ and plot the intercepts.
Solution:
$x \geq 3$, intercept: (3, 0)

x	3	4	7	12
y	0	-1	-2	-3

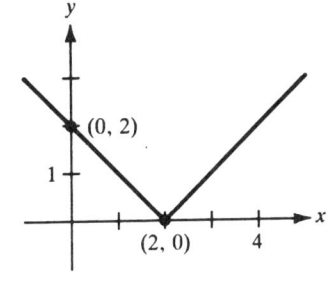

35. Sketch the graph of $y = |x - 2|$ and plot the intercepts.
Solution:
Intercepts: (2, 0) and (0, 2)

x	-1	0	1	2	3
y	3	2	1	0	1

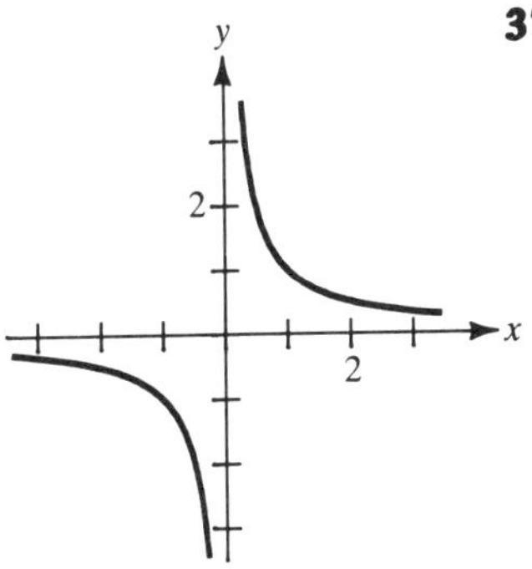

37. Sketch the graph of $y = 1/x$ and plot the intercepts.
Solution:
$x \neq 0$, no intercepts

x	-3	-2	-1	-1/2	1/2	1	2	3
y	-1/3	-1/2	-1	-2	2	1	1/2	1/3

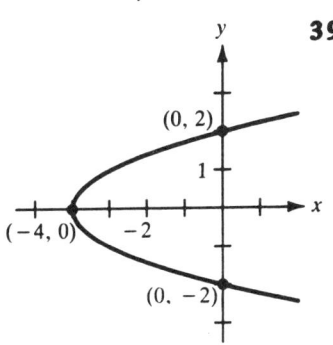

39. Sketch the graph of $x = y^2 - 4$ and plot the intercepts.
Solution: The graph is a parabola with vertex at (-4, 0) and intercepts at (-4, 0), (0, 2) and (0, -2).

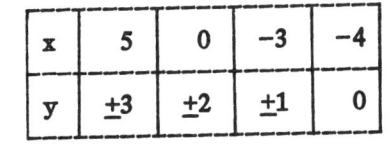

x	5	0	-3	-4
y	±3	±2	±1	0

Section 1.2

41. Write the general form of the equation of a circle with a center at (0, 0) and a radius of 3.
Solution:
$$(x - 0)^2 + (y - 0)^2 = 3^2$$
$$x^2 + y^2 = 9$$
$$x^2 + y^2 - 9 = 0$$

43. Write the general form of the equation of a circle with a center at (2, −1) and a radius of 4.
Solution:
$$(x - 2)^2 + (y + 1)^2 = 16$$
$$x^2 - 4x + 4 + y^2 + 2y + 1 = 16$$
$$x^2 + y^2 - 4x + 2y - 11 = 0$$

45. Write the general form of the equation of a circle with a center at (−1, 2) and solution point (0, 0).
Solution: Since the point (0, 0) lies on the circle, the radius must be the distance between (0, 0) and (−1, 2).

$$\text{Radius} = \sqrt{(0 + 1)^2 + (0 - 2)^2} = \sqrt{5}$$

$$(x + 1)^2 + (y - 2)^2 = 5$$
$$x^2 + y^2 + 2x - 4y = 0$$

47. Write the general form of the equation of a circle with endpoints of a diameter: (0, 0), (6, 8).
Solution:
Center = midpoint = (3, 4)
Radius = distance from the center to an endpoint
$$= \sqrt{(3 - 0)^2 + (4 - 0)^2} = 5$$

$$(x - 3)^2 + (y - 4)^2 = 25$$
$$x^2 + y^2 - 6x - 8y = 0$$

49. Use the process of completing the square to write
$$x^2 + y^2 - 2x + 6y + 6 = 0$$
in standard form. Sketch the graph of the circle.
Solution:
$$(x^2 - 2x + 1) + (y^2 + 6y + 9) = -6 + 1 + 9$$
$$(x - 1)^2 + (y + 3)^2 = 4$$

51. Use the process of completing the square to write
$$x^2 + y^2 - 2x + 6y + 10 = 0$$
in standard form. Sketch the graph of the circle.
Solution:
$$(x^2 - 2x + 1) + (y^2 + 6y + 9) = -10 + 1 + 9$$
$$(x - 1)^2 + (y + 3)^2 = 0$$

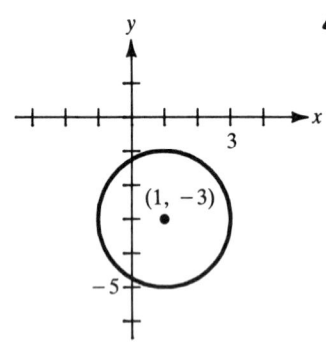

Section 1.2

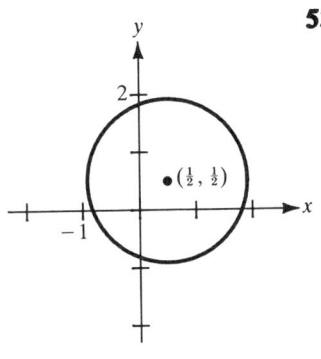

53. Use the process of completing the square to write $2x^2 + 2y^2 - 2x - 2y - 3 = 0$ in standard form. Sketch the graph of the circle.
Solution:
$$x^2 + y^2 - x - y = \frac{3}{2}$$
$$(x^2 - x + \frac{1}{4}) + (y^2 - y + \frac{1}{4}) = \frac{3}{2} + \frac{1}{4} + \frac{1}{4}$$
$$(x - \frac{1}{2})^2 + (y - \frac{1}{2})^2 = 2$$

55. Use the process of completing the square to write $16x^2 + 16y^2 + 16x + 40y - 7 = 0$ in standard form. Sketch the graph of the circle.
Solution:
$$x^2 + y^2 + x + \frac{5}{2}y = \frac{7}{16}$$
$$(x^2 + x + \frac{1}{4}) + (y^2 + \frac{5}{2}y + \frac{25}{16}) = \frac{7}{16} + \frac{1}{4} + \frac{25}{16}$$
$$(x + \frac{1}{2})^2 + (y + \frac{5}{4})^2 = \frac{9}{4}$$

57. Find the point of intersection of the graphs of $x + y = 2$ and $2x - y = 1$ and check your results.
Solution: Solving for y in the equation $x + y = 2$ yields $y = 2 - x$, and solving for y in the equation $2x - y = 1$ yields $y = 2x - 1$. Then setting these two y-values equal to each other, we have
$$2 - x = 2x - 1$$
$$3 = 3x$$
$$x = 1$$

The corresponding y-value is $y = 2 - 1 = 1$, so the point of intersection is $(1, 1)$.

59. Find the point of intersection of the graphs of $x + y = 7$ and $3x - 2y = 11$ and check your results.
Solution: Solving for y in the first equation yields $y = 7 - x$ and substituting this value in the second equation gives us
$$3x - 2(7 - x) = 11$$
$$5x = 25$$
$$x = 5$$

The corresponding y-value is $y = 7 - 5 = 2$, so the point of intersection is $(5, 2)$.

61. Find the points of intersection of the graphs of $x^2 + y^2 = 5$ and $x - y = 1$ and check your results.
Solution: Solving for x in the second equation yields $x = y + 1$, and substituting this into the first equation gives us

$$(y + 1)^2 + y^2 = 5$$
$$y^2 + 2y + 1 + y^2 = 5$$
$$2y^2 + 2y - 4 = 0$$
$$2(y - 1)(y + 2) = 0$$
$$y = 1, -2$$

The corresponding x-values are $x = 2$ and $x = -1$, so the points of intersection are $(2, 1)$ and $(-1, -2)$.

63. Find the points of intersection of the graphs of $y = x^3$ and $y = x$ and check your results.
    Solution: By equating the y-values for the two equations, we have

$$x^3 = x$$
$$x^3 - x = 0$$
$$x(x^2 - 1) = 0$$
$$x(x + 1)(x - 1) = 0$$
$$x = 0, -1, 1$$

The corresponding y-values are $y = 0$, $y = -1$, and $y = 1$, so the points of intersection are $(-1, -1)$, $(0, 0)$, and $(1, 1)$.

65. Find the points of intersection of the graphs of $y = x^4 - 2x^2 + 1$ and $y = 1 - x^2$ and check your results.
    Solution: By equating the y-values for the two equations, we have

$$x^4 - 2x^2 + 1 = 1 - x^2$$
$$x^4 - x^2 = 0$$
$$x^2(x + 1)(x - 1) = 0$$
$$x = 0, -1, 1$$

The corresponding y-values are $y = 1$, 0, and 0, so the points of intersection are $(-1, 0)$, $(0, 1)$, and $(1, 0)$.

67. A person setting up a part-time business makes an initial investment of $5,000. The unit cost of the product is $21.60, and the selling price is $34.10.
    (a) Find equations for the total cost C and total revenue R for x units.
    (b) Find the break-even point by finding the point of intersection of the cost and revenue equations of part (a).
    Solution:
    (a) $C = 21.6x + 5000$
        $R = 34.1x$
    (b) By equating R and C, we have

$$R = C$$
$$34.1x = 21.6x + 5000$$
$$12.5x = 5000$$
$$x = 5000/12.5 = 400 \text{ units}$$

Section 1.2   45

69. Find the sales necessary to break even (R = C) for the cost C of x units and the revenue R obtained from selling x units, when C = 8650x + 250,000 and R = 9950x. (Round your answer <u>up</u> to the nearest whole unit.)
Solution:
$$R = C$$
$$9950x = 8650x + 250,000$$
$$1300x = 250,000$$
$$x = 250,000/1300 \approx 193 \text{ units}$$

71. The following table gives the consumer price index (CPI) for the years 1969-1983. In the base year of 1967, CPI = 100.

Year	1969	1970	1971	1972	1973	1974	1975	1976
CPI	109.8	116.3	121.3	125.3	133.1	147.7	161.2	170.5

Year	1977	1978	1979	1980	1981	1982	1983
CPI	181.5	195.3	217.7	247.0	272.3	288.6	297.4

A mathematical model for the CPI during this period is $y = 0.79t^2 + 4.68t + 114.68$ where y represents the CPI and t the year, with t = 0 corresponding to 1970.
(a) Use a graph to compare the CPI with the model.
(b) Use the model to predict the CPI for 1985.
Solution:
(a) See accompanying graph.
(b) In 1985, t = 15, and the predicted CPI is

$$y = 0.79(15)^2 + 4.68(15) + 114.68 = 362.63$$

73. The average number of acres per farm in the United States is given in the following table:

Year	1950	1960	1965	1970	1975	1980	1984
Acres	213	297	340	374	391	427	437

A mathematical model for these data is given by $y = -0.09t^2 + 9.65t + 212.48$ where y represents the average number of acres per farm and t represents the year, with t = 0 corresponding to 1950.
(a) Use a graph to compare the actual number of acres per farm with that given by the model.
(b) Use the model to predict the average number of acres per farm in the United States in 1990.
Solution:
(a) See accompanying graph.
(b) In 1990, t = 40, and the predicted average number of acres is

$$y = -0.09(40)^2 + 9.65(40) + 212.48 = 454.48 \text{ acres}$$

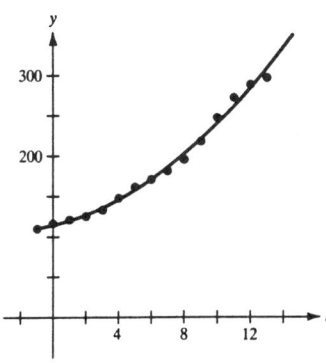

Section 1.3 Lines in the Plane, Slope

1. Estimate the slope of the given line from its graph.
   Solution: The slope is m = 1 since the line rises one unit vertically for each unit of horizontal change from left to right.

3. Estimate the slope of the given line from its graph.
   Solution: The slope is m = 0 since the line is horizontal.

5. Estimate the slope of the given line from its graph.
   Solution: The slope is m = -3 since the line falls three units vertically for each unit of horizontal change from left to right.

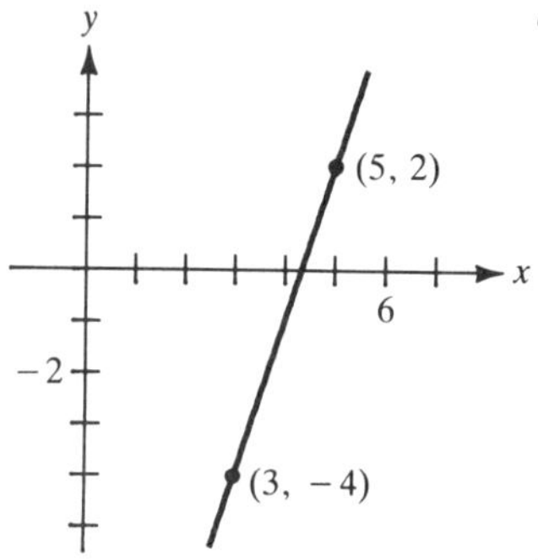

7. Plot the points (3, -4), (5, 2) and find the slope of the line passing through the points.
   Solution: The points are plotted in the accompanying graph and the slope is
   $$m = \frac{2 - (-4)}{5 - 3} = 3$$

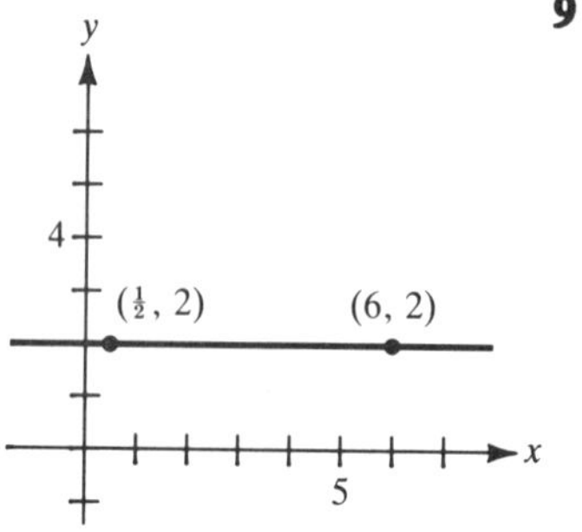

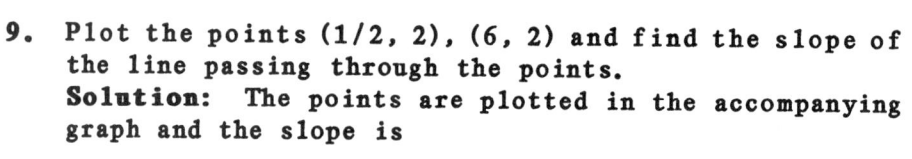

9. Plot the points (1/2, 2), (6, 2) and find the slope of the line passing through the points.
   Solution: The points are plotted in the accompanying graph and the slope is
   $$m = \frac{2 - 2}{6 - (1/2)} = 0$$
   Thus, the line is horizontal.

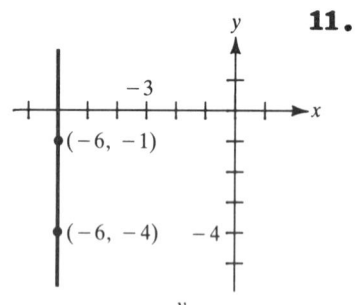

11. Plot the points (-6, -1), (-6, -4) and find the slope of the line passing through the points.
    Solution: The points are plotted on the accompanying graph. The slope is undefined since
    $$m = \frac{-4 - (-1)}{-6 - (-6)} \quad \text{(undefined slope)}$$
    Thus, the line is vertical.

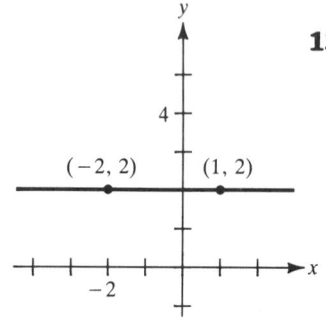

13. Plot the points (1, 2), (-2, 2) and find the slope of the line passing through the points.
    Solution: The points are plotted on the accompanying graph, and the slope is
    $$m = \frac{2 - 2}{-2 - 1} = 0$$
    Thus, the line is horizontal.

Section 1.3   47

15. A line with a slope of m = 0 passes through the point (2, 1). Find three additional points that the line passes through. (The solution is not unique.)
Solution: The equation of this horizontal line is y = 1. Therefore, three additional points are (0, 1), (1, 1), and (3, 1).

17. A line with a slope of m = 1 passes through the point (5, −6). Find three additional points that the line passes through. (The solution is not unique.)
Solution: The equation of this line is

$$y + 6 = 1(x - 5)$$
$$y = x - 11$$

Therefore, three additional points are (6, −5), (7, −4) and (8, −3).

19. A line with a slope of m = −3 passes through the point (1, 7). Find three additional points that the line passes through. (The solution is not unique.)
Solution: The equation of the line is

$$y - 7 = -3(x - 1)$$
$$y = -3x + 10$$

Therefore, three additional points are (0, 10), (2, 4), and (3, 1).

21. A line with an undefined slope passes through the point (−8, 1). Find three additional points that the line passes through. (The solution is not unique.)
Solution: The equation of this vertical line is x = −8. Therefore, three additional points are (−8, 0), (−8, 2), and (−8, 3).

23. Find the slope and y-intercept of the line given by x + 5y = 20.
Solution:
$$x + 5y = 20$$
$$y = -\frac{1}{5}x + 4$$

Therefore, the slope is m = −1/5, and the y-intercept is (0, 4).

25. Find the slope and y-intercept of the line given by 4x − 3y = 18.
Solution:
$$4x - 3y = 18$$
$$y = \frac{4}{3}x - 6$$

Therefore, the slope is m = 4/3, and the y-intercept is (0, −6).

48 Section 1.3

27. Find the slope and y-intercept (if possible) of the line given by x = 4.
Solution: Since the line is vertical, the slope is undefined and there is no y-intercept.

29. Find an equation for the line that passes through the points (2, 1) and (0, -3) and sketch the graph of the line.
Solution: The slope of the line is
$$m = \frac{-3 - 1}{0 - 2} = 2$$
Using the point-slope form, we have
$$y - 1 = 2(x - 2)$$
$$2x - y - 3 = 0$$

31. Find an equation for the line that passes through the points (0, 0) and (-1, 3) and sketch the graph of the line.
Solution: The slope of the line is
$$m = \frac{3 - 0}{-1 - 0} = -3$$
Using the point-slope, we have
$$y = -3x$$
$$3x + y = 0$$

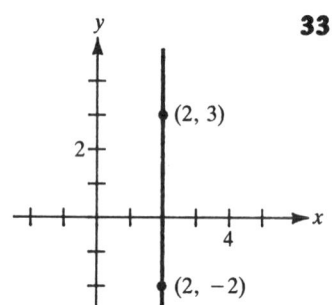

33. Find an equation for the line that passes through the points (2, 3) and (2, -2) and sketch the graph of the line.
Solution: The slope of the line is undefined, so the line is vertical and its equation is
$$x = 2$$
$$x - 2 = 0$$

35. Find an equation for the line that passes through the points (1, -2) and (3, -2) and sketch the graph of the line.
Solution: The slope of the line is m = 0, so the line is horizontal and its equation is
$$y = -2$$
$$y + 2 = 0$$

37. Find an equation of the line that passes through the point (0, 3) and has the slope m = 3/4. Sketch the line.
Solution: Using the point-slope form, we have
$$y - 3 = \frac{3}{4}(x - 0)$$
$$4y - 12 = 3x$$
$$3x - 4y + 12 = 0$$

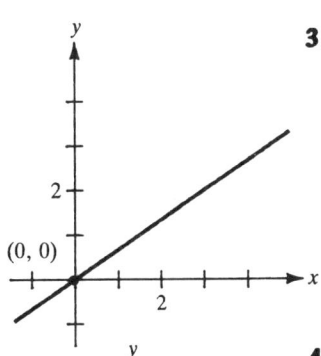

Section 1.3

39. Find an equation of the line that passes through the point (0, 0) and has the slope m = 2/3. Sketch the line.
Solution: Using the point-slope form, we have
$$y - 0 = \frac{2}{3}(x - 0)$$
$$2x - 3y = 0$$

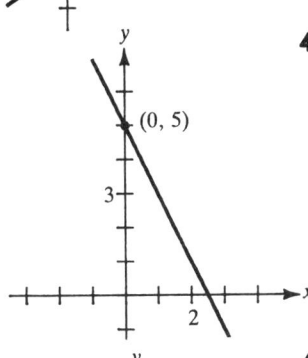

41. Find an equation of the line that passes through the point (0, 5) and has the slope m = -2. Sketch the line.
Solution: Using the point-slope form, we have
$$y - 5 = -2(x - 0)$$
$$2x + y - 5 = 0$$

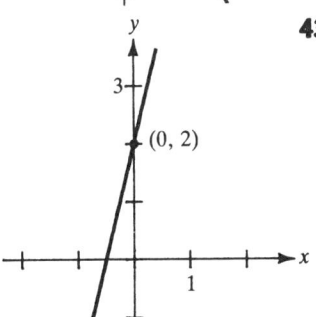

43. Find an equation of the line that passes through the point (0, 2) and has the slope m = 4. Sketch the line.
Solution: Using the point-slope form, we have
$$y - 2 = 4(x - 0)$$
$$4x - y + 2 = 0$$

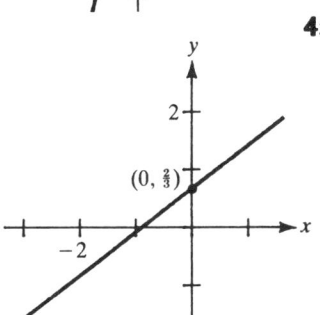

45. Find an equation of the line that passes through the point (0, 2/3) and has the slope m = 3/4. Sketch the line.
Solution: Using the point-slope form, we have
$$y - \frac{2}{3} = \frac{3}{4}(x - 0)$$
$$12y = 9x + 8$$
$$9x - 12y + 8 = 0$$

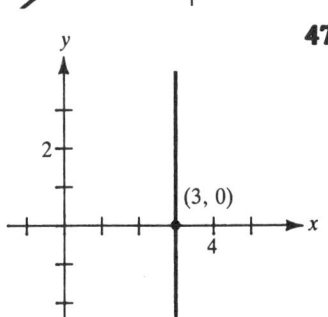

47. Find an equation of the vertical line with x-intercept at 3. Sketch the line.
Solution: Since the line is vertical, it has an undefined slope and its equation is
$$x = 3$$
$$x - 3 = 0$$

49. Write an equation of the line through the point (-3, 2) (a) parallel to the line x + y = 7 and (b) perpendicular to the line.
Solution:
(a) parallel, $m_2 = -1$ (b) perpendicular, $m_2 = 1$
    $y - 2 = -1(x + 3)$ $y - 2 = 1(x + 3)$
    $x + y + 1 = 0$ $x - y + 5 = 0$

51. Write an equation of the line through the point (-6, 4) (a) parallel to the line $3x + 4y = 7$ and (b) perpendicular to the line.
Solution:

(a) parallel, $m_2 = -\dfrac{3}{4}$ (b) perpendicular, $m_2 = \dfrac{4}{3}$

$y - 4 = -\dfrac{3}{4}(x + 6)$ $y - 4 = \dfrac{4}{3}(x + 6)$

$4y - 16 = -3x - 18$ $3y - 12 = 4x + 24$

$3x + 4y + 2 = 0$ $4x - 3y + 36 = 0$

53. Write an equation of the line through the point (-1, 0) (a) parallel to the line $y = -3$ and (b) perpendicular to the line.
Solution:
(a) $y = 0$ or the x-axis
(b) $x = -1$ or $x + 1 = 0$

55. Find the equation of the line giving the relationship between the temperature in degrees Celsius C and degrees Fahrenheit F. Use the fact that water freezes at 0° Celsius (32° Fahrenheit) and boils at 100° Celsius (212° Fahrenheit).
Solution:
$$m = \frac{212 - 32}{100 - 0} = \frac{9}{5}$$

$$F - 32 = \frac{9}{5}(C - 0)$$

$$F = \frac{9}{5}C + 32$$

57. A company reimburses its sales representatives $95 per day for lodging and meals plus $0.25 per mile driven. Write a linear equation giving the daily cost C to the company in terms of x, the number of miles driven.
Solution:

$$C = 0.25x + 95$$

59. A small business purchases a piece of equipment for $875. After 5 years the equipment will be obsolete and have no value. Write a linear equation giving the value y of the equipment during the five years it will be used. (Let t represent the time in years.)
Solution: The equipment depreciates

$$\frac{875}{5} = \$175 \text{ per year}$$

so the value is $y = 875 - 175t$ where $0 \leq t \leq 5$.

61. A real estate office handles an apartment complex with 50 units. When the rent is $380 per month, all 50 units are occupied. However, when the rent is $425, the average number of occupied units drops to 47. Assume that the relationship between the monthly rent p and the demand x is linear. (Note: Here we use the term <u>demand</u> to refer to the number of occupied units.)
  (a) Write a linear equation giving the quantity demanded x in terms of the rent p.
  (b) (Linear Extrapolation) Use this equation to predict the number of units occupied if the rent is raised to $455.
  (c) (Linear Interpolation) Predict the number of units occupied if the rent is lowered to $395.

Solution:
(a) (50, 380), (47, 425)

$$m = \frac{425 - 380}{47 - 50} = -15$$

$$p - 380 = -15(x - 50)$$
$$p = -15x + 1130 \quad \text{or} \quad x = \frac{1}{15}(1130 - p)$$

(b) $x = \frac{1}{15}(1130 - 455) = 45$ units

(c) $x = \frac{1}{15}(1130 - 395) = 49$ units

63. A contractor purchases a piece of equipment for $26,500. The equipment requires an average expenditure of $5.25 per hour for fuel and maintenance, and the operator is paid $9.50 per hour.
  (a) Write a linear equation giving the total cost C of operating this equipment t hours.
  (b) Given that customers are charged $25 per hour of machine use, write an equation for the revenue R derived from t hours of use.
  (c) Use the formula for profit (P = R − C) to write an equation for the profit derived from t hours of use.
  (d) Find the number of hours the equipment must be operated for the contractor to break even.

Solution:
(a) C = (5.25 + 9.50)t + 26,500
    = 14.75t + 26,500

(b) R = 25t

(c) P = R − C
    = 25t − (14.75t + 26,500)
    = 10.25t − 26,500

(d) R = C
    25t = 14.75t + 26,500
    10.25t = 26,500
    t ≈ 2585.4 hours

Section 1.4 Functions

1. Given $f(x) = 2x - 3$, find
  (a) $f(0)$ (b) $f(-3)$
  (c) $f(x - 1)$ (d) $f(1 + \Delta x)$
  Solution:
  (a) $f(0) = 2(0) - 3 = -3$
  (b) $f(-3) = 2(-3) - 3 = -9$
  (c) $f(x - 1) = 2(x - 1) - 3 = 2x - 5$
  (d) $f(1 + \Delta x) = 2(1 + \Delta x) - 3 = 2\Delta x - 1$

3. Given $f(x) = x^2$, find
  (a) $f(-2)$ (b) $f(6)$
  (c) $f(c)$ (d) $f(x + \Delta x)$
  Solution:
  (a) $f(-2) = (-2)^2 = 4$ (b) $f(6) = (6)^2 = 36$
  (c) $f(c) = (c)^2 = c^2$
  (d) $f(x + \Delta x) = (x + \Delta x)^2 = x^2 + 2x(\Delta x) + (\Delta x)^2$

5. Given $f(x) = |x|/x$, find
  (a) $f(2)$ (b) $f(-2)$
  (c) $f(x^2)$ (d) $f(x - 1)$
  Solution:
  (a) $f(2) = |2|/2 = 1$ (b) $f(-2) = |-2|/-2 = -1$
  (c) $f(x^2) = |x^2|/x^2 = 1$
  (d) $f(x - 1) = \dfrac{|x - 1|}{x - 1} = \begin{cases} -1, & x < 1 \\ 1, & x > 1 \end{cases}$

7. Given $f(x) = 3x - 1$, find and simplify

$$[f(x + \Delta x) - f(x)]/\Delta x$$
  Solution:

$$\frac{f(x + \Delta x) - f(x)}{\Delta x} = \frac{[3(x + \Delta x) - 1] - (3x - 1)}{\Delta x}$$

$$= \frac{3x + 3\Delta x - 1 - 3x + 1}{\Delta x}$$

$$= 3\Delta x/\Delta x = 3$$

9. Given $f(x) = x^2 - x + 1$, find and simplify

$$[f(2 + \Delta x) - f(2)]/\Delta x$$
  Solution:

$$\frac{f(2 + \Delta x) - f(2)}{\Delta x}$$

$$= \frac{(2 + \Delta x)^2 - (2 + \Delta x) + 1 - [(2)^2 - 2 + 1]}{\Delta x}$$

$$= \frac{4 + 4\Delta x + (\Delta x)^2 - 2 - \Delta x + 1 - 4 + 2 - 1}{\Delta x}$$

$$= \frac{3\Delta x + (\Delta x)^2}{\Delta x} = 3 + \Delta x$$

Section 1.4 53

11. Given $f(x) = x^3 - x$, find and simplify

$$\frac{f(x) - f(1)}{x - 1}$$

Solution:

$$\frac{f(x) - f(1)}{x - 1} = \frac{(x^3 - x) - ((1)^3 - 1)}{x - 1} = \frac{x^3 - x - 0}{x - 1}$$

$$= \frac{x(x + 1)(x - 1)}{x - 1} = x(x + 1)$$

13. Find the domain and range of $f(x) = 4 - 2x$. (Give your answer using interval notation.)
    Solution:
    Domain: $(-\infty, \infty)$
    Range: $(-\infty, \infty)$

15. Find the domain and range of $f(x) = x^2$. (Give your answer using interval notation.)
    Solution:
    Domain: $(-\infty, \infty)$
    Range: $[0, \infty)$

17. Find the domain and range of $f(x) = \sqrt{x - 1}$. (Give your answer using interval notation.)
    Solution:
    Domain: $[1, \infty)$
    Range: $[0, \infty)$

19. Find the domain and range of $f(x) = \sqrt{9 - x^2}$. (Give your answer using interval notation.)
    Solution:
    Domain: $[-3, 3]$
    Range: $[0, 3]$

21. Find the domain and range of $f(x) = |x - 2|$. (Give your answer using interval notation.)
    Solution:
    Domain: $(-\infty, \infty)$
    Range: $[0, \infty)$

23. Use the vertical line test to determine whether y is a function of x in $y = x^2$.
    Solution: y *is* a function of x.

25. Use the vertical line test to determine whether y is a function of x in $x - y^2 = 0$.
    Solution: y is *not* a function of x.

27. Use the vertical line test to determine whether y is a function of x in $\sqrt{x^2 - 4} - y = 0$.
    Solution: y *is* a function of x.

29. Use the vertical line test to determine whether y is a function of x in $x^2 = xy - 1$.
    Solution: y *is* a function of x.

Section 1.4

31. Use the vertical line test to determine whether y is a function of x in $x^2 - 4y^2 + 4 = 0$.
Solution: y is <u>not</u> a function of x.

33. Does the equation $x^2 + y^2 = 4$ determine y as a function of x?
Solution:
$$y = \pm\sqrt{4 - x^2}$$

y is <u>not</u> a function of x since there are two values of y for some x.

35. Does the equation $2x + 3y = 4$ determine y as a function of x?
Solution:
$$y = 4 - 2x/3$$

y <u>is</u> a function of x since there is only one value of y for each x.

37. Does the equation $x^2 + y = 4$ determine y as a function of x?
Solution:
$$y = 4 - x^2$$

y <u>is</u> a function of x since there is only one value of y for each x.

39. Does the equation $y^2 = x^2 - 1$ determine y as a function of x?
Solution:
$$y = \pm\sqrt{x^2 - 1}$$

y is <u>not</u> a function of x since there are two values of y for some x.

41. When $f(x) = x + 1$ and $g(x) = x - 1$, find
(a) $f(x) + g(x)$ (b) $f(x) \cdot g(x)$
(c) $f(x)/g(x)$ (d) $f(g(x))$
(e) $g(f(x))$
Solution:
(a) $f(x) + g(x) = (x + 1) + (x - 1) = 2x$
(b) $f(x) \cdot g(x) = (x + 1)(x - 1) = x^2 - 1$
(c) $f(x)/g(x) = (x + 1)/(x - 1)$, $x \neq 1$
(d) $f(g(x)) = f(x - 1) = (x - 1) + 1 = x$
(e) $g(f(x)) = g(x + 1) = (x + 1) - 1 = x$

43. When $f(x) = x^2$ and $g(x) = 1 - x$, find
(a) $f(x) + g(x)$ (b) $f(x) \cdot g(x)$
(c) $f(x)/g(x)$ (d) $f(g(x))$
(e) $g(f(x))$
Solution:
(a) $f(x) + g(x) = x^2 + (1 - x) = x^2 - x + 1$
(b) $f(x) \cdot g(x) = x^2(1 - x) = x^2 - x^3$
(c) $f(x)/g(x) = x^2/(1 - x)$, $x \neq 1$
(d) $f(g(x)) = f(1 - x) = (1 - x)^2$
(e) $g(f(x)) = g(x^2) = 1 - x^2$

Section 1.4   55

45. When $f(x) = x^2 + 5$ and $g(x) = \sqrt{1-x}$, find
  (a) $f(x) + g(x)$ (b) $f(x) \cdot g(x)$
  (c) $f(x)/g(x)$ (d) $f(g(x))$
  (e) $g(f(x))$
Solution:
  (a) $f(x) + g(x) = x^2 + 5 + \sqrt{1-x}$, $x \leq 1$
  (b) $f(x) \cdot g(x) = (x^2 + 5)\sqrt{1-x}$, $x \leq 1$
  (c) $f(x)/g(x) = (x^2 + 5)/\sqrt{1-x}$, $x < 1$
  (d) $f(g(x)) = f(\sqrt{1-x}) = (\sqrt{1-x})^2 + 5$
     $= |1 - x| + 5$
     $= \begin{cases} 6 - x, & x \leq 1 \\ 4 + x, & x > 1 \end{cases}$
  (e) $g(f(x))$ is not defined since the domain of g is $(-\infty, 1]$ and the range of f is $[5, \infty)$. The range of f is not in the domain of g.

47. When $f(x) = 1/x$ and $g(x) = 1/x^2$, find
  (a) $f(x) + g(x)$ (b) $f(x) \cdot g(x)$
  (c) $f(x)/g(x)$ (d) $f(g(x))$
  (e) $g(f(x))$
Solution:
  (a) $f(x) + g(x) = (1/x) + (1/x^2) = (x + 1)/x^2$
  (b) $f(x) \cdot g(x) = (1/x)(1/x^2) = 1/x^3$
  (c) $f(x)/g(x) = (1/x)/(1/x^2) = x$
  (d) $f(g(x)) = f(1/x^2) = 1/(1/x^2) = x^2$
  (e) $g(f(x)) = g(1/x) = 1/[(1/x)^2] = x^2$

49. When $f(x) = x^3$ and $g(x) = \sqrt[3]{x}$, (a) show that f and g are inverse functions by showing that $f(g(x)) = x$ and $g(f(x)) = x$, and (b) graph f and g on the same set of coordinate axes.
Solution:
  (a) $f(g(x)) = f(\sqrt[3]{x}) = (\sqrt[3]{x})^3 = x$
     $g(f(x)) = g(x^3) = \sqrt[3]{x^3} = x$
  (b) See accompanying graph.

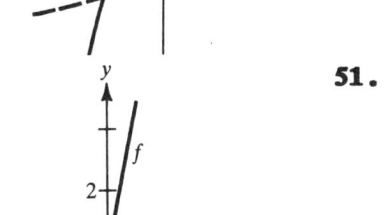

51. When $f(x) = 5x + 1$ and $g(x) = (x - 1)/5$, (a) show that f and g are inverse functions by showing that $f(g(x)) = x$ and $g(f(x)) = x$, and (b) graph f and g on the same set of coordinate axes.
Solution:
  (a) $f(g(x)) = f[(x - 1)/5] = 5[(x - 1)/5] + 1 = x$
     $g(f(x)) = g(5x + 1) = [(5x + 1) - 1]/5 = x$
  (b) See accompanying graph.

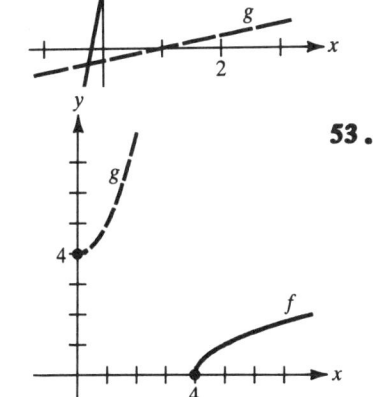

53. When $f(x) = \sqrt{x - 4}$, $x \geq 4$ and $g(x) = x^2 + 4$, $x \geq 0$, (a) show that f and g are inverse functions by showing that $f(g(x)) = x$ and $g(f(x)) = x$, and (b) graph f and g on the same set of coordinate axes.
Solution:
  (a) $f(g(x)) = f(x^2 + 4) = \sqrt{(x^2 + 4) - 4} = x$, $x \geq 0$
     $g(f(x)) = g(\sqrt{x - 4}) = (\sqrt{x - 4})^2 + 4 = x$, $x \geq 4$
  (b) See accompanying graph.

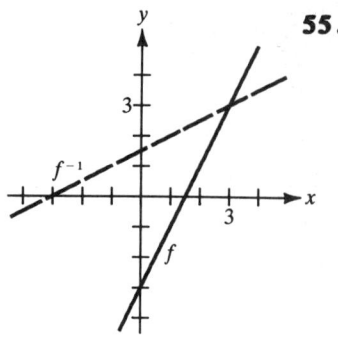

Section 1.4

55. When $f(x) = 2x - 3$, find the inverse of f. Then graph both f and f^{-1} on the same set of axes.
Solution:
$$f(x) = 2x - 3 = y$$
$$x = (y + 3)/2$$
$$f^{-1}(\ ) = \frac{(\ ) + 3}{2}$$
$$f^{-1}(x) = \frac{x + 3}{2}$$

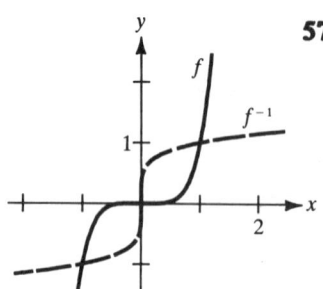

57. When $f(x) = x^5$, find the inverse of f. Then graph both f and f^{-1} on the same set of axes.
Solution:
$$f(x) = x^5 = y$$
$$x = \sqrt[5]{y}$$
$$f^{-1}(\ ) = \sqrt[5]{(\ )}$$
$$f^{-1}(x) = \sqrt[5]{x}$$

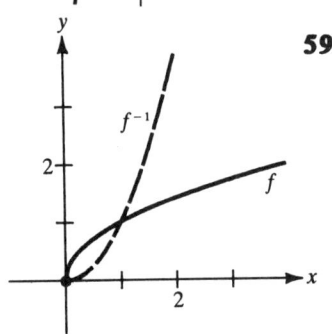

59. When $f(x) = \sqrt{x}$, find the inverse of f. Then graph both f and f^{-1} on the same set of axes.
Solution:
$$f(x) = \sqrt{x} = y$$
$$x = y^2$$
$$f^{-1}(\ ) = (\ )^2$$
$$f^{-1}(x) = x^2, \quad x \geq 0$$

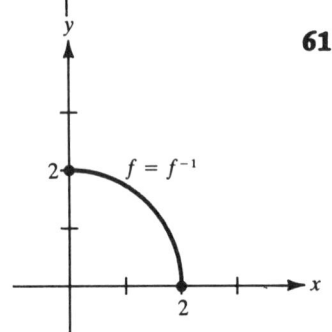

61. When $f(x) = \sqrt{4 - x^2}$, $0 \leq x \leq 2$, find the inverse of f. Then graph both f and f^{-1} on the same set of axes.
Solution:
$$f(x) = \sqrt{4 - x^2} = y, \quad 0 \leq x \leq 2$$
$$x = \sqrt{4 - y^2}$$
$$f^{-1}(\ ) = \sqrt{4 - (\ )^2}$$
$$f^{-1}(x) = \sqrt{4 - x^2}, \quad 0 \leq x \leq 2$$

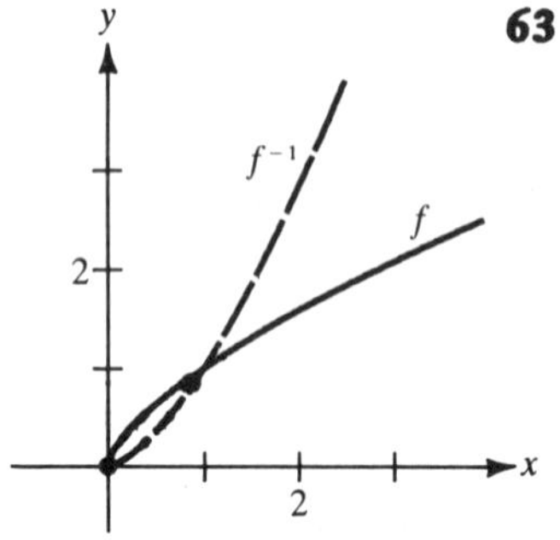

63. When $f(x) = x^{2/3}$, $x \geq 0$, find the inverse of f. Then graph both f and f^{-1} on the same set of axes.
Solution:
$$f(x) = x^{2/3} = y, \quad x \geq 0$$
$$x = y^{3/2}$$
$$f^{-1}(\ ) = (\ )^{3/2}$$
$$f^{-1}(x) = x^{3/2}$$

65. Determine whether $f(x) = ax + b$ is one-to-one, and if so, find its inverse.
Solution: Since $f(x) = ax + b = y$, f *is* one-to-one provided $a \neq 0$.
$$f^{-1}(x) = \frac{x - b}{a}$$

Section 1.4   57

67. Determine whether $f(x) = x^2$ is one-to-one.
   Solution:
   $$f(x) = x^2 = y$$

   f is <u>not</u> one-to-one since $f(1) = 1 = f(-1)$.

69. Determine whether $f(x) = |x - 2|$ is one-to-one.
   Solution:
   $$f(x) = |x - 2| = y$$

   f is <u>not</u> one-to-one since $f(0) = 2 = f(4)$.

71. Use the accompanying graph of $f(x) = \sqrt{x}$ to sketch the graph of each of the following. 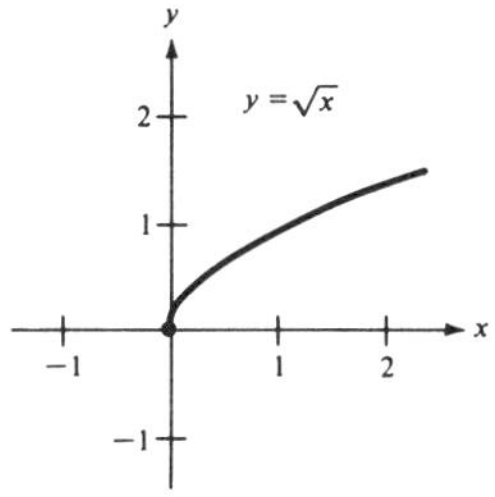
   (a) $y = \sqrt{x} + 2$ (b) $y = -\sqrt{x}$
   (c) $y = \sqrt{x - 2}$ (d) $y = \sqrt{x + 3}$
   (e) $y = \sqrt{x - 4}$ (f) $y = 2\sqrt{x}$
   Solution:
   (a)   (b) 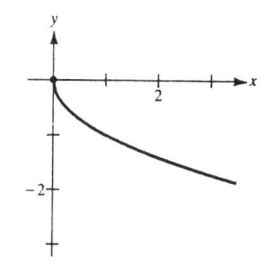
   (c) 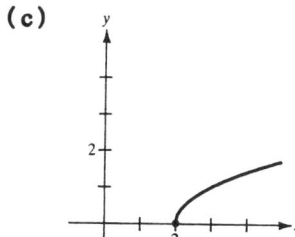   (d) 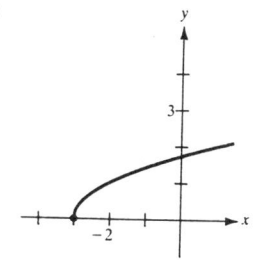
   (e) 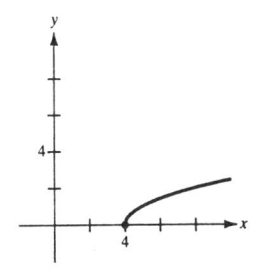   (f)

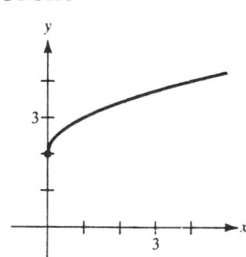

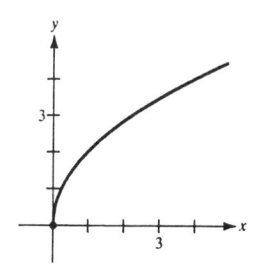

73. Use the given graph to write formulas for the functions whose graphs are shown in parts (a) through (d). 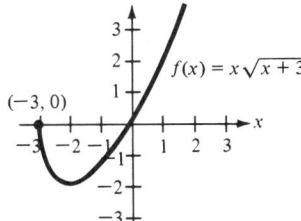
   (a) 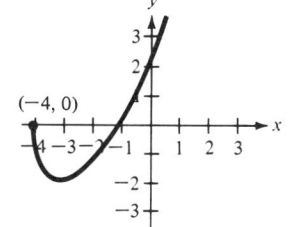   (b)

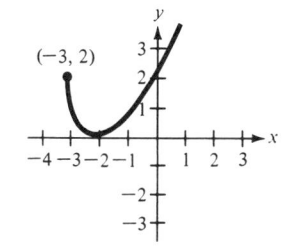

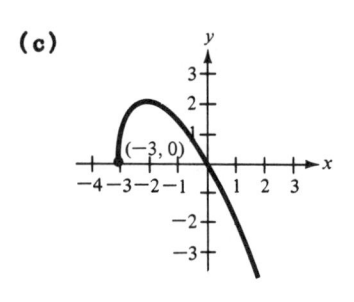

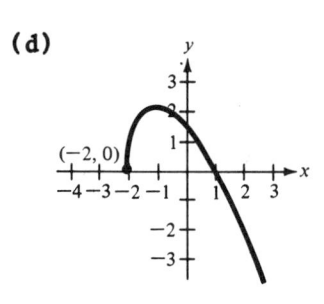

Solution:
$$f(x) = x\sqrt{x+3}$$

(a) Horizontal shift one unit to the left
$$y = (x - (-1))\sqrt{(x - (-1)) + 3} = (x + 1)\sqrt{x + 4}$$

(b) Vertical shift two units upward
$$y = x\sqrt{x + 3} + 2$$

(c) Negative of $f(x)$
$$y = -x\sqrt{x + 3}$$

(d) Negative of $f(x)$ and a horizontal shift one unit to the right
$$y = -(x - 1)\sqrt{(x - 1) + 3} = (1 - x)\sqrt{x + 2}$$

75. The rectangle shown has a perimeter of 100 feet. Express the area A of the rectangle as a function of x.

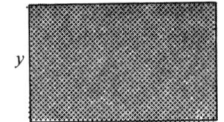

Solution:
$$2x + 2y = 100$$
$$y = \frac{100 - 2x}{2} = 50 - x$$
$$A = xy = x(50 - x) = 50x - x^2$$

77. Express the value V of a farm having $500,000 worth of buildings, livestock, and equipment in terms of the number of acres x on the farm if each acre is valued at $1,750.

Solution:
$$V = 1750x + 500{,}000, \quad x \geq 0$$

79. The demand function for a particular commodity is given by $p = 14.75/(1 + 0.01x)$, $x \geq 0$, where p is the price per unit and x is the number of units sold.
(a) Find x as a function of p.
(b) Use the result of part (a) to find the number of units sold when the price is $10.

Solution:

(a) $1 + 0.01x = 14.75/p$
$$x = [(14.75/p) - 1]/0.01$$
$$= (14.75 - p)/0.01p$$
$$= 100(14.75 - p)/p$$
$$= (1475/p) - 100$$

(b) $$x = \frac{100(14.75 - 10)}{10} = 47.5 \text{ units}$$

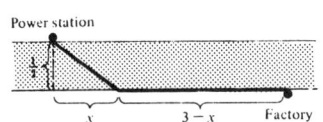

Section 1.5   59

81. A power station is on one side of a river that is one-half mile wide. A factory is 3 miles downstream on the other side of the river. It costs $10/ft to run the power lines on land and $15/ft to run them underwater. Write the cost C of running the line from the power station to the factory as a function of x.
Solution:
Cost = Cost on land + Cost underwater

$$= \left(\begin{array}{c}\text{Cost per}\\ \text{land ft}\end{array}\right)(\text{ft}) + \left(\begin{array}{c}\text{Cost per}\\ \text{water ft}\end{array}\right)(\text{ft})$$

$$= 10(3 - x) + 15\sqrt{x^2 + (1/4)}$$

83. Assume that the amount of money deposited in a bank is proportional to the square of the interest rate the bank pays on the money. that is, $d = kr^2$, where d is the total deposit, r is the interest rate, and k is the proportionality constant. Assuming the bank can reinvest the money for a return of 18%, write the bank's profit P as a function of the interest rate r.
Solution: Since $d = kr^2$, $C = rd = r(kr^2) = kr^3$, and $R = 0.18d = 0.18kr^2$, we have

$$P = R - C = 0.18kr^2 - kr^3 = kr^2(0.18 - r)$$

● **Section 1.5 Limits**

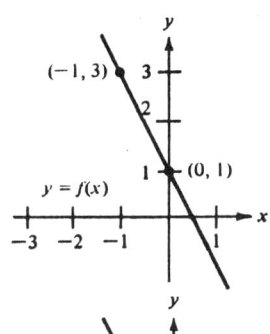

1. Use the graph to visually determine
(a) $\lim_{x \to 0} f(x)$ (b) $\lim_{x \to -1} f(x)$
Solution:
(a) $\lim_{x \to 0} f(x) = 1$
(b) $\lim_{x \to -1} f(x) = 3$

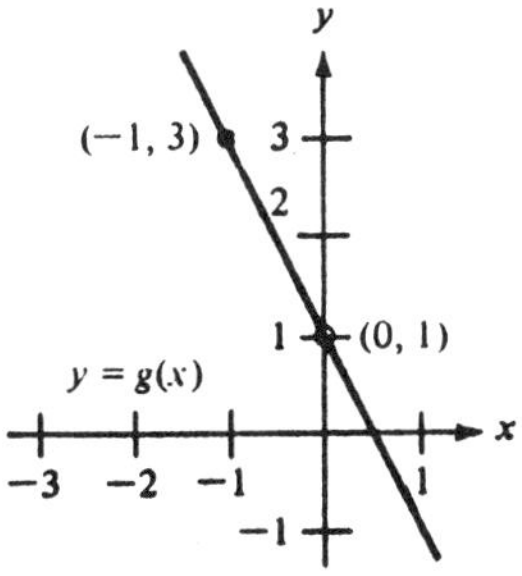

3. Use the graph to visually determine
(a) $\lim_{x \to 0} g(x)$ (b) $\lim_{x \to -1} g(x)$
Solution:
(a) $\lim_{x \to 0} g(x) = 1$
(b) $\lim_{x \to -1} g(x) = 3$

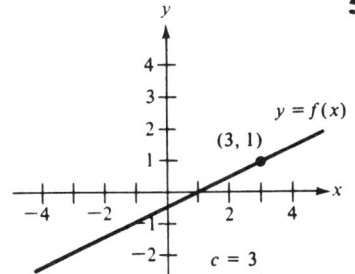

5. Use the graph to visually determine
(a) $\lim_{x \to 3^+} f(x)$ (b) $\lim_{x \to 3^-} f(x)$ (c) $\lim_{x \to 3} f(x)$
Solution:
(a) $\lim_{x \to 3^+} f(x) = 1$
(b) $\lim_{x \to 3^-} f(x) = 1$
(c) $\lim_{x \to 3} f(x) = 1$

7. Use the graph to visually determine
   (a) $\lim\limits_{x \to 3^+} f(x)$ (b) $\lim\limits_{x \to 3^-} f(x)$ (c) $\lim\limits_{x \to 3} f(x)$

   Solution:
   (a) $\lim\limits_{x \to 3^+} f(x) = 0$
   (b) $\lim\limits_{x \to 3^-} f(x) = 0$
   (c) $\lim\limits_{x \to 3} f(x) = 0$

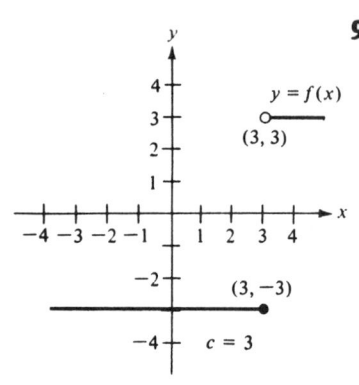

9. Use the graph to visually determine
   (a) $\lim\limits_{x \to 3^+} f(x)$ (b) $\lim\limits_{x \to 3^-} f(x)$ (c) $\lim\limits_{x \to 3} f(x)$
   if it exists.

   Solution:
   (a) $\lim\limits_{x \to 3^+} f(x) = 3$
   (b) $\lim\limits_{x \to 3^-} f(x) = -3$
   (c) $\lim\limits_{x \to 3} f(x)$ does not exist

11. Find $\lim\limits_{x \to 2} x^2$.

    Solution:
    $\lim\limits_{x \to 2} x^2 = 2^2 = 4$

13. Find $\lim\limits_{x \to 0} (2x - 1)$.

    Solution:
    $\lim\limits_{x \to 0} (2x - 1) = 2(0) - 1 = -1$

15. Find $\lim\limits_{x \to 2} (-x^2 + x - 2)$.

    Solution:
    $\lim\limits_{x \to 2} (-x^2 + x - 2) = -(2)^2 + 2 - 2 = -4$

17. Find $\lim\limits_{x \to 3} \sqrt{x + 1}$.

    Solution:
    $\lim\limits_{x \to 3} \sqrt{x + 1} = \sqrt{3 + 1} = 2$

19. Find $\lim\limits_{x \to 2} (1/x)$.

    Solution:
    $\lim\limits_{x \to 2} \dfrac{1}{x} = \dfrac{1}{2}$

Section 1.5   61

21. Use $\lim_{x \to c} f(x) = 2$ and $\lim_{x \to c} g(x) = 3$ to find

(a) $\lim_{x \to c} [f(x) + g(x)]$

(b) $\lim_{x \to c} [f(x)g(x)]$

(c) $\lim_{x \to c} [f(x)/g(x)]$

Solution:

(a) $\lim_{x \to c} [f(x) + g(x)] = \lim_{x \to c} f(x) + \lim_{x \to c} g(x)$
$= 2 + 3 = 5$

(b) $\lim_{x \to c} [f(x)g(x)] = [\lim_{x \to c} f(x)][\lim_{x \to c} g(x)]$
$= (2)(3) = 6$

(c) $\lim_{x \to c} \frac{f(x)}{g(x)} = \frac{\lim_{x \to c} f(x)}{\lim_{x \to c} g(x)} = \frac{2}{3}$

23. Find $\lim_{x \to -1} \frac{x^2 - 1}{x + 1}$.

Solution:
$\lim_{x \to -1} \frac{x^2 - 1}{x + 1} = \lim_{x \to -1} \frac{(x + 1)(x - 1)}{x + 1}$

$= \lim_{x \to -1} (x - 1) = -2$

25. Find $\lim_{x \to 3} \frac{x - 3}{x^2 - 9}$.

Solution:
$\lim_{x \to 3} \frac{x - 3}{x^2 - 9} = \lim_{x \to 3} \frac{x - 3}{(x + 3)(x - 3)}$

$= \lim_{x \to 3} \frac{1}{x + 3} = \frac{1}{6}$

27. Find $\lim_{t \to 5} \frac{t - 5}{t^2 - 25}$.

Solution:
$\lim_{t \to 5} \frac{t - 5}{t^2 - 25} = \lim_{t \to 5} \frac{t - 5}{(t + 5)(t - 5)}$

$= \lim_{t \to 5} \frac{1}{t + 5} = \frac{1}{10}$

29. Find $\lim_{x \to -2} \frac{x^3 + 8}{x + 2}$.

Solution:
$\lim_{x \to -2} \frac{x^3 + 8}{x + 2} = \lim_{x \to -2} \frac{(x + 2)(x^2 - 2x + 4)}{x + 2}$

$= \lim_{x \to -2} (x^2 - 2x + 4) = 12$

Section 1.5

31. Find $\lim\limits_{x \to 1^-} \dfrac{x}{x^2 - x}$.

    Solution:
    $$\lim_{x \to 1^-} \frac{x}{x^2 - x} = \lim_{x \to 1^-} \frac{1}{x - 1} = -\infty$$

33. Find $\lim\limits_{x \to -2^-} \dfrac{1}{x + 2}$.

    Solution:
    $$\lim_{x \to -2^-} \frac{1}{x + 2} = -\infty$$

35. Find $\lim\limits_{x \to 5} \dfrac{2}{(x - 5)^2}$.

    Solution:
    $$\lim_{x \to 5} \frac{2}{(x - 5)^2} = \infty$$

37. Find $\lim\limits_{x \to 0} \dfrac{|x|}{x}$.

    Solution:
    $$\lim_{x \to 0^-} \frac{|x|}{x} = -1, \qquad \lim_{x \to 0^+} \frac{|x|}{x} = 1$$

    Therefore, $\lim\limits_{x \to 0} |x|/x$ does not exist.

39. Find $\lim\limits_{x \to 3} f(x)$,

    where $f(x) = \begin{cases} (x + 2)/2, & x \leq 3 \\ (12 - 2x)/3, & x > 3 \end{cases}$

    Solution:
    $$\lim_{x \to 3^-} f(x) = \lim_{x \to 3^-} \frac{x + 2}{2} = \frac{5}{2}$$

    $$\lim_{x \to 3^+} f(x) = \lim_{x \to 3^+} \frac{12 - 2x}{3} = 2$$

    Therefore, $\lim\limits_{x \to 3} f(x)$ does not exist.

41. Find $\lim\limits_{x \to 1} f(x)$,

    where $f(x) = \begin{cases} x^3 + 1, & x < 1 \\ x + 1, & x \geq 1 \end{cases}$

    Solution:
    $$\lim_{x \to 1^-} f(x) = \lim_{x \to 1^-} (x^3 + 1) = 2$$

    $$\lim_{x \to 1^+} f(x) = \lim_{x \to 1^+} (x + 1) = 2$$

    Therefore, $\lim\limits_{x \to 1} f(x) = 2$.

Section 1.5

43. Find $\lim\limits_{\Delta x \to 0} \dfrac{2(x + \Delta x) - 2x}{\Delta x}$.

Solution:

$$\lim_{\Delta x \to 0} \frac{2(x + \Delta x) - 2x}{\Delta x} = \lim_{\Delta x \to 0} \frac{2x + 2\Delta x - 2x}{\Delta x}$$

$$= \lim_{\Delta x \to 0} 2 = 2$$

45. Find $\lim\limits_{\Delta t \to 0} \dfrac{(t + \Delta t)^2 + 3 - (t^2 + 3)}{\Delta t}$.

Solution:

$$\lim_{\Delta t \to 0} \frac{(t + \Delta t)^2 + 3 - (t^2 + 3)}{\Delta t}$$

$$= \lim_{\Delta t \to 0} \frac{t^2 + 2t\Delta t + (\Delta t)^2 + 3 - t^2 - 3}{\Delta t}$$

$$= \lim_{\Delta t \to 0} \frac{2t\Delta t + (\Delta t)^2}{\Delta t} = \lim_{\Delta t \to 0} (2t + \Delta t) = 2t$$

47. Find $\lim\limits_{\Delta x \to 0} \dfrac{(x + \Delta x)^2 - 2(x + \Delta x) + 1 - (x^2 - 2x + 1)}{\Delta x}$

Solution:

$$\lim_{\Delta x \to 0} \frac{(x + \Delta x)^2 - 2(x + \Delta x) + 1 - (x^2 - 2x + 1)}{\Delta x}$$

$$= \lim_{\Delta x \to 0} \frac{x^2 + 2x\Delta x + (\Delta x)^2 - 2x - 2\Delta x + 1 - x^2 + 2x - 1}{\Delta x}$$

$$= \lim_{\Delta x \to 0} \frac{2x\Delta x + (\Delta x)^2 - 2\Delta x}{\Delta x}$$

$$= \lim_{\Delta x \to 0} (2x + \Delta x - 2) = 2x - 2$$

49. Find $\lim\limits_{x \to 0} \dfrac{\sqrt{3 + x} - \sqrt{3}}{x}$.

Solution:

$$\lim_{x \to 0} \frac{\sqrt{3 + x} - \sqrt{3}}{x}$$

$$= \lim_{x \to 0} \frac{\sqrt{3 + x} - \sqrt{3}}{x} \cdot \frac{\sqrt{3 + x} + \sqrt{3}}{\sqrt{3 + x} + \sqrt{3}}$$

$$= \lim_{x \to 0} \frac{(3 + x) - 3}{x(\sqrt{3 + x} + \sqrt{3})}$$

$$= \lim_{x \to 0} \frac{1}{\sqrt{3 + x} + \sqrt{3}} = \frac{1}{2\sqrt{3}} = \frac{\sqrt{3}}{6}$$

51. Find $\lim\limits_{x \to 0} \dfrac{[1/(2+x)] - (1/2)}{x}$.

Solution:

$$\lim_{x \to 0} \dfrac{\dfrac{1}{2+x} - \dfrac{1}{2}}{x} = \lim_{x \to 0} \dfrac{\dfrac{2 - (2+x)}{2(2+x)}}{x}$$

$$= \lim_{x \to 0} \dfrac{-1}{2(2+x)} = -\dfrac{1}{4}$$

53. Find $\lim\limits_{x \to 4} \dfrac{\sqrt{x} - 2}{x - 4}$.

Solution:

$$\lim_{x \to 4} \dfrac{\sqrt{x} - 2}{x - 4} = \lim_{x \to 4} \dfrac{\sqrt{x} - 2}{x - 4} \cdot \dfrac{\sqrt{x} + 2}{\sqrt{x} + 2}$$

$$= \lim_{x \to 4} \dfrac{x - 4}{(x - 4)(\sqrt{x} + 2)}$$

$$= \lim_{x \to 4} \dfrac{1}{\sqrt{x} + 2} = \dfrac{1}{4}$$

55. Complete a table to estimate $\lim\limits_{x \to 2} (5x + 4)$.

Solution:

x	1.9	1.99	1.999	2	2.001	2.01	2.1
f(x)	13.5	13.95	13.995	14	14.005	14.05	14.5

$\lim\limits_{x \to 2} (5x + 4) = 14$

57. Complete a table to estimate $\lim\limits_{x \to 2} \dfrac{x - 2}{x^2 - 4}$.

Solution:

x	1.9	1.99	1.999	2	2.001	2.01	2.1
f(x)	0.2564	0.2506	0.2501	undef	0.2499	0.2494	0.2439

$\lim\limits_{x \to 2} \dfrac{x - 2}{x^2 - 4} = \dfrac{1}{4}$

59. Complete a table to estimate $\lim\limits_{x \to 0} \dfrac{\sqrt{x + 2} - \sqrt{2}}{x}$.

Solution:

x	-0.1	-0.01	-0.001	0	0.001	0.01	0.1
f(x)	0.358	0.354	0.354	undef	0.354	0.353	0.349

$\lim\limits_{x \to 0} \dfrac{\sqrt{x + 2} - \sqrt{2}}{x} = \dfrac{1}{2\sqrt{2}} \approx 0.354$

61. Complete a table to estimate

$$\lim_{x \to 0} \frac{[1/(2+x)] - (1/2)}{2x}$$

Solution:

x	−0.1	−0.01	−0.001	0	0.001	0.01	0.1
f(x)	−0.132	−0.126	−0.125	undef	−0.125	−0.124	−0.119

$$\lim_{x \to 0} \frac{[1/(2+x)] - (1/2)}{2x} = -\frac{1}{8} = -0.125$$

63. Given $4 - x^2 \leq f(x) \leq 4 + x^2$ for all x, find $\lim_{x \to 0} f(x)$.

Solution:
$$\lim_{x \to 0} (4 - x^2) \leq \lim_{x \to 0} f(x) \leq \lim_{x \to 0} (4 + x^2)$$

$$4 \leq \lim_{x \to 0} f(x) \leq 4$$

Therefore, $\lim_{x \to 0} f(x) = 4$.

Section 1.6 Continuity

1. Find the discontinuities (if any) for $f(x) = -x^3/2$.
Solution: Continuous on entire real line.

3. Find the discontinuities for $f(x) = \frac{x^2 - 1}{x + 1}$.
Solution:
$$f(x) = \frac{x^2 - 1}{x + 1} = \frac{(x-1)(x+1)}{x+1}$$
Removable discontinuity at $x = -1$.

5. Find the discontinuities for
$$f(x) = \begin{cases} x/2, & x < 1 \\ 2, & x = 1 \\ 2x - 1, & x > 1 \end{cases}$$
Solution: Nonremovable discontinuity at $x = 1$.

7. Find the discontinuities (if any) for
$$f(x) = x^2 - 2x + 1$$
Solution: Continuous on entire real line.

9. Find the discontinuities for $f(x) = 1/(x - 1)$. Which of the discontinuities are removable?
Solution: Nonremovable discontinuity at $x = 1$.

11. Find the discontinuities (if any) for
$$f(x) = \frac{x}{x^2 + 1}$$
Which of the discontinuities are removable?
Solution: Continuous on the entire real line.

13. Find the discontinuities (if any) for
$$f(x) = \frac{x + 2}{x^2 - 3x - 10}$$
Which of the discontinuities are removable?
Solution:
$$f(x) = \frac{x + 2}{x^2 - 3x - 10} = \frac{x + 2}{(x + 2)(x - 5)}$$
Removable discontinuity at $x = -2$ and a nonremovable discontinuity at $x = 5$.

15. Find the discontinuities (if any) for
$$f(x) = \begin{cases} x, & x \leq 1 \\ x^2, & x > 1 \end{cases}$$
Which of the discontinuities are removable?
Solution:

$$\lim_{x \to 1^-} f(x) = \lim_{x \to 1^-} x = 1$$
$$\lim_{x \to 1^+} f(x) = \lim_{x \to 1^+} x^2 = 1$$

Since $f(1) = 1$, f is continuous on the entire real line.

17. Find the discontinuities (if any) for
$$f(x) = \begin{cases} (x/2) + 1, & x \leq 2 \\ 3 - x, & x > 2 \end{cases}$$
Which of the discontinuities are removable?
Solution:

$$\lim_{x \to 2^-} f(x) = \lim_{x \to 2^-} \left(\frac{x}{2} + 1\right) = 2$$
$$\lim_{x \to 2^+} f(x) = \lim_{x \to 2^+} (3 - x) = 1$$

Since the limit of $f(x)$ as $x \to 2$ does not exist, there is a nonremovable discontinuity at $x = 2$.

19. Find the discontinuities (if any) for
$$f(x) = \begin{cases} 3 + x, & x \leq 2 \\ x^2 + 1, & x > 2 \end{cases}$$
Which of the discontinuities are removable?

Section 1.6

Solution:

$$\lim_{x \to 2^-} f(x) = \lim_{x \to 2^-} (3 + x) = 5$$
$$\lim_{x \to 2^+} f(x) = \lim_{x \to 2^+} (x^2 + 1) = 5$$

Since $f(2) = 5$, f is continuous on the entire real line.

21. Find the discontinuities (if any) for

$$f(x) = \frac{|x + 2|}{x + 2}$$

Which of the discontinuities are removable?
Solution:

$$\lim_{x \to -2^-} \frac{|x + 2|}{x + 2} = \lim_{x \to -2^-} \frac{-(x + 2)}{x + 2} = -1$$

$$\lim_{x \to -2^+} \frac{|x + 2|}{x + 2} = \lim_{x \to -2^+} \frac{x + 2}{x + 2} = 1$$

Since the limit of $|x + 2|/(x + 2)$ as $x \to -2$ does not exist, there is a nonremovable discontinuity at $x = -2$.

23. Find the discontinuities (if any) for

$$f(x) = [\![x - 1]\!]$$

Which of the discontinuities are removable?
Solution:

$$\lim_{x \to c^-} [\![x - 1]\!] = c - 2, \text{ for } c \text{ any integer}$$
$$\lim_{x \to c^+} [\![x - 1]\!] = c - 1, \text{ for } c \text{ any integer}$$

Since the limit of $[\![x - 1]\!]$ does not exist, f has a nonremovable discontinuity at every integer c.

25. Find the discontinuities (if any) for

$$h(x) = f(g(x)), \quad f(x) = \frac{1}{\sqrt{x}}, \quad g(x) = x - 1, \quad x > 1$$

Which of the discontinuities are removable?
Solution:

$$h(x) = f(g(x)) = f(x - 1) = \frac{1}{\sqrt{x - 1}}, \quad x > 1$$

Thus, h is continuous on its entire domain $(1, \infty)$.

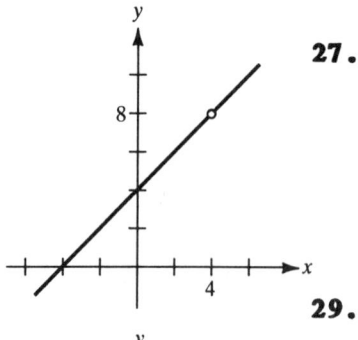

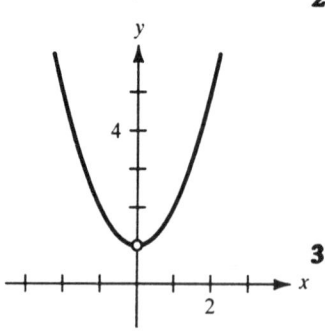

27. To determine any points of discontinuity, sketch the graph of $f(x) = (x^2 - 16)/(x - 4)$.
Solution:
$$f(x) = \frac{x^2 - 16}{x - 4} = \frac{(x + 4)(x - 4)}{x - 4} = x + 4, \quad x \neq 4$$
Removable discontinuity at $x = 4$.

29. To determine any points of discontinuity, sketch the graph of $f(x) = (x^3 + x)/x$.
Solution:
$$f(x) = \frac{x^3 + x}{x} = \frac{x(x^2 + 1)}{x} = x^2 + 1, \quad x \neq 0$$
Removable discontinuity at $x = 0$.

31. Determine the constant a so that the following function is continuous:
$$f(x) = \begin{cases} x^3, & x \leq 2 \\ ax^2, & x > 2 \end{cases}$$
Solution:
$$\lim_{x \to 2^-} f(x) = \lim_{x \to 2^-} x^3 = 8$$
$$\lim_{x \to 2^+} f(x) = \lim_{x \to 2^+} ax^2 = 4a$$

Therefore, $8 = 4a$ and $a = 2$.

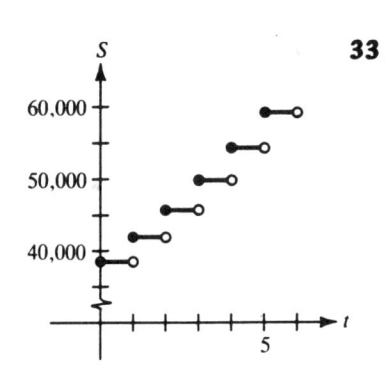

33. A union contract guarantees a 9% increase each year for five years. For a salary of $38,500, the salary S is given by
$$S = 38,500(1.09)^{[\![t]\!]}, \quad 0 \leq t < 6$$
where $t = 0$ corresponds to 1985. Sketch a graph of this function and discuss its continuity.
Solution: Nonremovable discontinuities at $t = 1, 2, 3, 4$, and 5.

35. The number of units in inventory in a small company is given by
$$N = 25\left(2 \left[\!\left[\frac{t + 2}{2} \right]\!\right] - t\right), \quad t \geq 0$$
where t is the time in months. Sketch the graph of this function and discuss its continuity. How often does this company replenish its inventory?
Solution: Nonremovable discontinuities at
$$t = 2, 4, 6, 8, \ldots$$
$N(t) = 0$ when $t = 2, 4, 6, 8, \ldots$, so the inventory is replenished every two months.

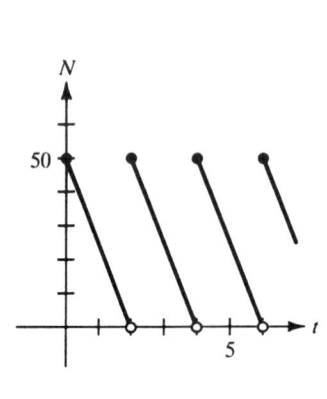

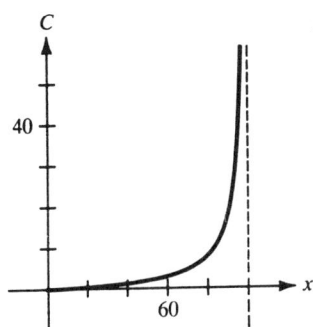

37. The cost (in millions of dollars) of removing x percent of the pollutants being emitted from the smokestack of a certain factory is given by

$$C = \frac{2x}{100 - x}, \quad 0 \leq x < 100$$

Sketch the graph of this function and discuss its continuity.
Solution: C is continuous on its entire domain [0, 100).

● Review Exercises for Chapter 1

1. Find (a) the distance between the points (0, 0), (6, 0), (b) the midpoint of the line segment between the two points, and (c) the general form of the equation of the line through the points.
   Solution:
   (a) Distance = $\sqrt{(6 - 0)^2 + (0 - 0)^2} = 6$

   (b) Midpoint = $(\frac{0 + 6}{2}, \frac{0 + 0}{2}) = (3, 0)$

   (c) Slope = $(0 - 0)/(6 - 0) = 0$

   Horizontal line: y = 0

3. Find (a) the distance between the points (-2, -1), (2, 2), (b) the midpoint of the line segment between the two points, and (c) the general form of the equation of the line through the points.
   Solution:
   (a) Distance = $\sqrt{(2 + 2)^2 + (2 + 1)^2} = 5$

   (b) Midpoint = $(\frac{-2 + 2}{2}, \frac{-1 + 2}{2}) = (0, \frac{1}{2})$

   (c) Slope = $(2 + 1)/(2 + 2) = 3/4$

   $$\begin{aligned} y - 2 &= (3/4)(x - 2) \\ 4y - 8 &= 3x - 6 \\ 0 &= 3x - 4y + 2 \end{aligned}$$

5. Find (a) the distance between the points (2, 1), (14, 6), (b) the midpoint of the line segment between the two points, and (c) the general form of the equation of the line through the points.
   Solution:
   (a) Distance = $\sqrt{(14 - 2)^2 + (6 - 1)^2} = 13$

   (b) Midpoint = $(\frac{2 + 14}{2}, \frac{1 + 6}{2}) = (8, \frac{7}{2})$

   (c) Slope = $(6 - 1)/(14 - 2) = 5/12$

   $$\begin{aligned} y - 1 &= (5/12)(x - 2) \\ 12y - 12 &= 5x - 10 \\ 0 &= 5x - 12y + 2 \end{aligned}$$

7. Find (a) the distance between the points (-1, 0), (6, 2), (b) the midpoint of the line segment between the two points, and (c) the general form of the equation of the line through the points.
Solution:
(a) Distance = $\sqrt{(6+1)^2 + (2-0)^2} = \sqrt{53}$

(b) Midpoint = $(\dfrac{-1+6}{2}, \dfrac{0+2}{2}) = (\dfrac{5}{2}, 1)$

(c) Slope = $(2-0)/(6+1) = 2/7$

$$y - 0 = (2/7)(x + 1)$$
$$7y = 2x + 2$$
$$0 = 2x - 7y + 2$$

9. Find (a) the distance between the points (1/3, 4/3), (2/3, 1/6), (b) the midpoint of the line segment between the two points, and (c) the general form of the equation of the line through the points.
Solution:
(a) Distance = $\sqrt{[(2/3) - (1/3)]^2 + [(1/6) - (4/3)]^2}$
   = $\sqrt{(1/9) + (49/36)} = \sqrt{53}/6$

(b) Midpoint = $(\dfrac{(1/3) + (2/3)}{2}, \dfrac{(4/3) + (1/6)}{2}) = (\dfrac{1}{2}, \dfrac{3}{4})$

(c) Slope = $\dfrac{(1/6) - (4/3)}{(2/3) - (1/3)} = \dfrac{-(7/6)}{1/3} = -\dfrac{7}{2}$

$$y - (4/3) = -(7/2)[x - (1/3)]$$
$$y - (4/3) = -(7/2)x + (7/6)$$
$$6y - 8 = -21x + 7$$
$$21x + 6y - 15 = 0$$

11. Determine the value of t so that the points (-2, 5), (0, t), (1, 1) are on the same line.
Solution:

$$m_1 = \dfrac{1-5}{1+2} = -\dfrac{4}{3}, \quad m_2 = \dfrac{1-t}{1-0} = 1 - t$$

For the points to be on the same line:

$$m_1 = m_2$$
$$-4/3 = 1 - t$$
$$t = 7/3$$

13. Find the general form of the line through the point (-2, 4) and satisfying the given characteristic.
(a) Slope is 7/16
(b) Parallel to the line $5x - 3y = 3$
(c) Passes through the origin
(d) Parallel to the y-axis
Solution:
(a) $\quad y - 4 = (7/16)(x + 2)$
$$16y - 64 = 7x + 14$$
$$0 = 7x - 16y + 78$$

Review Exercises for Chapter 1   71

(b) $y = (5/3)x - 1 \implies m = 5/3$

$$y - 4 = (5/3)(x + 2)$$
$$3y - 12 = 5x + 10$$
$$5x - 3y + 22 = 0$$

(c) $m = (0 - 4)/(0 + 2) = -2$

$$y = -2x$$
$$2x + y = 0$$

(d) Vertical line: $x = -2$

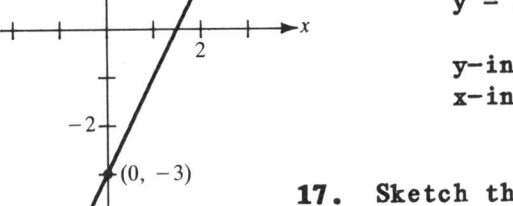

15. Sketch the graph of $4x - 2y = 6$.
Solution:
$$y = 2x - 3$$

y-intercept: $(0, -3)$
x-intercept: $(3/2, 0)$

17. Sketch the graph of $-(1/3)x + (5/6)y = 1$.
Solution:
$$-2x + 5y = 6$$
$$y = \frac{2}{5}x + \frac{6}{5}$$

y-intercept: $(0, 6/5)$
x-intercept: $(-3, 0)$

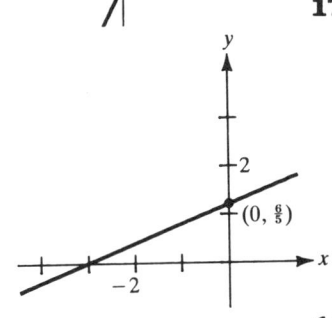

19. Determine the radius and center of the circle

$$x^2 + y^2 + 6x - 2y + 1 = 0$$

and sketch its graph.
Solution:
$$(x^2 + 6x + 9) + (y^2 - 2y + 1) = -1 + 9 + 1$$
$$(x + 3)^2 + (y - 1)^2 = 9$$

Center: $(-3, 1)$
Radius: 3

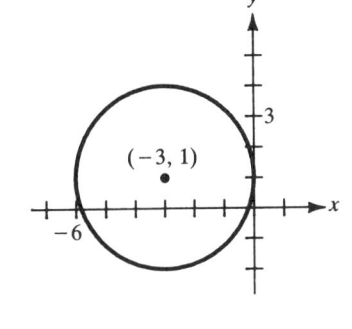

21. Determine the radius and center of the circle

$$x^2 + y^2 + 6x - 2y + 10 = 0$$

and sketch its graph.
Solution:
$$(x^2 + 6x + 9) + (y^2 - 2y + 1) = -10 + 9 + 1$$
$$(x + 3)^2 + (y - 1)^2 = 0$$

Single point: $(-3, 1)$

23. Find the general form of the equation of the circle with center (1, 2) and radius 3. Then determine if each of the points is inside, outside, or on the circle.
(a) (1, 5) (b) (0, 0)
(c) (-2, 1) (d) (0, 4)
Solution:
$$(x - 1)^2 + (y - 2)^2 = 3^2$$
$$x^2 - 2x + 1 + y^2 - 4y + 4 = 9$$
$$x^2 + y^2 - 2x - 4y - 4 = 0$$

(a) $(1)^2 + (5)^2 - 2(1) - 4(5) - 4 = 0$
On the circle
(b) $(0)^2 + (0)^2 - 2(0) - 4(0) - 4 = -4 < 0$
Inside the circle
(c) $(-2)^2 + (1)^2 - 2(-2) - 4(1) - 4 = 1 > 0$
Outside the circle
(d) $(0)^2 + (4)^2 - 2(0) - 4(4) - 4 = -4 < 0$
Inside the circle

25. Find the point of intersection of the graph of $3x - 4y = 8$, $x + y = 5$.
Solution: Solving for y in each equation produces

$$y = \frac{3}{4}x - 2 \quad \text{and} \quad y = 5 - x$$
$$\frac{3}{4}x - 2 = 5 - x$$
$$3x - 8 = 20 - 4x$$
$$7x = 28$$
$$x = 4$$
$$y = 1$$

Point of intersection: (4, 1)

27. Sketch the graph of $x - 2y = 0$ and use the vertical line test to determine if it expresses y as a function of x.
Solution:
$$y = \frac{1}{2}x$$
The graph is a line and y is a function of x.

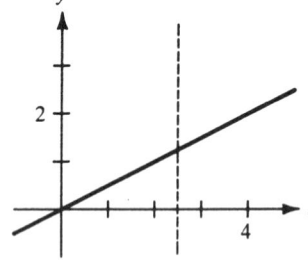

29. Sketch the graph of
$$x^2 + y^2 = 16$$
and use the vertical line test to determine if it expresses y as a function x.
Solution: The graph is a circle and y is not a function x.

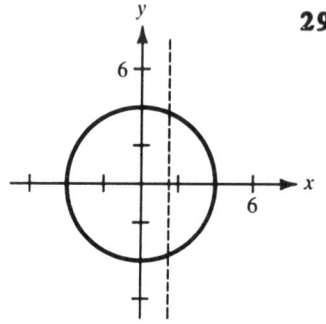

Review Exercises for Chapter 1

73

31. When $f(x) = 1 - x^2$ and $g(x) = 2x + 1$, use f and g to find the following:
   (a) $f(x) + g(x)$ (b) $f(x) - g(x)$
   (c) $f(x)g(x)$ (d) $f(x)/g(x)$
   (e) $f(g(x))$ (f) $g(f(x))$

   Solution:
   (a) $f(x) + g(x) = (1 - x^2) + (2x + 1) = -x^2 + 2x + 2$
   (b) $f(x) - g(x) = (1 - x^2) - (2x + 1) = -x^2 - 2x$
   (c) $f(x)g(x) = (1 - x^2)(2x + 1) = -2x^3 - x^2 + 2x + 1$
   (d) $f(x)/g(x) = (1 - x^2)/(2x + 1), \quad x \neq -(1/2)$
   (e) $f(g(x)) = f(2x + 1) = 1 - (2x + 1)^2$
       $= 1 - (4x^2 + 4x + 1) = -4x^2 - 4x$
   (f) $g(f(x)) = g(1 - x^2) = 2(1 - x^2) + 1 = -2x^2 + 3$

33. The sum of two positive numbers is 500. Let one of the numbers be x and express the product P of the two numbers as a function of x.

   Solution:
   $x + y = 500$
   $y = 500 - x$
   $P = xy = x(500 - x)$

35. When a wholesaler sold a certain product at $25 per unit, sales were 800 units per week. However, after a price increase of $5 per unit, the average number of units sold dropped to 775 units per week. Write the quantity demanded x as a linear function of the price p.

   Solution:
   $(25, 800), (30, 775)$
   $$m = \frac{775 - 800}{30 - 25} = -5$$
   $x - 800 = -5(p - 25)$
   $x = -5p + 925$
   $\quad = 5(185 - p)$

37. Find $\lim_{x \to 2} (5x - 3)$.

   Solution:
   $\lim_{x \to 2} (5x - 3) = 5(2) - 3 = 7$

39. Find $\lim_{x \to 2} (5x - 3)(2x + 9)$.

   Solution:
   $\lim_{x \to 2} (5x - 3)(2x + 9) = [5(2) - 3][2(2) + 9] = 91$

41. Find $\lim_{t \to 3} (t^2 + 1)/t$.

   Solution:
   $$\lim_{t \to 3} \frac{t^2 + 1}{t} = \frac{(3)^2 + 1}{3} = \frac{10}{3}$$

43. Find $\lim_{t \to 0} (t^2 + 1)/t$, if it exists.
Solution:

$$\lim_{t \to 0^-} \frac{t^2 + 1}{t} = -\infty$$

$$\lim_{t \to 0^+} \frac{t^2 + 1}{t} = \infty$$

$$\lim_{t \to 0} \frac{t^2 + 1}{t} \text{ does not exist.}$$

45. Find $\lim_{x \to -2} (x + 2)/(x^2 - 4)$.
Solution:

$$\lim_{x \to -2} \frac{x + 2}{x^2 - 4} = \lim_{x \to -2} \frac{x + 2}{(x + 2)(x - 2)}$$

$$= \lim_{x \to -2} \frac{1}{x - 2} = -\frac{1}{4}$$

47. Find $\lim_{x \to 0^+} [x - (1/x)]$.
Solution:

$$\lim_{x \to 0^+} \left(x - \frac{1}{x}\right) = \lim_{x \to 0^+} \frac{x^2 - 1}{x} = -\infty$$

49. Find

$$\lim_{x \to 0} \frac{[1/(x + 1)] - 1}{x}$$

Solution:

$$\lim_{x \to 0} \frac{[1/(x + 1)] - 1}{x} = \lim_{x \to 0} \frac{1 - (x + 1)}{x + 1} \cdot \frac{1}{x}$$

$$= \lim_{x \to 0} \frac{-1}{x + 1} = -1$$

51. Find $\lim_{\Delta x \to 0} [(x + \Delta x)^3 - (x + \Delta x) - (x^3 - x)]/\Delta x$.
Solution:

$$\lim_{\Delta x \to 0} \frac{(x + \Delta x)^3 - (x + \Delta x) - (x^3 - x)}{\Delta x}$$

$$= \lim_{\Delta x \to 0} \frac{x^3 + 3x^2 \Delta x + 3x(\Delta x)^2 + (\Delta x)^3 - x - \Delta x - x^3 + x}{\Delta x}$$

$$= \lim_{\Delta x \to 0} \frac{3x^2 \Delta x + 3x(\Delta x)^2 + (\Delta x)^3 - \Delta x}{\Delta x}$$

$$= \lim_{\Delta x \to 0} (3x^2 + 3x\Delta x + (\Delta x)^2 - 1) = 3x^2 - 1$$

Review Exercises for Chapter 1   75

53. Complete a table to estimate
$$\lim_{x \to 1^+} \frac{\sqrt{2x+1} - \sqrt{3}}{x-1}$$
Solution:

x	1.1	1.01	1.001	1.0001
f(x)	0.5680	0.5764	0.5773	0.5773

$$\lim_{x \to 1^+} \frac{\sqrt{2x+1} - \sqrt{3}}{x-1} = \frac{1}{\sqrt{3}} \approx 0.5774$$

55. Determine if the statement $\lim_{x \to 0} |x|/x = 1$ is true.
Solution: The statement is false since
$$\lim_{x \to 0^-} \frac{|x|}{x} = -1$$

57. Determine if the statement
$$\lim_{x \to 2} f(x) = 3, \quad f(x) = \begin{cases} 3, & x \leq 2 \\ 0, & x > 2 \end{cases}$$
is true or false.
Solution: The statement is false since
$$\lim_{x \to 2^+} f(x) = \lim_{x \to 2^+} 0 = 0$$

59. Determine if the statement $\lim_{x \to 0} \sqrt{x} = 0$ is true.
Solution: The statement is false since
$$\lim_{x \to 0^-} \sqrt{x} \text{ is undefined}$$

61. Determine the points of discontinuity (if any) of $f(x) = [\![x + 3]\!]$.
Solution: f is discontinuous at every integer. The discontinuities are nonremovable since
$$\lim_{x \to c^-} [\![x + 3]\!] = c + 2 \text{ for any integer } c$$
$$\lim_{x \to c^+} [\![x + 3]\!] = c + 3 \text{ for any integer } c$$

63. Determine the points of discontinuity (if any) of
$$f(x) = \begin{cases} \dfrac{3x^2 - x - 2}{x - 1}, & x \neq 1 \\ 0, & x = 1 \end{cases}$$

Solution:

$$\lim_{x \to 1} \frac{3x^2 - x - 2}{x - 1} = \lim_{x \to 1} \frac{(3x + 2)(x - 1)}{x - 1}$$

$$= \lim_{x \to 1} (3x + 2) = 5$$

Since $f(1) = 0$, $x = 1$ is a discontinuity, and this discontinuity is removable.

65. Determine the points of discontinuity (if any) of

$$f(x) = \frac{1}{(x - 2)^2}$$

Solution: f is discontinuous at $x = 2$ and since

$$\lim_{x \to 2} \frac{1}{(x - 2)^2} = \infty$$

the discontinuity is nonremovable.

67. Determine the points of discontinuity (if any) of

$$f(x) = \frac{3}{x + 1}$$

Solution: f is discontinuous at $x = -1$ and since

$$\lim_{x \to -1^-} \frac{3}{x + 1} = -\infty, \quad \lim_{x \to -1^+} \frac{3}{x + 1} = \infty$$

the discontinuity is nonremovable.

69. Determine the value of c so that

$$f(x) = \begin{cases} x + 3, & x \leq 2 \\ cx + 6, & x > 2 \end{cases}$$

is continuous.
Solution:

$$\lim_{x \to 2^-} f(x) = \lim_{x \to 2^-} (x + 3) = 5$$

$$\lim_{x \to 2^+} f(x) = \lim_{x \to 2^+} (cx + 6) = 2c + 6$$

Thus, $5 = 2c + 6$, and $c = -1/2$.

PRACTICE TEST FOR CHAPTER 1

1. Find the distance between (3, 7) and (4, −2).

2. Find the midpoint of the line segment joining (0, 5) and (2, 1).

3. Determine whether the points (0, −3), (2, 5), and (−3, −15) are collinear.

4. Find x so that the distance between (0, 3) and (x, 5) is 7.

5. Sketch the graph of $y = 4 - x^2$.

6. Sketch the graph of $y = \sqrt{x - 2}$.

7. Sketch the graph of $y = |x - 3|$.

8. Write the equation of the circle in standard form and sketch its graph: $x^2 + y^2 - 8x + 2y + 8 = 0$.

9. Find the points of intersection of the graphs of $x^2 + y^2 = 25$ and $x - 2y = 10$.

10. Find the general equation of the line passing through the points (7, 4) and (6, −2).

11. Find the general equation of the line passing through the point (−2, −1) with a slope of $m = 2/3$.

12. Find the general equation of the line passing through the point (6, −8) with undefined slope.

13. Find the general equation of the line passing through the point (0, 3) and perpendicular to the line given by $2x - 5y = 7$.

14. Given $f(x) = x^2 - 5$, find
    (a) $f(3)$ (b) $f(-6)$
    (c) $f(x - 5)$ (d) $f(x + \Delta x)$

15. Find the domain and range of $f(x) = \sqrt{3 - x}$.

16. Given $f(x) = 2x + 3$ and $g(x) = x^2 - 1$, find
    (a) $f(g(x))$ (b) $g(f(x))$

17. Given $f(x) = x^3 + 6$, find $f^{-1}(x)$.

18. Find $\lim\limits_{x \to -4} (2 - 5x)$.

19. Find $\lim\limits_{x \to 6} \dfrac{x^2 - 36}{x - 6}$.

20. Find $\lim\limits_{x \to -1} \dfrac{|x + 1|}{x + 1}$.

21. Find $\lim\limits_{x \to 0} \dfrac{\sqrt{x + 3} - \sqrt{3}}{x}$.

22. Find $\lim\limits_{x \to 1} f(x)$, where $f(x) = \begin{cases} 2x + 3, & x \leq 1 \\ x^2 + 4, & x > 1 \end{cases}$

23. Find the discontinuities of $f(x) = \dfrac{x - 8}{x^2 - 64}$. Which are removable?

24. Find the discontinuities of $f(x) = \dfrac{|x - 3|}{x - 3}$. Which are removable?

25. Sketch the graph of $f(x) = \dfrac{x^2 - 5x + 6}{x - 3}$.

Section 2.1

Chapter 2   Differentiation

Section 2.1   The Derivative and the Slope of a Curve

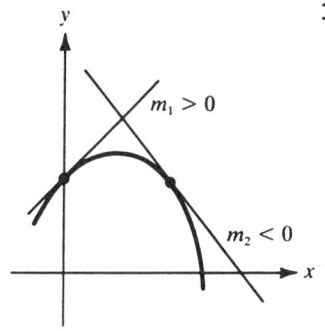

1. Trace the given curve on another piece of paper and sketch the tangent line at each of the points (x_1, y_1) and (x_2, y_2). Determine whether the slope of each tangent line is positive, negative, or zero.
Solution:
The tangent line at (x_1, y_1) has a positive slope.
The tangent line at (x_2, y_2) has a negative slope.

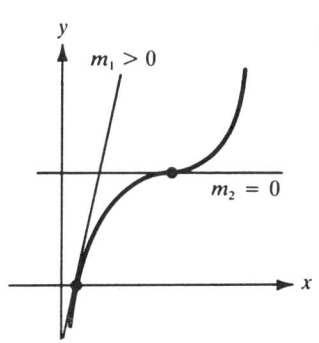

3. Trace the given curve on another piece of paper and sketch the tangent line at each of the points (x_1, y_1) and (x_2, y_2). Determine whether the slope of each tangent line is positive, negative, or zero.
Solution:
The tangent line at (x_1, y_1) has a positive slope.
The tangent line at (x_2, y_2) has zero slope.

5. Estimate the slope of the curve at the point (x, y).
Solution: The slope is $m = 1$.

7. Estimate the slope of the curve at the point (x, y).
Solution: The slope is $m = 0$.

9. Estimate the slope of the curve at the point (x, y).
Solution: The slope is $m = -1/3$.

11. Use the four-step process to find the derivative of $f(x) = 3$.
Solution:
1. $f(x + \Delta x) = 3$
2. $f(x + \Delta x) - f(x) = 0$
3. $[f(x + \Delta x) - f(x)]/\Delta x = 0$
4. $\lim\limits_{\Delta x \to 0} \dfrac{f(x + \Delta x) - f(x)}{\Delta x} = 0$

13. Use the four-step process to find the derivative of $f(x) = -5x + 3$.
Solution:
1. $f(x + \Delta x) = -5(x + \Delta x) + 3 = -5x - 5\Delta x + 3$
2. $f(x + \Delta x) - f(x) = -5\Delta x$
3. $[f(x + \Delta x) - f(x)]/\Delta x = -5$
4. $\lim\limits_{\Delta x \to 0} \dfrac{f(x + \Delta x) - f(x)}{\Delta x} = -5$

15. Use the four-step process to find the derivative of $f(x) = x^2$.
Solution:
1. $f(x + \Delta x) = (x + \Delta x)^2 = x^2 + 2x\Delta x + (\Delta x)^2$
2. $f(x + \Delta x) - f(x) = 2x\Delta x + (\Delta x)^2$
3. $\dfrac{f(x + \Delta x) - f(x)}{\Delta x} = 2x + \Delta x$
4. $\lim\limits_{\Delta x \to 0} \dfrac{f(x + \Delta x) - f(x)}{\Delta x} = 2x$

17. Use the four-step process to find the derivative of $f(x) = 2x^2 + x - 1$.
Solution:
1. $f(x + \Delta x) = 2(x + \Delta x)^2 + (x + \Delta x) - 1$
$\qquad = 2x^2 + 4x\Delta x + 2(\Delta x)^2 + x + \Delta x - 1$
2. $f(x + \Delta x) - f(x) = 4x\Delta x + 2(\Delta x)^2 + \Delta x$
$\qquad = \Delta x(4x + 2\Delta x + 1)$
3. $\dfrac{f(x + \Delta x) - f(x)}{\Delta x} = 4x + 2\Delta x + 1$
4. $\lim\limits_{\Delta x \to 0} \dfrac{f(x + \Delta x) - f(x)}{\Delta x} = 4x + 1$

19. Use the four-step process to find the derivative of $h(t) = \sqrt{t - 1}$.
Solution:
1. $h(t + \Delta t) = \sqrt{t + \Delta t - 1}$
2. $h(t + \Delta t) - h(t) = \sqrt{t + \Delta t - 1} - \sqrt{t - 1}$
$\qquad = \dfrac{\sqrt{t + \Delta t - 1} - \sqrt{t - 1}}{1} \cdot \dfrac{\sqrt{t + \Delta t - 1} + \sqrt{t - 1}}{\sqrt{t + \Delta t - 1} + \sqrt{t - 1}}$
$\qquad = \dfrac{\Delta t}{\sqrt{t + \Delta t - 1} + \sqrt{t - 1}}$
3. $\dfrac{h(t + \Delta t) - h(t)}{\Delta t} = \dfrac{1}{\sqrt{t + \Delta t - 1} + \sqrt{t - 1}}$
4. $\lim\limits_{\Delta t \to 0} \dfrac{h(t + \Delta t) - h(t)}{\Delta t} = \dfrac{1}{2\sqrt{t - 1}}$

21. Use the four-step process to find the derivative of $f(t) = t^3 - 12t$.
Solution:
1. $f(t + \Delta t) = (t + \Delta t)^3 - 12(t + \Delta t)$
$\qquad = t^3 + 3t^2\Delta t + 3t(\Delta t)^2 + (\Delta t)^3 - 12t - 12\Delta t$
2. $f(t + \Delta t) - f(t) = 3t^2\Delta t + 3t(\Delta t)^2 + (\Delta t)^3 - 12\Delta t$
3. $\dfrac{f(t + \Delta t) - f(t)}{\Delta t} = 3t^2 + 3t\Delta t + (\Delta t)^2 - 12$
4. $\lim\limits_{\Delta t \to 0} \dfrac{f(t + \Delta t) - f(t)}{\Delta t} = 3t^2 - 12$

Section 2.1   81

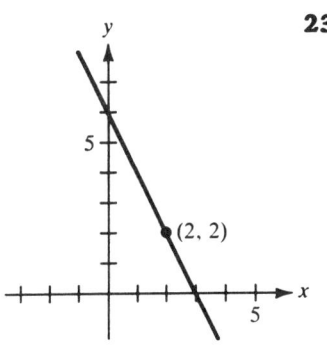

23. Find an equation of the tangent line to the graph of $f(x) = 6 - 2x$ at the point $(2, 2)$. Then verify your answer by sketching both the graph of f and the tangent line.
Solution:
1. $f(x + \Delta x) = 6 - 2(x + \Delta x) = 6 - 2x - 2\Delta x$
2. $f(x + \Delta x) - f(x) = -2\Delta x$
3. $[f(x + \Delta x) - f(x)]/\Delta x = -2$
4. $\lim_{\Delta x \to 0} \dfrac{f(x + \Delta x) - f(x)}{\Delta x} = -2$

At $(2, 2)$, the slope of the tangent line is $m = -2$. The equation of the tangent line is

$$y - 2 = -2(x - 2)$$
$$y = -2x + 6$$

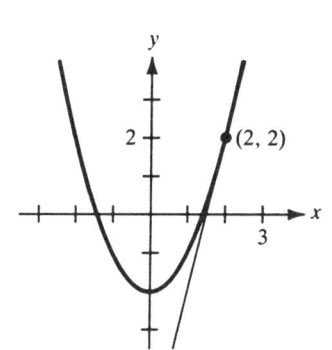

25. Find an equation of the tangent line to the graph of $f(x) = x^2 - 2$ at the point $(2, 2)$. Then verify your answer by sketching both the graph of f and the tangent line.
Solution:
1. $f(x + \Delta x) = (x + \Delta x)^2 - 2$
   $\qquad\qquad = x^2 + 2x\Delta x + (\Delta x)^2 - 2$
2. $f(x + \Delta x) - f(x) = 2x\Delta x + (\Delta x)^2$
3. $[f(x + \Delta x) - f(x)]/\Delta x = 2x + \Delta x$
4. $\lim_{\Delta x \to 0} \dfrac{f(x + \Delta x) - f(x)}{\Delta x} = 2x$

At $(2, 2)$, the slope of the tangent line is $m = 2(2) = 4$. The equation of the tangent line is

$$y - 2 = 4(x - 2)$$
$$y = 4x - 6$$

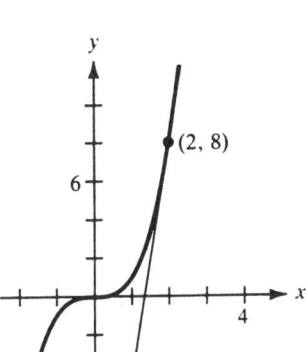

27. Find an equation of the tangent line to the graph of $f(x) = x^3$ at the point $(2, 8)$. Then verify your answer by sketching both the graph of f and the tangent line.
Solution:
1. $f(x + \Delta x) = x^3 + 3x^2\Delta x + 3x(\Delta x)^2 + (\Delta x)^3$
2. $f(x + \Delta x) - f(x) = 3x^2\Delta x + 3x(\Delta x)^2 + (\Delta x)^3$
3. $\dfrac{f(x + \Delta x) - f(x)}{\Delta x} = 3x^2 + 3x\Delta x + (\Delta x)^2$
4. $\lim_{\Delta x \to 0} \dfrac{f(x + \Delta x) - f(x)}{\Delta x} = 3x^2$

At $(2, 8)$, the slope of the tangent line is $m = 3(2)^2 = 12$. The equation of the tangent line is

$$y - 8 = 12(x - 2)$$
$$y = 12x - 16$$

29. Find an equation of the tangent line to the graph of $f(x) = \sqrt{x + 1}$ at the point (3, 2). Then verify your answer by sketching both the graph of f and the tangent line.
    Solution:
    1. $f(x + \Delta x) = \sqrt{x + \Delta x + 1}$
    2. $f(x + \Delta x) - f(x) = \sqrt{x + \Delta x + 1} - \sqrt{x + 1}$

    $$= \frac{\sqrt{x + \Delta x + 1} - \sqrt{x + 1}}{1} \cdot \frac{\sqrt{x + \Delta x + 1} + \sqrt{x + 1}}{\sqrt{x + \Delta x + 1} + \sqrt{x + 1}}$$

    $$= \frac{\Delta x}{\sqrt{x + \Delta x + 1} + \sqrt{x + 1}}$$

    3. $\dfrac{f(x + \Delta x) - f(x)}{\Delta x} = \dfrac{1}{\sqrt{x + \Delta x + 1} + \sqrt{x + 1}}$

    4. $\lim\limits_{\Delta x \to 0} \dfrac{f(x + \Delta x) - f(x)}{\Delta x} = \dfrac{1}{2\sqrt{x + 1}}$

    At (3, 2), the slope of the tangent line is m = 1/4. The equation of the tangent line is

    $$y - 2 = (1/4)(x - 3)$$
    $$y = (1/4)x + (5/4)$$
    $$4y = x + 5$$

31. Find an equation of the tangent line(s) to the curve $y = x^3$ parallel to the line $3x - y + 1 = 0$.
    Solution: From Exercise 27 we know that the slope of the tangent line is $f'(x) = 3x^2$. Since the slope of the given line is 3, we have $3x^2 = 3$ and $x = \pm 1$. Therefore, at the points (1, 1) and (-1, -1) the tangent lines are parallel to $3x - y + 1 = 0$. These lines have equations

    $$y - 1 = 3(x - 1) \quad \text{and} \quad y + 1 = 3(x + 1)$$
    $$y = 3x - 2 \qquad\qquad\qquad y = 3x + 2$$

33. Find the general form of the equations of the two tangent lines to the curve $y = 4x - x^2$ that pass through the point (2, 5).
    Solution: By the four-step process we have $dy/dx = 4 - 2x$. The equation of the tangent line is $y - 5 = (4 - 2x)(x - 2)$. Substituting the given equation for y, we have

    $$(4x - x^2) - 5 = (4 - 2x)(x - 2)$$
    $$4x - x^2 - 5 = -2x^2 + 8x - 8$$
    $$x^2 - 4x + 3 = 0$$
    $$(x - 1)(x - 3) = 0$$

    Therefore, the tangent lines intersect the parabola at (1, 3) and (3, 3) and their equations are

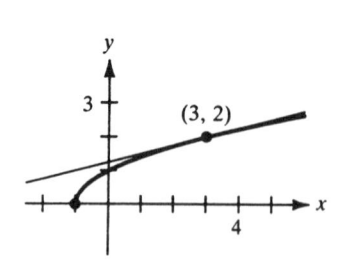

    $$y - 5 = 2(x - 2) \quad \text{and} \quad y - 5 = -2(x - 2)$$
    $$y = 2x + 1 \qquad\qquad\qquad y = -2x + 9$$

Section 2.2

35. Determine where the function $f(x) = |x + 3|$ is not differentiable.
Solution: f is not differentiable when x = -3. At (-3, 0), the graph has a node.

37. Determine where the function
$$f(x) = \begin{cases} 4 - x^2, & 0 < x \\ x^2 - 4, & x \leq 0 \end{cases}$$
is differentiable.
Solution: This function is differentiable for all values of x except x = 0.

● Section 2.2 Some Rules for Differentiation

1. Find the slope of the tangent line at the point (1, 1) to the graph of
  (a) $y = x^2$ (b) $y = x^{1/2}$
  Solution:
  (a) $y = x^2$ (b) $y = x^{1/2}$

  $y' = 2x$ $y' = \dfrac{1}{2}x^{-1/2} = \dfrac{1}{2\sqrt{x}}$

  At (1, 1), $y' = 2$.   At (1, 1), $y' = 1/2$.

3. Find the slope of the tangent line at the point (1, 1) to the graph of
  (a) $y = x^{-1}$ (b) $y = x^{-3/2}$
  Solution:
  (a) $y = x^{-1}$ (b) $y = x^{-3/2}$

  $y' = -x^{-2} = -\dfrac{1}{x^2}$ $y' = -\dfrac{3}{2}x^{-5/2} = -\dfrac{3}{2x^{5/2}}$

  At (1, 1), $y' = -1$.   At (1, 1), $y' = -3/2$.

5. Use the differentiation rules to find the derivative of $y = 3$.
  Solution:
    $y' = 0$

7. Use the differentiation rules to find the derivative of $f(x) = x + 1$.
  Solution:
    $f'(x) = 1$

9. Use the differentiation rules to find the derivative of $g(x) = x^2 + 4$.
  Solution:
    $g'(x) = 2x$

11. Use the differentiation rules to find the derivative of $f(t) = -2t^2 + 3t - 6$.
  Solution:
    $f'(t) = -4t + 3$

Section 2.2

13. Use the differentiation rules to find the derivative of $s(t) = t^3 - 2t + 4$.
Solution:
$$s'(t) = 3t^2 - 2$$

15. Use the differentiation rules to find the derivative of $y = 4t^{3/4}$.
Solution:
$$y' = 4\left(\frac{3}{4}\right)t^{(3/4)-1} = 3t^{-1/4} = \frac{3}{t^{1/4}}$$

17. Use the differentiation rules to find the derivative of $f(x) = 4\sqrt{x}$.
Solution:
$$f(x) = 4\sqrt{x} = 4x^{1/2}$$
$$f'(x) = 4\left(\frac{1}{2}\right)x^{(1/2)-1} = 2x^{-1/2} = \frac{2}{\sqrt{x}}$$

19. Use the differentiation rules to find the derivative of $y = 4x^{-2} + 2x^2$.
Solution:
$$y' = 4(-2)x^{-2-1} + 2(2)x^{2-1} = -8x^{-3} + 4x^1$$
$$= -\frac{8}{x^3} + 4x$$

21. Complete the table using Example 7 as a model.
Solution:

Function	Rewrite	Derivative	Simplify
$y = \dfrac{1}{3x^3}$	$y = \dfrac{1}{3}x^{-3}$	$y' = -x^{-4}$	$y' = -\dfrac{1}{x^4}$

23. Complete the table using Example 7 as a model.
Solution:

Function	Rewrite	Derivative	Simplify
$y = \dfrac{1}{(3x)^3}$	$y = \dfrac{1}{27}x^{-3}$	$y' = -\dfrac{1}{9}x^{-4}$	$y' = -\dfrac{1}{9x^4}$

25. Complete the table using Example 7 as a model.
Solution:

Function	Rewrite	Derivative	Simplify
$y = \dfrac{\sqrt{x}}{x}$	$y = x^{-(1/2)}$	$y' = -\dfrac{1}{2}x^{-(3/2)}$	$y' = -\dfrac{1}{2x^{3/2}}$

27. Find the value of the derivative of $f(x) = 1/x$ at $(1, 1)$.
Solution:
$$f(x) = 1/x = x^{-1}$$
$$f'(x) = -x^{-2} = -1/x^2 \implies f'(1) = -1$$

29. Find the value of the derivative of $f(t) = 3 - (3/5t)$ at the point $(3/5, 2)$.
Solution:
$$f(t) = 3 - (3/5)t^{-1}$$
$$f'(t) = 0 + \frac{3}{5}t^{-2} = \frac{3}{5t^2}$$
$$f'\left(\frac{3}{5}\right) = \frac{3}{5(9/25)} = \frac{5}{3}$$

Section 2.2

31. Find the value of the derivative of $y = (2x + 1)^2$ at the point $(0, 1)$.
Solution:
$$y = (2x + 1)^2 = 4x^2 + 4x + 1$$
$$y' = 8x + 4$$

At $(0, 1)$, $y' = 4$.

33. Find $f'(x)$ for $f(x) = x^2 - (4/x)$.
Solution:
$$f(x) = x^2 - \frac{4}{x} = x^2 - 4x^{-1}$$
$$f'(x) = 2x + 4x^{-2} = 2x + \frac{4}{x^2} = \frac{2(x^3 + 2)}{x^2}$$

35. Find $f'(x)$ for $f(x) = x^3 - 3x - (2/x^4)$.
Solution:
$$f(x) = x^3 - 3x - \frac{2}{x^4} = x^3 - 3x - 2x^{-4}$$
$$f'(x) = 3x^2 - 3 + 8x^{-5} = 3x^2 - 3 + \frac{8}{x^5}$$

37. Find $f'(x)$ for $f(x) = (x^3 - 3x^2 + 4)/x^2$.
Solution:
$$f(x) = \frac{x^3 - 3x^2 + 4}{x^2} = x - 3 + 4x^{-2}$$
$$f'(x) = 1 - 8x^{-3} = 1 - \frac{8}{x^3} = \frac{x^3 - 8}{x^3}$$

39. Find $f'(x)$ for $f(x) = x(x^2 + 1)$.
Solution:
$$f(x) = x(x^2 + 1) = x^3 + x$$
$$f'(x) = 3x^2 + 1$$

41. Find $f'(x)$ for $f(x) = x^{4/5}$.
Solution:
$$f'(x) = \frac{4}{5}x^{-1/5} = \frac{4}{5x^{1/5}}$$

43. Find $f'(x)$ for $f(x) = \sqrt[3]{x} + \sqrt[5]{x}$.
Solution:
$$f(x) = \sqrt[3]{x} + \sqrt[5]{x} = x^{1/3} + x^{1/5}$$
$$f'(x) = \frac{1}{3}x^{-2/3} + \frac{1}{5}x^{-4/5}$$
$$= \frac{1}{3x^{2/3}} + \frac{1}{5x^{4/5}}$$

45. Find an equation of the tangent line to the graph of $y = x^4 - 3x^2 + 2$ at the point $(1, 0)$.
Solution:
$$y' = 4x^3 - 6x$$

At $(1, 0)$, the slope is $m = y' = -2$. The equation of the tangent line is
$$y - 0 = -2(x - 1)$$
$$y = -2x + 2$$

47. Determine the point(s) at which $y = x^4 - 2x^2 + 2$ has a horizontal tangent line.
Solution:
$$y' = 4x^3 - 4x = 4x(x + 1)(x - 1) = 0$$
$$x = 0, \ x = -1, \ x = 1$$

The function has horizontal tangent lines at the points $(0, 2)$, $(1, 1)$ and $(-1, 1)$.

49. Determine the point(s) (if any) at which $f(x) = x^3 + x$ has a horizontal tangent line.
Solution:
$$f'(x) = 3x^2 + 1$$

$f'(x) \neq 0$ for any value of x. The graph of $f(x)$ has no horizontal tangent lines.

51. If $h(x) = f(x) + C$, use the Constant Rule and the Sum Rule to show that $h'(x) = f'(x)$.
Solution:
$$h'(x) = \frac{d}{dx}[h(x)] = \frac{d}{dx}[f(x) + C]$$
$$= \frac{d}{dx}[f(x)] + \frac{d}{dx}[C] = f'(x) + 0 = f'(x)$$

Therefore, $h'(x) = f'(x)$.

53. Use the Constant Rule, the Constant Multiple Rule, and the Sum Rule to find $h'(1)$ given that $f(x) = x^3$ and $f'(1) = 3$, where $h(x)$ is defined as follows:
(a) $h(x) = f(x) - 2$ (b) $h(x) = 2f(x)$
(c) $h(x) = -f(x)$ (d) $h(x) = 4 - f(x)$
Solution:
(a) $h(x) = f(x) - 2$
$h'(x) = f'(x) - 0 = f'(x)$
$h'(1) = f'(1) = 3$

(b) $h(x) = 2f(x)$
$h'(x) = 2f'(x)$
$h'(1) = 2f'(1) = 2(3) = 6$

(c) $h(x) = -f(x)$
$h'(x) = -f'(x)$
$h'(1) = -f'(1) = -3$

(d) $h(x) = 4 - f(x)$
$h'(x) = 0 - f'(x) = -f'(x)$
$h'(1) = -f'(1) = -3$

● Section 2.3 Rates of Change: velocity and marginals

1. The average hourly wage paid to U.S. workers for selected years from 1970 to 1983 is shown in the following table:

Year	1970	1975	1980	1982	1983
Wage	$3.23	$4.53	$6.66	$7.68	$8.02

Find the average rate of change in hourly wages per year for the following periods of time:
(a) 1970 to 1975   (b) 1975 to 1980
(c) 1980 to 1982   (c) 1982 to 1983
Solution:

(a) $\dfrac{4.53 - 3.23}{1975 - 1970} = \$0.26$

(b) $\dfrac{6.66 - 4.53}{1980 - 1975} = \$0.426$

(c) $\dfrac{7.68 - 6.66}{1982 - 1980} = \$0.51$

(d) $\dfrac{8.02 - 7.68}{1983 - 1982} = \$0.34$

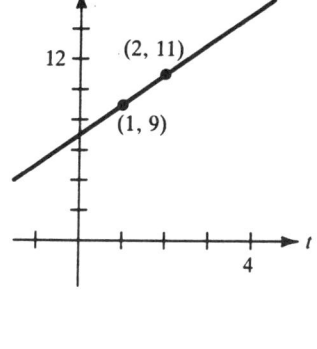

3. Sketch the graph of $f(t) = 2t + 7$ and find its average rate of change over the interval [1, 2]. Compare this to the instantaneous rates of change at the endpoints of the interval.
Solution:
$f'(t) = 2$

Average rate of change: $\dfrac{\Delta y}{\Delta x} = \dfrac{f(2) - f(1)}{2 - 1}$

$= \dfrac{11 - 9}{1} = 2$

Instantaneous rates of change: $f'(1) = 2$, $f'(2) = 2$

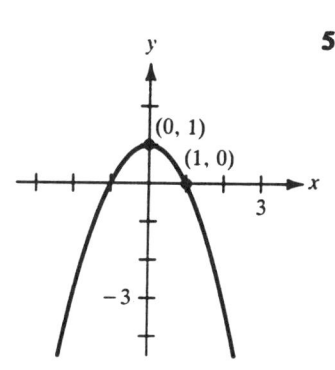

5. Sketch the graph of $h(x) = 1 - x^2$ and find its average rate of change over the interval [0, 1]. Compare this to the instantaneous rates of change at the endpoints of the interval.
Solution:
$h'(x) = -2x$

Average rate of change: $\dfrac{\Delta y}{\Delta x} = \dfrac{f(1) - f(0)}{1 - 0}$

$= \dfrac{0 - 1}{1 - 0} = -1$

Instantaneous rates of change: $h'(0) = 0$, $h'(1) = -2$

Section 2.3

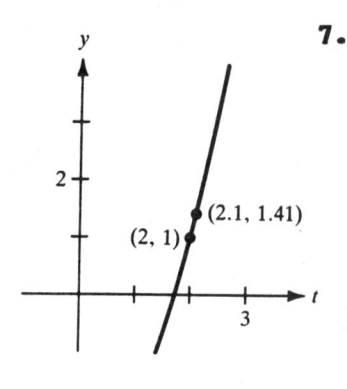

7. Sketch the graph of $f(t) = t^2 - 3$ and find its average rate of change over the interval [2, 2.1]. Compare this to the instantaneous rates of change at the endpoints of the interval.
Solution:
$$f'(t) = 2t$$

Average rate of change: $\dfrac{\Delta y}{\Delta x} = \dfrac{f(2.1) - f(2)}{2.1 - 2}$

$$= \dfrac{1.41 - 1}{0.1} = 4.1$$

Instantaneous rates of change: $f'(2) = 4$
$f'(2.1) = 4.2$

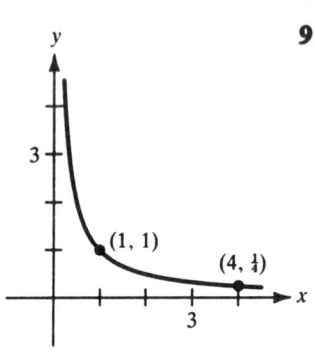

9. Sketch the graph of $f(x) = 1/x$ and find its average rate of change over the interval [1, 4]. Compare this to the instantaneous rates of change at the endpoints of the interval.
Solution:
$$f'(x) = -1/x^2$$

Average rate of change: $\dfrac{\Delta y}{\Delta x} = \dfrac{f(4) - f(1)}{4 - 1}$

$$= \dfrac{(1/4) - 1}{3} = -\dfrac{1}{4}$$

Instantaneous rates of change: $f'(1) = -1$
$f'(4) = -1/16$

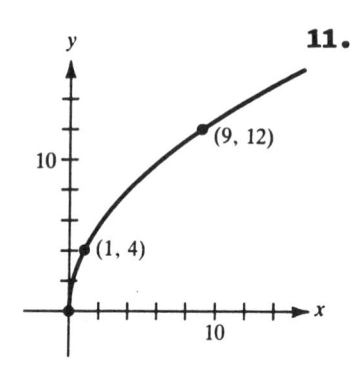

11. Sketch the graph of $g(x) = 4\sqrt{x}$ and find its average rate of change over the interval [1, 9]. Compare this to the instantaneous rates of change at the endpoints of the interval.
Solution:
$$g'(x) = 2/\sqrt{x}$$

Average rate of change: $\dfrac{\Delta y}{\Delta x} = \dfrac{g(9) - g(1)}{9 - 1} = \dfrac{12 - 4}{8} = 1$

Instantaneous rates of change: $g'(1) = 2$, $g'(9) = 2/3$

13. Suppose the effectiveness E (on a scale from 0 to 1) of a painkilling drug t hours after entering the bloodstream is given by

$$E(t) = \dfrac{1}{27}(9t + 3t^2 - t^3), \quad 0 \le t \le 4.5$$

Find the average rate of change of E over the indicated interval and compare this to the instantaneous rates of change at the endpoints of the interval.
(a) [0, 1]   (b) [1, 2]
(c) [2, 3]   (d) [3, 4]
Solution:

$$E'(t) = \dfrac{1}{27}(9 + 6t - 3t^2) = \dfrac{1}{9}(3 + 2t - t^2)$$

Section 2.3   89

(a) $\dfrac{E(1) - E(0)}{1 - 0} = \dfrac{(11/27) - 0}{1 - 0} = \dfrac{11}{27}$

   $E'(0) = 1/3, \quad E'(1) = 4/9$

(b) $\dfrac{E(2) - E(1)}{2 - 1} = \dfrac{(22/27) - (11/27)}{2 - 1} = \dfrac{11}{27}$

   $E'(1) = 4/9, \quad E'(2) = 1/3$

(c) $\dfrac{E(3) - E(2)}{3 - 2} = \dfrac{1 - (22/27)}{3 - 2} = \dfrac{5}{27}$

   $E'(2) = 1/3, \quad E'(3) = 0$

(d) $\dfrac{E(4) - E(3)}{4 - 3} = \dfrac{(20/27) - 1}{4 - 3} = -\dfrac{7}{27}$

   $E'(3) = 0, \quad E'(4) = -5/9$

15. The height s at time t of a silver dollar dropped from the World Trade Center is given by $s = -16t^2 + 1350$ where s is measured in feet and t is measured in seconds.
    (a) Find the average velocity on the interval [1, 2].
    (b) Find the instantaneous velocity when t = 1 and t = 2.
    (c) How long will it take the dollar to hit the ground?
    (d) Find the velocity of the dollar when it hits the ground.

    Solution:
       $s' = -32t$

    (a) $\dfrac{s(2) - s(1)}{2 - 1} = \dfrac{1286 - 1334}{2 - 1} = -48$ ft/sec

    (b) $s'(1) = -32$ ft/sec, $\quad s'(2) = -64$ ft/sec

    (c) $-16t^2 + 1350 = 0$
       $t^2 = 1350/16$
       $t = 15\sqrt{6}/4 \approx 9.2$ seconds

    (d) $s'(\dfrac{15\sqrt{6}}{4}) = -120\sqrt{6} \approx -293.9$ ft/sec

17. Suppose the position of an accelerating car is given by

    $$s(t) = 10t^{3/2}, \quad 0 \leq t \leq 10$$

    where s is measured in feet and t is measured in seconds. Find the velocity of the car when
    (a) t = 0   (b) t = 1
    (c) t = 4   (d) t = 9

    Solution:
       $s'(t) = 15\sqrt{t} = v(t)$

    (a) $v(0) = 0$ ft/sec
    (b) $v(1) = 15$ ft/sec
    (c) $v(4) = 30$ ft/sec
    (d) $v(9) = 45$ ft/sec

19. Use the cost function to find the marginal cost for producing x units when C = 4500 + 1.47x.
Solution:
$$\frac{dC}{dx} = 1.47$$

21. Use the cost function to find the marginal cost for producing x units when C = 55,000 + 470x − (1/4)x², 0 ≤ x ≤ 940.
Solution:
$$\frac{dC}{dx} = 470 - \frac{1}{2}x$$

23. Use the revenue function to find the marginal revenue for selling x units if R = 50x − (1/2)x².
Solution:
$$\frac{dR}{dx} = 50 - x$$

25. Use the revenue function to find the marginal revenue for selling x units if R = −6x³ + 8x² + 200x.
Solution:
$$\frac{dR}{dx} = -18x^2 + 16x + 200$$

27. Use the profit function to find the marginal profit for selling x units if P = −2x² + 72x − 145.
Solution:
$$\frac{dP}{dx} = -4x + 72$$

29. Use the profit function to find the marginal profit for selling x units if P = −(1/4000)x² + 12.2x − 25,000.
Solution:
$$\frac{dP}{dx} = -\frac{1}{2000}x + 12.2$$

31. The revenue from producing x units of a product is R = 12x − 0.001x².
    (a) Find the additional revenue if production is increased from 5000 to 5001 units.
    (b) Find the marginal revenue when 5000 units are produced.
Solution:
(a) R(5001) − R(5000) = 1.999
(b) dR/dx = 12 − 0.002x
When x = 5000, dR/dx = 12 − 0.002(5000) = 2.

33. If the average fuel cost is $1.35 per gallon, then the annual fuel cost of driving a car 15,000 mi/yr is approximately C = 20,250/x, where x is the number of miles per gallon.
    (a) Complete the table and use it to sketch the graph of the cost function.
    (b) Complete the table for the marginal cost.

(c) Find the change in annual fuel cost when x increases from 10 to 11 and compare this with the marginal cost when x = 10.
(d) Find the change in annual fuel cost when x increases from 30 to 31 and compare this with the marginal cost when x = 30.

Solution:

$$\frac{dC}{dx} = -\frac{20{,}250}{x^2}$$

(a)

x	10	15	20	25	30	35	40
C	2025	1350	1012.50	810	675	578.57	506.25

(b)

x	10	15	20	25	30	35	40
$\frac{dC}{dx}$	-202.50	-90.00	-50.63	-32.40	-22.50	-16.53	-12.66

(c) $C(11) - C(10) = -184.09$, $C'(10) = -202.50$
(d) $C(31) - C(30) = -21.77$, $C'(30) = -22.50$

35. The profit for producing x units of a product is given by $P = 2400 - 403.4x + 32x^2 - 0.664x^3$, $10 \le x \le 25$. Find the marginal profit when
(a) x = 10 (b) x = 20
(c) x = 23 (d) x = 25

Solution:

$$\frac{dP}{dx} = -403.4 + 64x - 1.992x^2$$

(a) $P'(10) = 37.40$ (b) $P'(20) = 79.80$
(c) $P'(23) = 14.83$ (d) $P'(25) = -48.40$

37. The monthly demand function and the cost function for x quarts of oil at a local service station are given, respectively, by

$$p = \frac{1100 - x}{400} \quad \text{and} \quad C = 65 + 1.25x$$

(a) Find the monthly revenue as a function of x.
(b) Find the profit as a function of x.
(c) Complete the table for marginal revenue, marginal profit, and profit.

Solution:

(a) $R = xp = \frac{1}{400}(1100x - x^2)$

(b) $P = R - C = \frac{1}{400}(1100x - x^2) - (65 + 1.25x)$

$$= -\frac{1}{400}x^2 + \frac{3}{2}x - 65$$

(c) $\frac{dR}{dx} = \frac{1}{400}(1100 - 2x)$, $\frac{dP}{dx} = -\frac{1}{200}x + \frac{3}{2}$

x	200	250	300	350	400
dR/dx	1.75	1.50	1.25	1.00	0.75
dP/dx	0.50	0.25	0	-0.25	-0.50
P	135.00	153.75	160.00	153.75	135.00

39. Since 1790 the center of population of the United States has been gradually moving westward. Use the figure to estimate the rate (in miles per year) at which the center of population was moving <u>westward</u> during the period
    (a) from 1790 to 1900   (b) from 1900 to 1970
    Solution:

    (a) $\frac{525 - 0}{1900 - 1790} \approx 4.77$ mi/yr

    (b) $\frac{735 - 525}{1970 - 1900} = 3$ mi/yr

● Section 2.4 The Product and Quotient Rules

1. Find the value of the derivative of $f(x) = x^2(3x^3 - 1)$ at the point (1, 2).
   Solution:
   $f'(x) = x^2(9x^2) + 2x(3x^3 - 1) = 15x^4 - 2x$
   $f'(1) = 13$

3. Find the value of the derivative of $f(x) = (1/3)(2x^3 - 4)$ at the point (0, -4/3).
   Solution:
   $f'(x) = \frac{1}{3}(6x^2) = 2x^2$
   $f'(0) = 0$

5. Find the value of the derivative of $g(x) = (x^2 + 3x - 1)(x + 3)$ at the point (-2, -3).
   Solution:
   $g'(x) = (x^2 + 3x - 1)(1) + (2x + 3)(x + 3)$
   $\quad\quad = 3x^2 + 12x + 8$
   $g'(-2) = -4$

7. Find the value of the derivative of $h(x) = x/(x - 5)$ at the point (6, 6).
   Solution:
   $h'(x) = \frac{(x - 5)(1) - (x)(1)}{(x - 5)^2} = \frac{-5}{(x - 5)^2}$
   $h'(6) = -5$

Section 2.4

9. Find the value of the derivative of $f(t) = (t^2 + 1)/4t$ at the point $(1, 1/2)$.
   Solution:
   $$f'(t) = \frac{(4t)(2t) - (t^2 + 1)(4)}{16t^2} = \frac{t^2 - 1}{4t^2}$$
   $$f'(1) = 0$$

11. Complete the table without using the Quotient Rule.
    Solution:

Function	Rewrite	Derivative	Simplify
$y = \dfrac{x^2 + 2x}{x}$	$y = x + 2$	$y' = 1$	$y' = 1$

13. Complete the table without using the Quotient Rule.
    Solution:

Function	Rewrite	Derivative	Simplify
$y = \dfrac{7}{3x^3}$	$y = \dfrac{7}{3}x^{-3}$	$y' = -7x^{-4}$	$y' = -\dfrac{7}{x^4}$

15. Complete the table without using the Quotient Rule.
    Solution:

Function	Rewrite	Derivative	Simplify
$y = \dfrac{3x^2 - 5}{7}$	$y = \dfrac{3}{7}x^2 - \dfrac{5}{7}$	$y' = \dfrac{6}{7}x$	$y' = \dfrac{6x}{7}$

17. Differentiate $f(x) = (x^3 - 3x)(2x^2 + 3x + 5)$.
    Solution:
    $$\begin{aligned}f'(x) &= (x^3 - 3x)(4x + 3) + (3x^2 - 3)(2x^2 + 3x + 5) \\ &= 4x^4 + 3x^3 - 12x^2 - 9x + 6x^4 + 9x^3 + 9x^2 - 9x - 15 \\ &= 10x^4 + 12x^3 - 3x^2 - 18x - 15\end{aligned}$$

19. Differentiate $h(t) = (3t + 4)(t^4 - 5t^2 + 2)$.
    Solution:
    $$\begin{aligned}h'(t) &= (3t + 4)(4t^3 - 10t) + 3(t^4 - 5t^2 + 2) \\ &= 12t^4 + 16t^3 - 30t^2 - 40t + 3t^4 - 15t^2 + 6 \\ &= 15t^4 + 16t^3 - 45t^2 - 40t + 6\end{aligned}$$

21. Differentiate $h(s) = (s^3 - 2)^2$.
    Solution:
    $$h'(s) = 6s^5 - 12s^2 = 6s^2(s^3 - 2)$$

23. Differentiate $f(x) = \sqrt[3]{x}(\sqrt{x} + 3)$.
    Solution:
    $$f(x) = \sqrt[3]{x}(\sqrt{x} + 3) = x^{1/3}(x^{1/2} + 3) = x^{5/6} + 3x^{1/3}$$
    $$f'(x) = \frac{5}{6}x^{-1/6} + x^{-2/3} = \frac{5}{6x^{1/6}} + \frac{1}{x^{2/3}}$$

25. Differentiate $f(x) = (3x - 2)/(2x - 3)$.
    Solution:
    $$f(x) = \frac{3x - 2}{2x - 3}$$
    $$f'(x) = \frac{(2x - 3)(3) - (3x - 2)(2)}{(2x - 3)^2} = \frac{-5}{(2x - 3)^2}$$

27. Differentiate $f(x) = (3 - 2x - x^2)/(x^2 - 1)$.
    Solution:
    $$f(x) = \frac{3 - 2x - x^2}{x^2 - 1}$$
    $$= \frac{(3 + x)(1 - x)}{(x + 1)(x - 1)} = \frac{-(3 + x)}{x + 1}$$
    $$f'(x) = \frac{(x + 1)(-1) + (3 + x)(1)}{(x + 1)^2} = \frac{2}{(x + 1)^2}$$

29. Differentiate $f(x) = (x^5 - 3x)(1/x^2)$.
    Solution:
    $$f(x) = (x^5 - 3x)\left(\frac{1}{x^2}\right) = x^3 - \frac{3}{x}$$
    $$f'(x) = 3x^2 + \frac{3}{x^2} = \frac{3x^4 + 3}{x^2} = \frac{3(x^4 + 1)}{x^2}$$

31. Differentiate $h(t) = (t + 1)/(t^2 + 2t + 2)$.
    Solution:
    $$h(t) = \frac{t + 1}{t^2 + 2t + 2}$$
    $$h'(t) = \frac{(t^2 + 2t + 2)(1) - (t + 1)(2t + 2)}{(t^2 + 2t + 2)^2}$$
    $$= \frac{-t^2 - 2t}{(t^2 + 2t + 2)^2}$$

33. Differentiate $f(x) = (x + 1)/\sqrt{x}$.
    Solution:
    $$f(x) = \frac{x + 1}{\sqrt{x}} = x^{1/2} + x^{-1/2}$$
    $$f'(x) = \frac{1}{2}x^{-1/2} - \frac{1}{2}x^{-3/2} = \frac{1}{2}\left(\frac{1}{x^{1/2}} - \frac{1}{x^{3/2}}\right)$$
    $$= \frac{1}{2}\left(\frac{x - 1}{x^{3/2}}\right) = \frac{x - 1}{2x^{3/2}}$$

35. Differentiate $g(x) = [(x + 1)/(x + 2)](2x - 5)$.
    Solution:
    $$g(x) = \left(\frac{x + 1}{x + 2}\right)(2x - 5)$$
    $$= \frac{2x^2 - 3x - 5}{x + 2}$$
    $$g'(x) = \frac{(x + 2)(4x - 3) - (2x^2 - 3x - 5)}{(x + 2)^2}$$
    $$= \frac{2x^2 + 8x - 1}{(x + 2)^2}$$

Section 2.4

37. Differentiate $f(x) = (3x^3 + 4x)(x - 5)(x + 1)$.
Solution:
$$f(x) = (3x^3 + 4x)(x - 5)(x + 1)$$
$$= (3x^3 + 4x)(x^2 - 4x - 5)$$
$$f'(x) = (3x^3 + 4x)(2x - 4) + (x^2 - 4x - 5)(9x^2 + 4)$$
$$= (6x^4 - 12x^3 + 8x^2 - 16x)$$
$$+ (9x^4 - 36x^3 - 41x^2 - 16x - 20)$$
$$= 15x^4 - 48x^3 - 33x^2 - 32x - 20$$

39. Differentiate $f(x) = (x^2 + c^2)/(x^2 - c^2)$, c is a constant.
Solution:
$$f'(x) = \frac{(x^2 - c^2)(2x) - (x^2 + c^2)(2x)}{(x^2 - c^2)^2}$$
$$= \frac{-4c^2 x}{(x^2 - c^2)^2}$$

41. Find an equation of the tangent line to the graph of $f(x) = x/(x - 1)$ at the point (2, 2).
Solution:
$$f(x) = \frac{x}{x - 1}$$
$$f'(x) = \frac{(x - 1)(1) - x(1)}{(x - 1)^2} = \frac{-1}{(x - 1)^2}$$

At (2, 2), the slope is $m = f'(2) = -1$. The equation of the tangent line is

$$y - 2 = -1(x - 2)$$
$$y = -x + 4$$

43. Find an equation of the tangent line to the graph of $f(x) = (x^3 - 3x + 1)(x + 2)$ at the point (1, -3).
Solution:
$$f'(x) = (x^3 - 3x + 1)(1) + (x + 2)(3x^2 - 3)$$
$$= 4x^3 + 6x^2 - 6x - 5$$

At (1, -3), the slope is $m = f'(1) = -1$. The equation of the tangent line is

$$y + 3 = -1(x - 1)$$
$$y = -x - 2$$

45. Determine the points where the graph of $f(x) = x^2/(x - 1)$ has a horizontal tangent.
Solution:
$$f'(x) = \frac{(x - 1)(2x) - x^2(1)}{(x - 1)^2} = \frac{x^2 - 2x}{(x - 1)^2}$$

$f'(x) = 0$ when $x^2 - 2x = x(x - 2) = 0$, which implies that $x = 0$ or $x = 2$. Thus, the horizontal tangent lines occur at (0, 0) and (2, 4).

47. Use the demand function to find the rate of change in the demand x for the price p when

$$x = 150\left(1 - \frac{p}{p+1}\right), \quad p = \$4$$

Solution:

$$\frac{dx}{dp} = 150\left[0 - \frac{(p+1)(1) - (p)(1)}{(p+1)^2}\right] = -\frac{150}{(p+1)^2}$$

When $p = 4$, $\frac{dx}{dp} = -\frac{150}{(4+1)^2} = -6$

49. The function

$$f(t) = \frac{t^2 - t + 1}{t^2 + 1}$$

measures the percentage of the normal level of oxygen in a pond, where t is the time in weeks after organic waste is dumped into the pond. Find the rate of change of f with respect to t when
(a) $t = 0.5$ (b) $t = 2$ (c) $t = 8$

Solution:

$$f'(t) = \frac{(t^2+1)(2t-1) - (t^2-t+1)(2t)}{(t^2+1)^2}$$

$$= \frac{t^2 - 1}{(t^2+1)^2}$$

(a) $f'(0.5) = -0.48$
(b) $f'(2) = 0.12$
(c) $f'(8) \approx 0.015$

51. A population of 500 bacteria is introduced into a culture. The number of bacteria in the population at any given time is given by

$$P = 500\left(1 + \frac{4t}{50 + t^2}\right)$$

where t is measured in hours. Find the rate of change of the population when $t = 2$.

Solution:

$$P' = 500\left(\frac{(50 + t^2)(4) - (4t)(2t)}{(50 + t^2)^2}\right)$$

$$= 500\left(\frac{200 - 4t^2}{(50 + t^2)^2}\right)$$

When $t = 2$,

$$P' = 500\left(\frac{184}{(54)^2}\right)$$

$$\approx 31.55 \text{ bacteria/hour}$$

Section 2.5

Section 2.5 The Chain Rule

1. Complete the table using Example 1 as a model.
   Solution:

$y = f(g(x))$	$u = g(x)$	$y = f(u)$
$y = (6x - 5)^4$	$u = 6x - 5$	$y = u^4$

3. Complete the table using Example 1 as a model.
   Solution:

$y = f(g(x))$	$u = g(x)$	$y = f(u)$
$y = (4 - x^2)^{-1}$	$u = 4 - x^2$	$y = u^{-1}$

5. Complete the table using Example 1 as a model.
   Solution:

$y = f(g(x))$	$u = g(x)$	$y = f(u)$
$y = \sqrt{x^2 - 1}$	$u = x^2 - 1$	$y = \sqrt{u}$

7. Complete the table using Example 1 as a model.
   Solution:

$y = f(g(x))$	$u = g(x)$	$y = f(u)$
$y = \dfrac{1}{3x + 1}$	$u = 3x + 1$	$y = \dfrac{1}{u}$

9. Find the derivative of $y = (2x - 7)^3$.
   Solution:
   $$y' = 3(2x - 7)^2(2) = 6(2x - 7)^2$$

11. Find the derivative of $f(x) = 2(x^2 - 1)^3$.
    Solution:
    $$f'(x) = 6(x^2 - 1)^2(2x) = 12x(x^2 - 1)^2$$

13. Find the derivative of $g(x) = (4 - 2x)^3$.
    Solution:
    $$g'(x) = 3(4 - 2x)^2(-2) = -6(4 - 2x)^2$$

15. Find the derivative of $h(x) = (6x - x^3)^2$.
    Solution:
    $$h'(x) = 2(6x - x^3)(6 - 3x^2) = 6x(6 - x^2)(2 - x^2)$$

17. Find the derivative of $f(x) = (x^2 - 9)^{2/3}$.
    Solution:
    $$f'(x) = \frac{2}{3}(x^2 - 9)^{-1/3}(2x) = \frac{4x}{3(x^2 - 9)^{1/3}}$$

19. Find the derivative of $f(t) = \sqrt{t + 1}$.
    Solution:
    $$f(t) = \sqrt{t + 1} = (t + 1)^{1/2}$$
    $$f'(t) = \frac{1}{2}(t + 1)^{-1/2}(1) = \frac{1}{2\sqrt{t + 1}}$$

21. Find the derivative of $s(t) = \sqrt{t^2 + 2t - 1}$.
    Solution:
    $$s(t) = \sqrt{t^2 + 2t - 1} = (t^2 + 2t - 1)^{1/2}$$
    $$s'(t) = \frac{1}{2}(t^2 + 2t - 1)^{-1/2}(2t + 2)$$
    $$= \frac{t + 1}{\sqrt{t^2 + 2t - 1}}$$

23. Find the derivative of $y = \sqrt[3]{9x^2 + 4}$.
    Solution:
    $$y = \sqrt[3]{9x^2 + 4} = (9x^2 + 4)^{1/3}$$
    $$y' = \frac{1}{3}(9x^2 + 4)^{-2/3}(18x) = \frac{6x}{(9x^2 + 4)^{2/3}}$$

25. Find the derivative of $y = 2\sqrt{4 - x^2}$.
    Solution:
    $$y = 2\sqrt{4 - x^2} = 2(4 - x^2)^{1/2}$$
    $$y' = 2(\frac{1}{2})(4 - x^2)^{-1/2}(-2x) = -\frac{2x}{\sqrt{4 - x^2}}$$

27. Find the derivative of $f(x) = (25 + x^2)^{-1/2}$.
    Solution:
    $$f'(x) = -\frac{1}{2}(25 + x^2)^{-3/2}(2x) = -\frac{x}{(25 + x^2)^{3/2}}$$

29. Find the derivative of $h(x) = (4 - x^3)^{-4/3}$.
    Solution:
    $$h'(x) = -\frac{4}{3}(4 - x^3)^{-7/3}(-3x^2) = \frac{4x^2}{(4 - x^3)^{7/3}}$$

31. Find the derivative of $y = 1/(x - 2)$.
    Solution:
    $$y = (x - 2)^{-1}$$
    $$y' = (-1)(x - 2)^{-2}(1) = -\frac{1}{(x - 2)^2}$$

33. Find the derivative of $f(t) = [1/(t - 3)]^2$.
    Solution:
    $$f(t) = (t - 3)^{-2}$$
    $$f'(t) = -2(t - 3)^{-3}(1) = -\frac{2}{(t - 3)^3}$$

35. Find the derivative of $f(x) = 3/(x^3 - 4)$.
    Solution:
    $$f(x) = 3(x^3 - 4)^{-1}$$
    $$f'(x) = -3(x^3 - 4)^{-2}(3x^2) = -\frac{9x^2}{(x^3 - 4)^2}$$

37. Find the derivative of $y = 1/\sqrt{x + 2}$.
    Solution:
    $$y = (x + 2)^{-1/2}$$
    $$y' = -\frac{1}{2}(x + 2)^{-3/2}(1) = -\frac{1}{2(x + 2)^{3/2}}$$

Section 2.5   99

39. Find the derivative of $g(x) = 3/\sqrt[3]{x^3 - 1}$.
   Solution:
   $$g(x) = 3(x^3 - 1)^{-1/3}$$
   $$g'(x) = 3(-\frac{1}{3})(x^3 - 1)^{-4/3}(3x^2) = -\frac{3x^2}{(x^3 - 1)^{4/3}}$$

41. Find the derivative of $y = x\sqrt{2x + 3}$.
   Solution:
   $$y = x(2x + 3)^{1/2}$$
   $$y' = x[\frac{1}{2}(2x + 3)^{-1/2}(2)] + (2x + 3)^{1/2}$$
   $$= \frac{x}{\sqrt{2x + 3}} + \frac{\sqrt{2x + 3}}{1} = \frac{3(x + 1)}{\sqrt{2x + 3}}$$

43. Find the derivative of $y = t^2\sqrt{t - 2}$.
   Solution:
   $$y = t^2(t - 2)^{1/2}$$
   $$y' = t^2[\frac{1}{2}(t - 2)^{-1/2}(1)] + 2t(t - 2)^{1/2}$$
   $$= \frac{t^2}{2\sqrt{t - 2}} + \frac{2t\sqrt{t - 2}}{1}$$
   $$= \frac{t^2 + 4t(t - 2)}{2\sqrt{t - 2}} = \frac{t(5t - 8)}{2\sqrt{t - 2}}$$

45. Find the derivative of $f(x) = x^2(x - 2)^4$.
   Solution:
   $$f'(x) = x^2[4(x - 2)^3(1)] + (x - 2)^4(2x)$$
   $$= (x - 2)^3[4x^2 + (x - 2)(2x)]$$
   $$= (x - 2)^3(6x^2 - 4x)$$
   $$= 2x(x - 2)^3(3x - 2)$$

47. Find the derivative of $f(t) = (t^2 - 9)\sqrt{t + 2}$.
   Solution:
   $$f(t) = (t^2 - 9)(t + 2)^{1/2}$$
   $$f'(t) = (t^2 - 9)[(1/2)(t + 2)^{-1/2}(1)] + (2t)(t + 2)^{1/2}$$
   $$= \frac{t^2 - 9}{2\sqrt{t + 2}} + \frac{2t\sqrt{t + 2}}{1}$$
   $$= \frac{t^2 - 9 + 4t(t + 2)}{2\sqrt{t + 2}} = \frac{5t^2 + 8t - 9}{2\sqrt{t + 2}}$$

49. Find the derivative of $f(x) = \sqrt{(x + 1)/x}$.
   Solution:
   $$f(x) = (\frac{x + 1}{x})^{1/2} = (1 + x^{-1})^{1/2}$$
   $$f'(x) = \frac{1}{2}(1 + x^{-1})^{-1/2}(-x^{-2})$$
   $$= \frac{-1}{2x^2}(\frac{x + 1}{x})^{-1/2} = -\frac{1}{2x^{3/2}\sqrt{x + 1}}$$

51. Find the derivative of $g(t) = 3t^2/\sqrt{t^2 + 2t - 1}$.
Solution:
$$g(t) = \frac{3t^2}{\sqrt{t^2 + 2t - 1}} = 3t^2(t^2 + 2t - 1)^{-1/2}$$
$$g'(t) = 3t^2[-(1/2)(t^2 + 2t - 1)^{-3/2}(2t + 2)]$$
$$+ (t^2 + 2t - 1)^{-1/2}(6t)$$
$$= 3t(t^2 + 2t - 1)^{-3/2}[-t(t + 1) + 2(t^2 + 2t - 1)]$$
$$= \frac{3t(t^2 + 3t - 2)}{(t^2 + 2t - 1)^{3/2}}$$

53. Find the derivative of $f(x) = \sqrt{x^2 + 1}/x$.
Solution:
$$f(x) = \frac{\sqrt{x^2 + 1}}{x}$$
$$f'(x) = \frac{x[(1/2)(x^2 + 1)^{-1/2}(2x)] - \sqrt{x^2 + 1}(1)}{x^2}$$
$$= \frac{(x^2/\sqrt{x^2 + 1}) - (\sqrt{x^2 + 1}/1)}{x^2} \left(\frac{\sqrt{x^2 + 1}}{\sqrt{x^2 + 1}}\right)$$
$$= \frac{x^2 - (x^2 + 1)}{x^2\sqrt{x^2 + 1}} = -\frac{1}{x^2\sqrt{x^2 + 1}}$$

55. Find an equation of the tangent line to the graph of f at point (3, 5) if $f(x) = \sqrt{3x^2 - 2}$.
Solution:
$$f(x) = (3x^2 - 2)^{1/2}$$
$$f'(x) = \frac{1}{2}(3x^2 - 2)^{-1/2}(6x) = \frac{3x}{\sqrt{3x^2 - 2}}$$

When $x = 3$, the slope is $f'(3) = 9/5$ and the equation of the tangent line is
$$y - 5 = \frac{9}{5}(x - 3)$$
$$y = \frac{9}{5}x - \frac{2}{5}$$

57. A sum of $1,000 is deposited in an account with an interest rate of r percent compounded monthly. At the end of 5 years the balance in the account is given by
$$A = 1000\left(1 + \frac{r}{1200}\right)^{60}$$
Find the rate of change of A with respect to r for the following rates:
(a) $r = 8\%$ (b) $r = 10\%$ (c) $r = 12\%$
(Compare these results with those of Example 9 to see the effect of increasing the frequency of compounding.)
Solution:
$$A' = 1000(60)\left(1 + \frac{r}{1200}\right)^{59}\left(\frac{1}{1200}\right) = 50\left(1 + \frac{r}{1200}\right)^{59}$$

(a) $A'(8) = 50(1 + \frac{8}{1200})^{59} \approx \$74.00$

(b) $A'(10) = 50(1 + \frac{10}{1200})^{59} \approx \$81.59$

(c) $A'(12) = 50(1 + \frac{12}{1200})^{59} \approx \$89.94$

59. Find the rate of change in the supply x of a product if the price is $150 and the supply function is given by

$$x = 10\sqrt{(p - 100)^2 + 25}, \quad p \geq \$100$$

Solution:

$$x' = 10(1/2)[(p - 100)^2 + 25]^{-1/2}[(2)(p - 100)(1)]$$

$$= \frac{10(p - 100)}{\sqrt{(p - 100)^2 + 25}}$$

When $p = 150$,

$$x' = \frac{10(150 - 100)}{\sqrt{(150 - 100)^2 + 25}} = \frac{500}{\sqrt{2525}}$$

$$= \frac{100}{\sqrt{101}} \approx 9.95$$

61. An assembly plant purchases small electric motors to install in one of their products. The plant management estimates that the cost per motor over the next five years is given by

$$C = 4(1.52t + 10)^{3/2}$$

where t is the time in years. Complete the table for the rate of change of the cost for each of the next five years.

Solution:

$$\frac{dC}{dt} = 4(\frac{3}{2})(1.52t + 10)^{1/2}(1.52)$$

$$= 9.12\sqrt{1.52t + 10}$$

t	1	2	3	4	5
dC/dt	$30.95	$32.93	$34.80	$36.57	$38.26

Section 2.6 Higher-Order Derivatives

1. Find the second derivative of $f(x) = 5 - 4x$.
   Solution:
   $f'(x) = -4$
   $f''(x) = 0$

3. Find the second derivative of $f(x) = x^2 - 2x + 1$.
   Solution:
   $f'(x) = 2x - 2$
   $f''(x) = 2$

5. Find the second derivative of
   $$g(t) = (1/3)t^3 - 4t^2 + 2t - 7$$
   Solution:
   $g'(t) = t^2 - 8t + 2$
   $g''(t) = 2t - 8$

7. Find the second derivative of $g(t) = t^{-1/3}$.
   Solution:
   $$g'(t) = -\frac{1}{3}t^{-4/3}$$
   $$g''(t) = \frac{4}{9}t^{-7/3} = \frac{4}{9t^{7/3}}$$

9. Find the second derivative of $f(x) = 4(x^2 - 1)^2$.
   Solution:
   $f'(x) = 8(x^2 - 1)(2x) = 16x(x^2 - 1) = 16x^3 - 16x$
   $f''(x) = 48x^2 - 16 = 16(3x^2 - 1)$

11. Find the second derivative of $f(x) = x\sqrt[3]{x}$.
    Solution:
    $$f(x) = x\sqrt[3]{x} = x^{4/3}$$
    $$f'(x) = \frac{4}{3}x^{1/3}$$
    $$f''(x) = \frac{4}{9}x^{-2/3} = \frac{4}{9x^{2/3}}$$

13. Find the second derivative of $f(x) = (x + 1)/(x - 1)$.
    Solution:
    $$f(x) = \frac{x + 1}{x - 1}$$
    $$f'(x) = \frac{(x - 1)(1) - (x + 1)(1)}{(x - 1)^2}$$
    $$= -\frac{2}{(x - 1)^2} = -2(x - 1)^{-2}$$
    $$f''(x) = 4(x - 1)^{-3}(1) = \frac{4}{(x - 1)^3}$$

Section 2.6   103

15. Find the second derivative of
$$f(x) = (x - 1)(x^2 - 3x + 2).$$
Solution:
$f(x) = (x - 1)(x^2 - 3x + 2) = x^3 - 4x^2 + 5x - 2$
$f'(x) = 3x^2 - 8x + 5$
$f''(x) = 6x - 8$

17. Find the third derivative of $f(x) = x^5 - 3x^4$.
Solution:
$f'(x) = 5x^4 - 12x^3$
$f''(x) = 20x^3 - 36x^2$
$f'''(x) = 60x^2 - 72x$

19. Find the third derivative of $f(x) = 2x(x - 1)^2$.
Solution:
$f(x) = 2x(x^2 - 2x + 1) = 2x^3 - 4x^2 + 2x$
$f'(x) = 6x^2 - 8x + 2$
$f''(x) = 12x - 8$
$f'''(x) = 12$

21. Find the third derivative of $f(x) = 3/(4x)^2$.
Solution:
$$f(x) = \frac{3}{(4x)^2} = \frac{3}{16}x^{-2}$$

$$f'(x) = -\frac{3}{8}x^{-3}$$

$$f''(x) = \frac{9}{8}x^{-4}$$

$$f'''(x) = -\frac{9}{2}x^{-5} = -\frac{9}{2x^5}$$

23. Find the third derivative of $f(x) = \sqrt{4 - x}$.
Solution:
$f(x) = (4 - x)^{1/2}$

$f'(x) = \frac{1}{2}(4 - x)^{-1/2}(-1) = -\frac{1}{2}(4 - x)^{-1/2}$

$f''(x) = \frac{1}{4}(4 - x)^{-3/2}(-1) = -\frac{1}{4}(4 - x)^{-3/2}$

$f'''(x) = \frac{3}{8}(4 - x)^{-5/2}(-1) = -\frac{3}{8(4 - x)^{5/2}}$

25. Given $f'(x) = 2x^2$, find $f''(x)$.
Solution:
$f''(x) = 4x$

27. Given $f''(x) = (2x - 2)/x$, find $f'''(x)$.
Solution:
$$f''(x) = \frac{2x - 2}{x} = 2 - \frac{2}{x}$$

$$f'''(x) = 0 - (-\frac{2}{x^2}) = \frac{2}{x^2}$$

Section 2.6

29. Given $f^{(4)}(x) = 2x + 1$, find $f^{(6)}(x)$.
Solution:
$$f^{(5)}(x) = 2$$
$$f^{(6)}(x) = 0$$

31. Find the second derivative and solve the equation $f''(x) = 0$ if $f(x) = x^3 - 9x^2 + 27x - 27$.
Solution:
$$f'(x) = 3x^2 - 18x + 27$$
$$f''(x) = 6x - 18 = 0$$

$f''(x) = 0$ when $x = 3$.

33. Find the second derivative and solve the equation $f''(x) = 0$ if $f(x) = (x + 1)(x - 2)(x - 5)$.
Solution:
$$f(x) = x^3 - 6x^2 + 3x + 10$$
$$f'(x) = 3x^2 - 12x + 3$$
$$f''(x) = 6x - 12 = 0$$

$f''(x) = 0$ when $x = 2$.

35. Find the second derivative and solve the equation $f''(x) = 0$ if $f(x) = x^4 - 8x^3 + 18x^2 - 16x + 2$.
Solution:
$$f(x) = x^4 - 8x^3 + 18x^2 - 16x + 2$$
$$f'(x) = 4x^3 - 24x^2 + 36x - 16$$
$$f''(x) = 12x^2 - 48x + 36 = 12(x^2 - 4x + 3)$$

$f''(x) = 0$ when

$$12(x^2 - 4x + 3) = 0$$
$$12(x - 1)(x - 3) = 0$$
$$x = 1, 3$$

37. Find the second derivative and solve the equation $f''(x) = 0$ if $f(x) = x/(x^2 + 3)$.
Solution:
$$f'(x) = \frac{(x^2 + 3)(1) - (x)(2x)}{(x^2 + 3)^2}$$

$$= \frac{3 - x^2}{(x^2 + 3)^2} = (3 - x^2)(x^2 + 3)^{-2}$$

$$f''(x) = (3 - x^2)[-2(x^2 + 3)^{-3}(2x)] + (x^2 + 3)^{-2}(-2x)$$

$$= -2x(x^2 + 3)^{-3}[2(3 - x^2) + (x^2 + 3)]$$

$$= \frac{-2x(9 - x^2)}{(x^2 + 3)^3} = \frac{2x(x^2 - 9)}{(x^2 + 3)^3}$$

$f''(x) = 0$ when

$$2x(x^2 - 9) = 0$$
$$x = 0, \pm 3$$

39. A ball is thrown upward from ground level, and its height at any time is given by

$$s(t) = -16t^2 + 48t$$

(a) Find expressions for the velocity and acceleration of the ball.
(b) Find the time when the ball is at its highest point by finding the time when the velocity is zero.
(c) Find the height at the time given in part (b).

Solution:
(a) $v(t) = s'(t) = -32t + 48$
    $a(t) = s''(t) = -32$

(b) $v(t) = -32t + 48$

   $v(t)$ is zero when

   $$-32t + 48 = 0, \quad t = \frac{48}{32} = \frac{3}{2} = 1.5 \text{ sec}$$

(c) $s(1.5) = -16(1.5)^2 + 48(1.5)$
         $= 36$ feet

41. The velocity of an automobile starting from rest is given by

$$\frac{ds}{dt} = \frac{90t}{t+10} \text{ ft/sec}$$

Complete the table showing the velocity and acceleration at 10-second intervals during the first minute of travel.

Solution:

$$\frac{d^2s}{dt^2} = \frac{(t+10)(90) - (90t)(1)}{(t+10)^2}$$

$$= \frac{900}{(t+10)^2}$$

t	0	10	20	30	40	50	60
$\frac{ds}{dt}$	0	45	60	67.5	72	75	77.14
$\frac{d^2s}{dt^2}$	9	2.25	1	0.56	0.36	0.25	0.18

Section 2.7 Implicit Differentiation

1. Find dy/dx by implicit differentiation and find the slope of the tangent line at the point (1, 4) on the graph for $3x^2 - 2y + 5 = 0$.
   Solution:
   $$3x^2 - 2y + 5 = 0$$
   $$6x - 2y' = 0$$
   $$y' = 3x$$

   At (1, 4), $y' = 3$.

3. Find dy/dx by implicit differentiation and find the slope of the tangent line at the point (4, −2) on the graph for $x - y^2 = 0$.
   Solution:
   $$x - y^2 = 0$$
   $$1 - 2yy' = 0$$
   $$y' = 1/2y$$

   At (4, −2), $y' = -1/4$.

5. Find dy/dx by implicit differentiation and find the slope of the tangent line at the point (1, 4) on the graph for $xy = 4$.
   Solution:
   $$xy = 4$$
   $$xy' + y = 0 \quad \text{(Product Rule)}$$
   $$y' = -y/x$$

   At (1, 4), $y' = -4$.

7. When $x^2 + y^2 = 25$, find dy/dx by implicit differentiation and evaluate the derivative at the point (3, 4).
   Solution:
   $$x^2 + y^2 = 25$$
   $$2x + 2yy' = 0$$
   $$y' = -x/y$$

   At (3, 4), $y' = -3/4$.

9. When $y + xy = 4$, find dy/dx by implicit differentiation and evaluate the derivative at the point (−5, −1).
   Solution:
   $$y + xy = 4$$
   $$y' + xy' + y = 0$$
   $$y'(1 + x) = -y$$
   $$y' = -y/(x + 1)$$

   At (−5, −1), $y' = -1/4$.

Section 2.7   107

11. When $x^2 - y^3 = 3$, find dy/dx by implicit differentiation and evaluate the derivative at the point $(2, 1)$.
   Solution:
$$x^2 - y^3 = 3$$
$$2x - 3y^2 y' = 0$$
$$y' = 2x/3y^2$$

   At $(2, 1)$, $y' = 4/3$.

13. When $x^3 - xy + y^2 = 4$, find dy/dx by implicit differentiation and evaluate the derivative at the point $(0, -2)$.
   Solution:
$$x^3 - xy + y^2 = 4$$
$$3x^2 - xy' - y + 2yy' = 0 \quad \text{(Product Rule)}$$
$$y'(2y - x) = y - 3x^2$$
$$y' = (y - 3x^2)/(2y - x)$$

   At $(0, -2)$, $y' = 1/2$.

15. When $x^3 y^3 - y = x$, find dy/dx by implicit differentiation and evaluate the derivative at the point $(0, 0)$.
   Solution:
$$x^3 y^3 - y = x$$
$$3x^3 y^2 y' + 3x^2 y^3 - y' = 1$$
$$y'(3x^3 y^2 - 1) = 1 - 3x^2 y^3$$
$$y' = (1 - 3x^2 y^3)/(3x^3 y^2 - 1)$$

   At $(0, 0)$, $y' = -1$.

17. When $x^{1/2} + y^{1/2} = 9$, find dy/dx by implicit differentiation and evaluate the derivative at the point $(16, 25)$.
   Solution:
$$x^{1/2} + y^{1/2} = 9$$
$$\frac{1}{2}x^{-1/2} + \frac{1}{2}y^{-1/2} y' = 0$$
$$x^{-1/2} + y^{-1/2} y' = 0$$
$$y' = \frac{-x^{-1/2}}{y^{-1/2}} = -\sqrt{\frac{y}{x}}$$

   At $(16, 25)$, $y' = -5/4$.

19. When $x^{2/3} + y^{2/3} = 5$, find dy/dx by implicit differentiation and evaluate the derivative at the point $(8, 1)$.
   Solution:
$$x^{2/3} + y^{2/3} = 5$$
$$\frac{2}{3}x^{-1/3} + \frac{2}{3}y^{-1/3} y' = 0$$
$$y' = \frac{-x^{-1/3}}{y^{-1/3}} = -\frac{y^{1/3}}{x^{1/3}} = -\sqrt[3]{\frac{y}{x}}$$

   At $(8, 1)$, $y' = -1/2$.

21. When $x^3 - 2x^2y + 3xy^2 = 38$, find dy/dx by implicit differentiation and evaluate the derivative at the point (2, 3).
Solution:
$$x^3 - 2x^2y + 3xy^2 = 38$$
$$3x^2 - 2x^2y' - 4xy + 6xyy' + 3y^2 = 0$$
$$y'(6xy - 2x^2) = 4xy - 3x^2 - 3y^2$$
$$y' = \frac{4xy - 3x^2 - 3y^2}{2x(3y - x)}$$

At (2, 3), $y' = \frac{24 - 12 - 27}{4(7)} = -\frac{15}{28}$.

23. Find dy/dx implicitly and explicitly (the explicit functions are shown on the graph) and show that the two results are equivalent. Find the slope of the tangent line at the point given on the graph for $x^2 + y^2 = 25$.
Solution:
Implicitly: $2x + 2yy' = 0$
$$y' = -x/y$$

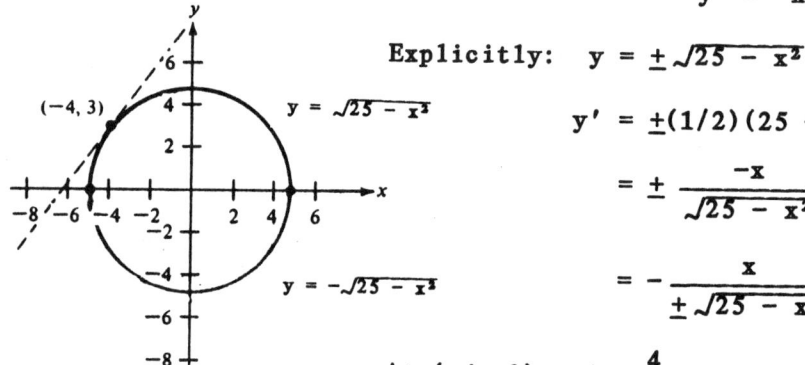

Explicitly: $y = \pm\sqrt{25 - x^2}$
$$y' = \pm(1/2)(25 - x^2)^{-1/2}(-2x)$$
$$= \pm\frac{-x}{\sqrt{25 - x^2}}$$
$$= -\frac{x}{\pm\sqrt{25 - x^2}} = -\frac{x}{y}$$

At (-4, 3), $y' = \frac{4}{3}$.

25. Find dy/dx implicitly and explicitly (the explicit functions are shown on the graph) and show that the two results are equivalent. Find the slope of the tangent line at the point given on the graph for $9x^2 + 16y^2 = 144$.
Solution:
Implicitly: $18x + 32yy' = 0$
$$y' = -9x/16y$$

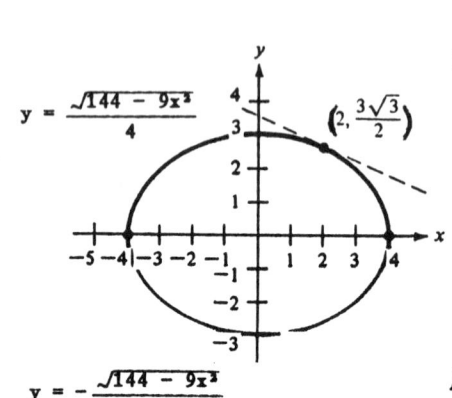

Explicitly: $y = \pm(1/4)\sqrt{144 - 9x^2}$
$$y' = \pm(1/8)(144 - 9x^2)^{-1/2}(-18x)$$
$$= \pm\frac{-9x}{4\sqrt{144 - 9x^2}}$$
$$= -\frac{9x}{16[\pm(1/4)\sqrt{144 - 9x^2}]} = -\frac{9x}{16y}$$

At $(2, \frac{3\sqrt{3}}{2})$, $y' = -\frac{\sqrt{3}}{4}$.

27. Find equations for the tangent lines to the circle $x^2 + y^2 = 169$ at points (5, 12) and (-12, 5).
Solution:
$$x^2 + y^2 = 169$$
$$2x + 2yy' = 0$$
$$y' = -x/y$$

At (5, 12): $\quad m = -5/12$
$$y - 12 = -(5/12)(x - 5)$$
$$5x + 12y - 169 = 0$$

At (-12, 5): $\quad m = 12/5$
$$y - 5 = (12/5)(x + 12)$$
$$12x - 5y + 169 = 0$$

29. The demand function for a certain commodity is given as $p = (200 - x)/2x$, $0 < x \le 200$. Use implicit differentiation to find dx/dp.
Solution:
$$p = \frac{200 - x}{2x} = \frac{100}{x} - \frac{1}{2} = 100x^{-1} - \frac{1}{2}$$

$$\frac{dp}{dx} = -100x^{-2} = -\frac{100}{x^2}$$

$$\frac{dx}{dp} = -\frac{x^2}{100}$$

31. Show that the tangent lines are perpendicular at the points of intersection for $2x^2 + y^2 = 6$ and $y^2 = 4x$.
Solution: To find the points of intersection, substitute $4x$ for y^2 in the equation $2x^2 + y^2 = 6$ to obtain

$$2(x^2 + 2x - 3) = 0$$
$$2(x + 3)(x - 1) = 0$$
$$x = -3, 1$$

y is undefined when $x = -3$ and $y = \pm 2$ when $x = 1$. Thus, the two points of intersection are (1, 2) and (1, -2). The slope of the tangent lines for each graph are given by

$$\begin{array}{ll} 2x^2 + y^2 = 6 & y^2 = 4x \\ 4x + 2yy' = 0 & 2yy' = 4 \\ y' = -2x/y & y' = 2/y \end{array}$$

At (1, 2), $y' = -1$ for $2x^2 + y^2 = 6$ and $y' = 1$ for $y^2 = 4x$. Thus, the tangent lines at (1, 2) are perpendicular.

At (1, -2), $y' = 1$ for $2x^2 + y^2 = 6$ and $y' = -1$ for $y^2 = 4x$. Thus, the tangent lines at (1, -2) are perpendicular.

33. The speed S of blood that is r centimeters from the center of an artery is given by $S = C(R^2 - r^2)$ where C is a constant, R is the radius of the artery, and S is measured in centimeters per second. Suppose a drug is administered and the artery begins dilating at the rate of dR/dt. At a constant distance r, find the rate at which S changes with respect to t for $C = 1.76 \times 10^5$, $R = 1.2 \times 10^{-2}$, and $dR/dt = 10^{-5}$.
Solution:

$$\frac{dS}{dt} = C\left(2R\frac{dR}{dt} - 0\right) = 2CR\frac{dR}{dt}$$

When $C = 1.76 \times 10^5$, $R = 1.2 \times 10^{-2}$, and $\frac{dR}{dt} = 10^{-5}$,

$$\frac{dS}{dt} = 2(1.76 \times 10^5)(1.2 \times 10^{-2})(10^{-5})$$

$$= 4.224 \times 10^{-2} = 0.04224$$

● **Section 2.8 Related Rates**

1. For $y = \sqrt{x}$, assume that x and y are differentiable functions of t. Find (a) dy/dt, given $x = 4$ and $dx/dt = 3$, and (b) dx/dt, given $x = 25$ and $dy/dt = 2$.
Solution:

$$y = \sqrt{x}, \quad \frac{dy}{dt} = \left(\frac{1}{2\sqrt{x}}\right)\frac{dx}{dt}, \quad \frac{dx}{dt} = 2\sqrt{x}\frac{dy}{dt}$$

(a) When $x = 4$ and $dx/dt = 3$,

$$\frac{dy}{dt} = \frac{1}{2\sqrt{4}}(3) = \frac{3}{4}$$

(b) When $x = 25$ and $dy/dt = 2$,

$$\frac{dx}{dt} = 2\sqrt{25}(2) = 20$$

3. For $xy = 4$ assume that x and y are both differentiable functions of t. Find (a) dy/dt, given $x = 8$ and $dx/dt = 10$, and (b) dx/dt, given $x = 1$ and $dy/dt = -6$.
Solution:

$$xy = 4, \quad x\frac{dy}{dt} + y\frac{dx}{dt} = 0$$

$$\frac{dy}{dt} = \left(-\frac{y}{x}\right)\frac{dx}{dt}, \quad \frac{dx}{dt} = \left(-\frac{x}{y}\right)\frac{dy}{dt}$$

(a) When $x = 8$, $y = 1/2$ and $dx/dt = 10$,

$$\frac{dy}{dt} = -\frac{1/2}{8}(10) = -\frac{5}{8}$$

(b) When $x = 1$, $y = 4$, and $dy/dt = -6$,

$$\frac{dx}{dt} = -\frac{1}{4}(-6) = \frac{3}{2}$$

5. The radius r of a circle is increasing at a rate of 2 in/min. Find the rate of change of the area when
   (a) r = 6 inches   (b) r = 24 inches
   Solution:
   $$A = \pi r^2, \quad \frac{dr}{dt} = 2, \quad \frac{dA}{dt} = 2\pi r \frac{dr}{dt}$$

   (a) When r = 6, dA/dt = 2π(6)(2) = 24π in²/min.

   (b) When r = 24, dA/dt = 2π(24)(2) = 96π in²/min.

7. Let A be the area of a circle of radius r that is changing with respect to time. If dr/dt is constant, is dA/dt constant? Explain why or why not.
   Solution:
   $$A = \pi r^2, \quad \frac{dA}{dt} = 2\pi r \frac{dr}{dt}$$

   If dr/dt is constant, dA/dt is proportional to r.

9. A spherical balloon is inflated with gas at the rate of 20 ft³/min. How fast is the radius of the balloon changing at the instant the radius is
   (a) 1 foot   (b) 2 feet
   Solution:
   $$V = \frac{4}{3}\pi r^3, \quad \frac{dV}{dt} = 20$$

   $$\frac{dV}{dt} = 4\pi r^2 \frac{dr}{dt}, \quad \frac{dr}{dt} = \left(\frac{1}{4\pi r^2}\right)\frac{dV}{dt}$$

   (a) When r = 1, $\frac{dr}{dt} = \frac{1}{4\pi(1)^2}(20) = \frac{5}{\pi}$ ft/min.

   (b) When r = 2, $\frac{dr}{dt} = \frac{1}{4\pi(2)^2}(20) = \frac{5}{4\pi}$ ft/min.

11. At a sand and gravel plant, sand is falling off a conveyer and onto a conical pile at the rate of 10 ft³/min. The diameter of the base of the cone is approximately three times the altitude. At what rate is the height of the pile changing when it is 15 feet high?
    Solution:
    $$V = \frac{1}{3}\pi r^2 h = \frac{1}{3}\pi \left(\frac{9}{4}h^2\right)h \quad \text{[since } 2r = 3h\text{]}$$

    $$= \frac{3\pi}{4}h^3, \quad \frac{dV}{dt} = 10$$

    $$\frac{dV}{dt} = \frac{9\pi}{4}h^2 \frac{dh}{dt} \implies \frac{dh}{dt} = \frac{4(dV/dt)}{9\pi h^2}$$

    When h = 15, $\frac{dh}{dt} = \frac{4(10)}{9\pi(15)^2} = \frac{8}{405\pi}$ ft/min.

Section 2.8

13. All edges of a cube are expanding at the rate of 3 cm/sec. How fast is the volume changing when each edge is (a) 1 centimeter and (b) 10 centimeters?
Solution:

$$V = x^3, \quad \frac{dx}{dt} = 3, \quad \frac{dV}{dt} = 3x^2 \frac{dx}{dt}$$

(a) When $x = 1$, $dV/dt = 3(1)^2(3) = 9$ cm³/sec.
(b) When $x = 10$, $dV/dt = 3(10)^2(3) = 900$ cm³/sec.

15. A point is moving along the graph of $y = x^2$ so that dx/dt is 2 cm/min. Find dy/dt when (a) $x = -3$, (b) $x = 0$, (c) $x = 1$, and (d) $x = 3$.
Solution:

$$y = x^2, \quad \frac{dx}{dt} = 2, \quad \frac{dy}{dt} = 2x \frac{dx}{dt}$$

(a) When $x = -3$, $dy/dt = 2(-3)(2) = -12$ cm/min.
(b) When $x = 0$, $dy/dt = 2(0)(2) = 0$ cm/min.
(c) When $x = 1$, $dy/dt = 2(1)(2) = 4$ cm/min.
(d) When $x = 3$, $dy/dt = 2(3)(2) = 12$ cm/min.

17. A ladder 25 feet long is leaning against a house, as shown in the figure. The base of the ladder is pulled away from the house wall at a rate of 2 ft/sec. How fast is the top of the ladder moving down the wall when the base of the ladder is (a) 7 feet, (b) 15 feet, and (c) 24 feet from the wall?
Solution:

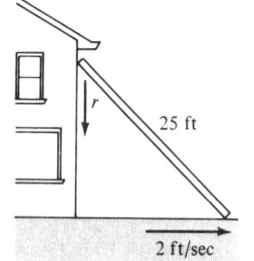

$$x^2 + y^2 = 25^2, \quad 2x \frac{dx}{dt} + 2y \frac{dy}{dt} = 0$$

$$\frac{dy}{dt} = \frac{-x}{y} \frac{dx}{dt} = \frac{-2x}{y} \text{ since } \frac{dx}{dt} = 2$$

(a) When $x = 7$, $y = \sqrt{576} = 24$

$$\frac{dy}{dt} = \frac{-2(7)}{24} = \frac{-7}{12} \text{ ft/sec}$$

(b) When $x = 15$, $y = \sqrt{400} = 20$

$$\frac{dy}{dt} = \frac{-2(15)}{20} = \frac{-3}{2} \text{ ft/sec}$$

(c) When $x = 24$, $y = 7$

$$\frac{dy}{dt} = \frac{-2(24)}{7} = \frac{-48}{7} \text{ ft/sec}$$

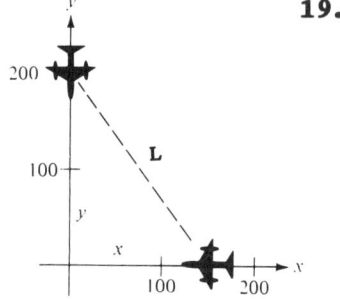

19. An air traffic controller spots two planes at the same altitude converging on a point as they fly at right angles to one another, as shown in the figure. One plane is 150 miles from the point and is moving 450 mi/hr. The other plane is 200 miles from the point and has a speed of 600 mi/hr.
(a) At what rate is the distance between the planes changing?
(b) How much time does the traffic controller have to get one of the planes on a different flight path?

Section 2.8   113

Solution:

(a) $L^2 = x^2 + y^2$, $dx/dt = -450$, $dy/dt = -600$, and

$$\frac{dL}{dt} = \frac{x(dx/dt) + y(dy/dt)}{L}$$

When $x = 150$ and $y = 200$,

$$\frac{dL}{dt} = \frac{150(-450) + 200(-600)}{250} = -750 \text{ mph}$$

(b) $t = 250/750 = 1/3 \text{ hr} = 20 \text{ min}$

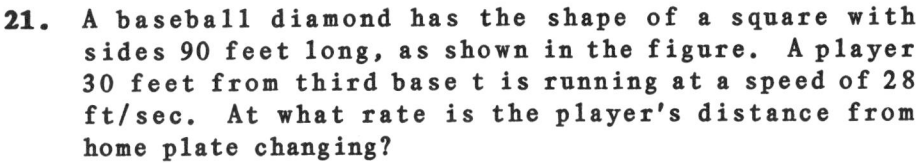

21. A baseball diamond has the shape of a square with sides 90 feet long, as shown in the figure. A player 30 feet from third base t is running at a speed of 28 ft/sec. At what rate is the player's distance from home plate changing?

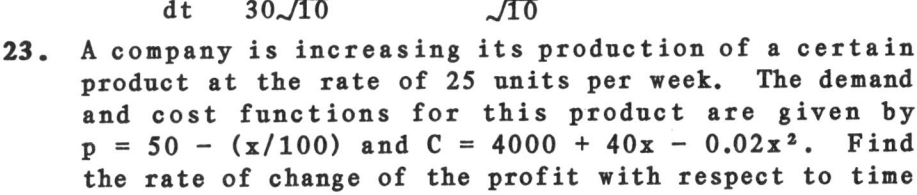

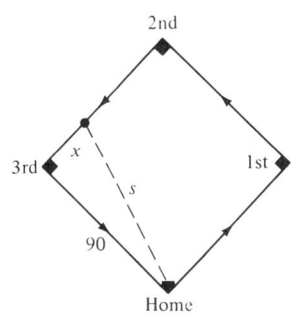

    Solution:
    $$s^2 = 90^2 + x^2, \quad x = 30, \quad dx/dt = -28$$

    $$2s\frac{ds}{dt} = 2x\frac{dx}{dt} \implies \frac{ds}{dt} = \frac{x}{s}\frac{dx}{dt}$$

    When $x = 30$, $s = \sqrt{90^2 + 30^2} = 30\sqrt{10}$

    $$\frac{ds}{dt} = \frac{30}{30\sqrt{10}}(-28) = -\frac{28}{\sqrt{10}} \approx -8.85 \text{ ft/sec}$$

23. A company is increasing its production of a certain product at the rate of 25 units per week. The demand and cost functions for this product are given by $p = 50 - (x/100)$ and $C = 4000 + 40x - 0.02x^2$. Find the rate of change of the profit with respect to time when weekly sales are $x = 800$ units.

    Solution:
    $$\begin{aligned} P &= R - C = xp - C \\ &= x[50 - (x/100)] - (4000 + 40x - 0.02x^2) \\ &= 50x - 0.01x^2 - 4000 - 40x + 0.02x^2 \\ &= 0.01x^2 + 10x - 4000 \end{aligned}$$

    $$\frac{dP}{dt} = 0.02x\frac{dx}{dt} + 10\frac{dx}{dt}$$

    When $x = 800$ and $dx/dt = 25$,

    $$\frac{dP}{dt} = 0.02(800)(25) + (10)(25) = \$650/\text{week}$$

25. An accident at an oil drilling platform in coastal waters is causing a circular oil slick to form. Engineers determine that the slick is 0.08 feet thick, and when the radius is 750 feet it is increasing at the rate of 1/2 ft/min. Estimate the rate at which oil is flowing from the site of the accident.

    Solution:
    $$V = \pi r^2 h, \quad h = 0.08, \quad V = 0.08\pi r^2$$

    $$\frac{dV}{dt} = 0.16\pi r \frac{dr}{dt}$$

    When $r = 750$ and $dr/dt = 1/2$,

    $$\frac{dV}{dt} = 0.16\pi(750)(1/2) = 60\pi \approx 188.5 \text{ ft}^3/\text{min}$$

Review Exercises for Chapter 2

1. Find the derivative of $f(x) = 7x + 3$ by the four-step process.
   Solution:
   1. $f(x + \Delta x) = 7(x + \Delta x) + 3 = 7x + 7\Delta x + 3$
   2. $f(x + \Delta x) - f(x) = 7\Delta x$
   3. $\dfrac{f(x + \Delta x) - f(x)}{\Delta x} = 7$
   4. $\lim\limits_{\Delta x \to 0} \dfrac{f(x + \Delta x) - f(x)}{\Delta x} = 7$

3. Find the derivative of $h(t) = \sqrt{t + 9}$ by the four-step process.
   Solution:
   1. $h(t + \Delta t) = \sqrt{t + \Delta t + 9}$
   2. $h(t + \Delta t) - h(t) = \sqrt{t + \Delta t + 9} - \sqrt{t + 9}$
   $= \dfrac{\sqrt{t + \Delta t + 9} - \sqrt{t + 9}}{1} \cdot \dfrac{\sqrt{t + \Delta t + 9} + \sqrt{t + 9}}{\sqrt{t + \Delta t + 9} + \sqrt{t + 9}}$
   $= \dfrac{\Delta t}{\sqrt{t + \Delta t + 9} + \sqrt{t + 9}}$
   3. $\dfrac{h(t + \Delta t) - h(t)}{\Delta t} = \dfrac{1}{\sqrt{t + \Delta t + 9} + \sqrt{t + 9}}$
   4. $\lim\limits_{\Delta t \to 0} \dfrac{h(t + \Delta t) - h(t)}{\Delta t} = \dfrac{1}{2\sqrt{t + 9}}$

5. Find the derivative of $f(x) = x^3 - 3x^2$.
   Solution:
   $f'(x) = 3x^2 - 6x = 3x(x - 2)$

7. Find the derivative of $f(x) = x^3 - 5 + 3x^{-3}$.
   Solution:
   $f'(x) = 3x^2 - 0 - 9x^{-4}$
   $= 3x^2 - \dfrac{9}{x^4} = 3(x^2 - \dfrac{3}{x^4})$

9. Find the derivative of $f(x) = x^{1/2} - x^{-1/2}$.
   Solution:
   $f'(x) = \dfrac{1}{2}x^{-1/2} + \dfrac{1}{2}x^{-3/2}$
   $= \dfrac{1}{2x^{1/2}} + \dfrac{1}{2x^{3/2}} = \dfrac{x + 1}{2x^{3/2}}$

11. Find the derivative of $f(x) = (3x^2 + 7)(x^2 - 2x + 3)$.
    Solution:
    $f'(x) = (3x^2 + 7)(2x - 2) + (6x)(x^2 - 2x + 3)$
    $= 6x^3 - 6x^2 + 14x - 14 + 6x^3 - 12x^2 + 18x$
    $= 12x^3 - 18x^2 + 32x - 14$
    $= 2(6x^3 - 9x^2 + 16x - 7)$

Review Exercises for Chapter 2

13. Find the derivative of $g(t) = 2/(3t^2)$.
Solution:
$$g(t) = \frac{2}{3t^2} = \frac{2}{3}t^{-2}$$
$$g'(t) = -\frac{4}{3}t^{-3} = -\frac{4}{3t^3}$$

15. Find the derivative of $f(x) = (x^2 + x - 1)/(x^2 - 1)$.
Solution:
$$f(x) = \frac{x^2 + x - 1}{x^2 - 1}$$
$$f'(x) = \frac{(x^2 - 1)(2x + 1) - (x^2 + x - 1)(2x)}{(x^2 - 1)^2}$$
$$= \frac{2x^3 + x^2 - 2x - 1 - 2x^3 - 2x^2 + 2x}{(x^2 - 1)^2}$$
$$= \frac{-x^2 - 1}{(x^2 - 1)^2} = -\frac{x^2 + 1}{(x^2 - 1)^2}$$

17. Find the derivative of $f(x) = 1/(4 - 3x^2)$.
Solution:
$$f(x) = \frac{1}{4 - 3x^2} = (4 - 3x^2)^{-1}$$
$$f'(x) = -(4 - 3x^2)^{-2}(-6x) = \frac{6x}{(4 - 3x^2)^2}$$

19. Find the derivative of $g(x) = 2/\sqrt{x + 1}$.
Solution:
$$g(x) = \frac{2}{\sqrt{x + 1}} = 2(x + 1)^{-1/2}$$
$$g'(x) = 2\left(-\frac{1}{2}\right)(x + 1)^{-3/2}(1) = -\frac{1}{(x + 1)^{3/2}}$$

21. Find the derivative of $f(x) = \sqrt{x^3 + 1}$.
Solution:
$$f(x) = \sqrt{x^3 + 1} = (x^3 + 1)^{1/2}$$
$$f'(x) = \frac{1}{2}(x^3 + 1)^{-1/2}(3x^2) = \frac{3x^2}{2\sqrt{x^3 + 1}}$$

23. Find the derivative of $g(x) = x\sqrt{x^2 + 1}$.
Solution:
$$g(x) = x\sqrt{x^2 + 1} = x(x^2 + 1)^{1/2}$$
$$g'(x) = x\left[\frac{1}{2}(x^2 + 1)^{-1/2}(2x)\right] + (1)(x^2 + 1)^{1/2}$$
$$= \frac{x^2}{\sqrt{x^2 + 1}} + \frac{\sqrt{x^2 + 1}}{1}$$
$$= \frac{2x^2 + 1}{\sqrt{x^2 + 1}}$$

25. Find the derivative of $f(t) = (t + 1)\sqrt[3]{t + 1}$.
Solution:
$$f(t) = (t + 1)(t + 1)^{1/3} = (t + 1)^{4/3}$$
$$f'(t) = \frac{4}{3}(t + 1)^{1/3}(1) = \frac{4}{3}\sqrt[3]{t + 1}$$

27. Find the derivative of $f(x) = -2(1 - 4x^2)^2$.
Solution:
$$f'(x) = -2(2)(1 - 4x^2)(-8x) = 32x(1 - 4x^2)$$

29. Find the derivative of $h(x) = [x^2(2x + 3)]^3$.
Solution:
$$h(x) = [x^2(2x + 3)]^3 = x^6(2x + 3)^3$$
$$h'(x) = x^6[3(2x + 3)^2(2)] + 6x^5(2x + 3)^3$$
$$= 6x^5(2x + 3)^2[x + (2x + 3)]$$
$$= 18x^5(2x + 3)^2(x + 1)$$

31. Find the second derivative of $f(x) = x^2 + 9$.
Solution:
$$f'(x) = 2x$$
$$f''(x) = 2$$

33. Find the second derivative of $f(t) = 5/(1 - t)^2$.
Solution:
$$f(t) = \frac{5}{(1 - t)^2} = 5(1 - t)^{-2}$$
$$f'(t) = 5(-2)(1 - t)^{-3}(-1) = 10(1 - t)^{-3}$$
$$f''(t) = 10(-3)(1 - t)^{-4}(-1) = \frac{30}{(1 - t)^4}$$

35. Find the second derivative of
$$f(x) = (3x^2 + 7)(x^2 - 2x + 3).$$
Solution:
$$f'(x) = (3x^2 + 7)(2x - 2) + (6x)(x^2 - 2x + 3)$$
$$= 6x^3 - 6x^2 + 14x - 14 + 6x^3 - 12x^2 + 18x$$
$$= 12x^3 - 18x^2 + 32x - 14$$
$$f''(x) = 36x^2 - 36x + 32 = 4(9x^2 - 9x + 8)$$

37. Find the second derivative of $f(x) = (1 - x^2)^4$.
Solution:
$$f'(x) = 4(1 - x^2)^3(-2x) = -8x(1 - x^2)^3$$
$$f''(x) = (-8x)[3(1 - x^2)^2(-2x)] + (-8)(1 - x^2)^3$$
$$= -8(1 - x^2)^2[-6x^2 + (1 - x^2)]$$
$$= -8(1 - x^2)^2(1 - 7x^2)$$

39. Find the second derivative of $f(x) = 18\sqrt[3]{x}$.
Solution:
$$f(x) = 18\sqrt[3]{x} = 18x^{1/3}$$
$$f'(x) = 18\left(\frac{1}{3}\right)x^{-2/3} = 6x^{-2/3}$$
$$f''(x) = 6\left(-\frac{2}{3}\right)x^{-5/3} = -\frac{4}{x^{5/3}}$$

Review Exercises for Chapter 2

41. Use implicit differentiation to find dy/dx for $x^2 + 3xy + y^3 = 10$.
Solution:
$$x^2 + 3xy + y^3 = 10$$
$$2x + 3xy' + 3y + 3y^2y' = 0$$
$$y'(3x + 3y^2) = -2x - 3y$$
$$y' = \frac{-2x - 3y}{3x + 3y^2} = -\frac{2x + 3y}{3(x + y^2)}$$

43. Use implicit differentiation to find dy/dx for $y^2 - x^2 = 25$.
Solution:
$$y^2 - x^2 = 25$$
$$2yy' - 2x = 0$$
$$y' = 2x/2y = x/y$$

45. Find an equation of the tangent line to the graph of $y = (x + 3)^3$ at the point $(-2, 1)$.
Solution:
$$y' = 3(x + 3)^2$$

At $(-2, 1)$, $m = y' = 3$, and the tangent line is

$$y - 1 = 3(x + 2)$$
$$y = 3x + 7$$

47. Find an equation of the tangent line to the graph of $x^2 + y^2 = 20$ at the point $(2, 4)$.
Solution:
$$x^2 + y^2 = 20$$
$$2x + 2yy' = 0$$
$$y' = -x/y$$

At $(2, 4)$, $m = y' = -1/2$, and the tangent line is

$$y - 4 = -\frac{1}{2}(x - 2)$$

$$y = -\frac{1}{2}x + 5$$

49. Find an equation of the tangent line to the graph of $y = \sqrt[3]{(x - 2)^2}$ at the point $(3, 1)$.
Solution:
$$y = \sqrt[3]{(x - 2)^2} = (x - 2)^{2/3}$$
$$y' = \frac{2}{3}(x - 2)^{-1/3} = \frac{2}{3\sqrt[3]{x - 2}}$$

At $(3, 1)$, $m = y' = 2/3$, and the tangent line is

$$y - 1 = \frac{2}{3}(x - 3)$$

$$y = \frac{2}{3}x - 1$$

51. Use the cost function to find the marginal cost for $C = 5000 + 650x$.
Solution:
$$\frac{dC}{dx} = 650$$

53. Use the revenue function to find the marginal revenue for $R = 50x/\sqrt{x - 2}$ when $x \geq 6$.
Solution:
$$R = \frac{50x}{\sqrt{x - 2}} = 50x(x - 2)^{-1/2}$$

$$\frac{dR}{dx} = 50x[-\frac{1}{2}(x - 2)^{-3/2}] + 50(x - 2)^{-1/2}$$

$$= 25(x - 2)^{-3/2}[-x + 2(x - 2)]$$

$$= \frac{25(x - 4)}{(x - 2)^{3/2}}$$

55. Use the profit function to find the marginal profit when $P = -0.0005x^3 + 5x^2 - x - 2500$.
Solution:
$$\frac{dP}{dx} = -0.0015x^2 + 10x - 1$$

57. Find the points on the graph of

$$f(x) = (1/3)x^3 + x^2 - x - 1$$

at which the slope is (a) -1, (b) 2, and (c) 0.
Solution:

$$f'(x) = x^2 + 2x - 1$$

(a) $x^2 + 2x - 1 = -1$
$x^2 + 2x = 0$
$x(x + 2) = 0$
$x = 0, -2$

The points are $(0, -1)$ and $(-2, 7/3)$.

(b) $x^2 + 2x - 1 = 2$
$x^2 + 2x - 3 = 0$
$(x + 3)(x - 1) = 0$
$x = -3, 1$

The points are $(-3, 2)$ and $(1, -2/3)$.

(c) $x^2 + 2x - 1 = 0$

$$x = \frac{-2 \pm \sqrt{(2)^2 - 4(1)(-1)}}{2(1)} = -1 \pm \sqrt{2}$$

The points are

$(-1 + \sqrt{2}, \frac{2 - 4\sqrt{2}}{3})$ and $(-1 - \sqrt{2}, \frac{2 + 4\sqrt{2}}{3})$

Review Exercises for Chapter 2   119

59. Derive the equations for the velocity and acceleration of a particle whose position function is $s(t) = t + [2/(t + 1)]$.

Solution:

$$s(t) = t + \frac{2}{t+1} = t + 2(t+1)^{-1}$$

$$v(t) = s'(t) = 1 - 2(t+1)^{-2} = 1 - \frac{2}{(t+1)^2}$$

$$a(t) = s''(t) = 4(t+1)^{-3} = \frac{4}{(t+1)^3}$$

61. Suppose that the temperature T of food placed in a freezer is given by $T = 700/(t^2 + 4t + 10)$ where t is the time in hours. Find the rate of change of T with respect to t when (a) $t = 1$, (b) $t = 3$, (c) $t = 5$, and (d) $t = 10$.

Solution:

$$T = 700(t^2 + 4t + 10)^{-1}$$

$$\frac{dT}{dt} = -700(t^2 + 4t + 10)^{-2}(2t + 4)$$

$$= -\frac{1400(t+2)}{(t^2 + 4t + 10)^2}$$

(a) When $t = 1$, $\dfrac{dT}{dt} = -\dfrac{1400(1+2)}{(1+4+10)^2} \approx -18.667°/\text{hr}$.

(b) When $t = 3$, $\dfrac{dT}{dt} = -\dfrac{1400(3+2)}{(9+12+10)^2} \approx -7.284°/\text{hr}$.

(c) When $t = 5$, $\dfrac{dT}{dt} = -\dfrac{1400(5+2)}{(25+20+10)^2} \approx -3.240°/\text{hr}$.

(d) When $t = 10$, $\dfrac{dT}{dt} = -\dfrac{1400(10+2)}{(100+40+10)^2} \approx -0.747°/\text{hr}$.

63. The **Doyle Log Rule** is a mathematical model for estimating the volume (in board feet) of a log of length L feet and diameter D inches at the small end. According to this model the volume is $V = [(D - 4)/4]^2 L$. Find the rate at which the volume is changing for a 12-foot log whose smallest diameter is (a) 8 inches, (b) 16 inches, (c) 24 inches, and (d) 36 inches.

Solution: When $L = 12$,

$$V = \frac{L}{16}(D-4)^2 = \frac{12}{16}(D-4)^2 = \frac{3}{4}(D-4)^2$$

$$V' = \frac{3}{2}(D-4) = (1.5)(D-4)$$

(a) When $D = 8$, $V' = (1.5)(8 - 4) = 6$ board ft/in.
(b) When $D = 16$, $V' = (1.5)(16 - 4) = 18$ board ft/in.
(c) When $D = 24$, $V' = (1.5)(24 - 4) = 30$ board ft/in.
(d) When $D = 36$, $V' = (1.5)(36 - 4) = 48$ board ft/in.

PRACTICE TEST FOR CHAPTER 2

1. Use the four-step process to find the derivative of $f(x) = 2x^2 + 3x - 5$.

2. Use the four-step process to find the derivative of $f(x) = 1/(x - 4)$.

3. Find the equation of the tangent line to the graph of $f(x) = \sqrt{x - 2}$ at the point $(6, 2)$.

4. Find $f'(x)$ for $f(x) = 5x^3 - 6x^2 + 15x - 9$.

5. Find $f'(x)$ for $f(x) = \dfrac{6x^2 - 4x + 1}{x^2}$.

6. Find $f'(x)$ for $f(x) = \sqrt[3]{x^2} + \sqrt[5]{x^3}$.

7. Find the average rate of change of $f(x) = x^3 - 11$ over the interval $[0, 2]$. Compare this to the instantaneous rate of change at the endpoints of the interval.

8. Given the cost function $C = 6200 + 4.31x - 0.0001x^2$, find the marginal cost of producing x units.

9. Find $f'(x)$ for $f(x) = (x^3 - 4x)(x^2 + 7x - 9)$.

10. Find $f'(x)$ for $f(x) = \dfrac{x + 7}{x^2 - 8}$.

11. Find $f'(x)$ for $f(x) = x^3 \left(\dfrac{x - 3}{x + 5}\right)$.

12. Find $f'(x)$ for $f(x) = \dfrac{\sqrt{x}}{x^2 + 4x - 1}$.

13. Find $f'(x)$ for $f(x) = (6x - 5)^{12}$.

14. Find $f'(x)$ for $f(x) = 8\sqrt{4 - 3x}$.

15. Find $f'(x)$ for $f(x) = -\dfrac{3}{(x^2 + 1)^3}$.

16. Find $f'(x)$ for $f(x) = \sqrt{\dfrac{10x}{x + 2}}$.

17. Find $f'''(x)$ for $f(x) = x^4 - 9x^3 + 17x^2 - 4x + 121$.

18. Find $f^{(4)}(x)$ for $f(x) = \sqrt{3 - x}$.

19. Use implicit differentiation to find dy/dx:

    $x^5 + y^5 = 100$.

20. Use implicit differentiation to find dy/dx:

    $x^2 y^3 + 2x - 3y + 11 = 0$.

Review Exercises for Chapter 2

21. Use implicit differentiation to find dy/dx:

$$\sqrt{xy + 4} = 5y - 4x.$$

22. Use implicit differentiation to find dy/dx:

$$y^3 = \frac{x^3 + 4}{x^3 - 4}.$$

23. Let $y = 3x^2$. Find dx/dt when $x = 2$ and dy/dt = 5.

24. The area A of a circle is increasing at a rate of 10 in²/min. Find the rate of change of the radius r when r = 4 inches.

25. The volume of a cone is $V = (1/3)\pi r^2 h$. Find the rate of change of the height when dV/dt is 200, h = r/2, and h = 20 inches.

Chapter 3 Applications of the Derivative

● Section 3.1 Increasing and Decreasing Functions

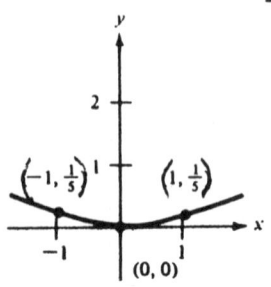

1. Evaluate the derivative of $f(x) = x^2/(x^2 + 4)$ at the indicated points on its graph. Observe the relationship between the sign of the derivative and the increasing or decreasing behavior of the graph.
   Solution:
   $$f'(x) = \frac{(x^2 + 4)(2x) - (x^2)(2x)}{(x^2 + 4)^2} = \frac{8x}{(x^2 + 4)^2}$$
   At $(-1, 1/5)$ f is decreasing since $f'(-1) = -8/25$.
   At $(0, 0)$ f has a critical number since $f'(0) = 0$.
   At $(1, 1/5)$ f is increasing $f'(1) = 8/25$.

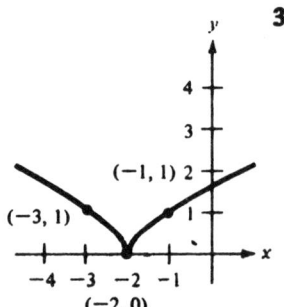

3. Evaluate the derivative of $f(x) = (x + 2)^{2/3}$ at the indicated points on its graph. Observe the relationship between the sign of the derivative and the increasing or decreasing behavior of the graph.
   Solution:
   $$f'(x) = \frac{2}{3}(x + 2)^{-1/3} = \frac{2}{3\sqrt[3]{x + 2}}$$
   At $(-3, 1)$, f is decreasing since $f'(-3) = -2/3$.
   At $(-2, 0)$ f has a critical number since f' is undefined.
   At $(-1, 1)$ f is increasing since $f'(-1) = 2/3$.

5. Find the open intervals on which $f(x) = x^2 - 6x + 8$ is increasing or decreasing.
   Solution:
   $$f'(x) = 2x - 6 = 2(x - 3)$$
   f has a critical number at $x = 3$. Moreover, f is increasing on $(3, \infty)$ and decreasing on $(-\infty, 3)$.

7. Find the open intervals on which $y = (x^3/4) - 3x$ is increasing or decreasing.
   Solution:
   $$y' = \frac{3}{4}x^2 - 3 = \frac{3}{4}(x^2 - 4)$$
   y has critical numbers at $x = \pm 2$. Moreover, y is increasing on $(-\infty, -2)$, $(2, \infty)$ and decreasing on $(-2, 2)$.

9. Find the open intervals on which $f(x) = 1/x^2$ is increasing or decreasing.
   Solution:
   $$f'(x) = -2/x^3$$
   f is undefined at $x = 0$. Moreover, f is increasing on $(-\infty, 0)$ and decreasing on $(0, \infty)$.

Section 3.1 123

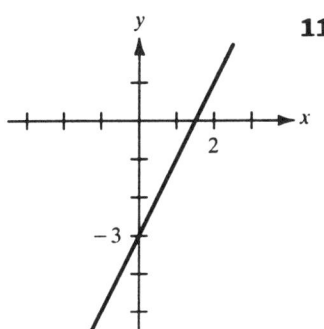

11. Find the critical numbers (if any) and the open intervals on which $f(x) = 2x - 3$ is increasing or decreasing. Sketch the graph of the function.
Solution:
$$f'(x) = 2$$

Since the derivative is positive for all x, the function is increasing for all x. Thus, there are no critical numbers.

13. Find the critical numbers and the open intervals on which $g(x) = -(x - 1)^2$ is increasing or decreasing. Sketch the graph of the function.
Solution:
$$g'(x) = -2(x - 1) = 0$$

Critical number: $x = 1$

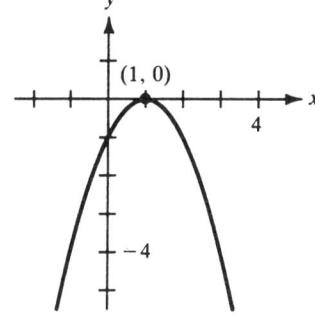

Interval	Sign of f'	Conclusion
$-\infty < x < 1$	$f' > 0$	Increasing
$1 < x < \infty$	$f' < 0$	Decreasing

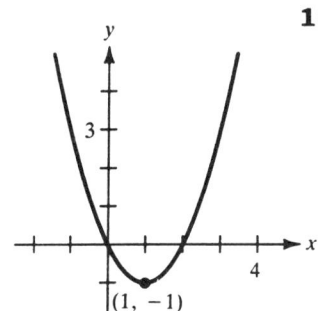

15. Find the critical numbers and the open intervals on which $y = x^2 - 2x$ is increasing or decreasing. Sketch the graph of the function.
Solution:
$$y' = 2x - 2 = 0$$

Critical number: $x = 1$

Interval	Sign of f'	Conclusion
$-\infty < x < 1$	$f' < 0$	Decreasing
$1 < x < \infty$	$f' > 0$	Increasing

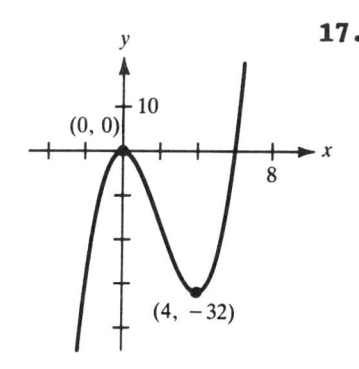

17. Find the critical numbers and the open intervals on which $y = x^3 - 6x^2$ is increasing or decreasing. Sketch the graph of the function.
Solution:
$$y' = 3x^2 - 12x = 3x(x - 4) = 0$$

Critical numbers: $x = 0$ and $x = 4$

Interval	Sign of f'	Conclusion
$-\infty < x < 0$	$f' > 0$	Increasing
$0 < x < 4$	$f' < 0$	Decreasing
$4 < x < \infty$	$f' > 0$	Increasing

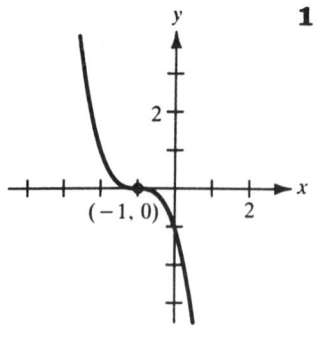

19. Find the critical numbers and the open intervals on which $f(x) = -(x + 1)^3$ is increasing or decreasing.
    Solution:
    $$f'(x) = -3(x + 1)^2 = 0$$

    Critical number: $x = -1$

Interval	Sign of f'	Conclusion
$-\infty < x < -1$	$f' < 0$	Decreasing
$-1 < x < \infty$	$f' < 0$	Decreasing

21. Find the critical numbers and open intervals on which $f(x) = -2x^2 + 4x + 3$ is increasing or decreasing.
    Solution:
    $$f'(x) = -4x + 4 = 0$$

    Critical number: $x = 1$

Interval	Sign of f'	Conclusion
$-\infty < x < 1$	$f' > 0$	Increasing
$1 < x < \infty$	$f' < 0$	Decreasing

23. Find the critical numbers and open intervals on which $f(x) = 2x^3 + 3x^2 - 12x$ is increasing or decreasing.
    Solution:
    $$f'(x) = 6x^2 + 6x - 12 = 6(x + 2)(x - 1) = 0$$

    Critical numbers: $x = -2$ and $x = 1$

Interval	Sign of f'	Conclusion
$-\infty < x < -2$	$f' > 0$	Increasing
$-2 < x < 1$	$f' < 0$	Decreasing
$1 < x < \infty$	$f' > 0$	Increasing

25. Find the critical numbers and the open intervals on which $h(x) = x^{2/3}$ is increasing or decreasing.
    Solution:
    $$h'(x) = (2/3)x^{-1/3} = 2/(3\sqrt[3]{x})$$

    Critical number: $x = 0$ (h' is undefined here)

Interval	Sign of h'	Conclusion
$-\infty < x < 0$	$h' < 0$	Decreasing
$0 < x < \infty$	$h' > 0$	Increasing

Section 3.1   125

27. Find the critical numbers and the open intervals on which $f(x) = x^4 - 2x^3$ is increasing or decreasing.
Solution:
$$f'(x) = 4x^3 - 6x^2 = 2x^2(2x - 3) = 0$$

Critical numbers: $x = 0$ and $x = 3/2$

Interval	Sign of f'	Conclusion
$-\infty < x < 0$	$f' < 0$	Decreasing
$0 < x < 3/2$	$f' < 0$	Decreasing
$3/2 < x < \infty$	$f' > 0$	Increasing

29. Find the critical numbers and the open intervals on which $f(x) = 2x\sqrt{3 - x}$ is increasing or decreasing.
Solution:
$$f'(x) = 2x[\frac{1}{2}(3 - x)^{-1/2}(-1)] + 2\sqrt{3 - x}$$
$$= -\frac{x}{\sqrt{3 - x}} + 2\sqrt{3 - x} = \frac{3(2 - x)}{\sqrt{3 - x}}$$

Domain: $(-\infty, 3]$
Critical numbers: $x = 2$

Interval	Sign of f'	Conclusion
$-\infty < x < 2$	$f' > 0$	Increasing
$2 < x < 3$	$f' < 0$	Decreasing

31. Find the critical numbers and the open intervals on which $y = x/(x^2 + 4)$ is increasing or decreasing.
Solution:
$$y' = \frac{(x^2 + 4)(1) - (x)(2x)}{(x^2 + 4)^2} = \frac{4 - x^2}{(x^2 + 4)^2}$$

Critical numbers: $x = -2$ and $x = 2$

Interval	Sign of y'	Conclusion
$-\infty < x < -2$	$y' < 0$	Decreasing
$-2 < x < 2$	$y' > 0$	Increasing
$2 < x < \infty$	$y' < 0$	Decreasing

33. Find the critical numbers and the open intervals on which the <u>discontinuous</u> function $f(x) = x + (1/x)$ is increasing or decreasing.

Solution:

$$f'(x) = 1 - \frac{1}{x^2} = \frac{x^2 - 1}{x^2}$$

Critical numbers: $x = -1$ and $x = 1$
Discontinuity: $x = 0$

Interval	Sign of f'	Conclusion
$-\infty < x < -1$	$f' > 0$	Increasing
$-1 < x < 0$	$f' < 0$	Decreasing
$0 < x < 1$	$f' < 0$	Decreasing
$1 < x < \infty$	$f' > 0$	Increasing

35. Find the critical numbers and the open intervals on which the <u>discontinuous</u> function $f(x) = x^2/(x^2 - 9)$ is increasing or decreasing.
    Solution:

$$f'(x) = \frac{(x^2 - 9)(2x) - (x^2)(2x)}{(x^2 - 9)^2} = -\frac{18x}{(x^2 - 9)^2}$$

Critical number: $x = 0$
Discontinuities: $x = -3$ and $x = 3$

Interval	Sign of f'	Conclusion
$-\infty < x < -3$	$f' > 0$	Increasing
$-3 < x < 0$	$f' > 0$	Increasing
$0 < x < 3$	$f' < 0$	Decreasing
$3 < x < \infty$	$f' < 0$	Decreasing

37. Find the critical numbers and the open intervals on which the following <u>discontinuous</u> function is increasing or decreasing.

$$f(x) = \begin{cases} 4 - x^2, & x \leq 0 \\ -2x - 2, & x > 0 \end{cases}$$

Solution: f is discontinuous when $x = 0$.

$$f'(x) = \begin{cases} -2x, & x < 0 \\ -2, & x > 0 \end{cases}$$

No critical numbers

Interval	Sign of f'	Conclusion
$-\infty < x < 0$	$f' > 0$	Increasing
$0 < x < \infty$	$f' < 0$	Decreasing

Section 3.1 127

39. The position function, $s(t) = 96t - 16t^2$, $0 \leq t \leq 6$, gives the height (in feet) of a ball, where the time t is measured in seconds. Find the time interval in which the ball is moving up and the interval in which it is moving down.
Solution: Since $s'(t) = 96 - 32t = 0$, the critical number is $t = 3$. Therefore, the ball is moving up on the interval $(0, 3)$ and moving down on $(3, 6)$.

41. A drug is administered to a patient. A model giving the drug concentration in the patient's bloodstream over two hours is $C = 0.29483t + 0.04253t^2 - 0.00035t^3$, $0 \leq t \leq 120$, where C is measured in milligrams and t is the time in minutes. Find the intervals on which C is increasing or decreasing.
Solution: Since $C' = 0.29483 + 0.08506t - 0.00105t^2$, we can use the Quadratic Formula to determine that the critical number is

$$t = \frac{-0.08506 \pm \sqrt{(0.08506)^2 + 4(0.00105)(0.29483)}}{2(-0.00105)}$$

$$= \frac{-0.08506 \pm 0.09205}{-0.00210} \approx 84.34$$

The other root is outside the domain. Thus, C is increasing on the interval $[0, 84.34)$ and decreasing on the interval $(84.34, 120]$.

43. After birth, an infant will normally lose weight for a few days and then start gaining. A model for the average weight W of infants over the first two weeks following birth is

$$W = 0.033t^2 - 0.3974t + 7.3032, \quad 0 \leq t \leq 14$$

where t is measured in days. Find the intervals on which W is increasing or decreasing.
Solution: Since $W' = 0.066t - 0.3974 = 0$, the critical number is $t = 0.3974/0.066 \approx 6.02$. Thus, W is decreasing on $[0, 6.02)$ and increasing on $(6.02, 14]$.

45. The sign of f' is given by

$f'(x) > 0$ on $(-\infty, -4)$
$f'(x) < 0$ on $(-4, 6)$
$f'(x) > 0$ on $(6, \infty)$.

For the function $g(x) = f(x) + 5$, supply the appropriate inequality for $g'(0)$ ___ 0.
Solution: Since $f'(0) < 0$, it follows that

$g'(x) = f'(x)$
$g'(0) = f'(0) \quad \Longrightarrow \quad g'(0) < 0$

47. The sign of f' is given by

$$f'(x) > 0 \text{ on } (-\infty, -4)$$
$$f'(x) < 0 \text{ on } (-4, 6)$$
$$f'(x) > 0 \text{ on } (6, \infty)$$

For the function $g(x) = -f(x)$, supply the appropriate inequality for $g'(-6)$ ___ 0.
Solution: Since $f'(-6) > 0$, it follows that

$$g'(x) = -f'(x)$$
$$g'(-6) = -f'(-6) \quad \Longrightarrow \quad g'(-6) < 0$$

49. The sign of f' is given by

$$f'(x) > 0 \text{ on } (-\infty, -4)$$
$$f'(x) < 0 \text{ on } (-4, 6)$$
$$f'(x) > 0 \text{ on } (6, \infty)$$

For the function $g(x) = f(x - 10)$, supply the appropriate inequality for $g'(0)$ ___ 0.
Solution: Since $f'(-10) > 0$, it follows that

$$g'(x) = f'(x - 10)\frac{d}{dx}[x - 10] = f'(x - 10)$$
$$g'(0) = f'(-10) \quad \Longrightarrow \quad g'(0) > 0$$

● **Section 3.2 Extrema and the First-Derivative Test**

1. Find all relative extrema of $f(x) = -2x^2 + 4x + 3$.
Solution:
$$f'(x) = 4 - 4x = 4(1 - x)$$

Critical number: $x = 1$

Interval	Sign of f'	f
$(-\infty, 1)$	$+$	Increasing
$(1, \infty)$	$-$	Decreasing

Relative maximum: $(1, 5)$

3. Find all relative extrema of $f(x) = x^2 - 6x$.
Solution:
$$f'(x) = 2x - 6 = 2(x - 3)$$

Critical number: $x = 3$

Interval	Sign of f'	f
$(-\infty, 3)$	$-$	Decreasing
$(3, \infty)$	$+$	Increasing

Relative minimum: $(3, -9)$

Section 3.2

5. Find all relative extrema of $g(x) = 2x^3 + 3x^2 - 12x$.
   Solution:
   $$g'(x) = 6x^2 + 6x - 12 = 6(x + 2)(x - 1)$$

   Critical numbers: $x = -2$ and $x = 1$

Interval	Sign of g'	g
$(-\infty, -2)$	+	Increasing
$(-2, 1)$	−	Decreasing
$(1, \infty)$	+	Increasing

   Relative maximum: $(-2, 20)$
   Relative minimum: $(1, -7)$

7. Find all relative extrema of $h(x) = -(x + 4)^3$.
   Solution:
   $$h'(x) = -3(x + 4)^2$$

   Critical number: $x = -4$

Interval	Sign of h'	h
$(-\infty, -4)$	−	Decreasing
$(-4, \infty)$	−	Decreasing

   No relative extrema

9. Find all relative extrema of $f(x) = x^3 - 6x^2 + 15$.
   Solution:
   $$f'(x) = 3x^2 - 12x = 3x(x - 4)$$

   Critical numbers: $x = 0, \quad x = 4$

Interval	Sign of f'	f
$(-\infty, 0)$	+	Increasing
$(0, 4)$	−	Decreasing
$(4, \infty)$	+	Increasing

   Relative maximum: $(0, 15)$
   Relative minimum: $(4, -17)$

11. Find all relative extrema of $f(x) = x^4 - 2x^3$.
    Solution:
    $$f'(x) = 2x^2(2x - 3)$$

    Critical numbers: $x = 0, \quad x = 3/2$

Section 3.2

Interval	Sign of f'	f
$(-\infty, 0)$	−	Decreasing
$(0, 3/2)$	−	Decreasing
$(3/2, \infty)$	+	Increasing

Relative minimum: $(3/2, -27/16)$

13. Find all relative extrema for $f(t) = t^{1/3} + 1$.
    Solution:
    $$f'(t) = \frac{1}{3}t^{-2/3} = \frac{1}{3\sqrt[3]{t^2}}$$

    Critical number: $t = 0$

Interval	Sign of f'	f
$(-\infty, 0)$	+	Increasing
$(0, \infty)$	+	Increasing

No relative extrema

15. Find all relative extrema for $g(t) = t^{2/3}$.
    Solution:
    $$g'(t) = \frac{2}{3}t^{-1/3} = \frac{2}{3\sqrt[3]{t}}$$

    Critical number: $t = 0$

Interval	Sign of g'	g
$(-\infty, 0)$	−	Decreasing
$(0, \infty)$	+	Increasing

Relative minimum: $(0, 0)$

17. Find all relative extrema for $f(x) = x + (1/x)$.
    Solution:
    $$f'(x) = 1 - \frac{1}{x^2} = \frac{x^2 - 1}{x^2}$$

    Critical numbers: $x = \pm 1$ ($x = 0$ is not in domain)

Interval	Sign of f'	f
$(-\infty, -1)$	+	Increasing
$(-1, 0)$	−	Decreasing
$(0, 1)$	−	Decreasing
$(1, \infty)$	+	Increasing

Relative maximum: $(-1, -2)$
Relative minimum: $(1, 2)$

Section 3.2   131

19. Find all relative extrema for $h(x) = 4/(x^2 + 1)$.
Solution:
$$h(x) = \frac{4}{x^2 + 1} = 4(x^2 + 1)^{-1}$$

$$h'(x) = -\frac{8x}{(x^2 + 1)^2}$$

Critical number: $x = 0$

Interval	Sign of h'	h
$(-\infty, 0)$	+	Increasing
$(0, \infty)$	−	Decreasing

Relative maximum: $(0, 4)$

21. Determine from the graph of f if f possesses a relative minimum in the interval (a, b).
Solution: Yes, this graph does have a relative minimum in the interval (a, b).

23. Determine from the graph of f if f possesses a relative minimum in the interval (a, b).
Solution: No, this graph does not have a relative minimum in the interval (a, b).

25. Determine from the graph of f if f possesses a relative minimum in the interval (a, b).
Solution: Yes, this graph does have a relative minimum in the interval (a, b).

27. Locate the extrema (if any exist) of $f(x) = 5 - x$ over the indicated interval.
(a) $[1, 4]$ (b) $[1, 4)$
(c) $(1, 4]$ (d) $(1, 4)$
Solution:
(a) Maximum: $(1, 4)$ Minimum: $(4, 1)$
(b) Maximum: $(1, 4)$
(c) Minimum: $(4, 1)$
(d) No extrema

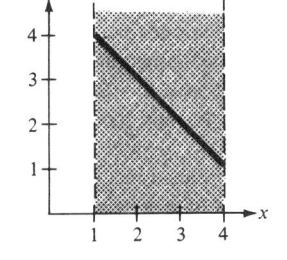

29. Locate the extrema (if any exist) of $f(x) = \sqrt{4 - x^2}$ over the indicated interval.
(a) $[-2, 2]$ (b) $[-2, 0)$
(c) $(-2, 2)$ (d) $[1, 2]$
Solution:
(a) Maximum: $(0, 2)$ Minima: $(-2, 0), (2, 0)$
(b) Minimum: $(-2, 0)$
(c) Maximum: $(0, 2)$
(d) Maximum: $(1, \sqrt{3})$

31. Locate the extrema of $f(x) = 2(3 - x)$ on the interval $[-1, 2]$.
Solution:
$$f'(x) = -2 \qquad \text{(No critical numbers)}$$

x-value	Endpoint $x = -1$	Endpoint $x = 2$
f(x)	8	2
Conclusion	Maximum	Minimum

33. Locate the extrema of $f(x) = -x^2 + 4x$ on the interval $[0, 3]$.
Solution:
$$f'(x) = -2x + 4$$

Critical number: $x = 2$

x-value	Endpoint $x = 0$	Critical $x = 2$	Endpoint $x = 3$
f(x)	0	4	3
Conclusion	Minimum	Maximum	

35. Locate the extrema of $f(x) = x^3 - 3x^2$ on $[-1, 3]$.
Solution:
$$f'(x) = 3x^2 - 6x$$

Critical numbers: $x = 0$ and $x = 2$

x-value	Endpoint $x = -1$	Critical $x = 0$	Critical $x = 2$	Endpoint $x = 3$
f(x)	-4	0	-4	0
Conclusion	Minimum	Maximum	Minimum	Maximum

37. Locate the extrema of $f(x) = 3x^{2/3} - 2x$ on the interval $[-1, 1]$.
Solution:
$$f'(x) = 2x^{-1/3} - 2 = \frac{2 - 2\sqrt[3]{x}}{\sqrt[3]{x}}$$

Critical numbers: $x = 1$ and $x = 0$

x-value	Endpoint $x = -1$	Critical $x = 0$	Endpoint $x = 1$
f(x)	5	0	1
Conclusion	Maximum	Minimum	

Section 3.2   133

39. Locate the extrema of $h(s) = 1/(s - 2)$ on the interval $[0, 1]$.
Solution:
$$h'(s) = \frac{-1}{(s - 2)^2} \quad \text{(No critical numbers)}$$

x-value	Endpoint $x = 0$	Endpoint $x = 1$
f(x)	$-1/2$	-1
Conclusion	Maximum	Minimum

41. For $f(x) = x^3(3x^2 - 10)$, find the maximum value of $|f''(x)|$ in the interval $[0, 1]$.
Solution:
$$f'(x) = 15x^4 - 30x^2$$
$$f''(x) = 60x^3 - 60x$$
$$f'''(x) = 180x^2 - 60 = 60(3x^2 - 1)$$

Critical numbers for f'' in $[0, 1]$: $x = 1/\sqrt{3}$

x-value	Endpoint $x = 0$	Critical $x = 1/\sqrt{3}$	Endpoint $x = 1$
$|f''(x)|$	0	$40/\sqrt{3}$	0
Conclusion		Maximum	

43. For $f(x) = 15x^4 - [(2x - 1)/2]^6$, find the maximum value of $|f^{(4)}(x)|$ in the interval $[0, 1]$.
Solution:
$$f'(x) = 60x^3 - 6\left(\frac{2x - 1}{2}\right)^5$$

$$f''(x) = 180x^2 - 30\left(\frac{2x - 1}{2}\right)^4$$

$$f'''(x) = 360x - 120\left(\frac{2x - 1}{2}\right)^3$$

$$f^{(4)}(x) = 360 - 360\left(\frac{2x - 1}{2}\right)^2$$

$$f^{(5)}(x) = -720\left(\frac{2x - 1}{2}\right)$$

Critical number of $f^{(4)}$: $x = 1/2$

x-value	Endpoint $x = 0$	Critical $x = 1/2$	Endpoint $x = 1$
$|f^{(5)}(x)|$	270	360	270
Conclusion		Maximum	

45. A retailer has determined that the cost C for ordering and storing x units of a certain product is

$$C = 2x + \frac{300,000}{x}, \qquad 0 < x \leq 300$$

Find the order size that will minimize cost given that the delivery truck can bring a maximum of 300 units per order.
Solution:

$$C' = 2 - \frac{300,000}{x^2} = \frac{2x^2 - 300,000}{x^2}$$

There are no critcal numbers in the interval (0, 300]. (Note that $x = \sqrt{300,000/2}$ is greater than 300.)

$$C(300) = 600 + 1000 = 1600$$

Since C is decreasing on (0, 300], x = 300 units gives the minimum cost.

47. Coughing forces the trachea (windpipe) to contract, which in turn affects the velocity v of the air through the trachea. Suppose the velocity of the air during coughing is

$$v = k(R - r)r^2, \qquad 0 \leq r < R$$

where k is a constant, R is the normal radius of the trachea, and r is the radius during coughing. What radius will produce the maximum air velocity?
Solution:

$$v = k(R - r)r^2 = k(Rr^2 - r^3)$$

$$\frac{dv}{dr} = k(2Rr - 3r^2) = kr(2R - 3r)$$

Maximum velocity occurs when dv/dr is zero.

$$kr(2R - 3r) = 0$$
$$r = 0 \text{ or } r = 2R/3$$

Since $v(0) = 0$ and $v(2R/3) = 4kR^3/27$, the maximum air velocity occurs when r = 2R/3.

● Section 3.3 Concavity and the Second-Derivative Test

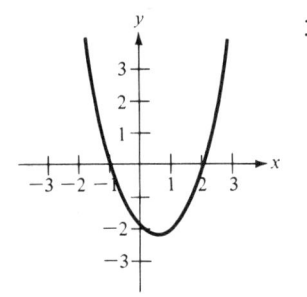

1. Find the intervals on which $y = x^2 - x - 2$ is concave upward and those on which it is concave downward.
Solution:

$$y' = 2x - 1$$
$$y'' = 2$$

Concave upward on $(-\infty, \infty)$

Section 3.3

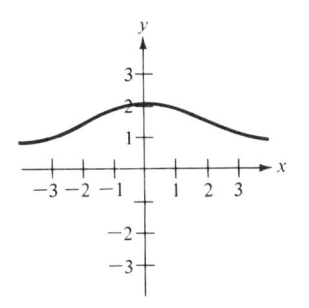

3. Find the intervals on which $f(x) = 24/(x^2 + 12)$ is concave upward and those on which it is concave downward.
   Solution:
   $$f(x) = 24(x^2 + 12)^{-1}$$
   $$f'(x) = -24(2x)(x^2 + 12)^{-2}$$
   $$f''(x) = -48[x(-2)(2x)(x^2 + 12)^{-3} + (x^2 + 12)^{-2}]$$
   $$= \frac{-48(-4x^2 + x^2 + 12)}{(x^2 + 12)^3}$$
   $$= \frac{144(x^2 - 4)}{(x^2 + 12)^3}$$

   $f''(x) = 0$ when $x = \pm 2$. Concave upward on $(-\infty, -2)$ and $(2, \infty)$. Concave downward on $(-2, 2)$.

5. Find the intervals on which $f(x) = (x^2 + 1)/(x^2 - 1)$ is concave upward and those on which it is concave downward.
   Solution:
   $$f'(x) = \frac{-4x}{(x^2 - 1)^2}$$
   $$f''(x) = \frac{4(3x^2 + 1)}{(x^2 - 1)^3}$$

   Concave upward on $(-\infty, -1)$ and $(1, \infty)$. Concave downward on $(-1, 1)$

7. For $f(x) = 6x - x^2$, identify all relative extrema. Use the Second-Derivative Test if applicable.
   Solution:
   $$f'(x) = 6 - 2x = 0$$

   Critical number: $x = 3$

   $$f''(x) = -2$$
   $$f''(3) = -2 < 0$$

   Thus, $(3, 9)$ is a relative maximum.

9. For $f(x) = (x - 5)^2$, identify all relative extrema. Use the Second-Derivative Test if applicable.
   Solution:
   $$f'(x) = 2(x - 5) = 0$$

   Critical number: $x = 5$

   $$f''(x) = 2$$
   $$f''(5) = 2 > 0$$

   Thus, $(5, 0)$ is a relative minimum.

11. For $f(x) = x^3 - 3x^2 + 3$, identify all relative extrema. Use the Second-Derivative Test if applicable.
    Solution:
    $$f'(x) = 3x^2 - 6x = 3x(x - 2)$$

    Critical numbers: $x = 0$, $x = 2$

    $$f''(x) = 6x - 6$$
    $$f''(0) = -6 < 0$$
    $$f''(2) = 6 > 0$$

    Thus, $(0, 3)$ is a relative maximum and $(2, -1)$ is a relative minimum.

13. For $f(x) = x^4 - 4x^3 + 2$, identify all relative extrema. Use the Second-Derivative Test if applicable.
    Solution:
    $$f'(x) = 4x^3 - 12x^2 = 4x^2(x - 3)$$

    Critical numbers: $x = 0$, $x = 3$

    $$f''(x) = 12x^2 - 24x$$
    $$f''(0) = 0 \quad \text{(Test fails)}$$
    $$f''(3) = 36 > 0$$

    Thus, $(3, -25)$ is a relative minimum.
    (Note: By the First-Derivative Test, $(0, 2)$ is not a relative extrema. In fact, it is an inflection point.)

15. For $f(x) = x^{2/3} - 3$, identify all relative extrema. Use the Second-Derivative Test if applicable.
    Solution:
    $$f'(x) = \frac{2}{3}x^{-1/3} = \frac{2}{3\sqrt[3]{x}}$$

    Critical number: $x = 0$
    The Second-Derivative Test does not apply, so we use the First-Derivative Test to conclude that $(0, -3)$ is a relative minimum.

17. For $f(x) = x + (4/x)$, identify all relative extrema. Use the Second-Derivative Test if applicable.
    Solution:
    $$f'(x) = 1 - \frac{4}{x^2} = \frac{x^2 - 4}{x^2}$$

    Critical numbers: $x = \pm 2$

    $$f''(x) = 8/x^3$$
    $$f''(2) = 1 > 0$$
    $$f''(-2) = -1 < 0$$

    Thus, $(2, 4)$ is a relative minimum and $(-2, -4)$ is a relative maximum.

Section 3.3

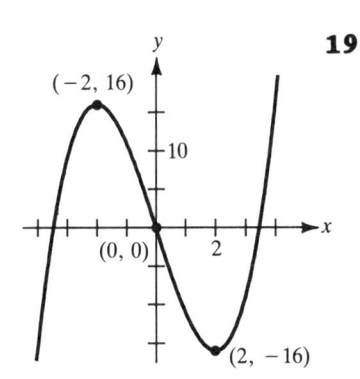

19. Sketch the graph of $f(x) = x^3 - 12x$ and identify all relative extrema and points of inflection.
Solution:
$$f'(x) = 3x^2 - 12 = 3(x^2 - 4)$$

Critical numbers: $x = \pm 2$

$$f''(x) = 6x$$
$$f''(2) = 12 > 0$$
$$f''(-2) = -12 < 0$$

Thus, $(-2, 16)$ is a relative maximum and $(2, -16)$ is a relative minimum.

$$f''(x) = 0 \text{ when } x = 0.$$
$$f''(x) < 0 \text{ on } (-\infty, 0)$$
$$f''(x) > 0 \text{ on } (0, \infty)$$

Thus, $(0, 0)$ is an inflection point.

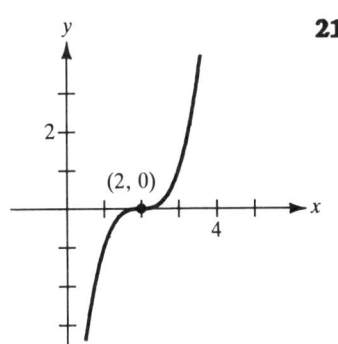

21. Sketch the graph of $f(x) = x^3 - 6x^2 + 12x - 8$ and identify all relative extrema and points of inflection.
Solution:
$$f'(x) = 3x^2 - 12x + 12 = 3(x - 2)^2$$

Critical number: $x = 2$

$$f''(x) = 6(x - 2)$$
$$f''(x) = 0 \text{ when } x = 2.$$

Since $f'(x) > 0$ when $x \neq 2$ and the concavity changes at $x = 2$, $(2, 0)$ is a point of inflection.

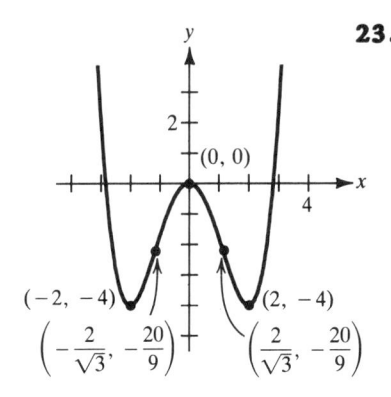

23. Sketch the graph of $f(x) = (1/4)x^4 - 2x^2$ and identify all relative extrema and points of inflection.
Solution:
$$f'(x) = x^3 - 4x = x(x + 2)(x - 2)$$

Critical numbers: $x = -2$, $x = 0$, and $x = 2$

$$f''(x) = 3x^2 - 4$$
$$f''(-2) = 8 > 0$$
$$f''(0) = -4 < 0$$
$$f''(2) = 8 > 0$$

Thus, $(0, 0)$ is a relative maximum and $(-2, -4)$ and $(2, -4)$ are relative minima.

$$f''(x) = 3x^2 - 4 = 0 \text{ when } x = \pm\frac{2\sqrt{3}}{3}$$
$$f''(x) > 0 \text{ on } (-\infty, -2\sqrt{3}/3)$$
$$f''(x) < 0 \text{ on } (-2\sqrt{3}/3, 2\sqrt{3}/3)$$
$$f''(x) > 0 \text{ on } (2\sqrt{3}/3, \infty)$$

Thus, $(-2\sqrt{3}/3, -20/9)$ and $(2\sqrt{3}/3, -20/9)$ are points of inflection.

Section 3.3

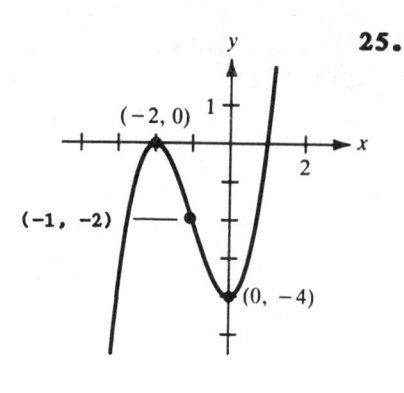

25. Sketch the graph of $g(x) = (x - 1)(x + 2)^2$ and identify all relative extrema and points of inflection.
Solution:
$$g'(x) = 3x^2 + 6x = 3x(x + 2)$$

Critical numbers: $x = -2$ and $x = 0$

$$g''(x) = 6x + 6$$
$$g''(-2) = -6 < 0$$
$$g''(0) = 6 > 0$$

Thus, $(-2, 0)$ is a relative maximum and $(0, -4)$ is a relative minimum.

$$g''(x) = 6x + 6 = 0 \text{ when } x = -1$$
$$g''(x) < 0 \text{ on } (-\infty, -1)$$
$$g''(x) > 0 \text{ on } (-1, \infty)$$

Thus, $(-1, -2)$ is an inflection point.

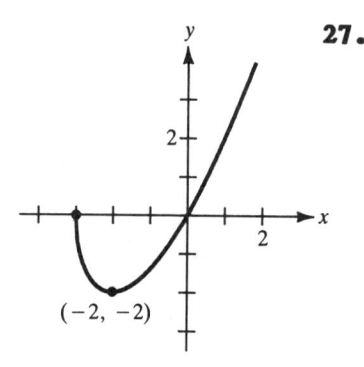

27. Sketch the graph of $g(x) = x\sqrt{x + 3}$ and identify all relative extrema and points of inflection.
Solution: The domain of g is $[-3, \infty)$.

$$g'(x) = x[\frac{1}{2}(x + 3)^{-1/2}] + \sqrt{x + 3} = \frac{3x + 6}{2\sqrt{x + 3}}$$

Critical numbers: $x = -3$ and $x = -2$
By the First-Derivative Test, $(-2, -2)$ is a relative minimum.

$$g''(x) = \frac{(2\sqrt{x + 3})(3) - (3x + 6)(1/\sqrt{x + 3})}{4(x + 3)}$$

$$= \frac{3(x + 4)}{4(x + 3)^{3/2}}$$

$x = -4$ is not in the domain of g.
On $[-3, \infty)$, $g''(x) > 0$ and is concave upward.

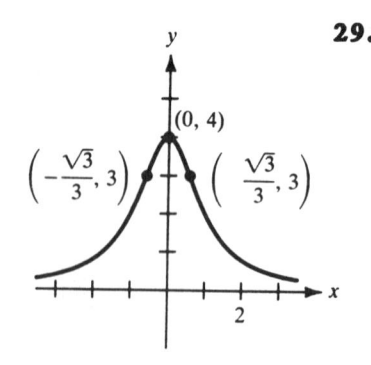

29. Sketch the graph of $f(x) = 4/(1 + x^2)$ and identify all relative extrema and points of inflection.
Solution:
$$f'(x) = \frac{-8x}{(1 + x^2)^2}$$

Critical number: $x = 0$

$$f''(x) = \frac{-8(1 - 3x^2)}{(1 + x^2)^3}$$
$$f''(0) = -8 < 0$$

Thus, $(0, 4)$ is a relative maximum.

$$f''(x) = 0 \text{ when } 1 - 3x^2 = 0, \quad x = \pm\sqrt{3}/3$$
$$f''(x) > 0 \text{ on } (-\infty, -\sqrt{3}/3)$$
$$f''(x) < 0 \text{ on } (-\sqrt{3}/3, \sqrt{3}/3)$$
$$f''(x) > 0 \text{ on } (\sqrt{3}/3, \infty)$$

Thus, $(\frac{\sqrt{3}}{3}, 3)$ and $(-\frac{\sqrt{3}}{3}, 3)$ are inflection points.

Section 3.3

31. Identify the point of diminishing returns for $R = -(4/9)(x^3 - 9x^2 - 27)$, $0 \leq x < 5$, where R is revenue and x is the amount spent on advertising. Assume that R and x are measured in 1000s of dollars.
Solution:
$$R' = -\frac{4}{9}(3x^2 - 18x)$$

$$R'' = -\frac{4}{9}(6x - 18) = 0 \text{ when } x = 3$$

$R'' > 0$ on $(0, 3)$
$R'' < 0$ on $(3, 5)$

Since $(3, 36)$ is a point of inflection it is the point of diminishing returns.

33. When $C = 0.5x^2 + 15x + 5000$ is the total cost function for producing x units, determine the production level that minimizes the average cost per unit. (The average cost per unit is given by $\bar{C} = C/x$.)
Solution:

$$C = 0.5x^2 + 15x + 5000$$

$$\bar{C} = 0.5x + 15 + \frac{5000}{x}$$

$$\bar{C}' = 0.5 - \frac{5000}{x^2}$$

Critical numbers: $x = \pm 100$

$x = 100$ units

35. Sketch a graph of function f having the given characteristics.

Function	First Derivative	Second Derivative
$f(2) = 0$	$f'(x) < 0, \ x < 3$	$f''(x) > 0$
$f(4) = 0$	$f'(3) = 0$	
	$f'(x) > 0, \ x > 3$	

Solution: The function has x-intercepts at $(2, 0)$ and $(4, 0)$. On $(-\infty, 3)$ f is decreasing and on $(3, \infty)$ f is increasing. A relative minimum occurs when $x = 3$. The graph of f is concave upward.

37. Use the graph of $f(x) = 4 - x^2$ to sketch the graph of f'. Find the intervals (if any) on which (a) $f'(x)$ is positive, (b) $f'(x)$ is negative, (c) f' is increasing, and (d) f' is decreasing. For each of these intervals describe the corresponding behavior of f.
Solution:
$$f'(x) = -2x$$

(a) $f'(x) > 0$ on $(-\infty, 0)$ where f is increasing.
(b) $f'(x) < 0$ on $(0, \infty)$ where f is decreasing.
(c) f' is not increasing. f is not concave upward.
(d) f' is decreasing on $(-\infty, \infty)$ where f is concave downward.

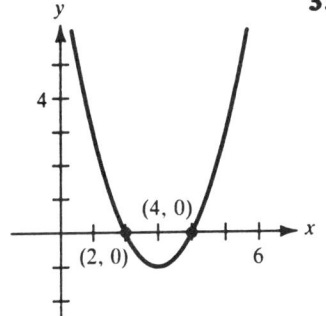

39. Show that the point of inflection of the graph of $f(x) = x(x - 6)^2$ lies midway between the relative extrema of f.
Solution:
$$f(x) = x(x - 6)^2 = x^3 - 12x^2 + 36x$$
$$f'(x) = 3x^2 - 24x + 36 = 3(x - 2)(x - 6)$$

Critical points: $x = 2$ and $x = 6$

$f''(x) = 6x - 24$
$f''(2) = -12 \Rightarrow (2, 32)$ is a relative maximum
$f''(6) = 12 \Rightarrow (6, 0)$ is a relative minimum

$f''(x) = 0$ when $x = 4$. Thus, $(4, 16)$ is an inflection point and lies midway between the relative extrema of f.

Section 3.4 Optimization Problems

1. Find two positive numbers when the sum is 110 and the product is maximum.
Solution: Let x be the first number and y be the second number. Then $x + y = 110$ and $y = 110 - x$. Thus, the product of x and y is given by

$$P = xy = x(110 - x)$$
$$P' = 110 - 2x$$

$P' = 0$ when $x = 55$. Since $P''(55) = -2 < 0$, the product is maximum when $x = 55$ and $y = 110 - 55 = 55$.

3. Find two positive numbers when the sum of the first and twice the second is 24 and the product is maximum.
Solution: Let x be the first number and y be the second number. Then $x + 2y = 24$ and $x = 24 - 2y$. The product of x and y is given by

$$P = xy = (24 - 2y)y$$
$$P' = 24 - 4y$$

$P' = 0$ when $y = 6$. Since $P'' = -4 < 0$, the product is maximum when $y = 6$ and $x = 24 - 2(6) = 12$.

5. Find two positive numbers when the product is 192 and the sum is minimum.
Solution: Let x be the first number and y be the second number. Then $xy = 192$ and $y = 192/x$. The sum of x and y is given by

$$S = x + y = x + \frac{192}{x}$$
$$S' = 1 - \frac{192}{x^2}$$

$S' = 0$ when $x = \sqrt{192}$. Since $S''(x) > 0$ when $x > 0$, S is minimum when $x = \sqrt{192}$ and $y = 192/\sqrt{192} = \sqrt{192}$.

Section 3.4

7. What positive number x minimizes the sum of x and its reciprocal?
   Solution:
   $$S = x + \frac{1}{x}, \quad x > 0$$
   $$S' = 1 - \frac{1}{x^2} = \frac{x^2 - 1}{x^2}$$
   Critical number: $x = 1$
   $$S'' = \frac{2}{x^3}$$
   Since $S''(1) = 2 > 0$, $(1, 2)$ is a relative minimum and the sum is a minimum when $x = 1$.

9. Find the length and width of a rectangle of maximum area with a perimeter of 100 feet.
   Solution: Let x be the length and y be the width of the rectangle. Then $2x + 2y = 100$ and $y = 50 - x$. The area is given by
   $$A = xy = x(50 - x)$$
   $$A' = 50 - 2x$$
   $A' = 0$ when $x = 25$. Since $A''(25) = -2 < 0$, A is maximum when $x = 25$ feet and $y = 50 - 25 = 25$ feet.

11. Find the length and width of a rectangle of minimum perimeter with an area of 64 square feet.
    Solution: Let x and y be the length and width of the rectangle. Then the area is $xy = 64$ and $y = 64/x$. Then perimeter is given by
    $$P = 2x + 2y = 2x + 2\left(\frac{64}{x}\right)$$
    $$P' = 2 - \frac{128}{x^2} = \frac{2(x^2 - 64)}{x^2}$$
    $P' = 0$ when $x = 8$. Since $P'' = 256/x^3$ and $P''(8) > 0$, the perimeter is a minimium when $x = 8$ feet and $y = 64/8 = 8$ feet.

13. A rancher has 200 feet of fencing to enclose two adjacent rectangular corrals, as shown in the figure. What dimensions should be used so that the enclosed area will be a maximum?
    Solution: Let x and y be the lengths shown in the figure. Then $4x + 3y = 200$ and $y = (200 - 4x)/3$. The area of the corrals is given by
    $$A = 2xy = 2x\left(\frac{200 - 4x}{3}\right) = \frac{8}{3}(50x - x^2)$$
    $$A' = \frac{8}{3}(50 - 2x)$$
    $A' = 0$ when $x = 25$. Since $A'' = -16/3 < 0$, A is maximum when $x = 25$ feet and $y = 100/3$ feet.

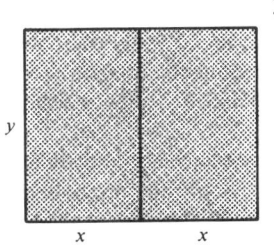

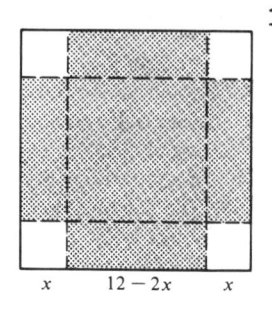

15. An open box is to be made from a square piece of material, 12 inches on a side, by cutting equal squares from each corner and turning up the sides, as shown in the figure. Find the volume of the largest box that can be made in this manner.
Solution: Let x be the length shown in the figure. Then the volume of the box is given by

$$V = x(12 - 2x)^2, \quad 0 < x < 6$$
$$V' = 12(6 - x)(2 - x)$$

$V' = 0$ when $x = 6$ and $x = 2$. Since $V = 0$ when $x = 6$ and $V = 128$ when $x = 2$, we conclude that the volume is maximum when $x = 2$. The corresponding volume is $V = 128$ cubic inches.

17. An open box is to be made from a rectangular piece of material by cutting equal squares from each corner and turning up the sides. Find the dimensions of the box of maximum volume if the material has dimensions 2 feet by 3 feet.
Solution: Let x be the length of the cut (the height of the box), then the length of the box is $3 - 2x$ and its width is $2 - 2x$. The volume of the box is given by

$$V = x(3 - 2x)(2 - 2x), \quad 0 < x < 1$$
$$= 4x^3 - 10x^2 + 6x$$
$$V' = 12x^2 - 20x + 6$$

$V' = 0$ when $x = (5 \pm \sqrt{7})/6$, but we do not consider $(5 + \sqrt{7})/6$ since it is not in the domain.

Length: $3 - 2(5 - \sqrt{7})/6 \approx 2.215$ feet
Width: $2 - 2(5 - \sqrt{7})/6 \approx 1.215$ feet
Height: $(5 - \sqrt{7})/6 \approx 0.392$ foot

19. An indoor physical fitness room consists of a rectangular region with a semicircle on each end. If the perimeter of the room is to be a 200-meter running track, find the dimensions that will make the area of the rectangular region as large as possible.
Solution: Let x and y be the length and width of the rectangle. The radius of the semicircle is $r = y/2$, and the perimeter of the track is

$$200 = 2x + 2\pi r = 2x + 2\pi(\frac{y}{2}) = 2x + \pi y$$

which implies that $y = (200 - 2x)/\pi$. The area of track is

$$A = xy = x[\frac{200 - 2x}{\pi}] = \frac{2}{\pi}(100x - x^2)$$
$$A' = \frac{2}{\pi}(100 - 2x)$$

$A' = 0$ when $x = 50$. Thus, A is maximum when $x = 50$ meters and $y = [200 - 2(50)]/\pi = 100/\pi$ meters.

Section 3.4

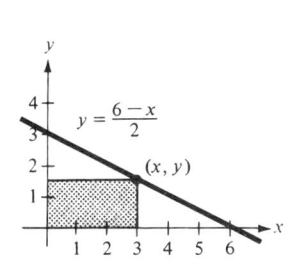

21. A rectangle is bounded by the x- and y-axes and the graph of y = (6 - x)/2, as shown in the figure. What length and width should the rectangle have so that its area is a maximum?
Solution: The area of the rectangle is

$$A = xy = x\left(\frac{6-x}{2}\right) = \frac{1}{2}(6x - x^2)$$

$$A' = \frac{1}{2}(6 - 2x)$$

A' = 0 when x = 3. Thus, A is maximum when x = 3 and y = (6 - 3)/2 = 3/2.

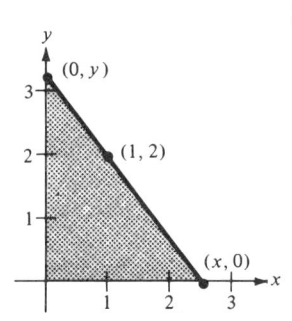

23. A right triangle is formed in the first quadrant by the x- and y-axes and a line through the point (1, 2). Find the vertices of the triangle so that the length of the hypotenuse is minimum.
Solution: The slope of the line connecting (0, y) and (1, 2) is m = (2 - y) and the slope of the line connecting (x, 0) and (1, 2) is m = 2/(1 - x). By equating these two slopes, we have

$$2 - y = \frac{2}{1-x} \quad \Longrightarrow \quad y = 2 + \frac{2}{x-1}$$

The length of the hypotenuse of the triangle is $d = \sqrt{x^2 + y^2}$. By minimizing $L = d^2$, we have

$$L = x^2 + \left(2 + \frac{2}{x-1}\right)^2, \quad x > 1$$

$$= x^2 + 4 + \frac{8}{x-1} + \frac{4}{(x-1)^2}$$

$$\frac{dL}{dx} = 2x - \frac{8}{(x-1)^2} - \frac{8}{(x-1)^3}$$

$$= \frac{2[x(x-1)^3 - 4(x-1) - 4]}{(x-1)^3}$$

$$= \frac{2x[(x-1)^3 - 4]}{(x-1)^3}$$

dL/dx = 0 when $x = 1 + \sqrt[3]{4}$ and $y = 2 + \sqrt[3]{2}$. Thus, the vertices occur at (0, 0), (2.587, 0), and (0, 3.260).

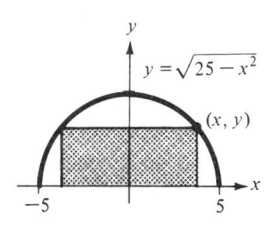

25. A rectangle is bounded by the x-axis and the semicircle $y = \sqrt{25 - x^2}$, as shown in the figure. What length and width should the rectangle have so that its area is a maximum?
Solution: The area is given by

$$A = 2xy = 2x\sqrt{25 - x^2}$$

$$A' = 2\frac{25 - 2x^2}{\sqrt{25 - x^2}}$$

A' = 0 when $x = 5/\sqrt{2}$. Thus, A is maximum when

$$x = \frac{5}{\sqrt{2}} \quad \text{and} \quad y = \sqrt{25 - (5/\sqrt{2})^2} = \frac{5}{\sqrt{2}} \approx 3.54.$$

27. Find the point on the graph of $y = \sqrt{x}$ that is closest to the point $(4, 0)$.
Solution: The distance between a point (x, y) on the graph and the point $(4, 0)$ is

$$d = \sqrt{(x-4)^2 + (y-0)^2}$$

and we can minimize d by minimizing its square, $L = d^2$.

$$L = (x-4)^2 + (\sqrt{x})^2 = x^2 - 7x + 16$$
$$L' = 2x - 7$$

$L' = 0$ when $x = 7/2$ and $y = \sqrt{7/2}$. Thus, the point nearest $(4, 0)$ is $(7/2, \sqrt{7/2})$.

29. A right circular cylinder is to be designed to hold 12 fluid ounces of a soft drink and to use a minimal amount of material in its construction. Find the dimensions for the container (1 fl oz ≈ 1.80469 in³).
Solution: The volume of the cylinder is

$$V = \pi r^2 h = 12(1.80469) \approx 21.66$$

which implies that $h = 21.66/\pi r^2$. The surface area of the cylinder is

$$S = 2\pi r^2 + 2\pi r h = 2\pi r^2 + 2\pi r \left(\frac{21.66}{\pi r^2}\right)$$
$$= 2\left(\pi r^2 + \frac{21.66}{r}\right)$$
$$S' = 2\left(2\pi r - \frac{21.66}{r^2}\right)$$

$S' = 0$ when $2\pi r^3 - 21.66 = 0$, which implies that

$$r = \sqrt[3]{21.66/2\pi} \approx 1.51 \text{ inches}$$
$$h = \frac{21.66}{\pi(1.51)^2} \approx 3.02 \text{ inches}$$

(Note that in the solution, $h = 2r$.)

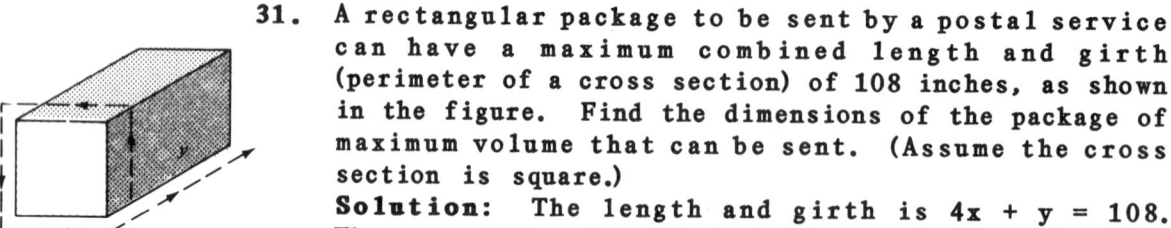

31. A rectangular package to be sent by a postal service can have a maximum combined length and girth (perimeter of a cross section) of 108 inches, as shown in the figure. Find the dimensions of the package of maximum volume that can be sent. (Assume the cross section is square.)
Solution: The length and girth is $4x + y = 108$. Thus, $y = 108 - 4x$. The volume is

$$V = x^2 y = x^2(108 - 4x) = 108x^2 - 4x^3$$
$$V' = 216x - 12x^2 = 12x(18 - x)$$

$V' = 0$ when $x = 0$ and $x = 18$. Thus, the maximum volume occurs when $x = 18$ inches and $y = 36$ inches. The dimensions are 18 in. by 18 in. by 36 in.

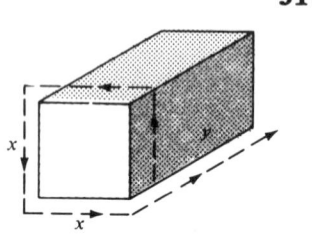

33. The combined perimeter of a circle and a square is 16. Find the dimensions of the circle and square that produce a minimum total area.
    Solution: Let x be the length of a side of the square and r the radius of the circle. Then the combined perimeter is $4x + 2\pi r = 16$, which implies that

    $$x = \frac{16 - 2\pi r}{4} = 4 - \frac{\pi r}{2}$$

    The combined area of the circle and square is

    $$A = x^2 + \pi r^2 = (4 - \frac{\pi r}{2})^2 + \pi r^2$$

    $$A' = 2(4 - \frac{\pi r}{2})(-\frac{\pi}{2}) + 2\pi r$$

    $$= \frac{1}{2}(\pi^2 r + 4\pi r - 8\pi)$$

    $A' = 0$ when $r = 8\pi/(\pi^2 + 4\pi) = 8/(\pi + 4)$ and the corresponding x-value is

    $$x = 4 - \frac{\pi[8/(\pi + 4)]}{2} = \frac{16}{\pi + 4}$$

35. A man is in a boat 2 miles from the nearest point on the coast. He is to go to a point Q, 3 miles down the coast and 1 mile inland, as shown in the figure. If he can row at 2 mi/hr and walk at 4 mi/hr, toward what point on the coast should he row in order to reach point Q in the least time?
    Solution: Using the formula Distance = (Rate)(Time), we have T = D/R.

    $$T = T_{rowed} + T_{walked} = \frac{D_{rowed}}{R_{rowed}} + \frac{D_{walked}}{R_{walked}}$$

    $$= \frac{\sqrt{x^2 + 4}}{2} + \frac{\sqrt{1 + (3 - x)^2}}{4}$$

    $$T' = \frac{x}{2\sqrt{x^2 + 4}} - \frac{3 - x}{4\sqrt{1 + (3 - x)^2}}$$

    By setting, T' = 0, we have

    $$\frac{x^2}{4(x^2 + 4)} = \frac{(3 - x)^2}{16[1 + (3 - x)^2]}$$

    $$\frac{x^2}{x^2 + 4} = \frac{9 - 6x + x^2}{4(10 - 6x + x^2)}$$

    $$4(x^4 - 6x^3 + 10x^2) = (x^2 + 4)(9 - 6x + x^2)$$

    $$x^4 - 6x^3 + 9x^2 + 8x - 12 = 0$$

    Possible rational roots: $\pm 1, \pm 2, \pm 3, \pm 4, \pm 6, \pm 12$
    By testing, we find that x = 1 mile.

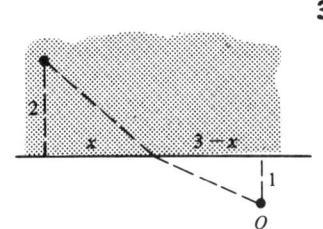

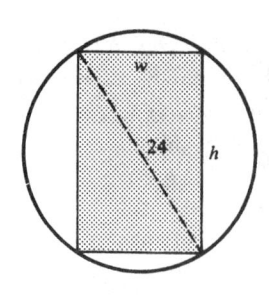

37. A wooden beam has a rectangular cross section of height h and width w, as shown in the figure. The strength S of the beam is directly proportional to the width and the square of the height. What are the dimensions of the strongest beam that can be cut from a round log of diameter 24 inches? (Hint: $S = kh^2w$, where k is the proportionality constant.)
Solution: Since $h^2 + w^2 = 24^2$, we have $h^2 = 24^2 - w^2$.

$$S = kh^2w = k(24^2 - w^2)w = k(576w - w^3)$$
$$S' = k(576 - 3w^2)$$

$S' = 0$ when $w = \sqrt{192} = 8\sqrt{3}$. Thus, S is maximum when

$$h^2 = 24^2 - 192 = 384 \implies h = \sqrt{384} = 8\sqrt{6}$$

● Section 3.5 Business and Economics Applications

1. Find the number of units x that produces a maximum revenue R when $R = 900x - 0.1x^2$.
   Solution:
   $$R' = 900 - 0.2x$$

   $R' = 0$ when $x = 900/0.2 = 4500$. Thus, R is maximum when $x = 4500$ units.

3. Find the number of units x that produces a maximum revenue R when $R = 1,000,000x/(0.02x^2 + 1800)$.
   Solution:

   $$R = 1,000,000 \left(\frac{x}{0.02x^2 + 1800} \right)$$

   $$R' = 1,000,000 \frac{(0.02x^2 + 1800)(1) - x(0.04x)}{(0.02x^2 + 1800)^2}$$

   $$= 1,000,000 \frac{1800 - 0.02x^2}{(0.02x^2 + 1800)^2}$$

   $R' = 0$ when $1800 - 0.02x^2 = 0$, which implies that

   $$x^2 = \frac{1800}{0.02} = 90,000$$

   Thus, R is maximum when $x = 300$ units.

5. When $C = 0.125x^2 + 20x + 5000$, find the number of units x that produces the minimum average cost per unit $\bar{C}$.
   Solution:

   $$\bar{C} = 0.125x + 20 + \frac{5000}{x}$$
   $$\bar{C}' = 0.125 - \frac{5000}{x^2}$$

   $\bar{C}' = 0$ when $0.125x^2 = 5000$, which implies that $x^2 = 40,000$ and $x = 200$ units.

7. For $C = 3000x - x^2\sqrt{300 - x}$, find the number of units x that produces the minimum average cost per unit $\bar{C}$.
   Solution:
   $$\bar{C} = 3000 - x\sqrt{300 - x}$$
   $$\bar{C}' = -x\frac{1}{2}(300 - x)^{-1/2}(-1) - \sqrt{300 - x}$$
   $$= \frac{x}{2\sqrt{300 - x}} - \sqrt{300 - x}$$
   $\bar{C}' = 0$ when $2(300 - x) = x$, which implies that $600 = 3x$ and $x = 200$ units.

9. When the cost function is $C = 100 + 30x$ and the demand function is $p = 90 - x$, find the price per unit p that produces the maximum profit P.
   Solution:
   $$P = xp - C = (90x - x^2) - (100 + 30x)$$
   $$= -x^2 + 60x - 100$$
   $$P' = -2x + 60$$

   $P' = 0$ when $x = 30$. Thus, the maximum profit occurs at $x = 30$ units and $p = 90 - 30 = \$60$.

11. When the cost function is $C = 4000 - 40x + 0.02x^2$ and the demand function is $p = 50 - (x/100)$, find the price per unit p that produces the maximum profit P.
    Solution:
    $$P = xp - C$$
    $$= (50x - 0.01x^2) - (4000 - 40x + 0.02x^2)$$
    $$= -0.03x^2 + 90x - 4000$$
    $$P' = -0.06x + 90$$

    $P' = 0$ when $x = 1500$. Thus, the maximum profit occurs at $x = 1500$ units and $p = 50 - (0.01)(1500) = \$35$.

13. For the cost function $C = 2x^2 + 5x + 18$, find the value of x for which the average cost is minimum. For this value of x, show that the marginal cost and average cost are equal.
    Solution:
    $$\bar{C} = 2x + 5 + \frac{18}{x}$$
    $$\bar{C}' = 2 - \frac{18}{x^2}$$
    $\bar{C}' = 0$ when $x = 3$. Thus, the average cost is minimum when $x = 3$ units and $\bar{C}(3) = \$17$ per unit.
    $$C' = \text{Marginal cost} = 4x + 5$$
    $$C'(3) = 17 = \bar{C}(3)$$

15. A given commodity has a demand function given by $p = 100 - (1/2)x^2$ and a total cost function given by $C = 40x + 37.5$.
    (a) What price gives the maximum profit?
    (b) What is the average cost per unit if production is set to give maximum profit?
    Solution:
    (a) $P = xp - C = x(100 - \frac{1}{2}x^2) - (40x + 37.5)$
    $$= -\frac{1}{2}x^3 + 60x - 37.5$$
    $$P' = -\frac{3}{2}x^2 + 60$$

    $P' = 0$ when $x = \sqrt{40} = 2\sqrt{10} \approx 6.32$ units. The price is $p = 100 - (1/2)(40) = \$80$.

    (b) $\bar{C} = 40 + (37.5/x)$. When $x = 2\sqrt{10}$, the average price is
    $$\bar{C}(2\sqrt{10}) = 40 + \frac{37.5}{2\sqrt{10}} \approx \$45.93$$

17. For $P = -2s^3 + 35s^2 - 100s + 200$, find the amount of advertising s (in 1000s of dollars) that yields the maximum profit P (in 1000s of dollars) and find the point of diminishing returns.
    Solution:
    $$P' = -6s^2 + 70s - 100 = -2(3s^2 - 35s + 50)$$
    $$= -2(3s - 5)(s - 10)$$

    Critical numbers: $s = 5/3$ and $s = 10$

    $P'' = -12s + 70$
    $P''(5/3) = 50 > 0 \implies$ Minimum
    $P''(10) = -50 < 0 \implies$ Maximum

    $P'' = -12s + 70 = 0$ when $s = 35/6$. The maximum profit occurs when $s = 10$ (or \$10,000) and the point of diminishing returns occurs at $s = 35/6$ (or \$5,833.33).

19. A manufacturer of radios charges \$90 per unit when the average production cost per unit is \$60. However, to encourage large orders from distributors, the manufacturer will reduce the charge by \$0.10 per unit for each unit ordered in excess of 100 (for example, there would be a charge of \$88 per radio for an order size of 120). Find the largest order the manufacturer should allow so as to realize maximum profit.
    Solution: Let x = number of units purchased, p = price per unit, and P = profit.

    $$p = 90 - (0.10)(x - 100) = 100 - 0.10x$$
    $$P = (p - 60)x = (40 - 0.10x)x = 40x - 0.10x^2$$
    $$P' = 40 - 0.20x$$

    $P' = 0$ when $x = 200$ radios.

Section 3.5   149

21. When a wholesaler sold a certain product at $25 per unit, sales were 800 units each week. However, after a price increase of $5, the average number of units sold dropped to 775 per week. Assuming that the demand function is linear, find the price that will maximize the total revenue.
Solution: The slope is m = (775 − 800)/(30 − 25) = −5 and the demand function is

$$x - 800 = -5(p - 25)$$
$$x = 800 - 5(p - 25) = 925 - 5p$$

Thus, the revenue is

$$R = xp = (925 - 5p)p = 925p - 5p^2$$
$$R' = 925 - 10p$$

R' = 0 when p = $92.50.

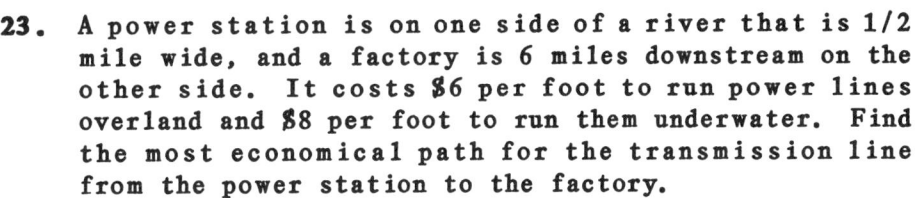

23. A power station is on one side of a river that is 1/2 mile wide, and a factory is 6 miles downstream on the other side. It costs $6 per foot to run power lines overland and $8 per foot to run them underwater. Find the most economical path for the transmission line from the power station to the factory.
Solution: Let T be the total cost.

$$T = 8(5280)\sqrt{x^2 + (1/4)} + 6(5280)(6 - x)$$
$$= 2(5280)[4\sqrt{x^2 + (1/4)} + 18 - 3x]$$
$$= 2(5280)[2\sqrt{4x^2 + 1} + 18 - 3x]$$

$$\frac{dT}{dx} = 2(5280)\left[\frac{2(8x)}{2\sqrt{4x^2 + 1}} - 3\right]$$

$$= 2(5280)\left(\frac{8x - 3\sqrt{4x^2 + 1}}{\sqrt{4x^2 + 1}}\right)$$

dT/dx = 0 when $8x = 3\sqrt{4x^2 + 1}$ which implies that

$$64x^2 = 9(4x^2 + 1)$$
$$28x^2 = 9$$
$$x^2 = 9/28$$
$$x = 3/(2\sqrt{7}) \approx 0.57 \text{ mile}$$

25. A small business has a minivan that it uses for making deliveries. The cost per hour for fuel for the van is $v^2/600$ where v is the speed in miles per hour. If the driver is paid $5 per hour, find the speed that minimizes cost on a 110-mile trip. (Assume there are no costs other than wages and fuel.)
Solution: The total cost per hour is

$$C = \frac{v^2}{600} + \frac{5(110)}{v}$$

$$C' = \frac{v}{300} - \frac{550}{v^2} = \frac{v^3 - 165,000}{300v^2}$$

C' = 0 when $v = 10\sqrt[3]{165} \approx 54.85$ mph.

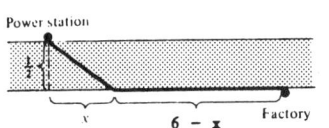

27. Find η (the price elasticity of demand) for p = 400 − 3x when x = 20. Is the demand elastic or inelastic (or neither) at the indicated x-value?
Solution: Since dp/dx = −3, the price elasticity of demand is

$$\eta = \frac{p/x}{dp/dx} = \frac{(400 - 3x)/x}{-3} = 1 - \frac{400}{3x}$$

When x = 20, we have

$$\eta = 1 - \frac{400}{3(20)} = -\frac{17}{3}$$

Since |η(20)| = 17/3 > 1, the demand is elastic.

29. Find η (the price elasticity of demand) for p = 400 − 0.5x² when x = 20. Is the demand elastic or inelastic (or neither) at the indicated x-value?
Solution: Since dp/dx = −x, the price elasticity of demand is

$$\eta = \frac{p/x}{dp/dx} = \frac{(400 - 0.5x^2)/x}{-x} = 0.5 - \frac{400}{x^2}$$

When x = 20, we have

$$\eta = 0.5 - \frac{400}{(20)^2} = -0.5$$

Since |η(20)| = 0.5 < 1, the demand is inelastic.

31. Find η (the price elasticity of demand) for p = (100/x²) + 2 when x = 10. Is the demand elastic or inelastic (or neither) at the indicated x-value?
Solution: Since dp/dx = −200/x³, the price elasticity of demand is

$$\eta = \frac{p/x}{dp/dx} = \frac{[(100/x^2) + 2]/x}{-(200/x^3)} = -\frac{1}{2} - \frac{x^2}{100}$$

When x = 10, we have

$$\eta = -\frac{1}{2} - \frac{(10)^2}{100} = -\frac{3}{2}$$

Since |η(10)| = 3/2 > 1, the demand is elastic.

33. The demand function for a product is x = 20 − 2p².
   (a) Consider a price of $2. If the price increases by 5%, determine the corresponding percentage change in quantity demanded.
   (b) Average elasticity of demand is defined to be the percentage change in quantity divided by the percentage change in price. Use the percentage of part (a) to find the average elasticity over the interval [2, 2.1].
   (c) Find the elasticity for a price of $2 and compare the result with that of part (b).
   (d) Find an expression for total revenue and find the values of x and p that maximize R.
   (e) For the value of x found in part (d), show that |η| = 1.

Solution:

(a) If $p = 2$, $x = 12$, and p increases by 5%, then

$$p = 2 + 2(0.05) = 2.1$$
$$x = 20 - 2(2.1)^2 = 11.18$$

The percentage increase in x is $(11.18 - 12)/12 = -41/600 \approx -6.83\%$.

(b) At $(2, 12)$, the average elasticity of demand is

$$\frac{\% \text{ change in } x}{\% \text{ change in } p} = \frac{-41/600}{0.05} = \frac{-41}{30} \approx -1.37$$

(c) The exact elasticity of demand at $(2, 12)$ is

$$\eta = (\frac{p}{x})(\frac{dx}{dp}) = (\frac{p}{x})(-4p) = (\frac{2}{12})[-4(2)] = -\frac{4}{3}$$

(d) The total revenue is

$$R = xp = (20 - 2p^2)p = 20p - 2p^3$$

$$\frac{dR}{dp} = 20 - 6p^2$$

$dR/dp = 0$ when $6p^2 = 20$ which implies that

$$p = \sqrt{10/3} \approx \$1.83$$
$$x = 20 - 2(\sqrt{10/3})^2 = 40/3 \text{ units}$$

(e) If $x = 40/3$ and $p = \sqrt{10/3}$ then

$$|\eta| = \left|\frac{p}{x}(-4p)\right| = \left|\frac{\sqrt{10/3}}{40/3}(-4\sqrt{10/3})\right| = 1$$

35. The demand function for a particular commodity is given by $p = (16 - x)^{1/2}$, $0 < x < 16$.
    (a) Find η when $x = 9$.
    (b) Find the values of x and p that maximize the total revenue.
    (c) Show that $|\eta| = 1$ for the value of x found in part (b).

Solution:

(a) $\dfrac{dp}{dx} = -\dfrac{1}{2\sqrt{16 - x}}$

$$\eta = \frac{p/x}{dp/dx} = \frac{(16 - x)^{1/2}/x}{-[1/(2\sqrt{16 - x})]} = -\frac{2(16 - x)}{x}$$

When $x = 9$, $\eta = -14/9$.

(b) $R = px = x(16 - x)^{1/2}$

$$R' = x\left(-\frac{1}{2\sqrt{16 - x}}\right) + (16 - x)^{1/2} = \frac{32 - 3x}{2\sqrt{16 - x}}$$

$R' = 0$ when $x = 32/3$ and $p = 4\sqrt{3}/3$

(c) $\left|\eta\left(\dfrac{32}{3}\right)\right| = \left|-\dfrac{2[16 - (32/3)]}{32/3}\right|$

$$= \left|-\frac{2(16/3)}{32/3}\right| = |-1| = 1$$

Section 3.6 Asymptotes

1. Find the vertical and horizontal asymptotes for $f(x) = 1/x^2$.
   Solution: A horizontal asymptote occurs at $y = 0$ since
   $$\lim_{x \to \infty} \frac{1}{x^2} = 0, \quad \lim_{x \to -\infty} \frac{1}{x^2} = 0$$
   A vertical asymptote occurs at $x = 0$ since
   $$\lim_{x \to 0^-} \frac{1}{x^2} = \infty, \quad \lim_{x \to 0^+} \frac{1}{x^2} = \infty$$

3. Find the vertical and horizontal asymptotes for $f(x) = (x^2 - 2)/(x^2 - x - 2)$.
   Solution: A horizontal asymptote occurs at $y = 1$ since
   $$\lim_{x \to \infty} \frac{x^2 - 2}{x^2 - x - 2} = 1, \quad \lim_{x \to -\infty} \frac{x^2 - 2}{x^2 - x - 2} = 1$$
   Vertical asymptotes occur at $x = -1$ and $x = 2$ since
   $$\lim_{x \to -1^-} \frac{x^2 - 2}{x^2 - x - 2} = -\infty, \quad \lim_{x \to -1^+} \frac{x^2 - 2}{x^2 - x - 2} = \infty$$
   $$\lim_{x \to 2^-} \frac{x^2 - 2}{x^2 - x - 2} = -\infty, \quad \lim_{x \to 2^+} \frac{x^2 - 2}{x^2 - x - 2} = \infty$$

5. Find the vertical and horizontal asymptotes for $f(x) = x^3/(x^2 - 1)$.
   Solution: There are no horizontal asymptotes. Vertical asymptotes occur at $x = -1$ and $x = 1$ since
   $$\lim_{x \to -1^-} \frac{x^3}{x^2 - 1} = -\infty, \quad \lim_{x \to -1^+} \frac{x^3}{x^2 - 1} = \infty$$
   $$\lim_{x \to 1^-} \frac{x^3}{x^2 - 1} = -\infty, \quad \lim_{x \to 1^+} \frac{x^3}{x^2 - 1} = \infty$$

7. Using asymptotes as an aid, match $f(x) = 3x^2/(x^2 + 2)$ with the correct graph.
   Solution: The graph of f has a horizontal asymptote at $y = 3$. It has no vertical asymptote, and it matches graph (f).

9. Using asymptotes as an aid, match $f(x) = x/(x^2 + 2)$ with the correct graph.
   Solution: The graph of f has a horizontal asymptote at $y = 0$. It has no vertical asymptote, and it matches graph (c).

Section 3.6   153

11. Match $f(x) = 5 - [1/(x^2 + 1)]$ with the correct graph.
Solution: f has a horizontal asymptote at $y = 5$. It has no vertical asymptote, and it matches graph (e).

13. Find $\lim\limits_{x \to -2^-} \dfrac{1}{(x + 2)^2}$.
Solution:
$$\lim_{x \to -2^-} \dfrac{1}{(x + 2)^2} = \infty$$

15. Find $\lim\limits_{x \to 2^+} \dfrac{x - 3}{x - 2}$.
Solution:
$$\lim_{x \to 2^+} \dfrac{x - 3}{x - 2} = -\infty$$

17. Find $\lim\limits_{x \to 4^-} \dfrac{x^2}{x^2 - 16}$.
Solution:
$$\lim_{x \to 4^-} \dfrac{x^2}{x^2 - 16} = -\infty$$

19. Find $\lim\limits_{x \to 0^-} \left(1 + \dfrac{1}{x}\right)$.
Solution:
$$\lim_{x \to 0^-} \left(1 + \dfrac{1}{x}\right) = -\infty$$

21. Find $\lim\limits_{x \to \infty} \dfrac{2x - 1}{3x + 2}$.
Solution:
$$\lim_{x \to \infty} \dfrac{2x - 1}{3x + 2} = \dfrac{2}{3}$$

23. Find $\lim\limits_{x \to \infty} \dfrac{x}{x^2 - 1}$.
Solution:
$$\lim_{x \to \infty} \dfrac{x}{x^2 - 1} = 0$$

25. Find $\lim\limits_{x \to -\infty} \dfrac{5x^2}{x + 3}$.
Solution:
$$\lim_{x \to -\infty} \dfrac{5x^2}{x + 3} = -\infty$$

27. Find $\lim\limits_{x \to \infty} \left(2x - \dfrac{1}{x^2}\right)$.
Solution:
$$\lim_{x \to \infty} \left(2x - \dfrac{1}{x^2}\right) = \lim_{x \to \infty} \dfrac{2x^3 - 1}{x^2} = \infty$$

29. Find $\lim\limits_{x \to -\infty} \left(\dfrac{2x}{x - 1} + \dfrac{3x}{x + 1}\right)$.
Solution:
$$\lim_{x \to -\infty} \left(\dfrac{2x}{x - 1} + \dfrac{3x}{x + 1}\right) = \lim_{x \to -\infty} \dfrac{5x^2 - x}{x^2 - 1} = 5$$

31. Complete the table and estimate the limit of f(x) as x approaches infinity for $f(x) = (x + 1)/(x\sqrt{x})$.
Solution:

x	1	10	10^2	10^3	10^4	10^5	10^6
f(x)	2.000	0.348	0.101	0.032	0.010	0.003	0.001

$$\lim_{x \to \infty} \frac{x + 1}{x\sqrt{x}} = 0$$

33. Complete the table and estimate the limit of f(x) as x approaches infinity for $f(x) = x^2 - x\sqrt{x(x - 1)}$.
Solution:

x	1	10	10^2	10^3	10^4	10^5	10^6
f(x)	1	5.1	50.1	500.1	5000.1	50000.1	500000.2

$$\lim_{x \to \infty} [x^2 - x\sqrt{x(x - 1)}] = \infty$$

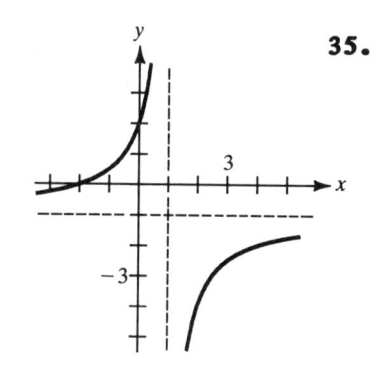

35. Sketch the graph of $y = (2 + x)/(1 - x)$, using intercepts, relative extrema, and asymptotes as sketching aids.
Solution:
x-intercept: (-2, 0)
y-intercept: (0, 2)
Horizontal asymptote: y = -1
Vertical asymptote: x = 1

$$y' = \frac{3}{(1 - x)^2}$$

No relative extrema

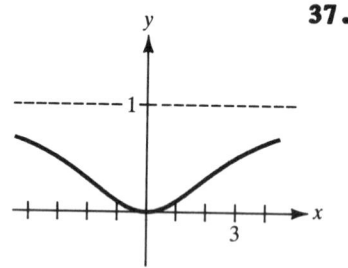

37. Sketch the graph of $f(x) = x^2/(x^2 + 9)$, using intercepts, relative extrema, and asymptotes as sketching aids.
Solution:
Intercept: (0, 0)
Horizontal asymptote: y = 1

$$f'(x) = \frac{18x}{(x^2 + 9)^2}$$

Relative minimum: (0, 0)

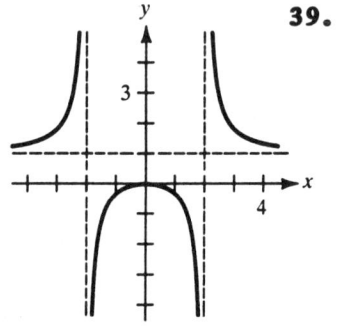

39. Sketch the graph of $g(x) = x^2/(x^2 - 4)$, using intercepts, relative extrema, and asymptotes as sketching aids.
Solution:
Intercept: (0, 0)
Horizontal asymptote: y = 1
Vertical asymptotes: x = ±2

$$g'(x) = -\frac{8x}{(x^2 - 4)^2}$$

Relative maximum: (0, 0)

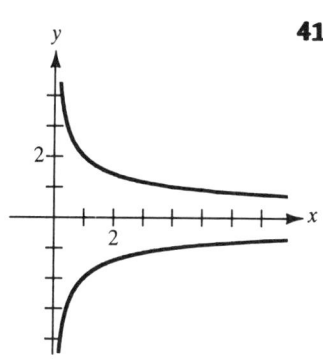

Section 3.6   155

41. Sketch the graph of $xy^2 = 4$, using intercepts, relative extrema, and asymptotes as sketching aids.
Solution:
$$x = \frac{4}{y^2}$$
No intercepts
Horizontal asymptote: $y = 0$
Vertical asymptote: $x = 0$
No relative extrema

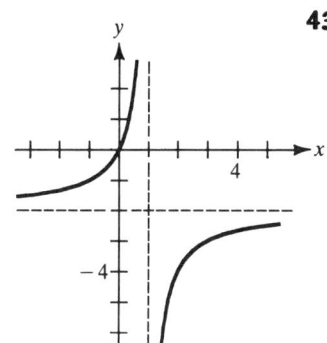

43. Sketch the graph of $y = 2x/(1 - x)$, using intercepts, relative extrema, and asymptotes as sketching aids.
Solution:
Intercept: $(0, 0)$
Horizontal asymptote: $y = -2$
$$y = \frac{2x}{1 - x}$$
Vertical asymptote: $x = 1$
$$y' = \frac{2}{(1 - x)^2}$$
No relative extrema

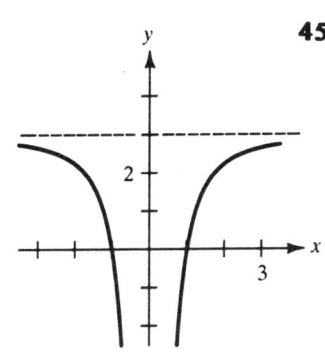

45. Sketch the graph of $y = 3[1 - (1/x^2)]$, using intercepts, relative extrema, and asymptotes as sketching aids.
Solution:
x-intercepts: $(-1, 0)$ and $(1, 0)$
Horizontal asymptote: $y = 3$
$$y = \frac{3(x^2 - 1)}{x^2}$$
Vertical asymptote: $x = 0$
$$y' = \frac{6}{x^3}$$
No relative extrema

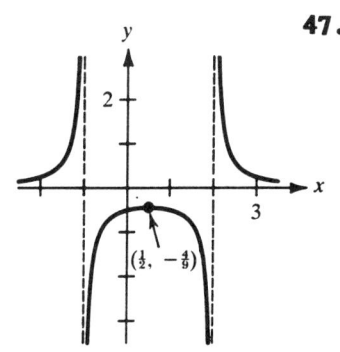

47. Sketch the graph of $f(x) = 1/(x^2 - x - 2)$, using intercepts, relative extrema, and asymptotes as sketching aids.
Solution:
y-intercept: $(0, -1/2)$
Horizontal asymptote: $y = 0$
$$f(x) = \frac{1}{(x + 1)(x - 2)}$$
Vertical asymptotes: $x = -1$, $x = 2$
$$f'(x) = -\frac{2x - 1}{(x^2 - x - 2)^2}$$
Relative maximum: $(1/2, -4/9)$

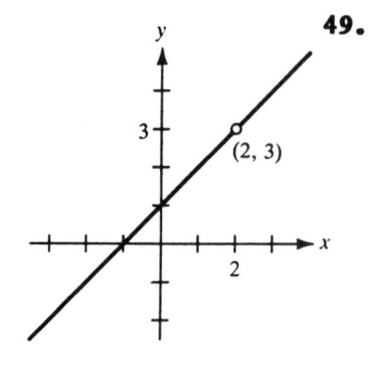

49. Sketch the graph of $g(x) = (x^2 - x - 2)/(x - 2)$, using intercepts, relative extrema, and asymptotes as sketching aids.
Solution:
$$g(x) = \frac{x^2 - x - 2}{x - 2} = x + 1 \text{ for } x \neq 2$$
x-intercept: (-1, 0)
y-intercept: (0, 1)
No asymptotes
No relative extrema

51. The cost function for a certain product is given by $C = 1.35x + 4570$ where C is measured in dollars and x is the number of units produced.
(a) Find the average cost per unit when x = 100 and when x = 1000.
(b) What is the limit of the average cost function as x approaches infinity?
Solution:
(a) $\bar{C} = 1.35 + (4570/x)$. When $x = 100$, $\bar{C} = \$47.05$. When $x = 1000$, $\bar{C} = \$5.92$.

(b) $\lim\limits_{x \to \infty} (1.35 + \frac{4570}{x}) = 1.35 + 0 = \$1.35$

53. The cost in millions of dollars for the federal government to seize p% of a certain illegal drug as it enters the country is given by $C = 528p/(100 - p)$, $0 \leq p < 100$.
(a) Find the cost of seizing 25%.
(b) Find the cost of seizing 50%.
(c) Find the cost of seizing 75%.
(d) Find the limit of C as $p \to 100^-$.
Solution:
(a) $C(25) = 528(25)/(100 - 25) = \$176$ million
(b) $C(50) = 528(50)/(100 - 50) = \$528$ million
(c) $C(75) = 528(75)/(100 - 75) = \$1584$ million

(d) $\lim\limits_{p \to 100^-} \frac{528p}{100 - p} = \infty$

55. Psychologists have developed mathematical models to predict performance as a function of the number of trials n for a certain task. One such model is

$$P = \frac{b + \theta a(n - 1)}{1 + \theta(n - 1)}$$

where P is the percentage of correct responses after n trials and a, b, and θ are constants depending on the actual learning situation. Find the limit of P as n approaches infinity.
Solution:

$$\lim_{n \to \infty} \frac{\theta a n - \theta a + b}{\theta n - \theta + 1} = \frac{\theta a}{\theta} = a$$

57. The game commission in a certain state introduces 50 deer into newly acquired state game lands. The population N of the herd is given by the model

$$N = \frac{10(5 + 3t)}{1 + 0.04t}$$

where t is time in years.
(a) Find the number in the herd when t is 5, 10, and 25.
(b) According to this model, what is the limiting size of the herd as time increases?

Solution:
(a) When $t = 5$, $N = 200/1.2 \approx 167$ deer.
When $t = 10$, $N = 350/1.4 = 250$ deer.
When $t = 25$, $N = 800/2 = 400$ deer.

(b) $\lim\limits_{t \to \infty} \dfrac{10(5 + 3t)}{1 + 0.04t} = \dfrac{30}{0.04} = 750$ deer

● Section 3.7 Curve Sketching: A Summary

1. Determine the sign of a for which the graph of $f(x) = ax^3 + bx^2 + cx + d$ will resemble the given graph.
Solution: Since the graph rises as $x \to -\infty$ and falls as $x \to \infty$, a must be negative.

3. Determine the sign of a for which the graph of $f(x) = ax^3 + bx^2 + cx + d$ will resemble the given graph.
Solution: Since the graph falls as $x \to -\infty$ and rises as $x \to \infty$, a must be positive.

5. Sketch the graph of $y = -x^2 - 2x + 3$. Choose a scale that allows all relative extrema and points of inflection to be identified on the sketch.
Solution:
$$y = -x^2 - 2x + 3 = -(x + 3)(x - 1)$$
$$y' = -2x - 2 = -2(x + 1)$$
$$y'' = -2$$

Intercepts: $(0, 3)$, $(1, 0)$, $(-3, 0)$
Relative maximum: $(-1, 4)$
Concave down

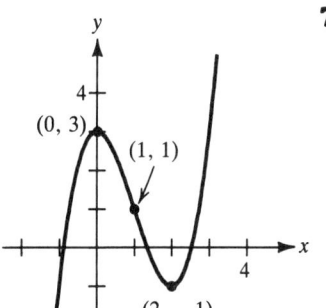

7. Sketch the graph of $y = x^3 - 3x^2 + 3$. Choose a scale that allows all relative extrema and points of inflection to be identified on the sketch.
Solution:
$$y = x^3 - 3x^2 + 3$$
$$y' = 3x^2 - 6x = 3x(x - 2)$$
$$y'' = 6x - 6 = 6(x - 1)$$

Relative maximum: $(0, 3)$
Relative minimum: $(2, -1)$
Point of inflection: $(1, 1)$

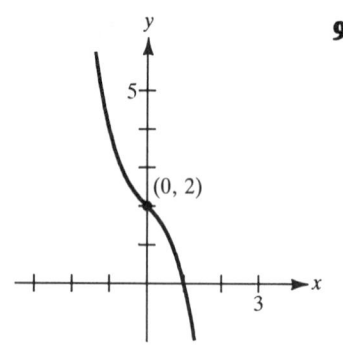

158 Section 3.7

9. Sketch the graph of $y = 2 - x - x^3$. Choose a scale that allows all relative extrema and points of inflection to be identified on the sketch.
Solution:
$$y = 2 - x - x^3$$
$$y' = -1 - 3x^2$$
$$y'' = -6x$$

No relative extrema
Point of inflection: (0, 2)

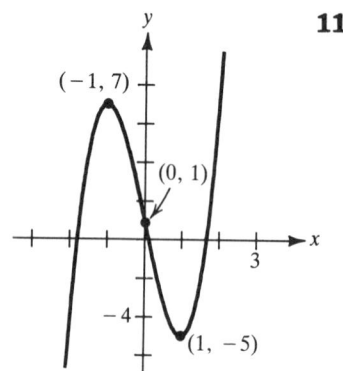

11. Sketch the graph of $f(x) = 3x^3 - 9x + 1$. Choose a scale that allows all relative extrema and points of inflection to be identified on the sketch.
Solution:
$$f(x) = 3x^3 - 9x + 1$$
$$f'(x) = 9x^2 - 9 = 9(x-1)(x+1)$$
$$f''(x) = 18x$$

Relative maximum: (-1, 7)
Relative minimum: (1, -5)
Point of inflection: (0, 1)

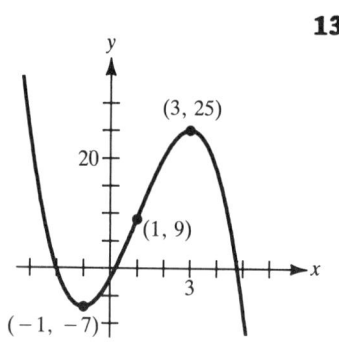

13. Sketch the graph of $f(x) = -x^3 + 3x^2 + 9x - 2$. Choose a scale that allows all relative extrema and points of inflection to be identified on the sketch.
Solution:
$$f(x) = -x^3 + 3x^2 + 9x - 2$$
$$f'(x) = -3x^2 + 6x + 9 = -3(x+1)(x-3)$$
$$f''(x) = -6x + 6 = -6(x-1)$$

Relative maximum: (3, 25)
Relative minimum: (-1, -7)
Point of inflection: (1, 9)

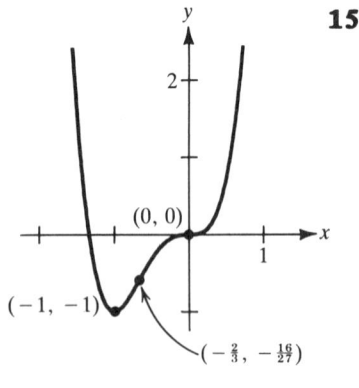

15. Sketch the graph of $y = 3x^4 + 4x^3$. Choose a scale that allows all relative extrema and points of inflection to be identified on the sketch.
Solution:
$$y = 3x^4 + 4x^3 = x^3(3x+4)$$
$$y' = 12x^3 + 12x^2 = 12x^2(x+1)$$
$$y'' = 36x^2 + 24x = 12x(3x+2)$$

Intercepts: (0, 0), (-4/3, 0)
Relative minimum: (-1, -1)
Points of inflection: (0, 0), (-2/3, -16/27)

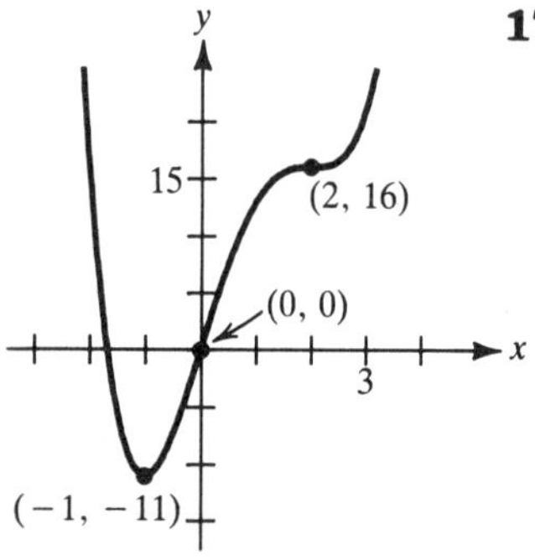

17. Sketch the graph of $f(x) = x^4 - 4x^3 + 16x$. Choose a scale that allows all relative extrema and points of inflection to be identified on the sketch.
Solution:
$$f(x) = x^4 - 4x^3 + 16x$$
$$f'(x) = 4x^3 - 12x^2 + 16 = 4(x+1)(x-2)^2$$
$$f''(x) = 12x^2 - 24x = 12x(x-2)$$

Relative minimum: (-1, -11)
Points of inflection: (0, 0), (2, 16)

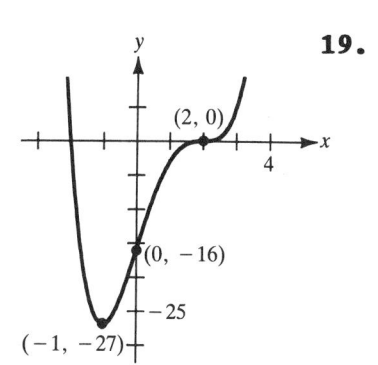

Section 3.7   159

19. Sketch the graph of $f(x) = x^4 - 4x^3 + 16x - 16$. Choose a scale that allows all relative extrema and points of inflection to be identified on the sketch.
    Solution:
    $$f(x) = x^4 - 4x^3 + 16x - 16$$
    $$f'(x) = 4x^3 - 12x^2 + 16 = 4(x + 1)(x - 2)^2$$
    $$f''(x) = 12x^2 - 24x = 12x(x - 2)$$

    Relative minimum: $(-1, -27)$
    Points of inflection: $(0, -16)$, $(2, 0)$

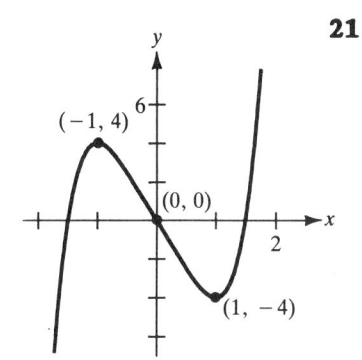

21. Sketch the graph of $y = x^5 - 5x$. Choose a scale that allows all relative extrema and points of inflection to be identified on the sketch.
    Solution:
    $$y = x^5 - 5x$$
    $$y' = 5x^4 - 5 = 5(x + 1)(x - 1)(x^2 + 1)$$
    $$y'' = 20x^3$$

    Intercepts: $(0, 0)$, $(\pm \sqrt[4]{5}, 0)$
    Relative maximum: $(-1, 4)$
    Relative minimum: $(1, -4)$
    Point of inflection: $(0, 0)$

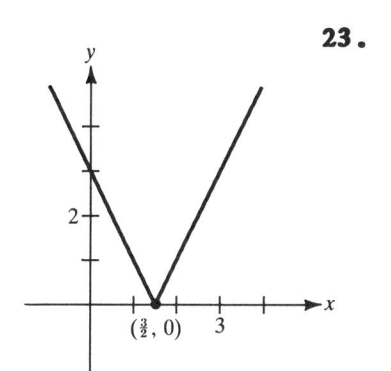

23. Sketch the graph of $y = |2x - 3|$. Choose a scale that allows all relative extrema and points of inflection to be identified on the sketch.
    Solution:
    $$y = |2x - 3|$$
    $$y' = (2)\frac{(2x - 3)}{|2x - 3|}$$
    $$y'' = 0$$

    Relative minimum: $(3/2, 0)$
    No points of inflection.

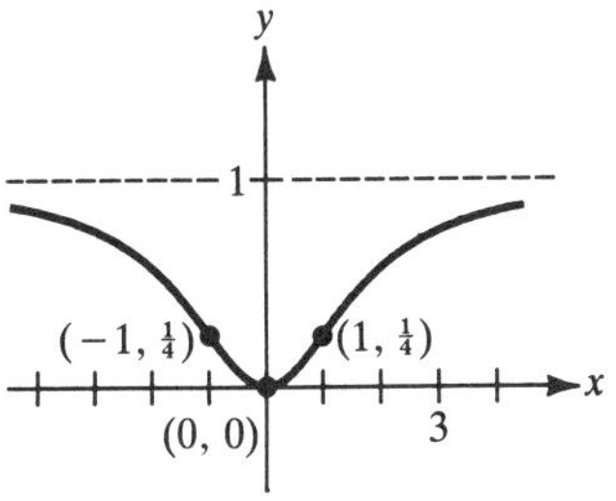

25. Sketch the graph of $y = x^2/(x^2 + 3)$. Choose a scale that allows all relative extrema and points of inflection to be identified on the sketch.
    Solution:
    $$y = \frac{x^2}{x^2 + 3}$$
    $$y' = \frac{6x}{(x^2 + 3)^2}$$
    $$y'' = \frac{6(x^2 + 3)^2 - 6x(2)(x^2 + 3)(2x)}{(x^2 + 3)^4}$$
    $$= \frac{18(1 + x)(1 - x)}{(x^2 + 3)^3}$$

    Relative minimum: $(0, 0)$
    Points of inflection: $(\pm 1, 1/4)$
    Horizontal asymptote: $y = 1$

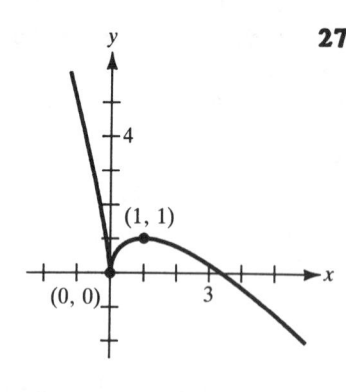

27. Sketch the graph of $y = 3x^{2/3} - 2x$. Choose a scale that allows all relative extrema and points of inflection to be identified on the sketch.
    Solution:
    $$y = 3x^{2/3} - 2x$$
    $$y' = 2x^{-1/3} - 2 = 2(x^{-1/3} - 1)$$
    $$y'' = -2/3 x^{4/3}$$

    Intercepts: $(0, 0)$, $(27/8, 0)$
    Relative maximum: $(1, 1)$
    Relative minimum: $(0, 0)$

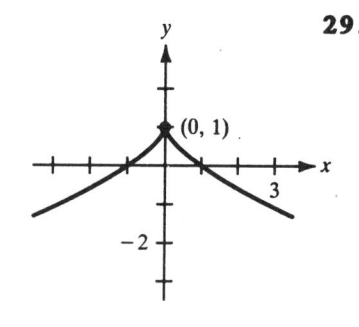

29. Sketch the graph of $y = 1 - x^{2/3}$. Choose a scale that allows all relative extrema and points of inflection to be identified on the sketch.
    Solution:
    $$y = 1 - x^{2/3}$$
    $$y' = -\frac{2}{3}x^{-1/3} = -\frac{2}{3\sqrt[3]{x}}$$
    $$y'' = \frac{2}{9}x^{-4/3} = \frac{2}{9\sqrt[3]{x^4}}$$

    Intercepts: $(0, 1)$, $(\pm 1, 0)$
    Relative maximum: $(0, 1)$

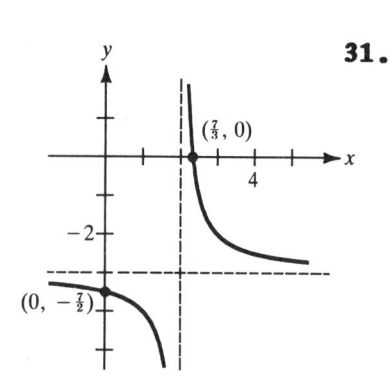

31. Sketch the graph of $y = [1/(x - 2)] - 3$. Label the intercepts, relative extrema, points of inflection, and asymptotes, and give the domain of the function.
    Solution:
    $$y = [1/(x - 2)] - 3$$
    $$y' = -1/(x - 2)^2$$
    $$y'' = 2/(x - 2)^3$$

    Intercepts: $(7/3, 0)$, $(0, -7/2)$
    No relative extrema.
    Horizontal asymptote: $y = -3$
    Vertical asymptote: $x = 2$
    Domain: $(-\infty, 2)$, $(2, \infty)$

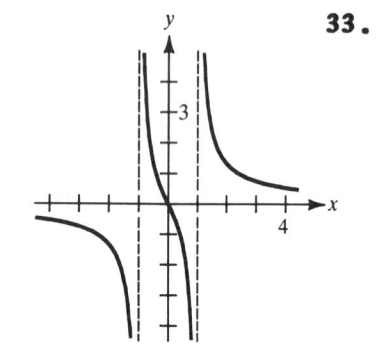

33. Sketch the graph of $y = 2x/(x^2 - 1)$. Label the intercepts, relative extrema, points of inflection, and asymptotes, and give the domain of the function.
    Solution:
    $$y = 2x/(x^2 - 1)$$
    $$y' = -2(x^2 + 1)/(x^2 - 1)^2$$
    $$y'' = 4x(x^2 + 3)/(x^2 - 1)^3$$

    Point of inflection: $(0, 0)$
    Intercept: $(0, 0)$
    Horizontal asymptote: $y = 0$
    Vertical asymptotes: $x = \pm 1$
    Symmetry with respect to the origin.
    Domain: $(-\infty, -1)$, $(-1, 1)$, $(1, \infty)$

Section 3.7

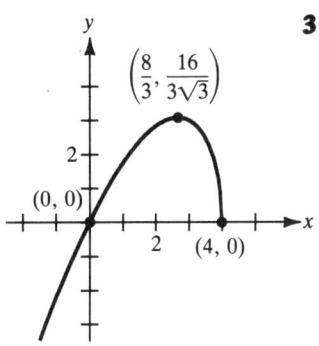

35. Sketch the graph of $y = x\sqrt{4-x}$. Label the intercepts, relative extrema, points of inflection, and asymptotes, and give the domain of the function.
Solution:
$$y = x\sqrt{4-x}$$
$$y' = \frac{8-3x}{2\sqrt{4-x}}$$
$$y'' = \frac{3x-16}{4(4-x)^{3/2}}$$

Intercepts: $(0, 0)$, $(4, 0)$
Relative maximum: $(8/3, 16/3\sqrt{3})$
Domain: $(-\infty, 4]$

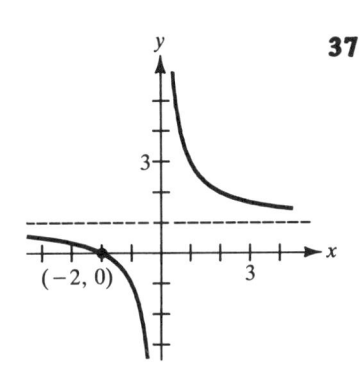

37. Sketch the graph of $y = (x+2)/x$. Label the intercepts, relative extrema, points of inflection, and asymptotes, and give the domain of the function.
Solution:
$$y = (x+2)/x = 1 + (2/x)$$
$$y' = -2/x^2$$
$$y'' = 4/x^3$$

Intercept: $(-2, 0)$
Horizontal asymptote: $y = 1$
Vertical asymptote: $x = 0$
Domain: $(-\infty, 0)$, $(0, \infty)$

39. Use the given graph of f' to sketch a graph of the function f. (The solution is not unique.)
Solution: Since $f'(x) = 2$, the graph of f is a line with a slope of 2.

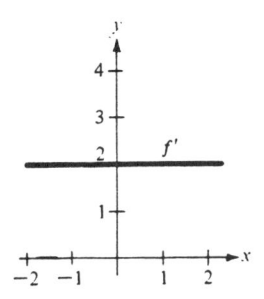

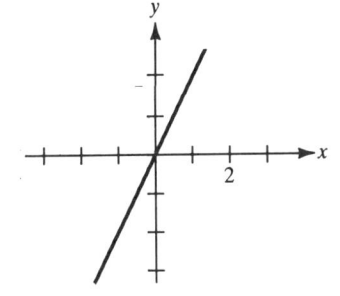

41. Use the given graph f'' to sketch a graph of the function f. (The solution is not unique.)
Solution: Since $f''(x) = 2$, the graph of f' is line with a slope of 2, and the graph of f is a parabola opening upward.

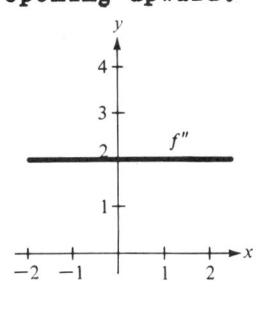

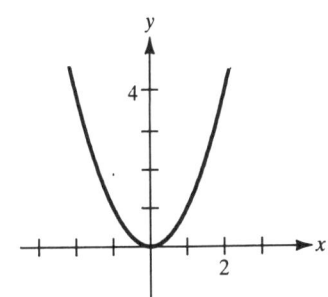

Section 3.8 Differentials

1. Find the differential dy of $y = 3x^2 - 4$.
   Solution:
   $$dy = 6x\, dx$$

3. Find the differential dy of $y = (2x + 5)^3$.
   Solution:
   $$dy = 6(2x + 5)^2\, dx$$

5. Find the differential dy of $y = \sqrt{x^2 + 1}$.
   Solution:
   $$dy = \frac{1}{2}(x^2 + 1)^{-1/2}(2x)\, dx = \frac{x}{\sqrt{x^2 + 1}}\, dx$$

7. Find the differential dy of $y = x\sqrt{1 - x^2}$.
   Solution:
   $$dy = [x(\frac{1}{2})(1 - x^2)^{-1/2}(-2x) + \sqrt{1 - x^2}\,]\, dx$$
   $$= \frac{1 - 2x^2}{\sqrt{1 - x^2}}\, dx$$

9. Find the differential dy of $y = (x + 1)/(2x - 1)$.
   Solution:
   $$dy = \frac{(2x - 1)(1) - (x + 1)(2)}{(2x - 1)^2}\, dx = -\frac{3}{(2x - 1)^2}\, dx$$

11. Use $y = x^2$ and the value of $\Delta x = dx$ to complete the table. (Let $x = 2$.)
   Solution:
   $$dy = 2x\, dx$$

$dx = \Delta x$	dy	Δy	$\Delta y - dy$	$dy/\Delta y$
1.0000	4.0000	5.0000	1.0000	0.8000
0.5000	2.0000	2.2500	0.2500	0.8889
0.1000	0.4000	0.4100	0.0100	0.9756
0.0100	0.0400	0.0401	0.0001	0.9975
0.0010	0.0040	0.0040	0.0000	0.9998

13. Use $y = x^5$ and the value of $\Delta x = dx$ to complete the table. (Let $x = 2$.)
   Solution:
   $$dy = 5x^4\, dx$$

$\Delta x = dx$	dy	Δy	$\Delta y - dy$	$dy/\Delta y$
1.0000	80.0000	211.0000	131.0000	0.3791
0.5000	40.0000	65.6562	25.6562	0.6092
0.1000	8.0000	8.8410	0.8410	0.9049
0.0100	0.8000	0.8080	0.0080	0.9901
0.0010	0.0800	0.0801	0.0001	0.9990

15. Approximate $\sqrt[4]{83}$ using differentials. Compare the result with the value obtained using a calculator.
Solution: Let $y = \sqrt[4]{x}$, then

$$dy = \frac{1}{4x^{3/4}} dx$$

By letting $x = 81$ and $dx = 2$, we have

$$dy = \frac{1}{4(81)^{3/4}}(2) = \frac{1}{54}$$

Thus, $\sqrt[4]{83} \approx \sqrt[4]{81} + (1/54) = 3 + (1/54) \approx 3.0185$. Using a calculator, we have $\sqrt[4]{83} \approx 3.0183$.

17. Approximate $1/\sqrt[3]{25}$ using differentials. Compare the result with the value obtained using a calculator.
Solution: Let $y = 1/\sqrt[3]{x}$, then

$$dy = -\frac{1}{3x^{4/3}} dx$$

By letting $x = 27$ and $dx = -2$, we have

$$dy = -\frac{1}{3(27)^{4/3}}(-2) = \frac{2}{243}.$$

Thus, $\dfrac{1}{\sqrt[3]{25}} \approx \dfrac{1}{\sqrt[3]{27}} + \dfrac{2}{243} = \dfrac{83}{243} \approx 0.3416$.

Using a calculator, we have $\dfrac{1}{\sqrt[3]{25}} \approx 0.3420$.

19. The area of a square of side x is given by $A = x^2$.
  (a) Compute dA and ΔA in terms of x and Δx.
  (b) Identify the region whose area is dA in the figure.
  (c) Identify the region whose area is $\Delta A - dA$ in the figure.
Solution:
(a) $dA = 2x\,dx = 2x\Delta x$
  $\Delta A = (x + \Delta x)^2 - x^2 = 2x\Delta x + (\Delta x)^2$

(b) See accompanying graph.

(c) $\Delta A - dA = (\Delta x)^2$ (See accompanying graph.)

21. The measurement of the radius of a circle is found to be 14 inches, with a possible error of 1/4 inch. Use differentials to approximate the possible error and the percentage error in computing the area of the circle.
Solution: Let $\Delta r = dr = 1/4$.

$$A = \pi r^2$$

$$dA = 2\pi r\, dr = 2\pi(14)(\pm \frac{1}{4}) = \pm 7\pi \text{ in}^2$$

When $r = 14$, $A = 196\pi$ and the percentage error is

$$\frac{dA}{A} = \frac{7\pi}{196\pi} = \frac{1}{28} \approx 3.57\%$$

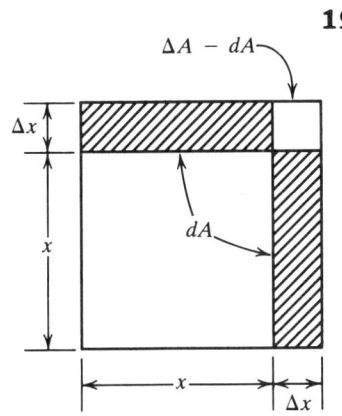

23. The radius of a sphere is measured to be 6 inches, with a possible error of 0.02 inch. Use differentials to approximate the possible error and the relative error in calculating the volume of the sphere.
Solution: Let $\Delta r = dr = 0.02$ inch.

$$V = \frac{4}{3}\pi r^3$$

$$dV = 4\pi r^2 \, dr = 4\pi(6)^2(\pm 0.02) = \pm 2.88\pi \text{ in}^3$$

When $r = 6$, the relative error is

$$\frac{dV}{V} = \frac{4\pi r^2 \, dr}{(4/3)\pi r^3} = \frac{3 \, dr}{r}$$

$$= \frac{3(0.02)}{6} = 0.01$$

25. The profit P for a company selling x units is given by $P = (500x - x^2) - [(1/2)x^2 - 77x + 3000]$. Approximate the change and the percentage change in profit as production changes from $x = 115$ to $x = 120$ units.
Solution: Since x changes from 115 to 120, we have $dx = 120 - 115 = 5$.

$$P = (500x - x^2) - (\frac{1}{2}x^2 - 77x + 3000)$$

$$dP = (500 - 2x - x + 77) \, dx = (577 - 3x) \, dx$$
$$= [577 - 3(115)](5) = \$1160$$

The approximate percentage change is

$$\frac{dP}{P} = \frac{1160}{43517.50} \approx 2.67\%$$

27. The game commission in a certain state introduces 50 deer into newly acquired state game lands. The population N of the herd is given by the model

$$N = \frac{10(5 + 3t)}{1 + 0.04t}$$

where t is time in years. Use differentials to approximate the change in the herd size from year $t = 5$ to year $t = 6$.
Solution:

$$N = \frac{10(5 + 3t)}{1 + 0.04t}$$

$$dN = \frac{(1 + 0.04t)(30) - 10(5 + 3t)(0.04)}{(1 + 0.04t)^2} \, dt$$

$$= \frac{28}{(1 + 0.04t)^2} \, dt$$

When $t = 5$ and $dt = 6 - 5 = 1$, we have

$$dN = \frac{28}{[1 + 0.04(5)]^2}(1) = \frac{28}{1.44} \approx 19.44$$

The change in herd size will be approximately 19 deer.

Review Exercises for Chapter 3

1. Make use of domain, range, asymptotes, and intercepts to match $f(x) = 4/(x^2 + 4)$ with the correct graph.
   Solution:
   Domain: all real numbers
   Range: (0, 1]
   Horizontal asymptote: $y = 0$
   y-intercept: (0, 1)
   Matches graph (b)

3. Make use of domain, range, asymptotes, and intercepts to match $g(x) = x/(x^2 - 4)$ with the correct graph.
   Solution:
   Domain: all real numbers except ± 2
   Range: all real numbers
   Horizontal asymptote: $y = 0$
   Vertical asymptotes: $x = -2$, $x = 2$
   Intercept: (0, 0)
   Matches graph (f)

5. Make use of domain, range, asymptotes, and intercepts to match $f(x) = 3x^2/(x^2 + 4)$ with the correct graph.
   Solution:
   Domain: all real numbers
   Range: [0, 3)
   Horizontal asymptote: $y = 3$
   Intercept: (0, 0)
   Matches graph (h)

7. Make use of domain, range, asymptotes, and intercepts to match $h(x) = \sqrt{x^2 - 4}$ with the correct graph.
   Solution:
   Domain: $(-\infty, -2]$, $[2, \infty)$
   Range: $[0, \infty)$
   No asymptotes
   Intercepts: (-2, 0), (2, 0)
   Matches graph (e)

9. Find $\lim\limits_{x \to 0^+} (x - \frac{1}{x^3})$.
   Solution:
   $\lim\limits_{x \to 0^+} (x - \frac{1}{x^3}) = -\infty$

11. Find $\lim\limits_{x \to 1/2} \frac{2x - 1}{6x - 3}$.
    Solution:
    $\lim\limits_{x \to 1/2} \frac{2x - 1}{6x - 3} = \lim\limits_{x \to 1/2} \frac{2x - 1}{3(2x - 1)} = \lim\limits_{x \to 1/2} \frac{1}{3} = \frac{1}{3}$

13. Find $\lim\limits_{x \to \infty} \frac{5x^2 + 3}{2x^2 - x + 1}$.
    Solution:
    $\lim\limits_{x \to \infty} \frac{5x^2 + 3}{2x^2 - x + 1} = \lim\limits_{x \to \infty} \frac{5 + (3/x^2)}{2 - (1/x) + (1/x^2)} = \frac{5}{2}$

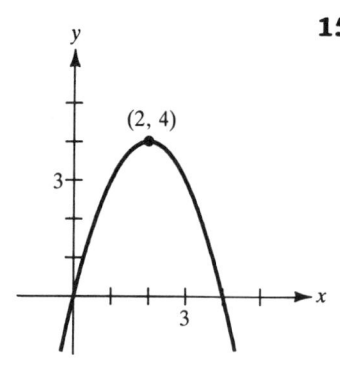

15. Make use of domain, range, asymptotes, intercepts, and relative extrema to sketch the graph of $f(x) = 4x - x^2$.
Solution:
$$f(x) = 4x - x^2 = x(4 - x)$$
$$f'(x) = 4 - 2x = -2(x - 2)$$
$$f''(x) = -2$$

Intercepts: $(0, 0)$, $(4, 0)$
Domain: $(-\infty, \infty)$
Range: $(-\infty, 4]$
Relative maximum: $(2, 4)$
Concave downward on $(-\infty, \infty)$

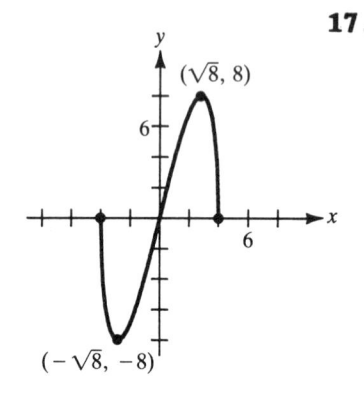

17. Make use of domain, range, asymptotes, intercepts, relative extrema, and points of inflection to sketch the graph of $f(x) = x\sqrt{16 - x^2}$.
Solution:
$$f(x) = x\sqrt{16 - x^2}$$
$$f'(x) = \frac{16 - 2x^2}{\sqrt{16 - x^2}}$$
$$f''(x) = \frac{2x(x^2 - 24)}{(16 - x^2)^{3/2}}$$

Domain: $[-4, 4]$
Range: $[-8, 8]$
Intercepts: $(0, 0)$, $(4, 0)$, $(-4, 0)$
Relative maximum: $(2\sqrt{2}, 8)$
Relative minimum: $(-2\sqrt{2}, -8)$
Point of inflection: $(0, 0)$
Concave upward on $(-4, 0)$
Concave downward on $(0, 4)$

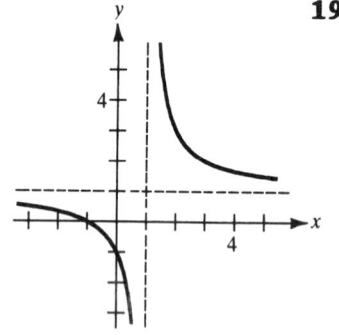

19. Make use of domain, range, asymptotes, intercepts, relative extrema, and points of inflection to sketch the graph of $f(x) = (x + 1)/(x - 1)$.
Solution:
$$f(x) = \frac{x + 1}{x - 1}$$
$$f'(x) = \frac{-2}{(x - 1)^2}$$
$$f''(x) = \frac{4}{(x - 1)^3}$$

Domain: all real numbers except 1
Range: all real numbers except 1
Intercepts: $(-1, 0)$, $(0, -1)$
Horizontal asymptote: $y = 1$
Vertical asymptote: $x = 1$
Concave upward on $(1, \infty)$
Concave downward on $(-\infty, 1)$

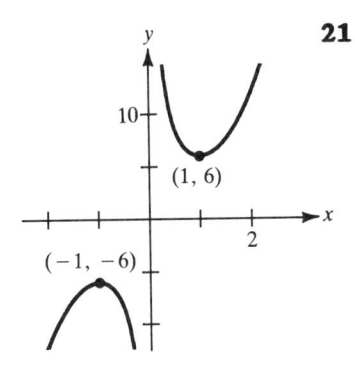

Review Exercises for Chapter 3

21. Make use of domain, range, asymptotes, intercepts, relative extrema, and points of inflection to sketch the graph of $f(x) = x^3 + x + (4/x)$.
Solution:

$$f(x) = x^3 + x + \frac{4}{x}$$

$$f'(x) = 3x^2 + 1 - \frac{4}{x^2} = \frac{3x^4 + x^2 - 4}{x^2}$$

$$f''(x) = 6x + \frac{8}{x^3} = \frac{6x^4 + 8}{x^3}$$

Domain: all real numbers except 0
Range: $(-\infty, -6] \cup [6, \infty)$
Relative maximum: $(-1, -6)$
Relative minimum: $(1, 6)$
Vertical asymptote: $x = 0$
Concave upward on $(0, \infty)$
Concave downward on $(-\infty, 0)$

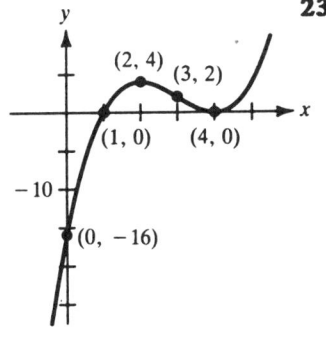

23. Make use of domain, range, asymptotes, intercepts, relative extrema, and points of inflection to sketch the graph of $f(x) = (x - 1)(x - 4)^2$.
Solution:
$$f(x) = (x - 1)(x - 4)^2$$
$$f'(x) = (x - 1)[2(x - 4)] + (x - 4)^2$$
$$= (x - 4)(3x - 6)$$
$$f''(x) = (x - 4)(3) + (3x - 6) = 6x - 18$$

Domain: $(-\infty, \infty)$
Range: $(-\infty, \infty)$
Intercepts: $(1, 0)$, $(4, 0)$, and $(0, -16)$
Relative maximum: $(2, 4)$
Relative minimum: $(4, 0)$
Point of inflection: $(3, 2)$
Concave upward on $(3, \infty)$
Concave downward on $(-\infty, 3)$

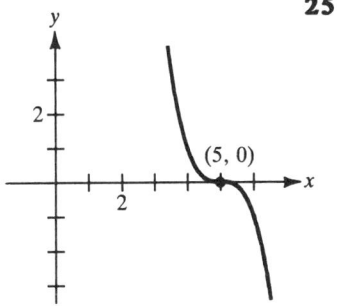

25. Make use of domain, range, asymptotes, intercepts, relative extrema, and points of inflection to sketch the graph of $f(x) = (5 - x)^3$.
Solution:
$$f(x) = (5 - x)^3$$
$$f'(x) = -3(5 - x)^2$$
$$f''(x) = 6(5 - x)$$

Domain: $(-\infty, \infty)$
Range: $(-\infty, \infty)$
Intercepts: $(5, 0)$, $(0, 125)$
No relative extrema.
Point of inflection: $(5, 0)$
Concave upward on $(-\infty, 5)$
Concave downward on $(5, \infty)$

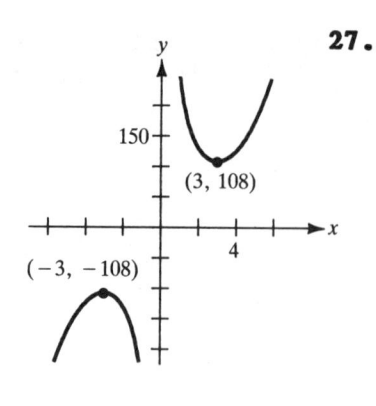

27. Make use of domain, range, asymptotes, intercepts, relative extrema, and points of inflection to sketch the graph of $f(x) = x^3 + (243/x)$.
Solution:
$$f(x) = x^3 + (243/x) = (x^4 + 243)/x$$
$$f'(x) = 3x^2 - (243/x^2) = 3(x^4 - 81)/x^2$$
$$f''(x) = 6x + (486/x^3) = 6(x^4 + 81)/x^3$$

Domain: all real numbers except 0
Range: $(-\infty, -108]$, $[108, \infty)$
Relative maximum: $(-3, -108)$
Relative minimum: $(3, 108)$
Vertical asymptote: $x = 0$
Concave downward on $(-\infty, 0)$
Concave upward on $(0, \infty)$

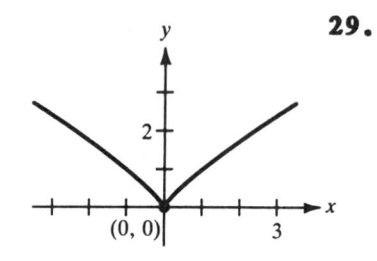

29. Make use of domain, range, asymptotes, intercepts, relative extrema, and points of inflection to sketch the graph of $f(x) = x^{4/5}$.
Solution:
$$f(x) = x^{4/5}$$
$$f'(x) = \frac{4}{5} x^{-1/5} = \frac{4}{5 \sqrt[5]{x}}$$
$$f''(x) = -\frac{4}{25} x^{-6/5} = -\frac{4}{25 x^{6/5}}$$

Domain: $(-\infty, \infty)$
Range: $[0, \infty)$
Intercept: $(0, 0)$
Relative minimum: $(0, 0)$
Concave downward on $(-\infty, 0)$ and $(0, \infty)$

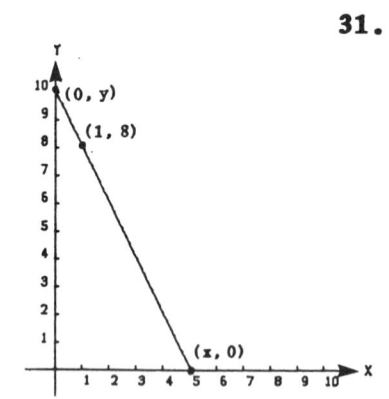

31. The sides of a right triangle (in the first quadrant) lie on the coordinate axes, and its hypotenuse passes through the point (1, 8). Find the vertices of the triangle so that the length of the hypotenuse is minimum.
Solution: The slope of the line passing through $(0, y)$ and $(1, 8)$ is $8 - y$ and the slope of the line passing through $(x, 0)$ and $(1, 8)$ is $8/(1 - x)$. By equating these two slopes, we have

$$m = 8 - y = \frac{8}{1-x} \implies y = \frac{8x}{x-1}$$

We can minimize the length of the hypotenuse by minimizing its square $L = d^2 = x^2 + y^2$.

$$L = x^2 + \left(\frac{8x}{x-1}\right)^2$$
$$\frac{dL}{dx} = 2x + 128\left(\frac{x}{x-1}\right)\left(\frac{(x-1)-x}{(x-1)^2}\right)$$
$$= 2\left[x - \frac{64x}{(x-1)^3}\right]$$

$dL/dx = 0$ when $x = 0$ and $x = 5$. By testing, we find the minimum occurs when $x = 5$ and $y = 10$. Thus, the vertices are $(0, 0)$, $(5, 0)$, and $(0, 10)$

Review Exercises for Chapter 3

169

33. Find the dimensions of the rectangle of maximum area, with sides parallel to the coordinate axes, that can be inscribed in the ellipse $(x^2/144) + (y^2/16) = 1$.

Solution: The area of the rectangle is

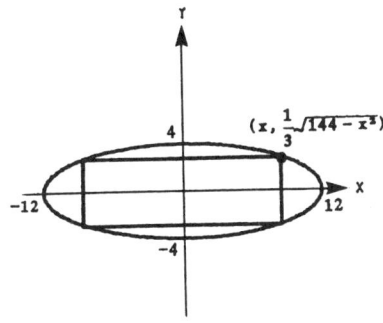

$$A = (2x)\left(\frac{2}{3}\sqrt{144 - x^2}\right) = \frac{4}{3}x\sqrt{144 - x^2}$$

$$\frac{dA}{dx} = \frac{4}{3}\left(\frac{-x^2}{\sqrt{144 - x^2}} + \sqrt{144 - x^2}\right)$$

$$= \frac{4}{3}\left(\frac{144 - 2x^2}{\sqrt{144 - x^2}}\right)$$

$dA/dx = 0$ when $x = \sqrt{72} = 6\sqrt{2}$. Thus, the dimensions of the rectangle are $12\sqrt{2}$ by $4\sqrt{2}$.

35. Find the maximum profit corresponding to a demand function of $p = 36 - 4x$ and a total cost function of $C = 2x^2 + 6$.

Solution: Since $R = xp$, the profit is given by

$$P = R - C = x(36 - 4x) - (2x^2 + 6) = -6(x^2 - 6x + 1)$$

$$\frac{dP}{dx} = -6(2x - 6)$$

$dP/dx = 0$ when $x = 3$ and $P = \$48$.

37. For groups of 80 or more, a charter bus company determines the rate per person according to the following formula: Rate = $\$8.00 - \$0.05(n - 80)$, $n \geq 80$. What number of passengers will give the bus company maximum revenue?

Solution: Let x be the number of people over 80 who go on the bus.

$$R = (\text{number who go})(\text{price/person})$$
$$= (80 + x)(8 - 0.05x)$$

$$\frac{dR}{dx} = (80 + x)(-0.05) + (8 - 0.05x) = 4 - 0.10x$$

$dR/dx = 0$ when $x = 40$. Thus, the revenue will be maximum when 120 people go on the bus.

39. The cost of inventory depends on ordering cost and storage cost. In the following inventory model, assume that sales occur at a constant rate, Q is the number of units sold per year, r is the cost of storing 1 unit for 1 year, s is the cost of placing an order, and x is the number of units per order. $C = (Q/x)s + (x/2)r$. Determine the order size that will minimize the cost.

Solution:

$$\frac{dC}{dx} = -\frac{Qs}{x^2} + \frac{r}{2}$$

$dC/dx = 0$ when $x^2 = 2Qs/r$ or $x = \sqrt{2Qs/r}$.

41. If a 1% error is made in measuring the edge of a cube, approximately what percentage error will be made in calculating the surface area and the volume of the cube?

Solution: For the surface area S of the cube, we have

$$S = 6x^2$$
$$dS = 12x\, dx$$

If $dx = 0.01x$, then the percentage error for the surface area is

$$\frac{dS}{S} = \frac{12x(0.01x)}{6x^2} = 0.02 = 2\%$$

For the volume of the cube, we have

$$V = x^3$$
$$dV = 3x^2\, dx$$

If $dx = 0.01x$, then the percentage error for the volume is

$$\frac{dV}{V} = \frac{3x^2(0.01x)}{x^3} = 0.03 = 3\%$$

43. A company finds that the demand for its commodity is given by $p = 75 - (1/4)x$. If x changes from 7 to 8, find the corresponding change in p. Compare the values of Δp and dp.

Solution:

$$p = 75 - \frac{1}{4}x$$

$$\Delta p = p(8) - p(7) = \left(75 - \frac{8}{4}\right) - \left(75 - \frac{7}{4}\right) = -\frac{1}{4}$$

$$dp = -\frac{1}{4}dx = -\frac{1}{4}(1) = -\frac{1}{4}$$

PRACTICE TEST FOR CHAPTER 3

1. Find the critical numbers and the intervals on which f is increasing or decreasing: $f(x) = x^3 - 6x^2 + 5$.

2. Find the critical numbers and the intervals on which f is increasing or decreasing: $f(x) = 2x\sqrt{1-x}$.

3. Find the relative extrema of $f(x) = x^4 - 32x + 3$.

4. Find the relative extrema of $f(x) = (x+3)^{4/3}$.

5. Find the extrema of $f(x) = x^2 - 4x - 5$ on $[0, 5]$.

6. Find the points of inflection of $f(x) = 3x^4 - 24x + 2$.

7. Find the points of inflection of $f(x) = \dfrac{x^2}{1+x^2}$.

8. Find two positive numbers whose product is 200 such that the sum of the first plus three times the second is a minimum.

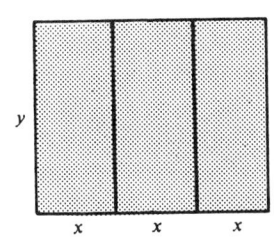

9. Three rectangular fields are to be enclosed by 3000 feet of fencing as shown in the accompanying figure. What dimensions should be used so that the enclosed area will be a maximum?

10. Find the number of units that produces a maximum revenue for $R = 400x^2 - 0.02x^3$.

11. Find the price per unit p that produces a maximum profit P given the cost function $C = 300x + 45{,}000$ and the demand function $p = 21{,}000 - 0.03x^2$.

12. Given the demand function $p = 600 - 0.02x^2$, find η (the price elasticity of demand) when $x = 100$.

13. Find $\lim\limits_{x \to 3^-} \dfrac{x+4}{x-3}$.

14. Find $\lim\limits_{x \to \infty} \dfrac{4x^3 - 9x^2 + 1}{1 - 2x^3}$.

15. Sketch the graph of $f(x) = \dfrac{x^2}{x^2 - 9}$.

16. Sketch the graph of $f(x) = \dfrac{x+2}{x^2+5}$.

17. Sketch the graph of $f(x) = x^3 + 3x^2 + 3x - 1$.

18. Sketch the graph of $f(x) = |4 - 2x|$.

19. Sketch the graph of $f(x) = (2-x)^{2/3}$.

20. Use differentials to approximate $\sqrt[3]{65}$.

Chapter 4 Integration

Section 4.1 Antiderivatives and the Indefinite Integral

1. Evaluate the indefinite integral and check your result by differentiation.

$$\int 6\, dx$$

Solution:

$$\int 6\, dx = 6x + C, \qquad \frac{d}{dx}[6x + C] = 6$$

3. Evaluate the indefinite integral and check your result by differentiation.

$$\int 3t^2\, dt$$

Solution:

$$\int 3t^2\, dt = t^3 + C, \qquad \frac{d}{dt}[t^3 + C] = 3t^2$$

5. Evaluate the indefinite integral and check your result by differentiation.

$$\int 3x^{-4}\, dx$$

Solution:

$$\int 3x^{-4}\, dx = -x^{-3} + C = -\frac{1}{x^3} + C$$

$$\frac{d}{dx}[-x^{-3} + C] = 3x^{-4}$$

7. Evaluate the indefinite integral and check your result by differentiation.

$$\int du$$

Solution:

$$\int du = u + C, \qquad \frac{d}{du}[u + C] = 1$$

9. Evaluate the indefinite integral and check your result by differentiation.

$$\int x^{3/2}\, dx$$

Solution:

$$\int x^{3/2}\, dx = \frac{2}{5}x^{5/2} + C$$

$$\frac{d}{dx}\left[\frac{2}{5}x^{5/2} + C\right] = x^{3/2}$$

Section 4.1

11. Complete the table using Example 2 as a model.
Solution:

Given	Rewrite	Integrate	Simplify
$\int \sqrt[3]{x}\, dx$	$\int x^{1/3}\, dx$	$\dfrac{x^{4/3}}{4/3} + C$	$\dfrac{3}{4} x^{4/3} + C$

13. Complete the table using Example 2 as a model.
Solution:

Given	Rewrite	Integrate	Simplify
$\int \dfrac{1}{x\sqrt{x}}\, dx$	$\int x^{-3/2}\, dx$	$\dfrac{x^{-1/2}}{-1/2} + C$	$-\dfrac{2}{\sqrt{x}} + C$

15. Complete the table using Example 2 as a model.
Solution:

Given	Rewrite	Integrate	Simplify
$\int \dfrac{1}{2x^3}\, dx$	$\dfrac{1}{2}\int x^{-3}\, dx$	$\dfrac{1}{2}\left(\dfrac{x^{-2}}{-2}\right) + C$	$-\dfrac{1}{4x^2} + C$

17. Evaluate the indefinite integral and check your result by differentiation.

$$\int (x^3 + 2)\, dx$$

Solution:

$$\int (x^3 + 2)\, dx = \frac{x^4}{4} + 2x + C$$

$$\frac{d}{dx}\left[\frac{x^4}{4} + 2x + C\right] = x^3 + 2$$

19. Evaluate the indefinite integral and check your result by differentiation.

$$\int (x^{3/2} + 2x + 1)\, dx$$

Solution:

$$\int (x^{3/2} + 2x + 1)\, dx = \frac{2}{5} x^{5/2} + x^2 + x + C$$

$$\frac{d}{dx}\left[\frac{2}{5} x^{5/2} + x^2 + x + C\right] = (x^{3/2} + 2x + 1)$$

21. Evaluate the indefinite integral and check your result by differentiation.

$$\int \sqrt[3]{x^2}\, dx$$

Solution:

$$\int \sqrt[3]{x^2}\, dx = \int x^{2/3}\, dx = \frac{3}{5} x^{5/3} + C$$

$$\frac{d}{dx}\left[\frac{3}{5} x^{5/3} + C\right] = \sqrt[3]{x^2}$$

23. Evaluate the indefinite integral and check your result by differentiation.

$$\int \frac{1}{x^3}\, dx$$

Solution:

$$\int \frac{1}{x^3}\, dx = \int x^{-3}\, dx = \frac{x^{-2}}{-2} + C = -\frac{1}{2x^2} + C$$

$$\frac{d}{dx}[-\frac{1}{2x^2} + C] = \frac{1}{x^3}$$

25. Evaluate the indefinite integral and check your result by differentiation.

$$\int \frac{1}{4x^2}\, dx$$

Solution:

$$\int \frac{1}{4x^2}\, dx = \frac{1}{4}\int x^{-2}\, dx = \frac{1}{4}(\frac{x^{-1}}{-1}) + C = -\frac{1}{4x} + C$$

$$\frac{d}{dx}[-\frac{1}{4x} + C] = \frac{1}{4x^2}$$

27. Evaluate the indefinite integral and check your result by differentiation.

$$\int \frac{t^2 + 2}{t^2}\, dt$$

Solution:

$$\int \frac{t^2 + 2}{t^2}\, dt = \int (1 + 2t^{-2})\, dt$$

$$= t + 2(\frac{t^{-1}}{-1}) + C = t - \frac{2}{t} + C$$

$$\frac{d}{dt}[t - \frac{2}{t} + C] = 1 + \frac{2}{t^2} = \frac{t^2 + 2}{t^2}$$

29. Evaluate the indefinite integral and check your result by differentiation.

$$\int u(3u^2 + 1)\, du$$

Solution:

$$\int u(3u^2 + 1)\, du = \int (3u^3 + u)\, du = \frac{3}{4}u^4 + \frac{1}{2}u^2 + C$$

$$\frac{d}{du}[\frac{3}{4}u^4 + \frac{1}{2}u^2 + C] = 3u^3 + u = u(3u^2 + 1)$$

31. Evaluate the indefinite integral and check your result by differentiation.

$$\int (x + 1)(3x - 2)\, dx$$

Section 4.1

Solution:

$$\int (x+1)(3x-2)\,dx = \int (3x^2 + x - 2)\,dx$$

$$= x^3 + \frac{x^2}{2} - 2x + C$$

$$\frac{d}{dx}\left[x^3 + \frac{x^2}{2} - 2x + C\right] = 3x^2 + x - 2$$

33. Evaluate the indefinite integral and check your result by differentiation.

$$\int y^2 \sqrt{y}\,dy$$

Solution:

$$\int y^2 \sqrt{y}\,dy = \int y^{5/2}\,dy = \frac{2}{7}y^{7/2} + C$$

$$\frac{d}{dy}\left[\frac{2}{7}y^{7/2} + C\right] = y^{5/2} = y^2\sqrt{y}$$

35. Find the equation of the curve, given the derivative $dy/dx = 2x - 1$ and the point $(1, 1)$ on the curve.

Solution:

$$y = \int (2x - 1)\,dx = x^2 - x + C$$

At $(1, 1)$, $1 = 1 - 1 + C$, which implies that $C = 1$. Thus, $y = x^2 - x + 1$.

37. Find $y = f(x)$ satisfying the conditions $f''(x) = 2$, $f'(2) = 5$, $f(2) = 10$.

Solution:

$$f'(x) = \int 2\,dx = 2x + C_1$$

Since $f'(2) = 4 + C_1 = 5$, we know that $C_1 = 1$. Thus, $f'(x) = 2x + 1$.

$$f(x) = \int (2x + 1)\,dx = x^2 + x + C_2$$

Since $f(2) = 4 + 2 + C_2 = 10$, we know that $C_2 = 4$. Thus, $f(x) = x^2 + x + 4$.

39. Find $y = f(x)$ satisfying the conditions $f''(x) = x^{-3/2}$, $f'(4) = 2$, $f(0) = 0$.

Solution:

$$f'(x) = \int x^{-3/2}\,dx = -2x^{-1/2} + C_1$$

Since $f'(4) = -1 + C_1 = 2$, we know that $C_1 = 3$. Thus, $f'(x) = -2x^{-1/2} + 3$.

$$f(x) = \int (-2x^{-1/2} + 3)\,dx = -4x^{1/2} + 3x + C_2$$

Since $f(0) = C_2 = 0$, we have $f(x) = -4\sqrt{x} + 3x$.

Section 4.1

41. Find the revenue and demand functions for the marginal revenue $dR/dx = 500 - 5x$. (Use the fact that $R = 0$ when $x = 0$.)
Solution:
$$R = \int (500 - 5x)\, dx = 500x - \frac{5}{2}x^2 + C$$

Since $R = 0$ when $x = 0$, it follows that $C = 0$. Thus, $R = 500x - (5/2)x^2$ and the demand function is $p = R/x = 500 - (5/2)x$

43. A company produces a product for which the marginal cost of producing x units is $dC/dx = 2x - 12$ and fixed costs are $125.
 (a) Find the total cost function and the average cost function.
 (b) Find the total cost of producing 50 units.
Solution:

(a) $C(x) = \int (2x - 12)\, dx = x^2 - 12x + C_1$

Since $C(0) = 125$, it follows that $C_1 = 125$. Thus, $C(x) = x^2 - 12x + 125$ and the average cost is $C/x = x - 12 + (125/x)$.

(b) $C(50) = \$2025$

45. An evergreen nursery usually sells a certain type of shrub after 6 years of growth and shaping. The growth rate after t years is given by $dh/dt = 0.5t + 2$ where $t = 0$ represents the time when the shrubs are 5-inch seedlings ($h = 5$ when $t = 0$).
 (a) Find the height h after t years.
 (b) How tall are the shrubs when they are sold?
Solution:

(a) $h(t) = \int (0.5t + 2)\, dt = \frac{t^2}{4} + 2t + C$

Since $h = 5$ when $t = 0$, it follows that $C = 5$. Thus, $h(t) = (t^2/4) + 2t + 5$.

(b) $h(6) = 26$ inches

47. Use $a(t) = -32$ ft/s^2 as the acceleration due to gravity. A ball is thrown upward with an initial velocity of 60 ft/sec. How high will the ball go?
Solution:
$$v(t) = \int -32\, dt = -32t + C_1$$

Since $v(0) = 60$, it follows that $C_1 = 60$. Thus, we have $v(t) = -32t + 60$.

$$s(t) = \int (-32t + 60)\, dt = -16t^2 + 60t + C_2$$

Since $s(0) = 0$, it follows that $C_2 = 0$. Therefore, the position function is $s(t) = -16t^2 + 60t$. Now, since $v(t) = 0$ when $t = 60/32 = 1.875$ seconds, the maximum height of the ball is $s(1.875) = 56.25$ feet.

49. Use $a(t) = -32$ ft/s² as the acceleration due to gravity. With what initial velocity must an object be thrown upward from the ground to reach the height of the Washington Monument (550 ft)?
Solution:
$$v(t) = \int -32 \, dt = -32t + C_1$$
Letting v_0 be the initial velocity, we have $C_1 = v_0$.
$$s(t) = \int (-32t + v_0) \, dt = -16t^2 + v_0 t + C_2$$
Since $s(0) = 0$, we have $C_2 = 0$. Therefore, the position function is $s(t) = -16t^2 + v_0 t$. At the highest point, the velocity is zero. Therefore, we have $v(t) = -32t + v_0 = 0$, and $t = v_0/32$ seconds. Finally, substituting this value into the position funciton, we have
$$s(v_0/32) = -16(v_0/32)^2 + v_0(v_0/32) = 550$$
which implies that $v_0^2 = 35,200$ and the initial velocity should be $v_0 = 40\sqrt{22} \approx 187.617$ ft/sec.

● **Section 4.2   The General Power Rule**

1. Complete the table by identifying u and du/dx for the given integral.
$$\int (5x^2 + 1)^2 (10x) \, dx$$
Solution:
$u = 5x^2 + 1$, $du/dx = 10x$

3. Complete the table by identifying u and du/dx for the given integral.
$$\int \sqrt{1 - x^2}(-2x) \, dx$$
Solution:
$u = 1 - x^2$, $du/dx = -2x$

5. Complete the table by identifying u and du/dx for the given integral.
$$\int \left(4 + \frac{1}{x^2}\right)\left(\frac{-2}{x^3}\right) dx$$
Solution:
$$u = 4 + \frac{1}{x^2}, \quad \frac{du}{dx} = -\frac{2}{x^3}$$

Section 4.2

7. Evaluate $\int (1 + 2x)^4(2)\, dx$.

   Solution:
   $$\int (1 + 2x)^4(2)\, dx = \frac{(1 + 2x)^5}{5} + C$$

9. Evaluate $\int \sqrt{3x^2 + 4}(6x)\, dx$.

   Solution:
   $$\int \sqrt{3x^2 + 4}(6x)\, dx = \int (3x^2 + 4)^{1/2}(6x)\, dx$$
   $$= \frac{2}{3}(3x^2 + 4)^{3/2} + C$$

11. Evaluate $\int x^2(x^3 - 1)^4\, dx$.

    Solution:
    $$\int x^2(x^3 - 1)^4\, dx = \frac{1}{3}\int (x^3 - 1)^4(3x^2)\, dx$$
    $$= \frac{1}{3}\frac{(x^3 - 1)^5}{5} + C$$
    $$= \frac{(x^3 - 1)^5}{15} + C$$

13. Evaluate $\int x(x^2 - 1)^7\, dx$.

    Solution:
    $$\int x(x^2 - 1)^7\, dx = \frac{1}{2}\int (x^2 - 1)^7(2x)\, dx$$
    $$= \frac{1}{2}\frac{(x^2 - 1)^8}{8} + C$$
    $$= \frac{(x^2 - 1)^8}{16} + C$$

15. Evaluate $\int \frac{x^2}{(1 + x^3)^2}\, dx$.

    Solution:
    $$\int \frac{x^2}{(1 + x^3)^2}\, dx = \frac{1}{3}\int (1 + x^3)^{-2}(3x^2)\, dx$$
    $$= \frac{1}{3}\frac{(1 + x^3)^{-1}}{-1} + C$$
    $$= -\frac{1}{3(1 + x^3)} + C$$

Section 4.2

17. Evaluate $\int \dfrac{x+1}{(x^2+2x-3)^2}\,dx$.

Solution:

$$\int \dfrac{x+1}{(x^2+2x-3)^2}\,dx = \dfrac{1}{2}\int (x^2+2x-3)^{-2}\,2(x+1)\,dx$$

$$= \dfrac{1}{2}\dfrac{(x^2+2x-3)^{-1}}{-1} + C$$

$$= -\dfrac{1}{2(x^2+2x-3)} + C$$

19. Evaluate $\int \dfrac{x-4}{\sqrt{x^2-8x+1}}\,dx$.

Solution:

$$\int \dfrac{x-4}{\sqrt{x^2-8x+1}}\,dx = \dfrac{1}{2}\int (x^2-8x+1)^{-1/2}\,2(x-4)\,dx$$

$$= \dfrac{1}{2}(2)(x^2-8x+1)^{1/2} + C$$

$$= \sqrt{x^2-8x+1} + C$$

21. Evaluate $\int 5x\sqrt[3]{1+x^2}\,dx$.

Solution:

$$\int 5x\sqrt[3]{1+x^2}\,dx = 5\left(\dfrac{1}{2}\right)\int (1+x^2)^{1/3}(2x)\,dx$$

$$= \dfrac{5}{2}\left(\dfrac{3}{4}\right)(1+x^2)^{4/3} + C$$

$$= \dfrac{15(1+x^2)^{4/3}}{8} + C$$

23. Evaluate $\int \dfrac{4x}{\sqrt{1+x^2}}\,dx$.

Solution:

$$\int \dfrac{4x}{\sqrt{1+x^2}}\,dx = 4\left(\dfrac{1}{2}\right)\int (1+x^2)^{-1/2}(2x)\,dx$$

$$= 2(2)(1+x^2)^{1/2} + C$$

$$= 4\sqrt{1+x^2} + C$$

Section 4.2

25. Evaluate $\int \dfrac{-3}{\sqrt{2x+3}}\, dx$.

   Solution:
   $$\int \dfrac{-3}{\sqrt{2x+3}}\, dx = -\dfrac{3}{2}\int (2x+3)^{-1/2}(2)\, dx$$
   $$= -\dfrac{3}{2}(2)(2x+3)^{1/2} + C$$
   $$= -3\sqrt{2x+3} + C$$

27. Evaluate $\int \dfrac{x^3}{\sqrt{1+x^4}}\, dx$.

   Solution:
   $$\int \dfrac{x^3}{\sqrt{1+x^4}}\, dx = \dfrac{1}{4}\int (1+x^4)^{-1/2}(4x^3)\, dx$$
   $$= \dfrac{1}{4}(2)(1+x^4)^{1/2} + C$$
   $$= \dfrac{\sqrt{1+x^4}}{2} + C$$

29. Evaluate $\int \dfrac{1}{2\sqrt{x}}\, dx$.

   Solution:
   $$\int \dfrac{1}{2\sqrt{x}}\, dx = \dfrac{1}{2}\int x^{-1/2}\, dx$$
   $$= \dfrac{1}{2}(2x^{1/2}) + C = \sqrt{x} + C$$

31. Evaluate $\int \dfrac{1}{\sqrt{2x}}\, dx$.

   Solution:
   $$\int \dfrac{1}{\sqrt{2x}}\, dx = \dfrac{1}{\sqrt{2}}\int x^{-1/2}\, dx$$
   $$= \dfrac{1}{\sqrt{2}}(2x^{1/2}) + C = \sqrt{2x} + C$$

33. Evaluate $\int t^2\left(t - \dfrac{2}{t}\right) dt$.

   Solution:
   $$\int t^2\left(t - \dfrac{2}{t}\right) dt = \int (t^3 - 2t)\, dt$$
   $$= \dfrac{1}{4}t^4 - t^2 + C$$

Section 4.2
181

35. Evaluate $\int (9 - y)\sqrt{y}\, dy$.

Solution:

$$\int (9 - y)\sqrt{y}\, dy = \int (9y^{1/2} - y^{3/2})\, dy$$

$$= 9(\frac{2}{3})y^{3/2} - \frac{2}{5}y^{5/2} + C$$

$$= 6y^{3/2} - \frac{2}{5}y^{5/2} + C$$

37. Perform the integration in two ways and explain the difference in appearance of the answers.

$$\int (2x - 1)^2\, dx$$

Solution:

$$\int (2x - 1)^2\, dx = \frac{1}{2} \int (2x - 1)^2 (2)\, dx$$

$$= \frac{1}{2} \frac{(2x - 1)^3}{3} + C_1$$

$$= \frac{1}{6}(2x - 1)^3 + C_1$$

$$= \frac{1}{6}(8x^3 - 12x^2 + 6x - 1) + C_1$$

$$= \frac{4}{3}x^3 - 2x^2 + x - \frac{1}{6} + C_1$$

$$\int (2x - 1)^2\, dx = \int (4x^2 - 4x + 1)\, dx$$

$$= \frac{4}{3}x^3 - 2x^2 + x + C_2$$

The two answers differ by a constant.

39. Use formal substitution to find the indefinite integral.

$$\int x(3x^2 - 5)^3\, dx$$

Solution: Let $u = 3x^2 - 5$ then $du = 6x\, dx$, which implies that $x\, dx = (1/6)\, du$.

$$\int x(3x^2 - 5)^3\, dx = \int (3x^2 - 5)^3 (x)\, dx$$

$$= \int u^3 (\frac{1}{6}\, du) = \frac{1}{6}(\frac{u^4}{4}) + C$$

$$= \frac{1}{24}u^4 + C = \frac{1}{24}(3x^2 - 5)^4 + C$$

Section 4.2

41. Use formal substitution to find the indefinite integral.

$$\int x^2(2-3x^3)^{3/2}\,dx$$

Solution: Let $u = 2 - 3x^3$, then $du = -9x^2\,dx$, which implies that $x^2\,dx = -(1/9)\,du$.

$$\int x^2(2-3x^3)^{3/2}\,dx = \int (2-3x^3)^{3/2}(x^2)\,dx$$

$$= \int u^{3/2}\left(-\frac{1}{9}\right)du$$

$$= -\frac{1}{9}\left(\frac{2}{5}\right)u^{5/2} + C$$

$$= -\frac{2}{45}(2-3x^3)^{5/2} + C$$

43. Use formal substitution to find the indefinite integral.

$$\int \frac{x}{\sqrt{x^2+25}}\,dx$$

Solution: Let $u = x^2 + 25$ then $du = 2x\,dx$, which implies that $x\,dx = (1/2)\,du$.

$$\int \frac{x}{\sqrt{x^2+25}}\,dx = \int (x^2+25)^{-1/2}(x)\,dx$$

$$= \int u^{-1/2}\left(\frac{1}{2}\right)du$$

$$= \frac{1}{2}(2u^{1/2}) + C$$

$$= \sqrt{u} + C = \sqrt{x^2+25} + C$$

45. Use formal substitution to find the indefinite integral.

$$\int \frac{x^2+1}{\sqrt{x^3+3x+4}}\,dx$$

Solution: Let $u = x^3 + 3x + 4$, then $du = (3x^2+3)\,dx = 3(x^2+1)\,dx$, and $(x^2+1)\,dx = (1/3)\,du$.

$$\int \frac{x^2+1}{\sqrt{x^3+3x+4}}\,dx = \int (x^3+3x+4)^{-1/2}(x^2+1)\,dx$$

$$= \int u^{-1/2}\left(\frac{1}{3}\right)du$$

$$= \left(\frac{1}{3}\right)2u^{1/2} + C = \frac{2}{3}\sqrt{u} + C$$

$$= \frac{2}{3}\sqrt{x^3+3x+4} + C$$

Section 4.2   183

47. Find the equation of the function f whose graph passes through the point (0, 4/3) and whose derivative is $f'(x) = x\sqrt{1-x^2}$.
Solution:

$$f(x) = \int x\sqrt{1-x^2}\, dx = -\frac{1}{2}\int (1-x^2)^{1/2}(-2x)\, dx$$

$$= -\frac{1}{2}\left(\frac{2}{3}\right)(1-x^2)^{3/2} + C$$

$$= -\frac{1}{3}(1-x^2)^{3/2} + C$$

Since $f(0) = 4/3$, it follows that $C = 5/3$, and we have

$$f(x) = -\frac{1}{3}(1-x^2)^{3/2} + \frac{5}{3} = \frac{1}{3}[5 - (1-x^2)^{3/2}]$$

49. A company has determined the marginal cost for a particular product to be

$$\frac{dC}{dx} = \frac{4}{\sqrt{x+1}}$$

(a) Find the cost function if $C = 50$ when $x = 15$.
(b) Graph the marginal cost function and the cost function on the same set of axes.
Solution:

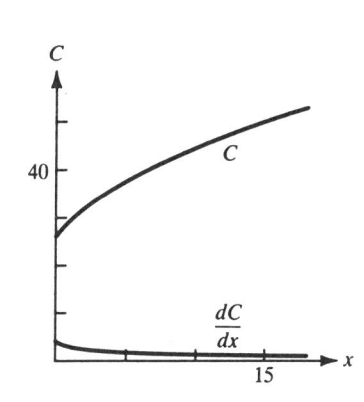

(a) $\quad C = \int \frac{4}{\sqrt{x+1}}\, dx = 4\int (x+1)^{-1/2}\, dx$

$$= 4(2)(x+1)^{1/2} + K$$

$$= 8\sqrt{x+1} + K$$

Since $C(15) = 50$, it follows that $K = 18$, and we have $C = 8\sqrt{x+1} + 18$.
(b) See accompanying graph.

51. Find the supply function $x = f(p)$ that satisfies the given conditions.

$$\frac{dx}{dp} = p\sqrt{p^2 - 16}, \quad x = 50 \text{ when } p = \$5$$

Solution:

$$x = \int p\sqrt{p^2 - 16}\, dp = \frac{1}{2}\int (p^2 - 16)^{1/2}(2p)\, dp$$

$$= \frac{1}{2}\left(\frac{2}{3}\right)(p^2 - 16)^{3/2} + C$$

$$= \frac{1}{3}(p^2 - 16)^{3/2} + C$$

Since $x = 50$ when $p = 5$, it follows that $C = 41$. Therefore, we have

$$x = \frac{1}{3}(p^2 - 16)^{3/2} + 41$$

53. Find the demand function x = f(p) that satisfies the given conditions.

$$\frac{dx}{dp} = -\frac{6000p}{(p^2 - 16)^{3/2}}, \quad x = 5000 \text{ when } p = \$5$$

Solution:

$$x = \int -\frac{6000p}{(p^2 - 16)^{3/2}}\, dp$$

$$= -\frac{6000}{2}\int (p^2 - 16)^{-3/2}(2p)\, dp$$

$$= -3000(-2)(p^2 - 16)^{-1/2} + C$$

$$= \frac{6000}{\sqrt{p^2 - 16}} + C$$

Since x = 5000 when p = 5, it follows that C = 3000 and we have

$$x = \frac{6000}{\sqrt{p^2 - 16}} + 3000$$

55. A lumber company is seeking a model that yields the average weight loss W per ponderosa pine log as a function of the number of days of drying time t. The model is to be reliable up to 100 days after the log is cut. Based on the weight loss during the first 30 days, it was determined that

$$\frac{dW}{dt} = \frac{12}{\sqrt{16t + 9}}, \quad 0 \leq t \leq 100$$

(a) Find W as a function of t. Note that no weight loss occurs until the tree is cut.
(b) Find the total weight loss after 100 days.

Solution:

(a) $W(t) = \int \frac{12}{\sqrt{16t + 9}}\, dt$

$$= \frac{12}{16}\int (16t + 9)^{-1/2}(16)\, dt$$

$$= \frac{3}{4}(2\sqrt{16t + 9}) + C$$

$$= \frac{3\sqrt{16t + 9}}{2} + C$$

Since W = 0 when t = 0, it follows that C = -9/2, and we have

$$W(t) = \frac{3\sqrt{16t + 9} - 9}{2} = \frac{3}{2}(\sqrt{16t + 9} - 3)$$

(b) $W(100) \approx 55.67$ lb.

Section 4.3   Area and the Fundamental Theorem of Calculus

1. Evaluate $\int_0^1 2x\,dx$.

   Solution:
   $$\int_0^1 2x\,dx = x^2\Big]_0^1 = 1 - 0 = 1$$

3. Evaluate $\int_{-1}^0 (x - 2)\,dx$.

   Solution:
   $$\int_{-1}^0 (x - 2)\,dx = \left[\frac{x^2}{2} - 2x\right]_{-1}^0$$
   $$= 0 - \frac{5}{2} = -\frac{5}{2}$$

5. Evaluate $\int_{-1}^1 (t^2 - 2)\,dt$.

   Solution:
   $$\int_{-1}^1 (t^2 - 2)\,dt = \left[\frac{t^3}{3} - 2t\right]_{-1}^1$$
   $$= -\frac{5}{3} - \frac{5}{3} = -\frac{10}{3}$$

7. Evaluate $\int_0^1 (2t - 1)^2\,dt$.

   Solution:
   $$\int_0^1 (2t - 1)^2\,dt = \frac{1}{2}\int_0^1 (2t - 1)^2(2)\,dt$$
   $$= \frac{1}{6}(2t - 1)^3\Big]_0^1$$
   $$= \frac{1}{6} - \left(-\frac{1}{6}\right) = \frac{1}{3}$$

9. Evaluate $\int_1^2 \left(\frac{3}{x^2} - 1\right)dx$.

   Solution:
   $$\int_1^2 \left(\frac{3}{x^2} - 1\right)dx = \int_1^2 (3x^{-2} - 1)\,dx$$
   $$= \left[3\left(\frac{x^{-1}}{-1}\right) - x\right]_1^2 = \left[-\frac{3}{x} - x\right]_1^2$$
   $$= -\frac{7}{2} - (-4) = \frac{1}{2}$$

11. Evaluate $\int_1^2 (5x^4 + 5)\, dx$.

    Solution:
    $$\int_1^2 (5x^4 + 5)\, dx = \left[x^5 + 5x\right]_1^2$$
    $$= 42 - 6 = 36$$

13. Evaluate $\int_{-1}^1 (\sqrt[3]{t} - 2)\, dt$.

    Solution:
    $$\int_{-1}^1 (\sqrt[3]{t} - 2)\, dt = \left[\frac{3}{4}t^{4/3} - 2t\right]_{-1}^1$$
    $$= -\frac{5}{4} - \frac{11}{4} = -4$$

15. Evaluate $\int_1^4 \frac{u - 2}{\sqrt{u}}\, du$.

    Solution:
    $$\int_1^4 \frac{u - 2}{\sqrt{u}}\, du = \int_1^4 (u^{1/2} - 2u^{-1/2})\, du$$
    $$= \left[\frac{2}{3}u^{3/2} - 4u^{1/2}\right]_1^4$$
    $$= \left(\frac{16}{3} - 8\right) - \left(\frac{2}{3} - 4\right) = \frac{2}{3}$$

17. Evaluate $\int_0^1 \frac{x - \sqrt{x}}{3}\, dx$.

    Solution:
    $$\int_0^1 \frac{x - \sqrt{x}}{3}\, dx = \frac{1}{3}\left(\frac{x^2}{2} - \frac{2}{3}x^{3/2}\right)\Big]_0^1$$
    $$= \frac{1}{3}\left(\frac{1}{2} - \frac{2}{3}\right) - 0$$
    $$= -\frac{1}{18}$$

19. Evaluate $\int_{-1}^0 (t^{1/3} - t^{2/3})\, dt$.

    Solution:
    $$\int_{-1}^0 (t^{1/3} - t^{2/3})\, dt = \left[\frac{3}{4}t^{4/3} - \frac{3}{5}t^{5/3}\right]_{-1}^0$$
    $$= 0 - \left(\frac{3}{4} + \frac{3}{5}\right)$$
    $$= -\frac{27}{20}$$

Section 4.3

21. Evaluate $\int_0^4 \dfrac{1}{\sqrt{2x+1}}\, dx$.

Solution:
$$\int_0^4 \frac{1}{\sqrt{2x+1}}\, dx = \frac{1}{2}\int_0^4 (2x+1)^{-1/2}(2)\, dx$$
$$= \left[\frac{1}{2}(2)(2x+1)^{1/2}\right]_0^4 = \sqrt{2x+1}\,\Big]_0^4$$
$$= 3 - 1 = 2$$

23. Evaluate $\int_{-1}^1 x(x^2+1)^3\, dx$.

Solution:
$$\int_{-1}^1 x(x^2+1)^3\, dx = \frac{1}{2}\int_{-1}^1 (x^2+1)^3(2x)\, dx$$
$$= \frac{1}{2}\left(\frac{1}{4}\right)(x^2+1)^4\,\Big]_{-1}^1$$
$$= \frac{1}{8}(x^2+1)^4\,\Big]_{-1}^1$$
$$= 2 - 2 = 0$$

25. Evaluate $\int_{-2}^2 x\sqrt[3]{4+x^2}\, dx$.

Solution:
$$\int_{-2}^2 x\sqrt[3]{4+x^2}\, dx = \frac{1}{2}\int_{-2}^2 (4+x^2)^{1/3}(2x)\, dx$$
$$= \frac{1}{2}\left(\frac{3}{4}\right)(4+x^2)^{4/3}\,\Big]_{-2}^2$$
$$= \frac{3}{8}(4+x^2)^{4/3}\,\Big]_{-2}^2$$
$$= \frac{3}{8}(8)^{4/3} - \frac{3}{8}(8)^{4/3} = 0$$

27. Evaluate $\int_{-1}^1 |x|\, dx$.

Solution:
$$\int_{-1}^1 |x|\, dx = \int_{-1}^0 -x\, dx + \int_0^1 x\, dx$$
$$= -\frac{x^2}{2}\,\Big]_{-1}^0 + \frac{x^2}{2}\,\Big]_0^1$$
$$= \left(0 + \frac{1}{2}\right) + \left(\frac{1}{2} - 0\right)$$
$$= 1$$

Section 4.3

29. Evaluate $\int_1^2 (x - 1)(2 - x)\, dx$.

Solution:

$$\int_1^2 (x - 1)(2 - x)\, dx = \int_1^2 (-x^2 + 3x - 2)\, dx$$

$$= \left[-\frac{1}{3}x^3 + \frac{3}{2}x^2 - 2x\right]_1^2$$

$$= \left(-\frac{8}{3} + 6 - 4\right) - \left(-\frac{1}{3} + \frac{3}{2} - 2\right)$$

$$= 4 - \frac{7}{3} - \frac{3}{2} = \frac{1}{6}$$

31. Evaluate $\int_0^3 x(x - 3)^2\, dx$.

Solution:

$$\int_0^3 x(x - 3)^2\, dx = \int_0^3 x(x^2 - 6x + 9)\, dx$$

$$= \int_0^3 (x^3 - 6x^2 + 9x)\, dx$$

$$= \left[\frac{1}{4}x^4 - 2x^3 + \frac{9}{2}x^2\right]_0^3$$

$$= \left(\frac{81}{4} - 54 + \frac{81}{2}\right) - 0 = \frac{27}{4}$$

33. Evaluate $\int_0^7 x\sqrt[3]{x^2 + 1}\, dx$.

Solution:

$$\int_0^7 x\sqrt[3]{x^2 + 1}\, dx = \frac{1}{2}\int_0^7 (x^2 + 1)^{1/3}(2x)\, dx$$

$$= \frac{1}{2}\left(\frac{3}{4}\right)(x^2 + 1)^{4/3}\Big]_0^7$$

$$= \frac{3}{8}(x^2 + 1)^{4/3}\Big]_0^7 = \frac{3}{8}(50)^{4/3} - \frac{3}{8}(1)$$

$$= \frac{3}{8}[50\sqrt[3]{50} - 1] \approx 68.7$$

35. Determine the area of the region bounded by the graphs of $y = x - x^2$ and $y = 0$.

Solution:

$$A = \int_0^1 (x - x^2)\, dx = \left[\frac{x^2}{2} - \frac{x^3}{3}\right]_0^1$$

$$= \frac{1}{6} \text{ square units}$$

Section 4.3   189

37. Determine the area of the region bounded by the graphs of $y = 1 - x^4$ and $y = 0$.
Solution:
$$A = 2\int_0^1 (1 - x^4)\, dx = 2\left(x - \frac{x^5}{5}\right)\Big]_0^1$$
$$= \frac{8}{5} \text{ square units}$$

39. Determine the area of the region bounded by the graphs of $y = \sqrt[3]{2x}$, $x = 0$, $x = 4$, and $y = 0$.
Solution:
$$A = \int_0^4 \sqrt[3]{2x}\, dx = \frac{1}{2}\int_0^4 (2x)^{1/3}(2)\, dx$$
$$= \frac{1}{2}\left(\frac{3}{4}\right)(2x)^{4/3}\Big]_0^4 = \frac{3}{8}(\sqrt[3]{2x})^4\Big]_0^4$$
$$= \frac{3}{8}(16) - \frac{3}{8}(0) = 6 \text{ square units}$$

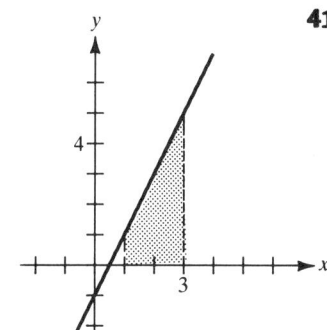

41. Evaluate the definite integral and make a sketch of the region whose area is given by the integral.
$$\int_1^3 (2x - 1)\, dx$$
Solution:
$$\int_1^3 (2x - 1)\, dx = \left[x^2 - x\right]_1^3$$
$$= 6 - 0 = 6$$

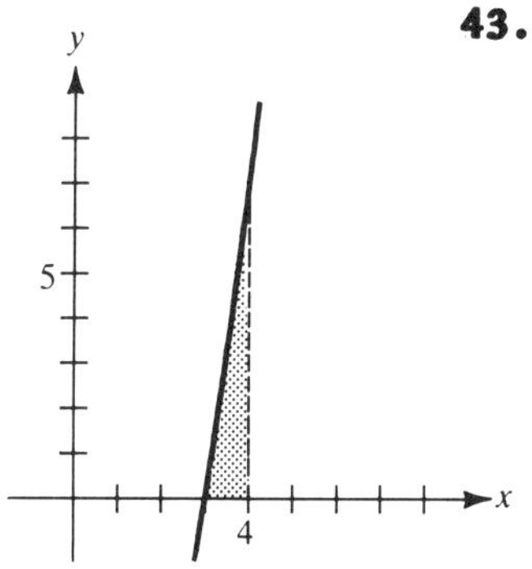

43. Evaluate the definite integral and make a sketch of the region whose area is given by the integral.
$$\int_3^4 (x^2 - 9)\, dx$$
Solution:
$$\int_3^4 (x^2 - 9)\, dx = \left[\frac{x^3}{3} - 9x\right]_3^4$$
$$= \left(\frac{64}{3} - 36\right) - (9 - 27) = \frac{10}{3}$$

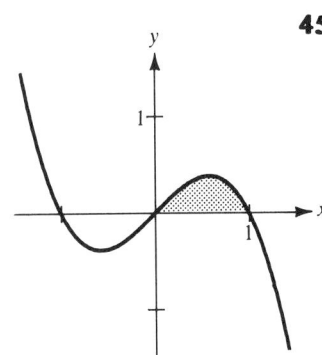

45. Evaluate the definite integral and make a sketch of the region whose area is given by the integral.
$$\int_0^1 (x - x^3)\, dx$$
Solution:
$$\int_0^1 (x - x^3)\, dx = \left[\frac{x^2}{2} - \frac{x^4}{4}\right]_0^1$$
$$= \left(\frac{1}{2} - \frac{1}{4}\right) - 0 = \frac{1}{4}$$

47. Determine the area of the region bounded by the graphs of $y = 3x^2 + 1$, $x = 0$, $x = 2$, and $y = 0$.
Solution:
$$A = \int_0^2 (3x^2 + 1)\, dx = \left[x^3 + x\right]_0^2 = 10 \text{ sq. units}$$

49. Determine the area of the region bounded by the graphs of $y = x^3 + x$, $x = 2$, and $y = 0$.
Solution:
$$A = \int_0^2 (x^3 + x)\, dx = \left[\frac{1}{4}x^4 + \frac{1}{2}x^2\right]_0^2 = 6 \text{ sq. units}$$

51. Find the average value of $f(x) = 4 - x^2$ on the interval $[-2, 2]$ and find all values of x where the function equals its average. Sketch your result.
Solution:
$$\text{Average Value} = \frac{1}{2 - (-2)} \int_{-2}^2 (4 - x^2)\, dx$$
$$= \frac{1}{4}\left(4x - \frac{x^3}{3}\right)\Big]_{-2}^2 = \frac{4}{3} + \frac{4}{3} = \frac{8}{3}$$

To find the x-values for which $f(x) = 8/3$, we let $4 - x^2 = 8/3$, and solve for x to obtain
$$x = \pm\sqrt{4/3} = \pm\frac{2\sqrt{3}}{3} \approx \pm 1.155$$

53. Find the average value of $f(x) = x\sqrt{4 - x^2}$ on the interval $[0, 2]$ and find all values of x where the function equals its average. Sketch your result.
Solution:
$$\text{Average Value} = \frac{1}{2 - 0}\int_0^2 x\sqrt{4 - x^2}\, dx$$
$$= \frac{1}{2}\left(-\frac{1}{2}\right)\int_0^2 (4 - x^2)^{1/2}(-2x)\, dx$$
$$= \frac{1}{2}\left(-\frac{1}{2}\right)\left(\frac{2}{3}\right)(4 - x^2)^{3/2}\Big]_0^2$$
$$= -\frac{1}{6}(4 - x^2)^{3/2}\Big]_0^2 = 0 + \frac{4}{3} = \frac{4}{3}$$

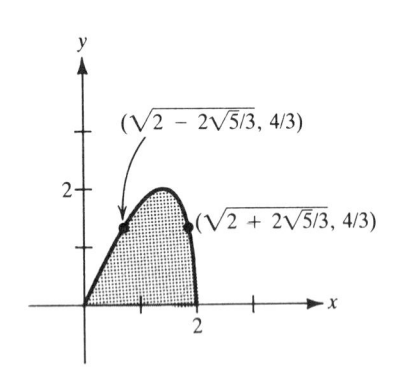

To find the x-values for which $f(x) = 4/3$, we let $x\sqrt{4 - x^2} = 4/3$, and solve for x to obtain
$$x^2(4 - x^2) = 16/9$$
$$36x^2 - 9x^4 = 16$$
$$9x^4 - 36x^2 + 16 = 0$$
$$x^2 = \frac{36 \pm \sqrt{720}}{18} = 2 \pm \frac{2\sqrt{5}}{3}$$
$$x = \sqrt{2 \pm \frac{2\sqrt{5}}{3}}$$

55. Find the average value of $f(x) = x - 2\sqrt{x}$ on the interval $[0, 4]$ and find all values of x where the function equals its average. Sketch your result.
   Solution:

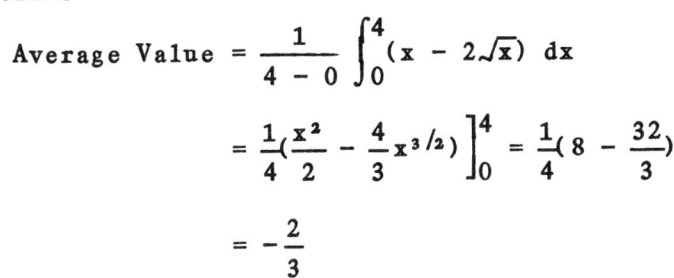

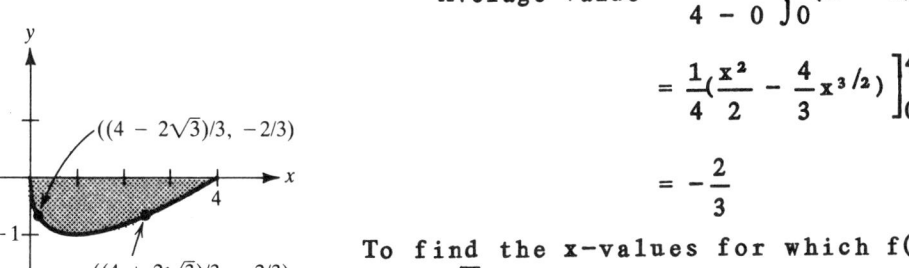

$$\text{Average Value} = \frac{1}{4-0}\int_0^4 (x - 2\sqrt{x})\, dx$$

$$= \frac{1}{4}\left(\frac{x^2}{2} - \frac{4}{3}x^{3/2}\right)\Big]_0^4 = \frac{1}{4}\left(8 - \frac{32}{3}\right)$$

$$= -\frac{2}{3}$$

To find the x-values for which $f(x) = -2/3$, we let $x - 2\sqrt{x} = -2/3$ and solve for x to obtain

$$x + (2/3) = 2\sqrt{x}$$
$$x^2 + (4/3)x + (4/9) = 4x$$
$$x^2 - (8/3)x + (4/9) = 0$$
$$9x^2 - 24x + 4 = 0$$

$$x = \frac{24 \pm \sqrt{432}}{18} = \frac{4 \pm 2\sqrt{3}}{3}$$

57. Use the fact that

$$\int_0^2 x^2\, dx = \frac{8}{3}$$

to evaluate the following definite integrals without using the Fundamental Theorem of Calculus:

(a) $\int_{-2}^0 x^2\, dx$ (b) $\int_{-2}^2 x^2\, dx$

(c) $\int_0^2 -x^2\, dx$ (d) $\int_{-2}^0 3x^2\, dx$

Solution:

(a) $\int_{-2}^0 x^2\, dx = \int_0^2 x^2\, dx = \frac{8}{3}$

(b) $\int_{-2}^2 x^2\, dx = 2\int_0^2 x^2\, dx = \frac{16}{3}$

(c) $\int_0^2 -x^2\, dx = -\int_0^2 x^2\, dx = -\frac{8}{3}$

(d) $\int_{-2}^0 3x^2\, dx = 3\int_0^2 x^2\, dx = 8$

59. The total cost of purchasing and maintaining a piece of equipment for x years is given by

$$C = 5000\left(25 + 3\int_0^x t^{1/4}\, dt\right)$$

Find (a) $C(1)$, (b) $C(5)$, and (c) $C(10)$.

Solution:

$$C(x) = 5000\left(25 + 3\int_0^x t^{1/4}\, dt\right)$$

$$= 5000\left(25 + \frac{12}{5}t^{5/4}\Big]_0^x\right) = 5000\left(25 + \frac{12}{5}x^{5/4}\right)$$

(a) $C(1) = 5000[25 + (12/5)] = \$137{,}000.00$
(b) $C(5) = 5000[25 + (12/5)(5)^{5/4}] \approx \$214{,}720.93$
(c) $C(10) = 5000[25 + (12/5)(10)^{5/4}] \approx \$338{,}393.53$

61. A company purchases a new machine for which the rate of depreciation is given by

$$\frac{dV}{dt} = 10{,}000(t - 6), \quad 0 \le t \le 5$$

where V is the value of the machine after t years. Set up and evaluate the definite integral that yields the total loss of value of the machine over the first 3 years.

Solution:

$$V = \int_0^3 10{,}000(t - 6)\, dt = 10{,}000\left(\frac{t^2}{2} - 6t\right)\Big]_0^3$$

$$= 10{,}000\left(\frac{9}{2} - 18\right) = -\$135{,}000.00$$

63. The air temperature during a period of 12 hours is given by the model $T = 53 + 5t - 0.3t^2$, $0 \le t \le 12$, where t is measured in hours and T in degrees Fahrenheit. Find the average temperature during (a) the first 6-hour period and (b) the entire period.

Solution:

(a) $\dfrac{1}{6 - 0}\displaystyle\int_0^6 (53 + 5t - 0.3t^2)\, dt$

$$= \frac{1}{6}\left(53t + \frac{5}{2}t^2 - 0.1t^3\right)\Big]_0^6$$

$$= \frac{1}{6}(318 + 90 - 21.6)$$

$$= 64.4°F$$

(b) $\dfrac{1}{12 - 0}\displaystyle\int_0^{12} (53 + 5t - 0.3t^2)\, dt$

$$= \frac{1}{12}\left(53t + \frac{5}{2}t^2 - 0.1t^3\right)\Big]_0^{12}$$

$$= \frac{1}{12}(636 + 360 - 172.8)$$

$$= 68.6°F$$

Section 4.4

● Section 4.4 The Area of a Region Between Two Curves

1. Sketch the region whose area is given by
$$\int_0^4 x \, dx$$
   Solution: The region is bounded by the graphs of
   $$y = x$$
   $y = 0$, $x = 0$, and $x = 4$, as shown in the accompanying figure.

3. Sketch the region whose area is given by
$$\int_{-3}^3 \sqrt{9 - x^2} \, dx$$
   Solution: The region is bounded by the graphs of
   $$y = \sqrt{9 - x^2}$$
   $y = 0$, $x = -3$, and $x = 3$, as shown in the accompanying figure.

5. Sketch the region whose area is given by
$$\int_0^4 \left[(x + 1) - \frac{x}{2}\right] dx$$
   Solution: The region is bounded by the graphs of
   $$y = x + 1 \quad \text{and} \quad y = \frac{x}{2}$$
   $x = 0$, and $x = 4$, as shown in the accompanying figure.

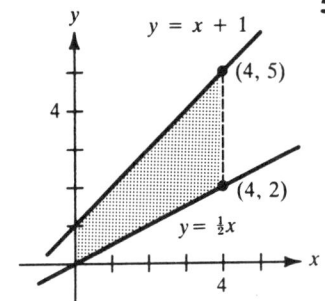

7. Find the area of the region bounded by the graphs of $f(x) = x^2 - 6x$ and $g(x) = 0$.
   Solution:
   $$A = \int_0^6 -(x^2 - 6x) \, dx = -\left(\frac{x^3}{3} - 3x^2\right)\Big]_0^6 = 36$$

9. Find the area of the region bounded by the graphs of $f(x) = x^2 - 4x + 3$ and $g(x) = -x^2 + 2x + 3$.
   Solution:
   $$A = \int_0^3 [(-x^2 + 2x + 3) - (x^2 - 4x + 3)] \, dx$$
   $$= \int_0^3 (-2x^2 + 6x) \, dx$$
   $$= \left[\frac{-2x^3}{3} + 3x^2\right]_0^3 = 9$$

Section 4.4

11. Find the area of the region bounded by the graphs of $f(x) = 3(x^3 - x)$ and $g(x) = 0$.
Solution:
$$A = 2\int_0^1 -3(x^3 - x)\, dx = -6\left(\frac{x^4}{4} - \frac{x^2}{2}\right)\Big]_0^1 = \frac{3}{2}$$

13. Sketch the region bounded by the graphs of $f(x) = x^2 - 4x$ and $g(x) = 0$, and find the area of the region.
Solution: The points of intersection of f and g are found by setting $f(x) = g(x)$ and solving for x.

$$x^2 - 4x = 0$$
$$x(x - 4) = 0$$
$$x = 0,\ 4$$

$$A = \int_0^4 -(x^2 - 4x)\, dx = -\left(\frac{x^3}{3} - 2x^2\right)\Big]_0^4 = \frac{32}{3}$$

15. Sketch the region bounded by the graphs of $f(x) = x^2 + 2x + 1$ and $g(x) = 3x + 3$, and find the area of the region.
Solution: The points of intersection of f and g are found by setting $f(x) = g(x)$ and solving for x.

$$x^2 + 2x + 1 = 3x + 3$$
$$x^2 - x - 2 = 0$$
$$(x + 1)(x - 2) = 0$$
$$x = -1,\ 2$$

$$A = \int_{-1}^2 [(3x + 3) - (x^2 + 2x + 1)]\, dx$$
$$= \int_{-1}^2 (-x^2 + x + 2)\, dx = \left[-\frac{x^3}{3} + \frac{x^2}{2} + 2x\right]_{-1}^2$$
$$= \left(-\frac{8}{3} + 6\right) - \left(\frac{1}{3} + \frac{1}{2} - 2\right) = \frac{9}{2}$$

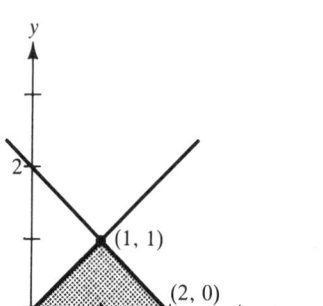

17. Sketch the region bounded by the graphs of $y = x$, $y = 2 - x$, and $y = 0$, and find the area of the region.
Solution: The point of intersection of the two graphs is found by equating y-values for $y = x$ and $y = 2 - x$ and solving for x.

$$x = 2 - x$$
$$2x = 2$$
$$x = 1$$

$$A = \int_0^1 x\, dx + \int_1^2 (2 - x)\, dx = \frac{x^2}{2}\Big]_0^1 + \left[2x - \frac{x^2}{2}\right]_1^2$$
$$= \frac{1}{2} + \left(2 - \frac{3}{2}\right) = 1$$

Section 4.4

19. Sketch the region bounded by the graphs of $f(x) = x^2 - x$ and $g(x) = 2(x + 2)$, and find the area of the region.

Solution: The points of intersection of f and g are found by setting $f(x) = g(x)$ and solving for x.

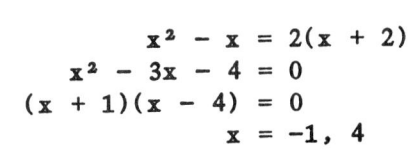

$$x^2 - x = 2(x + 2)$$
$$x^2 - 3x - 4 = 0$$
$$(x + 1)(x - 4) = 0$$
$$x = -1, 4$$

$$A = \int_{-1}^{4} [2(x + 2) - (x^2 - x)] \, dx$$

$$= \int_{-1}^{4} [-x^2 + 3x + 4] \, dx = \left[-\frac{x^3}{3} + \frac{3x^2}{2} + 4x\right]_{-1}^{4}$$

$$= \left(-\frac{64}{3} + 24 + 16\right) - \left(\frac{1}{3} + \frac{3}{2} - 4\right) = \frac{125}{6}$$

21. Sketch the region bounded by the graphs of $y = x^3 - 2x + 1$, $y = -2x$, and $x = 1$, and find the area of the region.

Solution: The point of intersection of the two graphs is found by equating y-values and solving for x.

$$x^3 - 2x + 1 = -2x$$
$$x^3 + 1 = 0$$
$$x = -1$$

$$A = \int_{-1}^{1} [(x^3 - 2x + 1) - (-2x)] \, dx$$

$$= \int_{-1}^{1} (x^3 + 1) \, dx = \left[\frac{x^4}{4} + x\right]_{-1}^{1} = 2$$

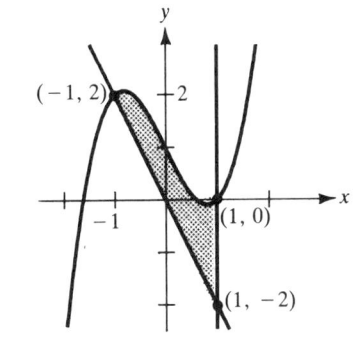

23. Sketch the region bounded by the graphs of $f(x) = \sqrt{3x} + 1$, $g(x) = x + 1$, and find the area of the region.

Solution: The points of intersection of f and g are found by setting $f(x) = g(x)$ and solving for x.

$$\sqrt{3x} + 1 = x + 1$$
$$\sqrt{3x} = x$$
$$3x = x^2$$
$$x^2 - 3x = 0$$
$$x(x - 3) = 0$$
$$x = 0, 3$$

$$A = \int_{0}^{3} [(\sqrt{3x} + 1) - (x + 1)] \, dx = \int_{0}^{3} (\sqrt{3x} - x) \, dx$$

$$= \left[\frac{2}{9}(3x)^{3/2} - \frac{x^2}{2}\right]_{0}^{3} = \frac{3}{2}$$

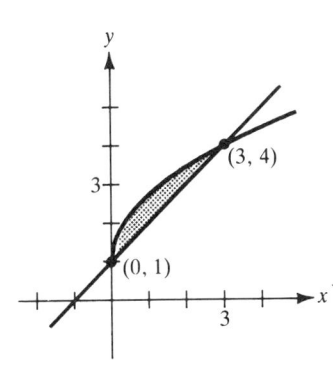

25. Sketch the region bounded by the graphs of $y = x^2 - 4x + 3$ and $y = 3 + 4x - x^2$, and find the area of the region.

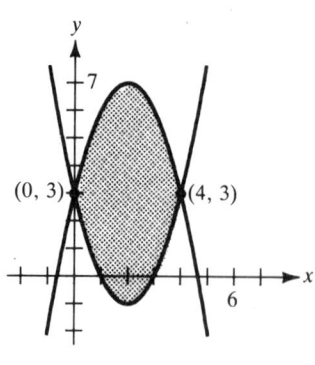

Solution: The points of intersection of the two graphs are found by equating y-values and solving for x.

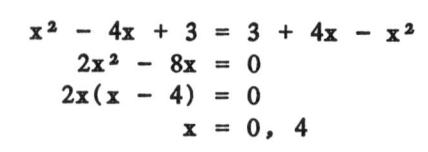

$$x^2 - 4x + 3 = 3 + 4x - x^2$$
$$2x^2 - 8x = 0$$
$$2x(x - 4) = 0$$
$$x = 0, 4$$

$$A = \int_0^4 [(3 + 4x - x^2) - (x^2 - 4x + 3)]\, dx$$

$$= \int_0^4 (-2x^2 + 8x)\, dx = \left[-\frac{2x^3}{3} + 4x^2\right]_0^4 = \frac{64}{3}$$

27. Sketch the region bounded by the graphs of $f(y) = y^2$, and $g(y) = y + 2$, and find the area of the region.

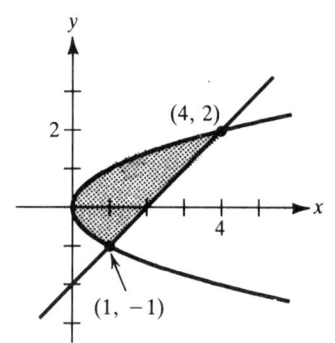

Solution: The points of intersection are found by setting $f(y)$ equal to $g(y)$ and solving for y.

$$y^2 = y + 2$$
$$y^2 - y - 2 = 0$$
$$(y + 1)(y - 2) = 0$$
$$y = -1, 2$$

$$A = \int_{-1}^2 [(y + 2) - y^2]\, dy = \left[\frac{y^2}{2} + 2y - \frac{y^3}{3}\right]_{-1}^2$$

$$= (2 + 4 - \frac{8}{3}) - (\frac{1}{2} - 2 + \frac{1}{3}) = \frac{9}{2}$$

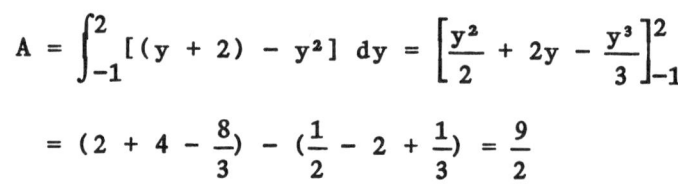

29. Sketch the region bounded by the graphs of $x = y^2 + 1$, $x = 0$, $y = -1$, and $y = 2$, and find the area of the region.

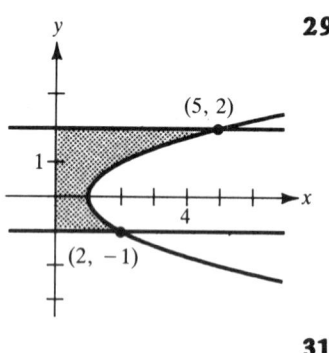

Solution:

$$A = \int_{-1}^2 (y^2 + 1)\, dy = \left[\frac{y^3}{3} + y\right]_{-1}^2$$

$$= (\frac{8}{3} + 2) - (-\frac{1}{3} - 1) = 6$$

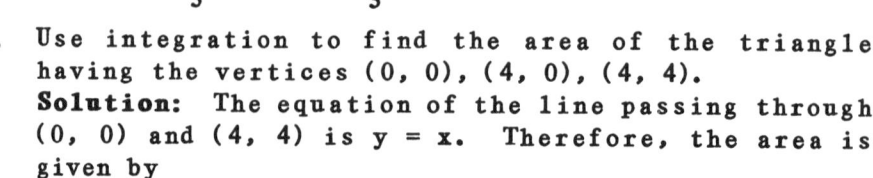

31. Use integration to find the area of the triangle having the vertices (0, 0), (4, 0), (4, 4).

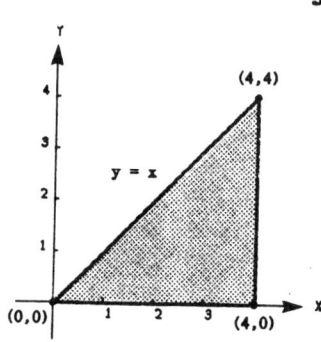

Solution: The equation of the line passing through (0, 0) and (4, 4) is $y = x$. Therefore, the area is given by

$$A = \int_0^4 x\, dx = \left.\frac{x^2}{2}\right]_0^4$$

$$= 8$$

Section 4.4   197

33. Two models $R_1 = 7.21 + 0.58t$ and $R_2 = 7.21 + 0.45t$ are given for revenue (in billions of dollars) for a large corporation. Model R_1 gives projected annual revenues from 1985 to 1990, with $t = 0$ corresponding to 1985, and model R_2 gives projected annual revenues if there is a decrease in growth of corporate sales over the period. Approximate the total reduction in revenue if corporate sales are actually closer to model R_2.
Solution: The reduction in revenue is given by

$$\int_0^5 (R_1 - R_2)\, dt$$

$$= \int_0^5 [(7.21 + 0.58t) - (7.21 + 0.45t)]\, dt$$

$$= \int_0^5 0.13t\, dt = 0.065t^2 \Big]_0^5 = \$1.625 \text{ billion}$$

35. The total consumption of beef (in billions of pounds) in the U.S. from 1950 to 1970 followed a growth pattern given by $f(t) = 23.703 + 1.002t + 0.015t^2$, with $t = 0$ representing 1970. From 1970 to 1980, the growth pattern was more closely approximated by $g(t) = 22.93 + 0.678t - 0.037t^2$. Estimate the total reduction in consumption of beef from 1970 to 1980.
Solution: The reduction is given by

$$\int_0^{10} [f(t) - g(t)]\, dt$$

$$= \int_0^{10} [(23.703 + 1.002t + 0.015t^2)$$

$$\qquad\qquad - (22.93 + 0.678t - 0.037t^2)]\, dt$$

$$= \int_0^{10} (0.052t^2 + 0.324t + 0.773)\, dt$$

$$= \left[\frac{0.052t^3}{3} + \frac{0.324t^2}{2} + 0.773t\right]_0^{10}$$

$$= 41.2633 \text{ billion pounds}$$

37. Find the consumer surplus and producer surplus for the demand function $p_1(x) = 50 - 0.5x$ and supply function $p_2(x) = 0.125x$.
Solution: The point of equilibrium is found by equating $50 - 0.5x = 0.125x$ to obtain $x = 80$ and $p = 10$.

$$CS = \int_0^{80} [(50 - 0.5x) - 10]\, dx = \left[-\frac{0.5x^2}{2} + 40x\right]_0^{80}$$

$$= 1600$$

$$PS = \int_0^{80} (10 - 0.125x)\, dx = \left[10x - \frac{0.125x^2}{2}\right]_0^{80}$$

$$= 400$$

Section 4.4

39. Find the consumer surplus and producer surplus for the demand function $p_1(x) = 300 - x$ and the supply function $p_2(x) = 100 + x$.

Solution: The point of equilibrium is found by equating $300 - x = 100 + x$ to obtain $x = 100$ and $p = 200$.

$$CS = \int_0^{100} [(300 - x) - 200] \, dx = \left[100x - \frac{x^2}{2}\right]_0^{100}$$
$$= 5000$$

$$PS = \int_0^{100} [200 - (100 + x)] \, dx = \left[100x - \frac{x^2}{2}\right]_0^{100}$$
$$= 5000$$

41. Find the consumer surplus and producer surplus for the given the demand function $p_1(x) = 300 - 0.01x^2$ and the supply function $p_2(x) = 100 + x$.

Solution: The point of equilibrium is found by equating the demand and supply functions.

$$300 - 0.01x^2 = 100 + x$$
$$0.01x^2 + x - 200 = 0$$
$$x^2 + 100x - 20,000 = 0$$
$$(x + 200)(x - 100) = 0$$
$$x = 100 \quad \text{and} \quad p = 200$$

$$CS = \int_0^{100} \left[(300 - \frac{1}{100}x^2) - 200\right] dx$$

$$= \left[100x - \frac{1}{300}x^3\right]_0^{100} \approx 6667$$

$$PS = \int_0^{100} [200 - (100 + x)] \, dx$$

$$= \left[100x - \frac{x^2}{2}\right]_0^{100} = 5000$$

43. Find the consumer surplus and producer surplus for the demand function $p_1(x) = 0.125(-x + 400)$ and the supply function $p_2(x) = 10 + 0.025x + 0.00025x^2$.

Solution: The point of equilibrium is found by equating the demand and supply functions.

$$0.125(-x + 400) = 10 + 0.025x + 0.00025x^2$$
$$500(-x + 400) = 40,000 + 100x + x^2$$
$$x^2 + 600x - 160,000 = 0$$
$$(x + 800)(x - 200) = 0$$
$$x = 200 \quad \text{and} \quad p = 25$$

Section 4.4   199

$$CS = \int_0^{200} [\tfrac{1}{8}(-x + 400) - 25]\, dx$$

$$= \left[-\frac{1}{16}x^2 + 25x\right]_0^{200} = 2500$$

$$PS = \int_0^{200} [25 - (10 + \frac{1}{40}x + \frac{1}{4000}x^2)]\, dx$$

$$= \left[15x - \frac{1}{80}x^2 - \frac{1}{12{,}000}x^3\right]_0^{200} \approx 1833$$

45. Find the consumer surplus and producer surplus for the given supply and demand curves.

<u>Demand Function</u>   <u>Supply Function</u>

$$p_1(x) = \frac{10{,}000}{\sqrt{x + 100}} \qquad p_2(x) = 100\sqrt{0.05x + 10}$$

Solution: The point of equilibrium is found by equating the demand and supply functions.

$$\frac{10{,}000}{\sqrt{x+100}} = 100\sqrt{0.05x + 10}$$

$$100 = \sqrt{(x + 100)(0.05x + 10)}$$

$$10{,}000 = 0.05x^2 + 15x + 1000$$

$$5x^2 + 1500x - 900{,}000 = 0$$

$$5(x^2 + 300x - 180{,}000) = 0$$

$$5(x + 600)(x - 300) = 0$$

$$x = 300 \quad \text{and} \quad p = 500$$

$$CS = \int_0^{300} \left(\frac{10{,}000}{\sqrt{x+100}} - 500\right) dx$$

$$= \left[20{,}000\sqrt{x + 100} - 500x\right]_0^{300}$$

$$= 250{,}000 - 200{,}000 = 50{,}000$$

$$PS = \int_0^{300} (500 - 100\sqrt{0.05x + 10})\, dx$$

$$= \left[500x - \frac{4000}{3}(0.05x + 10)^{3/2}\right]_0^{300}$$

$$= \frac{-50{,}000}{3} + \frac{40{,}000\sqrt{10}}{3}$$

$$= \frac{10{,}000}{3}(4\sqrt{10} - 5) \approx 25{,}497.$$

● Section 4.5 The Definite Integral as the Limit of a Sum

1. Use the Midpoint Rule with n = 4 to approximate the area of the region bounded by the graph of $f(x) = -2x + 3$ and the x-axis on the interval [0, 1]. Compare this result with the exact area obtained by using the definite integral.
   Solution: The midpoints of the four intervals are 1/8, 3/8, 5/8, and 7/8. The approximate area is
   $$A \approx \frac{1-0}{4}[f(\tfrac{1}{8}) + f(\tfrac{3}{8}) + f(\tfrac{5}{8}) + f(\tfrac{7}{8})]$$
   $$= \frac{1}{4}[\tfrac{11}{4} + \tfrac{9}{4} + \tfrac{7}{4} + \tfrac{5}{4}] = 2$$
   The exact area is
   $$A = \int_0^1 (-2x + 3)\, dx = \left[-x^2 + 3x\right]_0^1 = 2$$

3. Use the Midpoint Rule with n = 4 to approximate the area of the region bounded by the graph of $y = \sqrt{x}$ and the x-axis on the interval [0, 1]. Compare this result with the exact area obtained by using the definite integral.
   Solution: The midpoints of the four intervals are 1/8, 3/8, 5/8, and 7/8. The approximate area is
   $$A \approx \frac{1-0}{4}\left[\sqrt{\tfrac{1}{8}} + \sqrt{\tfrac{3}{8}} + \sqrt{\tfrac{5}{8}} + \sqrt{\tfrac{7}{8}}\right]$$
   $$= \frac{1}{4}\left[\frac{\sqrt{2}}{4} + \frac{\sqrt{6}}{4} + \frac{\sqrt{10}}{4} + \frac{\sqrt{14}}{4}\right] \approx 0.6730$$
   The exact area is
   $$A = \int_0^1 \sqrt{x}\, dx = \tfrac{2}{3}x^{3/2}\Big]_0^1 = \tfrac{2}{3} \approx 0.6667$$

5. Use the Midpoint Rule with n = 4 to approximate the area of the region bounded by the graph of $y = x^2 + 2$ and the x-axis on the interval [-1, 1]. Compare this result with the exact area obtained by using the definite integral. Sketch the region.
   Solution: The midpoints of the four intervals are -3/4, -1/4, 1/4, and 3/4. The approximate area is
   $$A \approx \frac{1-(-1)}{4}\left[\tfrac{41}{16} + \tfrac{33}{16} + \tfrac{33}{16} + \tfrac{41}{16}\right]$$
   $$= \frac{1}{2}\left[\tfrac{148}{16}\right] = \tfrac{37}{8} = 4.625$$
   The exact area is
   $$A = \int_{-1}^1 (x^2 + 2)\, dx = \left[\tfrac{x^3}{3} + 2x\right]_{-1}^1 = \tfrac{14}{3} \approx 4.6667$$

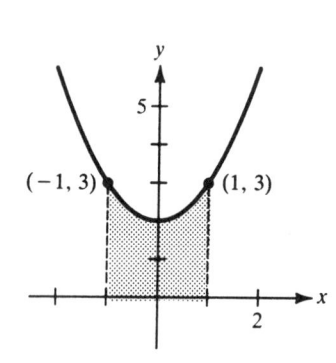

Section 4.5

7. Use the Midpoint Rule with n = 4 to approximate the area of the region bounded by the graph of $g(x) = 2x^2$ and the x-axis on the interval [1, 3]. Compare this result with the exact area obtained by using the definite integral. Sketch the region. 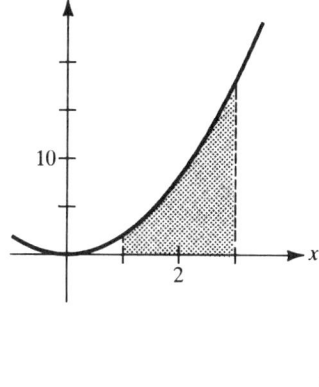
Solution: The midpoints of the four intervals are 5/4, 7/4, 9/4, and 11/4. The approximate area is

$$A \approx \frac{3-1}{4}[g(\frac{5}{4}) + g(\frac{7}{4}) + g(\frac{9}{4}) + g(\frac{11}{4})]$$

$$= \frac{1}{2}[\frac{25}{8} + \frac{49}{8} + \frac{81}{8} + \frac{121}{8}] = \frac{69}{4} = 17.25$$

The exact area is

$$A = \int_1^3 2x^2 \, dx = \frac{2x^3}{3}\Big]_1^3 = \frac{52}{3} \approx 17.3333$$

9. Use the Midpoint Rule with n = 4 to approximate the area of the region bounded by the graph of $f(x) = 1 - x^3$ and the x-axis on the interval [0, 1]. Compare this result with the exact area obtained by using the definite integral. Sketch the region. 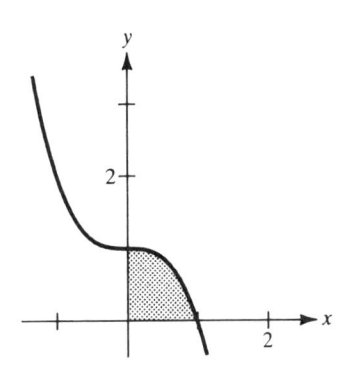
Solution: The midpoints of the four intervals are 1/8, 3/8, 5/8, and 7/8. The approximate area is

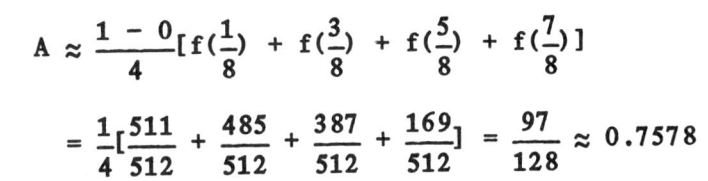

$$A \approx \frac{1-0}{4}[f(\frac{1}{8}) + f(\frac{3}{8}) + f(\frac{5}{8}) + f(\frac{7}{8})]$$

$$= \frac{1}{4}[\frac{511}{512} + \frac{485}{512} + \frac{387}{512} + \frac{169}{512}] = \frac{97}{128} \approx 0.7578$$

The exact area is

$$A = \int_0^1 (1 - x^3) \, dx = \left[x - \frac{x^4}{4}\right]_0^1$$

$$= \frac{3}{4} = 0.75$$

11. Use the Midpoint Rule with n = 4 to approximate the area of the region bounded by the graph of $y = x^2 - x^3$ and the x-axis on the interval [-1, 0]. Compare this result with the exact area obtained by using the definite integral. Sketch the region. 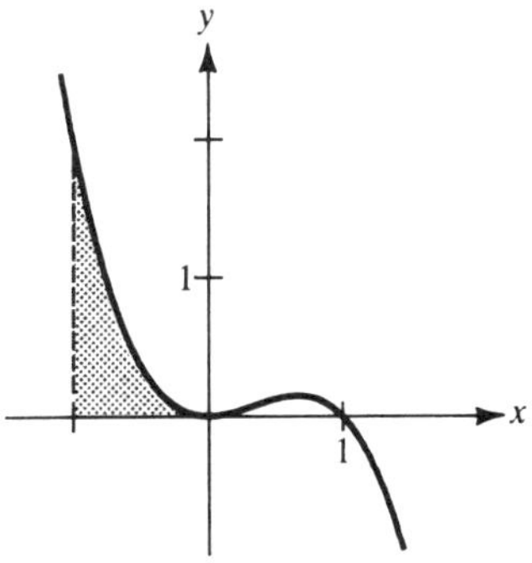
Solution: The midpoints of the four intervals are -7/8, -5/8, -3/8, and -1/8. The approximate area is

$$A \approx \frac{0 - (-1)}{4}[\frac{735}{512} + \frac{325}{512} + \frac{99}{512} + \frac{9}{512}]$$

$$= \frac{73}{128} \approx 0.5703$$

The exact area is

$$A = \int_{-1}^0 (x^2 - x^3) \, dx = \left[\frac{x^3}{3} - \frac{x^4}{4}\right]_{-1}^0$$

$$= \frac{7}{12} \approx 0.5833$$

13. Use the Midpoint Rule with n = 4 to approximate the area of the region bounded by the graph of $y = x(1 - x)^2$ and the x-axis on the interval [0, 1]. Compare this result with the exact area obtained by using the definite integral. Sketch the region. 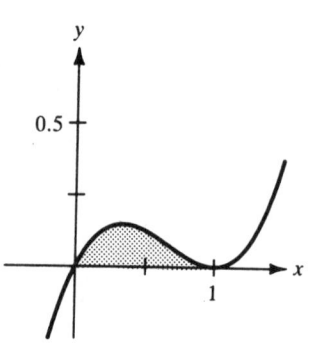
Solution: The midpoints of the four intervals are 1/8, 3/8, 5/8, and 7/8. The approximate area is

$$A \approx \frac{1-0}{4}\left[\frac{49}{512} + \frac{75}{512} + \frac{45}{512} + \frac{7}{512}\right] = \frac{11}{128} \approx 0.0859$$

The exact area is

$$A = \int_0^1 x(1 - x)^2 \, dx = \int_0^1 (x^3 - 2x^2 + x) \, dx$$

$$= \left[\frac{x^4}{4} - \frac{2x^3}{3} + \frac{x^2}{2}\right]_0^1 = \frac{1}{12} \approx 0.0833$$

15. Use the Midpoint Rule with n = 4 to approximate the area of the region between the graph of $f(y) = 3y$ and the y-axis on the interval [0, 2]. Compare this result with the exact area obtained by using the definite integral.
Solution: The midpoints of the four intervals are 1/4, 3/4, 5/4, and 7/4. The approximate area is

$$A \approx \frac{2-0}{4}\left[f\left(\frac{1}{4}\right) + f\left(\frac{3}{4}\right) + f\left(\frac{5}{4}\right) + f\left(\frac{7}{4}\right)\right]$$

$$= \frac{1}{2}\left[\frac{3}{4} + \frac{9}{4} + \frac{15}{4} + \frac{21}{4}\right] = 6$$

The exact area is

$$A = \int_0^2 3y \, dy = \frac{3y^2}{2}\Big]_0^2 = 6$$

17. Use the Midpoint Rule to approximate the given definite integral using n = 2, 4, and 8. (The exact value of the integral is 4π.)

$$\int_0^4 \sqrt{16 - x^2} \, dx$$

Solution: For n = 2 the midpoints are 1 and 3 and the approximation is

$$\int_0^4 \sqrt{16 - x^2} \, dx \approx 13.0375$$

For n = 4 the midpoints are 1/2, 3/2, 5/2, and 7/2 and the approximation is

$$\int_0^4 \sqrt{16 - x^2} \, dx \approx 12.7357$$

For n = 8 the midpoints are 1/4, 3/4, 5/4, 7/4, 9/4, 11/4, 13/4, and 15/4 and the approximation is

$$\int_0^4 \sqrt{16 - x^2} \, dx \approx 12.6267$$

Section 4.6

19. Use the Midpoint Rule to approximate the given definite integral using n = 2, 4, and 8.

$$\int_0^2 \sqrt{1 + x^3}\, dx$$

Solution: For n = 2 the midpoints are 1/2 and 3/2 and the approximation is

$$\int_0^2 \sqrt{1 + x^3}\, dx \approx 3.1523$$

For n = 4 the midpoints are 1/4, 3/4, 5/4, and 7/4 and the approximation is

$$\int_0^2 \sqrt{1 + x^3}\, dx \approx 3.2202$$

For n = 8 the midpoints are 1/8, 3/8, 5/8, 7/8, 9/8, 11/8, 13/8, and 15/8 and the approximation is

$$\int_0^2 \sqrt{1 + x^3}\, dx \approx 3.2361$$

● Section 4.6 Volumes of Solids of Revolution

1. Find the volume of the solid formed by revolving the region bounded by the graph of y = −x + 1 about the x-axis.
   Solution:
   $$V = \pi \int_0^1 (-x + 1)^2\, dx = \pi \int_0^1 (x^2 - 2x + 1)\, dx$$
   $$= \pi \left(\frac{x^3}{3} - x^2 + x\right)\Big]_0^1 = \frac{\pi}{3}$$

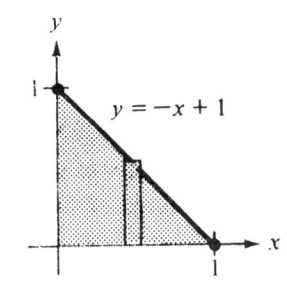

3. Find the volume of the solid formed by revolving the region bounded by the graph of $y = \sqrt{4 - x^2}$ about the x-axis.
   Solution:
   $$V = \pi \int_0^2 (\sqrt{4 - x^2})^2\, dx = \pi \int_0^2 (4 - x^2)\, dx$$
   $$= \pi\left(4x - \frac{x^3}{3}\right)\Big]_0^2 = \frac{16\pi}{3}$$

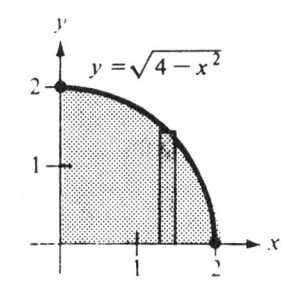

5. Find the volume of the solid formed by revolving the region bounded by the graph of $y = \sqrt{x}$ about the x-axis.
   Solution:
   $$V = \pi \int_1^4 (\sqrt{x})^2\, dx = \pi \int_1^4 x\, dx = \pi\left(\frac{x^2}{2}\right)\Big]_1^4$$
   $$= 8\pi - \frac{\pi}{2} = \frac{15\pi}{2}$$

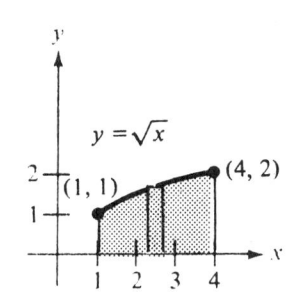

Section 4.6

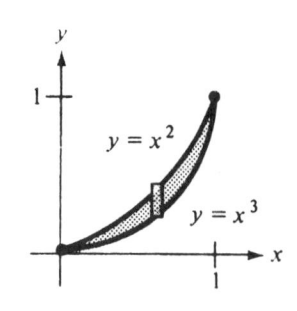

7. Find the volume of the solid formed by revolving the region bounded by the graphs of $y = x^2$ and $y = x^3$ about the x-axis.
Solution:
$$V = \pi \int_0^1 [(x^2)^2 - (x^3)^2]\, dx = \pi \int_0^1 (x^4 - x^6)\, dx$$
$$= \pi\left(\frac{x^5}{5} - \frac{x^7}{7}\right)\Big]_0^1 = \frac{2\pi}{35}$$

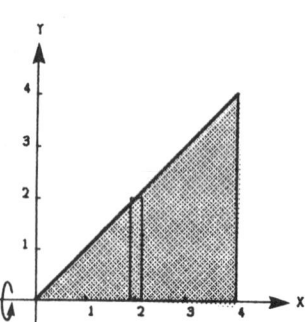

9. Find the volume of the solid formed by revolving the region bounded by the graphs of $y = x$, $y = 0$, and $x = 4$ about the x-axis.
Solution:
$$V = \pi \int_0^4 x^2\, dx = \pi\left(\frac{x^3}{3}\right)\Big]_0^4$$
$$= \frac{64\pi}{3}$$

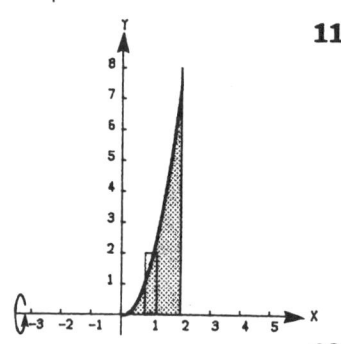

11. Find the volume of the solid formed by revolving the region bounded by the graphs of $y = 2x^2$, $y = 0$, and $x = 2$ about the x-axis.
Solution:
$$V = \pi \int_0^2 (2x^2)^2\, dx = \pi \int_0^2 4x^4\, dx = \pi\left(\frac{4x^5}{5}\right)\Big]_0^2$$
$$= \frac{128\pi}{5}$$

13. Find the volume of the solid formed by revolving the region bounded by the graphs of $y = 1/x$, $y = 0$, $x = 1$, and $x = 3$ about the x-axis.
Solution:
$$V = \pi \int_1^3 \left(\frac{1}{x}\right)^2 dx = \pi \int_1^3 \frac{1}{x^2}\, dx = -\frac{\pi}{x}\Big]_1^3$$
$$= \frac{2\pi}{3}$$

15. Find the volume of the solid formed by revolving the region bounded by the graphs of $y = 6 - 2x - x^2$ and $y = x + 6$ about the x-axis.
Solution: The points of intersection of the two graphs occur when $6 - 2x - x^2 = x + 6$, which implies that $x = 0$ or $x = -3$.
$$V = \pi \int_{-3}^0 [(6 - 2x - x^2)^2 - (x + 6)^2]\, dx$$
$$= \pi \int_{-3}^0 (x^4 + 4x^3 - 9x^2 - 36x)\, dx$$
$$= \pi\left(\frac{x^5}{5} + x^4 - 3x^3 - 18x^2\right)\Big]_{-3}^0 = \frac{243\pi}{5}$$

Section 4.6   205

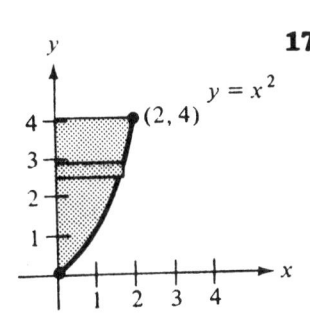

17. Find the volume of the solid formed by revolving the region bounded by the graphs of $y = x^2$, $y = 4$, and $x = 0$ about the y-axis.
Solution: The points of intersection of the two graphs occur when $y = 0$ and $y = 4$.
$$V = \pi \int_0^4 (\sqrt{y})^2 \, dy = \pi \int_0^4 y \, dy = \pi \left(\frac{y^2}{2}\right)\Big]_0^4$$
$$= 8\pi$$

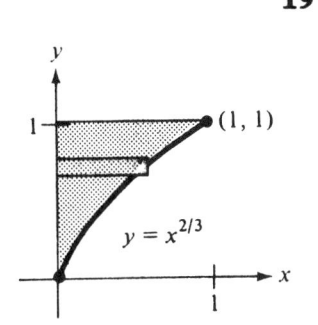

19. Find the volume of the solid formed by revolving the region bounded by the graphs $y = x^{2/3}$, $y = 1$, and $x = 0$ about the y-axis.
Solution: The points of intersection of the two graphs occur when $y = 0$ and $y = 1$.
$$V = \pi \int_0^1 (y^{3/2})^2 \, dy = \pi \int_0^1 y^3 \, dy = \pi \left(\frac{y^4}{4}\right)\Big]_0^1$$
$$= \frac{\pi}{4}$$

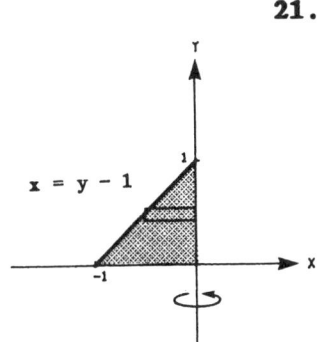

21. Find the volume of the solid formed by revolving the region bounded by the graphs of $x = y - 1$, $x = 0$, and $y = 0$ about the y-axis.
Solution:
$$V = \pi \int_0^1 (y - 1)^2 \, dy = \pi \int_0^1 (y^2 - 2y + 1) \, dy$$
$$= \pi \left[\frac{y^3}{3} - y^2 + y\right]_0^1 = \frac{\pi}{3}$$

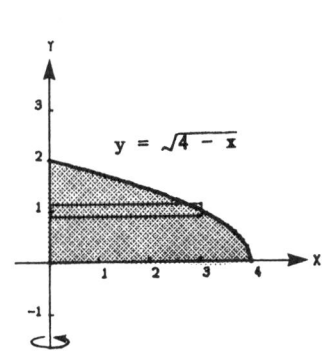

23. Find the volume of the solid formed by revolving the region bounded by the graphs of $y = \sqrt{4 - x}$, $y = 0$, and $x = 0$ about the y-axis.
Solution: The points of intersection of the three graphs occur when $y = 0$ and $y = 2$.
$$V = \pi \int_0^2 (4 - y^2)^2 \, dy = \pi \int_0^2 (16 - 8y^2 + y^4) \, dy$$
$$= \pi \left[16y - \frac{8y^3}{3} + \frac{y^5}{5}\right]_0^2 = \pi \left[32 - \frac{64}{3} + \frac{32}{5}\right]$$
$$= \frac{256\pi}{15}$$

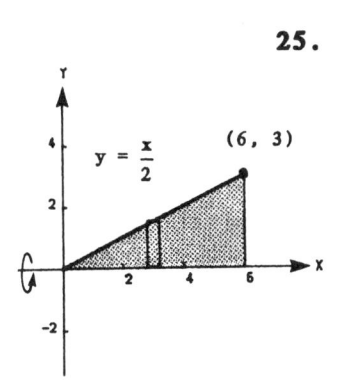

25. If the portion of the line $y = (1/2)x$ lying in the first quadrant is revolved about the x-axis, a cone is generated. Find the volume of the cone formed by the line extending from $x = 0$ to $x = 6$.
Solution:
$$V = \pi \int_0^6 \left(\frac{1}{2}x\right)^2 dx = \frac{\pi}{4} \int_0^6 x^2 \, dx$$
$$= \frac{\pi}{4}\left(\frac{x^3}{3}\right)\Big]_0^6 = 18\pi$$

27. Use the Disc Method to verify that the volume of a sphere of radius r is

$$V = (4/3)\pi r^3$$

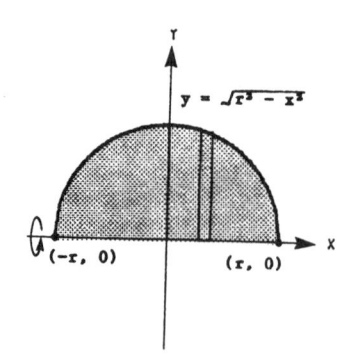

Solution: A sphere of radius r can be formed by revolving the graph of $y = \sqrt{r^2 - x^2}$ about the x-axis.

$$V = \pi \int_{-r}^{r} (\sqrt{r^2 - x^2})^2 \, dx = \pi \int_{-r}^{r} (r^2 - x^2) \, dx$$

$$= \pi \left[r^2 x - \frac{x^3}{3} \right]_{-r}^{r}$$

$$= \pi \left[(r^3 - \frac{r^3}{3}) - (-r^3 + \frac{r^3}{3}) \right]$$

$$= \frac{4\pi r^3}{3}$$

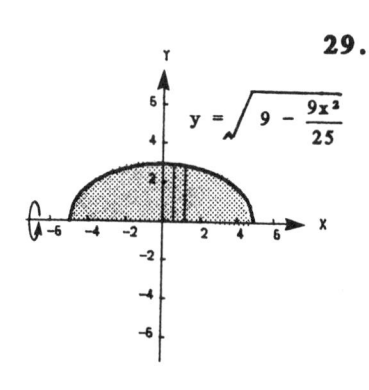

29. The upper half of the ellipse $9x^2 + 25y^2 = 225$ is revolved about the x-axis to form a prolate spheroid (shaped like a football). Find the volume of the spheroid.

Solution: The upper half of the ellipse is given by $y = \sqrt{9 - (9/25)x^2}$.

$$V = \pi \int_{-5}^{5} (9 - \frac{9}{25} x^2) \, dx = 18\pi \int_{0}^{5} (1 - \frac{1}{25} x^2) \, dx$$

$$= 18\pi (x - \frac{x^3}{75}) \Big]_{0}^{5} = 60\pi$$

31. Solve the problem given in Example 3 of this section when the pond is 15 feet deep at its center and has a radius of 150 feet.

Solution: We can find the volume by revolving the region bounded by the graphs of

$$y = 15[(\frac{x}{150})^2 - 1]$$

and $x = 0$ about the y-axis. By solving for x^2, we have $x^2 = 1{,}500(y + 15)$.

$$V = \pi \int_{-15}^{0} 1500(y + 15) \, dy = 1500\pi \left[\frac{y^2}{2} + 15y \right]_{-15}^{0}$$

$$= 1500\pi [0 - (\frac{225}{2} - 225)]$$

$$= 168{,}750\pi \text{ cubic feet}$$

Therefore, the maximum number of fish is

$$168{,}750\pi/500 \approx 1060 \text{ fish}$$

Review Exercises for Chapter 4

1. Evaluate $\int (2x^2 + x - 1)\, dx$.

Solution:
$$\int (2x^2 + x - 1)\, dx = \frac{2}{3}x^3 + \frac{1}{2}x^2 - x + C$$

3. Evaluate $\int \frac{x^2 + 3}{x^2}\, dx$.

Solution:
$$\int \frac{x^2 + 3}{x^2}\, dx = \int [1 + 3x^{-2}]\, dx$$
$$= x + \frac{3x^{-1}}{-1} + C = x - \frac{3}{x} + C$$

5. Evaluate $\int \frac{x^2}{(x^3 - 1)^2}\, dx$.

Solution:
$$\int \frac{x^2}{(x^3 - 1)^2}\, dx = \frac{1}{3}\int (x^3 - 1)^{-2}(3x^2)\, dx$$
$$= (\frac{1}{3})\frac{(x^3 - 1)^{-1}}{-1} + C$$
$$= -\frac{1}{3(x^3 - 1)} + C$$

7. Evaluate $\int \frac{x^3 - 2x^2 + 1}{x^2}\, dx$.

Solution:
$$\int \frac{x^3 - 2x^2 + 1}{x^2}\, dx = \int (x - 2 + x^{-2})\, dx$$
$$= \frac{1}{2}x^2 - 2x + \frac{x^{-1}}{-1} + C$$
$$= \frac{1}{2}x^2 - 2x - \frac{1}{x} + C$$

9. Evaluate $\int \frac{2}{3\sqrt[3]{x}}\, dx$.

Solution:
$$\int \frac{2}{3\sqrt[3]{x}}\, dx = \frac{2}{3}\int x^{-1/3}\, dx = \frac{2}{3}(\frac{x^{2/3}}{2/3}) + C$$
$$= x^{2/3} + C$$

11. Evaluate $\int \dfrac{(1+x)^2}{\sqrt{x}}\, dx$.

Solution:

$$\int \dfrac{(1+x)^2}{\sqrt{x}}\, dx = \int x^{-1/2}(1 + 2x + x^2)\, dx$$

$$= \int (x^{-1/2} + 2x^{1/2} + x^{3/2})\, dx$$

$$= \dfrac{x^{1/2}}{1/2} + \dfrac{2x^{3/2}}{3/2} + \dfrac{x^{5/2}}{5/2} + C$$

$$= 2\sqrt{x} + \dfrac{4}{3}x\sqrt{x} + \dfrac{2}{5}x^2\sqrt{x} + C$$

$$= \dfrac{2\sqrt{x}}{15}(15 + 10x + 3x^2) + C$$

13. Evaluate $\int \dfrac{x^2}{\sqrt{x^3 + 3}}\, dx$.

Solution:

$$\int \dfrac{x^2}{\sqrt{x^3 + 3}}\, dx = \dfrac{1}{3}\int (x^3 + 3)^{-1/2}(3x^2)\, dx$$

$$= \left(\dfrac{1}{3}\right)\dfrac{(x^3 + 3)^{1/2}}{1/2} + C$$

$$= \dfrac{2\sqrt{x^3 + 3}}{3} + C$$

15. Evaluate $\int \dfrac{3x}{\sqrt{1 - 2x^2}}\, dx$.

Solution:

$$\int \dfrac{3x}{\sqrt{1 - 2x^2}}\, dx = -\dfrac{3}{4}\int (1 - 2x^2)^{-1/2}(-4x)\, dx$$

$$= \left(-\dfrac{3}{4}\right)\dfrac{(1 - 2x^2)^{1/2}}{1/2} + C$$

$$= -\dfrac{3}{2}\sqrt{1 - 2x^2} + C$$

17. Evaluate $\int (x^2 + 1)^3\, dx$.

Solution:

$$\int (x^2 + 1)^3\, dx = \int (x^6 + 3x^4 + 3x^2 + 1)\, dx$$

$$= \dfrac{x^7}{7} + \dfrac{3x^5}{5} + x^3 + x + C$$

Review Exercises for Chapter 4

19. Evaluate $\int \sqrt{x}(x + 3)\, dx$.

Solution:

$$\int \sqrt{x}(x + 3)\, dx = \int (x^{3/2} + 3x^{1/2})\, dx$$

$$= \frac{x^{5/2}}{5/2} + \frac{3x^{3/2}}{3/2} + C = \frac{2}{5}x^2\sqrt{x} + 2x\sqrt{x} + C$$

$$= \frac{2x^{3/2}}{5}(x + 5) + C$$

21. Evaluate $\int_0^4 (2 + x)\, dx$.

Solution:

$$\int_0^4 (2 + x)\, dx = \left[2x + \frac{x^2}{2} \right]_0^4 = 16$$

23. Evaluate $\int_{-1}^1 (4t^3 - 2t)\, dt$.

Solution:

$$\int_{-1}^1 (4t^3 - 2t)\, dt = \left[t^4 - t^2 \right]_{-1}^1 = 0$$

25. Evaluate $\int_0^3 \frac{1}{\sqrt{1 + x}}\, dx$.

Solution:

$$\int_0^3 \frac{1}{\sqrt{1 + x}}\, dx = \int_0^3 (1 + x)^{-1/2}\, dx = 2\sqrt{1 + x}\, \Big]_0^3 = 2$$

27. Evaluate $\int_4^9 x\sqrt{x}\, dx$.

Solution:

$$\int_4^9 x\sqrt{x}\, dx = \int_4^9 x^{3/2}\, dx = \frac{2}{5}x^{5/2}\, \Big]_4^9$$

$$= \frac{2}{5}[243 - 32] = \frac{422}{5}$$

29. Evaluate $2\pi \int_{-1}^0 x^2(x + 1)^2\, dx$

Solution:

$$2\pi \int_{-1}^0 x^2(x + 1)^2\, dx = 2\pi \int_{-1}^0 x^2(x^2 + 2x + 1)\, dx$$

$$= 2\pi \int_{-1}^0 (x^4 + 2x^3 + x^2)\, dx$$

$$= 2\pi \left[\frac{x^5}{5} + \frac{x^4}{2} + \frac{x^3}{3} \right]_{-1}^0 = \frac{\pi}{15}$$

31. Set up a definite integral that yields the area of the region bounded by the graphs of $f(x) = 3$ and $y = 0$ on the interval $[0, 5]$.
Solution:
$$A = \int_0^5 3 \, dx$$

33. Set up a definite integral that yields the area of the region bounded by graphs of $f(x) = x^2$ and $y = 0$ on the interval $[0, 2]$.
Solution:
$$A = \int_0^2 x^2 \, dx$$

35. Set up a definite integral that yields the area of the region bounded by the graphs of $f(x) = 4 - x^2$ and $y = 0$.
Solution:
$$A = \int_{-2}^2 (4 - x^2) \, dx = 2 \int_0^2 (4 - x^2) \, dx$$

37. Set up a definite integral that yields the area of the region bounded by the graphs of $f(y) = y^3$ and $x = 0$ on the interval $[0, 2]$.
Solution:
$$A = \int_0^2 y^3 \, dy$$

39. Sketch the graph of the region whose area is given by the integral, and find the area.
$$\int_0^2 (2 - x) \, dx$$
Solution:
$$\int_0^2 (2 - x) \, dx = \left[2x - \frac{x^2}{2} \right]_0^2 = 2$$

41. Sketch the graph of the region whose area is given by the integral, and find the area.
$$\int_1^4 \frac{1}{\sqrt{x}} \, dx$$
Solution:
$$\int_1^4 \frac{1}{\sqrt{x}} \, dx = \int_1^4 x^{-1/2} \, dx = 2\sqrt{x} \,\Big]_1^4 = 2$$

43. Sketch the graph of the region whose area is given by the integral, and find the area.
$$\int_0^4 \frac{y}{2} \, dy$$

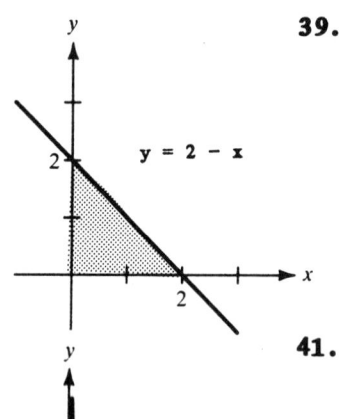

Solution:

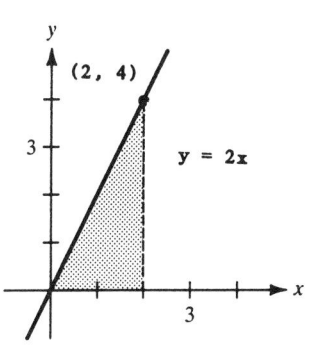

$$\int_0^4 \frac{y}{2}\, dy = \frac{y^2}{4}\Big]_0^4 = 4$$

45. If $\int_2^6 f(x)\, dx = 10$ and $\int_2^6 g(x)\, dx = -2$, find

(a) $\int_2^6 [f(x) + g(x)]\, dx$ (b) $\int_2^6 [g(x) - f(x)]\, dx$

(c) $\int_2^6 [2f(x) - 3g(x)]\, dx$ (d) $\int_2^6 3f(x)\, dx$

Solution:

(a) $\int_2^6 [f(x) + g(x)]\, dx = \int_2^6 f(x)\, dx + \int_2^6 g(x)\, dx$
$$= 10 + (-2) = 8$$

(b) $\int_2^6 [g(x) - f(x)]\, dx = \int_2^6 g(x)\, dx - \int_2^6 f(x)\, dx$
$$= -2 - 10 = -12$$

(c) $\int_2^6 [2f(x) - 3g(x)]\, dx = 2\int_2^6 f(x)\, dx - 3\int_2^6 g(x)\, dx$
$$= 2(10) - 3(-2) = 26$$

(d) $\int_2^6 3f(x)\, dx = 3\int_2^6 f(x)\, dx = 3(10) = 30$

47. Approximate the definite integral by the Midpoint Rule, letting n = 4.

$$\int_0^2 x^3\, dx$$

Solution:

$$\int_0^2 x^3\, dx \approx \frac{2-0}{4}[f(\tfrac{1}{4}) + f(\tfrac{3}{4}) + f(\tfrac{5}{4}) + f(\tfrac{7}{4})]$$

$$= \frac{1}{2}[\frac{1}{64} + \frac{27}{64} + \frac{125}{64} + \frac{343}{64}] = \frac{31}{8} = 3.875$$

49. Approximate the definite integral by the Midpoint Rule, letting n = 4.

$$\int_0^1 \sqrt{1 - x^2}\, dx$$

Solution:

$$\int_0^1 \sqrt{1 - x^2}\, dx \approx \frac{1-0}{4}[f(\tfrac{1}{8}) + f(\tfrac{3}{8}) + f(\tfrac{5}{8}) + f(\tfrac{7}{8})]$$

$$= \frac{1}{4}[\frac{\sqrt{63}}{8} + \frac{\sqrt{55}}{8} + \frac{\sqrt{39}}{8} + \frac{\sqrt{15}}{8}]$$

$$\approx 0.7960$$

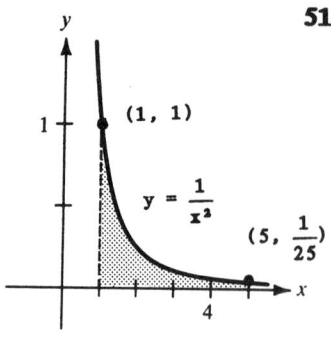

51. Sketch the region bounded by the graphs of $y = 1/x^2$, $y = 0$, $x = 1$, and $x = 5$ and determine the area of the region.
Solution:
$$A = \int_1^5 \frac{1}{x^2}\, dx = -\frac{1}{x}\Big]_1^5$$
$$= \frac{4}{5}$$

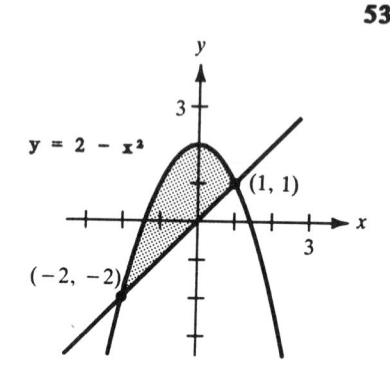

53. Sketch the region bounded by the graphs of $y = x$ and $y = 2 - x^2$ and determine the area of the region.
Solution: The points of intersection of the two graphs occur when $x = 2 - x^2$ which implies that $x = -2$ and $x = 1$.
$$A = \int_{-2}^1 [(2 - x^2) - x]\, dx = \left[2x - \frac{x^3}{3} - \frac{x^2}{2}\right]_{-2}^1$$
$$= \left(2 - \frac{1}{3} - \frac{1}{2}\right) - \left(-4 + \frac{8}{3} - 2\right) = \frac{9}{2}$$

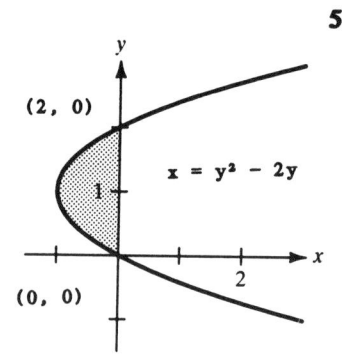

55. Sketch the region bounded by the graphs of $x = y^2 - 2y$ and $x = 0$ and determine the area of the region.
Solution: The points of intersection occur when $y = 0$ and $y = 2$.
$$A = \int_0^2 [0 - (y^2 - 2y)]\, dy$$
$$= \left[-\frac{y^3}{3} + y^2\right]_0^2 = \frac{4}{3}$$

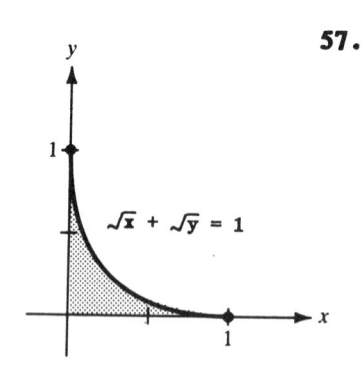

57. Sketch the region bounded by the graphs of $\sqrt{x} + \sqrt{y} = 1$, $y = 0$, and $x = 0$ and determine the area of the region.
Solution: The first equation yields $y = (1 - \sqrt{x})^2$.
$$A = \int_0^1 (1 - \sqrt{x})^2\, dx = \int_0^1 (1 - 2\sqrt{x} + x)\, dx$$
$$= \left[x - \frac{4}{3}x^{3/2} + \frac{1}{2}x^2\right]_0^1 = \frac{1}{6}$$

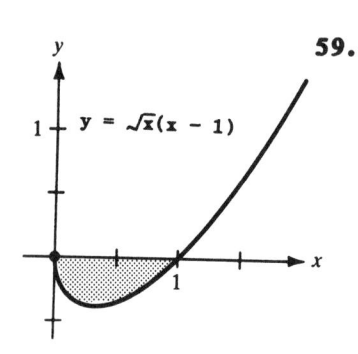

59. Sketch the region bounded by the graphs of $y = \sqrt{x}(x - 1)$ and $y = 0$ and determine the area of the region.
Solution:
$$A = \int_0^1 [0 - \sqrt{x}(x - 1)]\, dx = \int_0^1 [-x^{3/2} + x^{1/2}]\, dx$$
$$= \left[-\frac{2}{5}x^{5/2} + \frac{2}{3}x^{3/2}\right]_0^1 = \frac{4}{15}$$

Review Exercises for Chapter 4

61. Find the average value of $f(x) = 1/\sqrt{x - 1}$ over the interval $[5, 10]$. Find the values of x where the function assumes its mean value and sketch the graph of the function.
Solution:

$$\text{Average Value} = \frac{1}{10 - 5} \int_5^{10} \frac{1}{\sqrt{x - 1}} \, dx$$

$$= \frac{2}{5} \sqrt{x - 1} \Big]_5^{10} = \frac{2}{5}$$

To find the x-values for which $f(x) = 2/5$, we let $1/\sqrt{x - 1} = 2/5$ and solve for x.

$$\sqrt{x - 1} = 5/2$$
$$x - 1 = 25/4$$
$$x = 29/4$$

63. Find the volume of the solid generated by revolving the region bounded by the graphs of $y = \sqrt{16 - x}$, $y = 0$, and $x = 0$ about (a) the x-axis and (b) the y-axis.
Solution:

(a) $V = \pi \int_0^{16} (\sqrt{16 - x})^2 \, dx = \pi \int_0^{16} (16 - x) \, dx$

$$= \pi \left[16x - \frac{x^2}{2} \right]_0^{16} = 128\pi$$

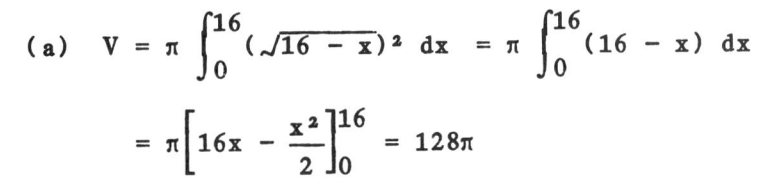

(b) $y = \sqrt{16 - x}$ implies that $x = 16 - y^2$.

$$V = \pi \int_0^4 (16 - y^2)^2 \, dy = \pi \int_0^4 (256 - 32y^2 + y^4) \, dy$$

$$= \pi \left[256y - \frac{32y^3}{3} + \frac{y^5}{5} \right]_0^4$$

$$= \frac{8192\pi}{15}$$

65. Find the function f whose derivative is $f'(x) = -2x$ and whose graph passes through the point $(-1, 1)$.
Solution:

$$f(x) = \int -2x \, dx = -x^2 + C$$

Since $f(-1) = 1$, it follows that $C = 2$. Therefore, we have

$$f(x) = -x^2 + 2 = 2 - x^2$$

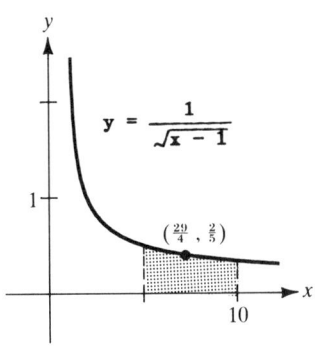

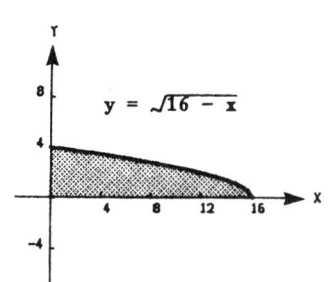

67. An airplane taking off from a runway travels 3600 feet before lifting off. If it starts from rest, moves with constant acceleration, and makes the run in 30 seconds, with what velocity does it lift off?
Solution: Since the acceleration is constant, we have $s''(t) = a(t) = a$, and the velocity is

$$s'(t) = v(t) = \int a(t)\, dt = \int a\, dt = at + C_1$$

Since it starts from rest, we have $v(0) = 0$, which implies that $C_1 = 0$ and $s'(t) = at$. Therefore, the position function is

$$s(t) = \int at\, dt = \frac{at^2}{2} + C_2$$

Since $s(0) = 0$, we have $C_2 = 0$ and $s(t) = at^2/2$. When $t = 30$, the position is

$$s(30) = 450a = 3600$$

which implies that $a = 8$ ft/sec². Thus $v(t) = at = 8t$ and $v(30) = 240$ ft/sec

69. Suppose that gasoline is increasing in price according to the equation

$$p = 1.00 + 0.1t + 0.02t^2$$

where p is the dollar price per gallon and $t = 0$ represents the year 1985. If an automobile is driven 15,000 miles a year and gets M miles per gallon, then the annual fuel cost is

$$C = \frac{15{,}000}{M} \int_t^{t+1} p\, dt$$

Find the annual fuel cost for the years
(a) 1987   (b) 1990
Solution:
(a) For 1987, $t = 2$

$$C = \frac{15{,}000}{M} \int_2^3 (1.00 + 0.1t + 0.02t^2)\, dt$$

$$= \frac{15{,}000}{M} \left[1.00t + 0.05t^2 + \frac{0.02}{3} t^3 \right]_2^3$$

$$= \frac{20{,}650}{M}$$

(b) For 1990, $t = 5$

$$C = \frac{15{,}000}{M} \left[1.00t + 0.05t^2 + \frac{0.02}{3} t^3 \right]_5^6$$

$$= \frac{32{,}350}{M}$$

PRACTICE TEST FOR CHAPTER 4

1. Evaluate $\int (3x^2 - 8x + 5)\, dx$.

2. Evaluate $\int (x + 7)(x^2 - 4)\, dx$.

3. Evaluate $\int \dfrac{x^3 - 9x^2 + 1}{x^2}\, dx$.

4. Evaluate $\int x^3 \sqrt[4]{1 - x^4}\, dx$.

5. Evaluate $\int \dfrac{3}{\sqrt[3]{7x}}\, dx$.

6. Evaluate $\int \sqrt{6 - 11x}\, dx$.

7. Evaluate $\int (\sqrt[4]{x} + \sqrt[6]{x})\, dx$

8. Evaluate $\int \left(\dfrac{1}{x^4} - \dfrac{1}{x^5}\right) dx$.

9. Evaluate $\int (1 - x^2)^3\, dx$.

10. Evaluate $\int \dfrac{5x}{(1 + 3x^2)^3}\, dx$.

11. Evaluate $\int_0^3 (x^2 - 4x + 2)\, dx$.

12. Evaluate $\int_1^8 x\sqrt[3]{x}\, dx$

13. Evaluate $\int_2^{\sqrt{13}} \dfrac{x}{\sqrt{x^2 - 4}}\, dx$

14. Sketch the region bounded by the graphs of $f(x) = x^2 - 6x$ and $g(x) = 0$ and find the area of the region.

15. Sketch the region bounded by the graphs of $f(x) = x^3 + 1$ and $g(x) = x + 1$ and find the area of the region.

16. Sketch the region bounded by the graphs of $f(y) = 1/y^2$, $x = 0$, $y = 1$, and $y = 3$ and find the area of the region.

17. Approximate the definite integral by the Midpoint Rule using n = 4.

$$\int_0^1 \sqrt{x^3 + 2} \, dx$$

18. Approximate the definite integral by the Midpoint Rule using n = 4.

$$\int_3^4 \frac{1}{x^2 - 5} \, dx$$

19. Find the volume of the solid generated by revolving the region bounded by the graphs of

$$f(x) = \frac{1}{\sqrt[3]{x}}, \quad x = 1, \quad x = 8, \quad y = 0$$

about the x-axis.

20. Find the volume of the solid generated by revolving the region bounded by the graphs of

$$y = \sqrt{25 - x}, \quad y = 0, \quad x = 0$$

about the y-axis.

Chapter 5   Exponential and Logarithmic Functions

Section 5.1   Exponential Functions

1. Evaluate
   (a) $5(5^3)$ (b) $27^{2/3}$
   (c) $64^{3/4}$ (d) $81^{1/2}$
   (e) $25^{3/2}$ (f) $32^{2/5}$

   Solution:
   (a) $5(5^3) = 5^4 = 625$
   (b) $27^{2/3} = (\sqrt[3]{27})^2 = 3^2 = 9$
   (c) $64^{3/4} = (2^6)^{3/4} = 2^{9/2} = \sqrt{2^9} = 2^4 \sqrt{2} = 16\sqrt{2}$
   (d) $81^{1/2} = \sqrt{81} = 9$
   (e) $25^{3/2} = (\sqrt{25})^3 = 5^3 = 125$
   (f) $32^{2/5} = (\sqrt[5]{32})^2 = 2^2 = 4$

3. Use the properties of exponents to simplify
   (a) $(5^2)(5^3)$ (b) $(5^2)(5^{-3})$
   (c) $(5^2)^2$ (d) 5^{-3}

   Solution:
   (a) $(5^2)(5^3) = 5^5 = 3125$
   (b) $(5^2)(5^{-3}) = 5^{-1} = \dfrac{1}{5}$
   (c) $(5^2)^2 = 5^4 = 625$
   (d) $5^{-3} = \dfrac{1}{5^3} = \dfrac{1}{125}$

5. Use the properties of exponents to simplify
   (a) $\dfrac{5^3}{25^2}$ (b) $(9^{2/3})(3)(3^{2/3})$
   (c) $[(25^{1/2})(25^2)]^{1/5}$ (d) $(8^2)(4^3)$

   Solution:
   (a) $\dfrac{5^3}{25^2} = \dfrac{5^3}{(5^2)^2} = \dfrac{5^3}{5^4} = \dfrac{1}{5}$
   (b) $(9^{2/3})(3)(3^{2/3}) = (3^2)^{2/3}(3)(3^{2/3}) = (3^{4/3})(3^{5/3})$
       $= 3^{9/3} = 3^3 = 27$
   (c) $[(25^{1/2})(25^2)]^{1/5} = [25^{5/2}]^{1/5} = 25^{1/2} = \sqrt{25} = 5$
   (d) $(8^2)(4^3) = (64)(64) = 4096$

7. Use the properties of exponents to simplify
   (a) $e^2(e^4)$ (b) $(e^3)^4$
   (c) $(e^3)^{-2}$ (d) $\dfrac{e^5}{e^3}$

   Solution:
   (a) $e^2(e^4) = e^6$ (b) $(e^3)^4 = e^{12}$
   (c) $(e^3)^{-2} = e^{-6} = \dfrac{1}{e^6}$ (d) $\dfrac{e^5}{e^3} = e^2$

9. If $3^x = 81$, solve for x.
   Solution:
   $$3^x = 81 = 3^4$$
   $$x = 4$$

11. If $(1/3)^{x-1} = 27$, solve for x.
    Solution:
    $$\left(\frac{1}{3}\right)^{x-1} = 27$$
    $$3^{-(x-1)} = 3^3$$
    $$-(x - 1) = 3$$
    $$-x + 1 = 3$$
    $$x = -2$$

13. If $4^3 = (x + 2)^3$, solve for x.
    Solution:
    $$4^3 = (x + 2)^3$$
    $$4 = x + 2$$
    $$x = 2$$

15. If $x^{3/4} = 8$, solve for x.
    Solution:
    $$x = 8^{4/3} = (\sqrt[3]{8})^4 = 2^4 = 16$$

17. If $e^{-2x} = e^5$, solve for x.
    Solution:
    $$-2x = 5$$
    $$x = -5/2$$

19. Match $f(x) = 3^x$ with the correct graph.
    Solution: The graph is an exponential curve with the following characteristics.

    Passes through (0, 1), (1, 3), (-1, 1/3)
    Horizontal asymptote: x-axis

    Therefore, it matches graph (e).

21. Match $f(x) = -3^x$ with the correct graph.
    Solution: The graph of $f(x) = -3^x = (-1)(3^x)$ is an exponential curve with the following characteristics.

    Passes through (0, -1), (1, -3), (-1, -1/3)
    Horizontal asymptote: x-axis

    Therefore, it matches graph (a).

23. Match $f(x) = 3^{-x} - 1$ with the correct graph.
    Solution: The graph of $f(x) = 3^{-x} - 1 = (1/3)^x - 1$ is an exponential curve with the following characteristics.

    Passes through (0, 0), (1, -2/3), (-1, 2)
    Horizontal asymptote: $y = -1$

    Therefore, it matches graph (d).

Section 5.1

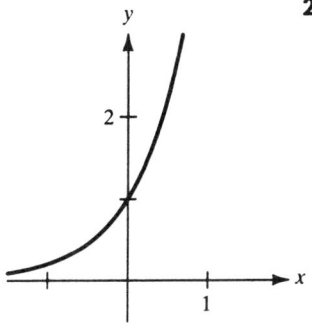

25. Sketch the graph of $f(x) = 5^x$.
Solution:

x	-2	-1	0	1	2
y	1/25	1/5	1	5	25

27. Sketch the graph of $f(x) = (1/5)^x = 5^{-x}$.
Solution:

x	-2	-1	0	1	2
y	25	5	1	1/5	1/25

29. Sketch the graph of $y = 3^{-x^2}$.
Solution:

x	-2	-1	0	1	2
y	1/81	1/3	1	1/3	1/81

31. Sketch the graph of $y = 3^{-|x|}$.
Solution:

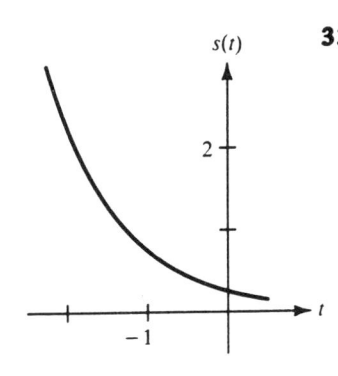

x	-2	-1	0	1	2
y	1/9	1/3	1	1/3	1/9

33. Sketch the graph of $s(t) = (3^{-t})/4$.
Solution:

$$s(t) = \frac{3^{-t}}{4} = \frac{1}{4(3^t)}$$

t	-2	-1	0	1	2
s(t)	9/4	3/4	1/4	1/12	1/36

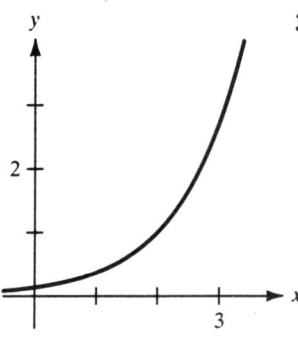

35. Sketch the graph of $h(x) = e^{x-2}$.
Solution:

x	-1	0	1	2	3
h(x)	0.050	0.135	0.368	1	2.178

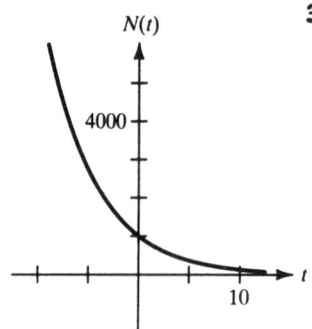

37. Sketch the graph of $N(t) = 1000e^{-0.2t}$.
Solution:

t	-5	0	5	10	20
N(t)	2718.3	1000	367.9	135.3	18.3

39. Sketch the graph of $g(x) = 2/(1 + e^{x^2})$.
Solution:

x	-2	-1	0	1	2
g(x)	0.036	0.538	1	0.538	0.036

41. Find the amount of an investment of $1000 invested at 10% for 10 years if the interest is compounded (a) annually, (b) semiannually, (c) quarterly, (d) monthly, (e) daily, and (f) continuously.
Solution:
(a) $A = 1000[1 + (0.10/1)]^{1(10)} = \$2593.74$
(b) $A = 1000[1 + (0.10/2)]^{2(10)} = \$2653.30$
(c) $A = 1000[1 + (0.10/4)]^{4(10)} = \$2685.06$
(d) $A = 1000[1 + (0.10/12)]^{12(10)} = \$2707.04$
(e) $A = 1000[1 + (0.10/365)]^{365(10)} = \$2717.91$
(f) $A = 1000e^{0.10(10)} = \$2718.28$

43. Find the investment that would be required at 12% compounded continuously to yield an amount of $100,000 in (a) 1 year, (b) 10 years, (c) 20 years, and (d) 50 years.
Solution:

$$A = 100,000, \quad r = 0.12, \quad P = \frac{A}{e^{rt}}$$

(a) $P = 100,000/e^{0.12(1)} = \$88,692.04$
(b) $P = 100,000/e^{0.12(10)} = \$30,119.42$
(c) $P = 100,000/e^{0.12(20)} = \$9,071.80$
(d) $P = 100,000/e^{0.12(50)} = \$247.88$

45. The demand function for a certain product is given by

$$p = 5000\left(1 - \frac{4}{4 + e^{-0.002x}}\right)$$

Find the price of the product if the quantity demanded is (a) x = 100 units, and (b) x = 500 units.
Solution:

(a) $p(100) = 5000\left(1 - \frac{4}{4 + e^{-0.002(100)}}\right) \approx \$849.53$

(b) $p(500) = 5000\left(1 - \frac{4}{4 + e^{-0.002(500)}}\right) \approx \$421.12$

47. The average time between incoming calls at a switchboard is 3 minutes. If a call has just come in, the probability that the next call will come within the next t minutes is given by $P(t) = 1 - e^{-t/3}$. Find (a) P(1/2), (b) P(2), and (c) P(5).
Solution:
(a) $P(1/2) = 1 - e^{-1/6} \approx 0.1535 = 15.35\%$
(b) $P(2) = 1 - e^{-2/3} \approx 0.4866 = 48.66\%$
(c) $P(5) = 1 - e^{-5/3} \approx 0.8111 = 81.11\%$

49. The population of a bacterial culture is given by the logistics growth function $y = 850/(1 + e^{-0.2t})$ where y is the number of bacteria and t is the time in days.
(a) Find the limit of this function as t approaches infinity.
(b) Sketch the graph of this function.
Solution:

(a) $\lim_{t \to \infty} \frac{850}{1 + e^{-0.2t}} = \frac{850}{1 + 0} = 850$

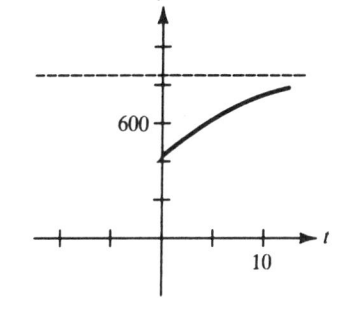

(b)

t	0	5	10	15
y	425	621.4	748.7	809.7

51. In a group project in learning theory, a mathematical model for the proportion P of correct responses after n trials was found to be

$$P = \frac{0.83}{1 + e^{-0.2n}}$$

(a) Find the proportion of correct responses after 10 trials.
(b) Find the limiting proportion of correct responses as n approaches infinity.
Solution:
(a) $P(10) = \frac{0.83}{1 + e^{-0.2(10)}} \approx 0.731$

(b) $\lim_{n \to \infty} \frac{0.83}{1 + e^{-0.2n}} = \frac{0.83}{1 + 0} = 0.83$

Section 5.2 Differentiation and Integration of Exponential Functions

1. Find the slope of the tangent line to $y = e^{3x}$ at the point $(0, 1)$.
Solution:
$$y' = 3e^{3x}$$
$$y'(0) = 3$$

3. Find the slope of the tangent line to $y = e^x$ at the point $(0, 1)$.
Solution:
$$y' = e^x$$
$$y'(0) = 1$$

5. Find the slope of the tangent line to $y = e^{-2x}$ at the point $(0, 1)$.
Solution:
$$y' = -2e^{-2x}$$
$$y'(0) = -2$$

7. Find the derivative of $y = e^{2x}$.
Solution:
$$y' = 2e^{2x}$$

9. Find the derivative of $y = e^{-2x+x^2}$.
Solution:
$$y' = (-2 + 2x)e^{-2x+x^2} = 2(x - 1)e^{-2x+x^2}$$

11. Find the derivative of $f(x) = e^{1/x}$.
Solution:
$$f'(x) = (-\frac{1}{x^2})e^{1/x} = -\frac{e^{1/x}}{x^2}$$

13. Find the derivative of $g(x) = e^{\sqrt{x}}$.
Solution:
$$g'(x) = (\frac{1}{2}x^{-1/2})e^{\sqrt{x}} = \frac{e^{\sqrt{x}}}{2\sqrt{x}}$$

15. Find the derivative of $f(x) = (x + 1)e^{3x}$.
Solution:
$$f'(x) = (x + 1)(3e^{3x}) + (1)e^{3x}$$
$$= e^{3x}[(x + 1)(3) + 1] = e^{3x}(3x + 4)$$

17. Find the derivative of $y = (e^{-x} + e^x)^3$.
Solution:
$$y' = 3(e^{-x} + e^x)^2(-e^{-x} + e^x)$$
$$= 3(e^x - e^{-x})(e^{-x} + e^x)^2$$

19. Find the derivative of $f(x) = 2/(e^x + e^{-x})$.
Solution:
$$f(x) = \frac{2}{e^x + e^{-x}} = 2(e^x + e^{-x})^{-1}$$
$$f'(x) = -2(e^x + e^{-x})^{-2}(e^x - e^{-x}) = -\frac{2(e^x - e^{-x})}{(e^x + e^{-x})^2}$$

Section 5.2

21. Find the derivative of $y = xe^x - e^x$.
Solution:
$$y' = xe^x + e^x - e^x = xe^x$$

23. Find the second derivative of $f(x) = 2e^{3x} + 3e^{-2x}$.
Solution:
$$f'(x) = 6e^{3x} - 6e^{-2x}$$
$$f''(x) = 18e^{3x} + 12e^{-2x} = 6(3e^{3x} + 2e^{-2x})$$

25. Find the second derivative of $g(x) = (1 + 2x)e^{4x}$.
Solution:
$$g'(x) = (1 + 2x)(4e^{4x}) + 2e^{4x}$$
$$= 2e^{4x}[(1 + 2x)(2) + 1] = 2e^{4x}(4x + 3)$$
$$g''(x) = 2e^{4x}(4) + 8e^{4x}(4x + 3)$$
$$= 8e^{4x}[1 + (4x + 3)] = 32e^{4x}(x + 1)$$

27. Find the extrema and the points of inflection and sketch the graph of $f(x) = 2/(1 + e^{-x})$.
Solution:
$$f'(x) = \frac{2e^{-x}}{(1 + e^{-x})^2}$$

Since $f'(x) > 0$, f is increasing on $(-\infty, \infty)$ and there are no extrema.

$$f''(x) = \frac{-2e^{-x}(1 - e^{-x})}{(1 + e^{-x})^3}$$

$f''(x) = 0$ when $x = 0$. Therefore, $(0, 1)$ is an inflection point. Horizontal asymptotes occur at $y = 0$ and $y = 2$.

x	-2	-1	0	1	2
f(x)	0.238	0.538	1	1.462	1.762

29. Find the extrema and the points of inflection and sketch the graph of $f(x) = x^2 e^{-x}$.
Solution:
$$f'(x) = -x^2 e^{-x} + 2xe^{-x} = xe^{-x}(2 - x)$$

$f'(x) = 0$ when $x = 0$ and $x = 2$.

$$f''(x) = x^2 e^{-x} - 2xe^{-x} - 2xe^{-x} + 2e^{-x}$$
$$= e^{-x}(x^2 - 4x + 2)$$

Since $f''(0) > 0$ and $f''(2) < 0$, we have

Relative minimum: $(0, 0)$, Relative maximum: $(2, 4/e^2)$

Since $f''(x) = 0$ when $x = 2 \pm \sqrt{2}$, the inflection points occur at $(2 - \sqrt{2}, 0.191)$ and $(2 + \sqrt{2}, 0.384)$.

x	-2	-1	0	1	2	3
f(x)	29.556	2.718	0	0.368	0.541	0.448

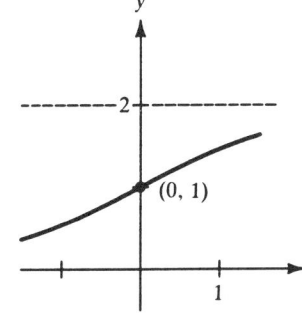

224   Section 5.2

31. Find an equation of the line tangent to the graph of $y = e^{-x}$ at $(0, 1)$.
    Solution:
    $$y' = -e^{-x}$$

    At $(0, 1)$, $y' = -1$. Therefore, the equation of the tangent line is
    $$y - 1 = -1(x - 0)$$
    $$y = -x + 1$$

33. Find the rate at which newly purchased equipment is depreciating (a) $t = 1$ year and (b) $t = 5$ years after it is purchased, if its value V is given by
    $$V = 15,000e^{-0.6286t}$$
    Solution:
    $$V' = -9429e^{-0.6286t}$$

    (a) $V'(1) = -9429e^{-0.6286} \approx -5028.84$
    (b) $V'(5) = -9429e^{-0.6286(5)} \approx -406.89$

35. A lake is stocked with 500 fish, and their population is given by the **logistics curve**
    $$p = \frac{10,000}{1 + 19e^{-t/5}}$$
    where t is measured in months. At what rate is the fish population changing at the end of 1 month and at the end of 10 months? After how many months is the population increasing most rapidly?
    Solution:
    $$\frac{dp}{dt} = \frac{38,000e^{-t/5}}{(1 + 19e^{-t/5})^2}$$
    $$p'(1) = 113.506$$
    $$p'(10) = 403.204$$

    To maximize $\frac{dp}{dt}$, set $\frac{d^2p}{dt^2} = 0$.
    $$p'' = \frac{-38,000e^{-t/5}(1 - 19e^{-t/5})}{5(1 + 19e^{-t/5})^3} = 0$$
    $$1 = 19e^{-t/5}$$
    $$\ln 1 = \ln 19 - \frac{t}{5} \ln e$$
    $$0 = \ln 19 - \frac{t}{5}$$
    $$t = 5 \ln 19 \approx 14.7 \text{ months}$$

37. The yield V (in millions of cubic feet per acre) for a forest stand at age t is given by $V = 6.7e^{-48.1/t}$ where t is measured in years. Find the rate at which the yield is changing when (a) $t = 15$ years, (b) $t = 20$ years, and (c) $t = 60$ years.

Section 5.2 225

Solution:
$$V' = \frac{322.27}{t^2} e^{-48.1/t}$$

(a) $V'(15) = 0.058$ million ft³/year
(b) $V'(20) = 0.073$ million ft³/year
(c) $V'(60) = 0.040$ million ft³/year

39. The balance in a certain savings account is given by $A = (5000)e^{0.08t}$ where A is measured in dollars and t is measured in years. Find the rate at which the balance is changing when (a) t = 1 year, (b) t = 10 years, and (c) t = 50 years.

Solution:
$$A' = 400e^{0.08t}$$

(a) $A'(1) = \$433.31$ per year
(b) $A'(10) = \$890.22$ per year
(c) $A'(50) = \$21,839.26$ per year

41. Evaluate $\int e^{5x}(5)\, dx$.

Solution:
$$\int e^{5x}(5)\, dx = e^{5x} + C$$

43. Evaluate $\int e^{-x^4}(-4x^3)\, dx$.

Solution:
$$\int e^{-x^4}(-4x^3)\, dx = e^{-x^4} + C$$

45. Evaluate $\int_0^1 e^{-2x}\, dx$.

Solution:
$$\int_0^1 e^{-2x}\, dx = -\frac{1}{2}e^{-2x}\Big]_0^1$$
$$= -\frac{e^{-2}}{2} + \frac{1}{2}$$
$$= \frac{1}{2}(1 - e^{-2}) \approx 0.432$$

47. Evaluate $\int_0^2 (x^2 - 1)e^{x^3 - 3x + 1}\, dx$

Solution:
$$\int_0^2 (x^2 - 1)e^{x^3 - 3x + 1}\, dx = \frac{1}{3}(e^{x^3 - 3x + 1})\Big]_0^2$$
$$= \frac{1}{3}(e^3 - e)$$
$$\approx 5.789$$

49. Evaluate $\int xe^{ax^2}\,dx$.

Solution:
$$\int xe^{ax^2}\,dx = \frac{1}{2a}\int e^{ax^2}(2ax)\,dx = \frac{1}{2a}e^{ax^2} + C$$

51. Evaluate $\int_1^3 \frac{e^{3/x}}{x^2}\,dx$.

Solution:
$$\int_1^3 \frac{e^{3/x}}{x^2}\,dx = -\frac{1}{3}\int_1^3 e^{3/x}\left(-\frac{3}{x^2}\right)dx$$
$$= -\frac{1}{3}e^{3/x}\Big]_1^3 = -\frac{1}{3}(e - e^3)$$
$$= \frac{e^3 - e}{3} \approx 5.789$$

53. Evaluate $\int \frac{e^{2x} + 2e^x + 1}{e^x}\,dx$.

Solution:
$$\int \frac{e^{2x} + 2e^x + 1}{e^x}\,dx = \int (e^x + 2 + e^{-x})\,dx$$
$$= e^x + 2x - e^{-x} + C$$

55. Evaluate $\int e^x\sqrt{1 - e^x}\,dx$.

Solution:
$$\int e^x\sqrt{1 - e^x}\,dx = -\int (1 - e^x)^{1/2}(-e^x)\,dx$$
$$= -\frac{2}{3}(1 - e^x)^{3/2} + C$$

57. Find the area of the region bounded by the graphs of $y = e^x$, $y = 0$, $x = 0$, and $x = 5$.

Solution:
$$A = \int_0^5 e^x\,dx = e^x\Big]_0^5 = e^5 - 1 \approx 147.413$$

59. Find the area of the region bounded by the graphs of $y = xe^{-x^2}$, $y = 0$, $x = 0$, $x = 4$.

Solution:
$$A = \int_0^4 xe^{-x^2}\,dx = -\frac{1}{2}\int_0^4 e^{-x^2}(-2x)\,dx$$
$$= -\frac{1}{2}e^{-x^2}\Big]_0^4$$
$$= -\frac{1}{2}(e^{-16} - 1) \approx 0.500$$

61. Find the volume of the solid generated by revolving the region bounded by the graphs of $y = e^x$, $y = 0$, $x = 0$, and $x = 1$ about the x-axis.
Solution:
$$V = \pi \int_0^1 (e^x)^2 \, dx = \pi \int_0^1 e^{2x} \, dx$$
$$= \frac{\pi}{2} \int_0^1 e^{2x}(2) \, dx = \frac{\pi}{2} e^{2x} \Big]_0^1$$
$$= \frac{\pi(e^2 - 1)}{2} \approx 10.036$$

63. A deposit of $2,500 is made in a savings account at an annual percentage rate of 12% compounded continuously. Find the average balance in the account during the first 5 years.
Solution:
$$\text{Average balance} = \frac{1}{5-0} \int_0^5 2500 e^{0.12t} \, dt$$
$$= \frac{1}{5}\left(\frac{1}{0.12}\right)(2500) \int_0^5 e^{0.12t}(0.12) \, dt$$
$$= \frac{12500}{3} e^{0.12} \Big]_0^5 = \frac{12500}{3}[e^{0.60} - 1]$$
$$\approx \$3425.50$$

● **Section 5.3 The Natural Logarithmic Function**

1. Write the logarithmic equation $\ln 2 = 0.6931\ldots$ as an exponential equation.
Solution:
$e^{0.6931\ldots} = 2$

3. Write the logarithmic equation $\ln 0.5 = -0.6931\ldots$ as an exponential equation.
Solution:
$e^{-0.6931\ldots} = 0.5$

5. Write the exponential equation $e^0 = 1$ as a logarithmic equation.
Solution:
$\ln 1 = 0$

7. Write the exponential equation $e^{-2} = 0.1353\ldots$ as a logarithmic equation.
Solution:
$\ln 0.1353\ldots = -2$

228   Section 5.3

9. Use the graph of $y = \ln x$ to match $f(x) = \ln x + 2$ with the correct graph.
    Solution: The graph is a logarithmic curve that passes through the point (1, 2) with a vertical asymptote at $x = 0$. Therefore, it matches graph (c).

11. Use the graph of $y = \ln x$ to match $f(x) = \ln (x + 2)$ with the correct graph.
    Solution: The graph is a logarithmic curve that passes through the point (-1, 0) with vertical asymptote at $x = -2$. Therefore, it matches graph (b).

13. Sketch the graph of $y = \ln (x - 1)$.
    Solution:

x	1.5	2	3	4	5
y	-0.69	0	0.69	1.10	1.39

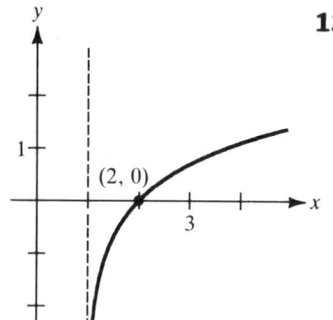

15. Sketch the graph of $y = \ln 2x$.
    Solution:

x	0.25	0.5	1	3	5
y	-0.69	0	0.69	1.79	2.30

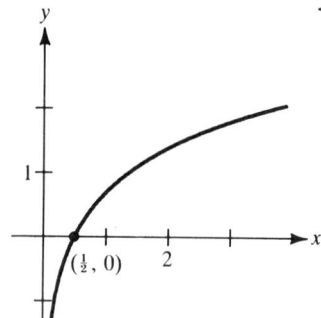

17. Sketch the graph of $y = 3 \ln x$.
    Solution:

x	0.5	1	2	3	4
y	-2.08	0	2.08	3.30	4.16

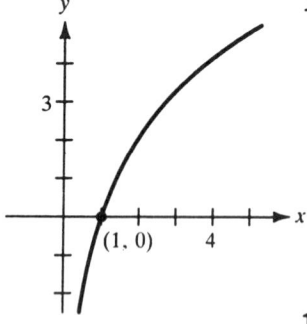

19. Show that $f(x) = e^{2x}$ and $g(x) = \ln \sqrt{x}$ are inverses of each other and sketch their graphs on the same coordinate axes.
    Solution:
    $$g(x) = \ln \sqrt{x} = \frac{1}{2} \ln x$$
    $$f(g(x)) = f(\frac{1}{2} \ln x) = e^{2(1/2 \ln x)} = e^{\ln x} = x$$
    $$g(f(x)) = g(e^{2x}) = \frac{1}{2} \ln e^{2x} = \frac{1}{2}(2x) \ln e = x$$

Section 5.3

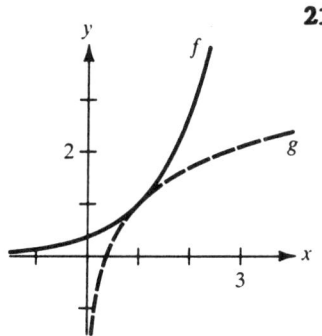

21. Show that $f(x) = e^{x-1}$ and $g(x) = 1 + \ln x$ are inverses of each other and sketch their graphs on the same coordinate axes.
    Solution:
    $$\begin{aligned} f(g(x)) &= f(1 + \ln x) \\ &= e^{(1+\ln x)-1} \\ &= e^{\ln x} = x \end{aligned}$$

    $$\begin{aligned} g(f(x)) &= g(e^{x-1}) \\ &= 1 + \ln e^{x-1} \\ &= 1 + (x - 1) \ln e = x \end{aligned}$$

23. Apply the inverse properties of $\ln x$ and e^x to simplify $\ln e^{x^2}$.
    Solution:
    $$\ln e^{x^2} = x^2$$

25. Apply the inverse properties of $\ln x$ and e^x to simplify $e^{\ln(5x+2)}$.
    Solution:
    $$e^{\ln(5x+2)} = 5x + 2$$

27. Apply the inverse properties of $\ln x$ and e^x to simplify $e^{\ln\sqrt{x}}$.
    Solution:
    $$e^{\ln\sqrt{x}} = \sqrt{x}$$

29. Use the properties of logarithms and the fact that $\ln 2 \approx 0.6931$ and $\ln 3 \approx 1.0986$ to approximate

    (a) $\ln 6$ (b) $\ln(2/3)$
    (c) $\ln 81$ (d) $\ln \sqrt{3}$
    Solution:
    (a) $\ln 6 = \ln(2 \cdot 3) = \ln 2 + \ln 3$
    $ = 0.6931 + 1.0986 = 1.7917$
    (b) $\ln(2/3) = \ln 2 - \ln 3 = 0.6931 - 1.0986 = -0.4055$
    (c) $\ln 81 = \ln 3^4 = 4 \ln 3 = 4(1.0986) = 4.3944$
    (d) $\ln \sqrt{3} = (1/2) \ln 3 = (1/2)(1.0986) = 0.5493$

31. Use the properties of logarithms to write $\ln(2/3)$ as a sum, difference, or multiple of logarithms.
    Solution:
    $$\ln \frac{2}{3} = \ln 2 - \ln 3$$

33. Use the properties of logarithms to write $\ln xyz$ as a sum, difference, or multiple of logarithms.
    Solution:
    $$\ln xyz = \ln x + \ln y + \ln z$$

35. Use the properties of logarithms to write $\ln \sqrt{a-1}$ as a sum, difference, or multiple of logarithms.
    Solution:
    $$\ln \sqrt{a-1} = \frac{1}{2} \ln(a-1)$$

37. Use the properties of logarithms to write

$$\ln \frac{2x}{\sqrt{x^2 - 1}}$$

as a sum, difference, or multiple of logarithms.
Solution:

$$\ln \frac{2x}{\sqrt{x^2 - 1}} = \ln 2x - \ln \sqrt{x^2 - 1}$$

$$= \ln 2 + \ln x - \frac{1}{2} \ln [(x + 1)(x - 1)]$$

$$= \ln 2 + \ln x - \frac{1}{2}[\ln (x + 1) + \ln (x - 1)]$$

$$= \ln 2 + \ln x - \frac{1}{2} \ln (x + 1) - \frac{1}{2} \ln (x - 1)$$

39. Use the properties of logarithms to write

$$\ln \left(\frac{x^2 - 1}{x^3}\right)^3$$

as a sum, difference, or multiple of logarithms.
Solution:

$$\ln \left(\frac{x^2 - 1}{x^3}\right)^3 = 3[\ln (x^2 - 1) - 3 \ln x]$$

$$= 3[\ln (x + 1) + \ln (x - 1) - 3 \ln x]$$

41. Write $\ln (x - 2) - \ln (x + 2)$ as a single logarithm.
Solution:

$$\ln (x - 2) - \ln (x + 2) = \ln \frac{x - 2}{x + 2}$$

43. Write $3 \ln x + 2 \ln y - 4 \ln z$ as a single logarithm.
Solution:

$$3 \ln x + 2 \ln y - 4 \ln z = \ln x^3 + \ln y^2 - \ln z^4$$

$$= \ln \left(\frac{x^3 y^2}{z^4}\right)$$

45. Write $2[\ln x - \ln (x + 1) - \ln (x - 1)]$ as a single logarithm.
Solution:

$$2[\ln x - \ln (x + 1) - \ln (x - 1)] = 2 \ln \left(\frac{x}{(x + 1)(x - 1)}\right)$$

$$= \ln \left(\frac{x}{x^2 - 1}\right)^2$$

47. Write $(3/2)[\ln x(x^2 + 1) - \ln (x + 1)]$ as a single logarithm.
Solution:

$$\frac{3}{2}[\ln x(x^2 + 1) - \ln (x + 1)] = \frac{3}{2} \ln \frac{x(x^2 + 1)}{x + 1}$$

$$= \ln \left[\frac{x(x^2 + 1)}{x + 1}\right]^{3/2}$$

Section 5.3

49. When $e^{\ln x} = 4$, solve for x.
Solution:
$$x = 4$$

51. When $\ln x = 0$, solve for x.
Solution:
$$x = e^0 = 1$$

53. When $e^{x+1} = 4$, solve for x.
Solution:
$$x + 1 = \ln 4$$
$$x = (\ln 4) - 1 \approx 0.3863$$

55. When $500e^{-0.11t} = 600$, solve for t.
Solution:
$$e^{-0.11t} = 6/5$$
$$-0.11t = \ln 6 - \ln 5$$
$$t = (\ln 6 - \ln 5)/(-0.11) \approx -1.6575$$

57. When $5^{2x} = 15$, solve for x.
Solution:
$$2x \ln 5 = \ln 15$$
$$x = \frac{\ln 15}{2 \ln 5} \approx 0.8413$$

59. When $500(1.07)^t = 1000$, solve for t.
Solution:
$$t \ln 1.07 = \ln 2$$
$$t = \frac{\ln 2}{\ln 1.07} \approx 10.2448$$

61. A deposit of $1,000 is made into a fund with an annual interest rate of 11%. Find the time for the investment to double if the interest is compounded
(a) annually   (b) monthly
(c) daily   (d) continuously
Solution:
$$P = 1000, \quad r = 0.11, \quad A = 2000$$

(a) $2000 = 1000(1 + 0.11)^t$

$$t = \frac{\ln 2}{\ln 1.11} \approx 6.64 \text{ years}$$

(b) $2000 = 1000(1 + \frac{0.11}{12})^{12t}$

$$t = \frac{\ln 2}{12 \ln 1.0092} \approx 6.33 \text{ years}$$

(c) $2000 = 1000(1 + \frac{0.11}{365})^{365t}$

$$t = \frac{\ln 2}{365 \ln 1.0003} \approx 6.30 \text{ years}$$

(d) $2000 = 1000e^{0.11t}$

$$t = \frac{\ln 2}{0.11} \approx 6.30 \text{ years}$$

63. Complete the table for the time t necessary for P dollars to triple if interest is compounded continuously at the rate r.
Solution:
$$3P = Pe^{rt}$$
$$3 = e^{rt}$$
$$\ln 3 = rt$$
$$t = \frac{\ln 3}{r}$$

r	2%	4%	6%	8%	10%	12%
t	54.93	27.47	18.31	13.73	10.99	9.16

65. The demand function for a certain product is given by

$$p = 500 - 0.5e^{0.004x}$$

Find the quantity x demanded for a price of (a) p = $350, and (b) p = $300.
Solution:
(a) $\qquad 350 = 500 - 0.5e^{0.004x}$
$\qquad e^{0.004x} = 300$

$$x = \frac{\ln 300}{0.004} \approx 1426$$

(b) $\qquad 300 = 500 - 0.5e^{0.004x}$
$\qquad e^{0.004x} = 400$

$$x = \frac{\ln 400}{0.004} \approx 1498$$

67. Use a calculator to demonstrate that

$$\frac{\ln x}{\ln y} \neq \ln \frac{x}{y} = \ln x - \ln y$$

by completing the table.
Solution:

x	y	$\frac{\ln x}{\ln y}$	$\ln \frac{x}{y}$	$\ln x - \ln y$
1	2	0.0000	-0.6931	-0.6931
3	4	0.7925	-0.2877	-0.2877
10	5	1.4307	0.6931	0.6931
4	0.5	-2.0000	2.0794	2.0794

Section 5.4 Logarithmic Functions: Differentiation and Integration

1. Find the slope of the tangent line to the graph of $y = \ln x^3$ at the point $(1, 0)$.
   Solution:
   $$y = \ln x^3 = 3 \ln x$$
   $$y' = \frac{3}{x}$$
   $$y'(1) = 3$$

3. Find the slope of the tangent line to the graph of $y = \ln x^2$ at the point $(1, 0)$.
   Solution:
   $$y = \ln x^2 = 2 \ln x$$
   $$y' = \frac{2}{x}$$
   $$y'(1) = 2$$

5. Find the slope of the tangent line to the graph of $y = \ln x^{3/2}$ at the point $(1, 0)$.
   Solution:
   $$y = \ln x^{3/2} = \frac{3}{2} \ln x$$
   $$y' = \frac{3}{2x}$$
   $$y'(1) = 3/2$$

7. Find the derivative of $y = \ln x^2$.
   Solution:
   $$y = \ln x^2 = 2 \ln x$$
   $$y' = \frac{2}{x}$$

9. Find the derivative of $f(x) = \ln 2x$.
   Solution:
   $$f'(x) = \frac{2}{2x} = \frac{1}{x}$$

11. Find the derivative of $y = \ln \sqrt{x^4 - 4x}$.
    Solution:
    $$y = \ln \sqrt{x^4 - 4x} = \frac{1}{2} \ln(x^4 - 4x)$$
    $$y' = \frac{1}{2}\left(\frac{4x^3 - 4}{x^4 - 4x}\right) = \frac{2(x^3 - 1)}{x(x^3 - 4)}$$

13. Find the derivative of $y = (\ln x)^4$.
    Solution:
    $$y' = 4(\ln x)^3 \left(\frac{1}{x}\right) = \frac{4(\ln x)^3}{x}$$

15. Find the derivative of $y = x \ln x$.
    Solution:
    $$y' = x(\frac{1}{x}) + \ln x = 1 + \ln x$$

17. Find the derivative of $y = \ln x \sqrt{x^2 - 1}$.
    Solution:
    $$y = \ln x \sqrt{x^2 - 1} = \ln x + \frac{1}{2} \ln(x^2 - 1)$$
    $$y' = \frac{1}{x} + \frac{1}{2}(\frac{2x}{x^2 - 1}) = \frac{2x^2 - 1}{x(x^2 - 1)}$$

19. Find the derivative of $y = \ln[x/(x^2 + 1)]$.
    Solution:
    $$y = \ln(\frac{x}{x^2 + 1}) = \ln x - \ln(x^2 + 1)$$
    $$y' = \frac{1}{x} - \frac{2x}{x^2 + 1} = \frac{1 - x^2}{x(x^2 + 1)}$$

21. Find the derivative of $y = \ln[(x + 1)/(x - 1)]$.
    Solution:
    $$y = \ln\frac{x + 1}{x - 1} = \ln(x + 1) - \ln(x - 1)$$
    $$y' = \frac{1}{x + 1} - \frac{1}{x - 1} = -\frac{2}{x^2 - 1} = \frac{2}{1 - x^2}$$

23. Find the derivative of $y = \ln\sqrt{(x + 1)/(x - 1)}$.
    Solution:
    $$y = \ln\sqrt{\frac{x + 1}{x - 1}} = \frac{1}{2}[\ln(x + 1) - \ln(x - 1)]$$
    $$y' = \frac{1}{2}(\frac{1}{x + 1} - \frac{1}{x - 1}) = \frac{-1}{x^2 - 1} = \frac{1}{1 - x^2}$$

25. Find the derivative of $y = (\ln x)/x^2$.
    Solution:
    $$y = \frac{\ln x}{x^2}$$
    $$y' = \frac{x^2(1/x) - (\ln x)(2x)}{x^4} = \frac{1 - 2\ln x}{x^3}$$

27. Find the derivative of $y = \ln(\sqrt{4 + x^2}/x)$.
    Solution:
    $$y = \ln\frac{\sqrt{4 + x^2}}{x} = \frac{1}{2}\ln(4 + x^2) - \ln x$$
    $$y' = \frac{1}{2}(\frac{2x}{4 + x^2}) - \frac{1}{x} = -\frac{4}{x(4 + x^2)}$$

29. Find the derivative of $y = \ln \sqrt{x^2 - 4}$.
Solution:
$$y = \ln \sqrt{x^2 - 4} = \frac{1}{2} \ln(x^2 - 4)$$
$$y' = \frac{1}{2}\left(\frac{2x}{x^2 - 4}\right) = \frac{x}{x^2 - 4}$$

31. Find the derivative of $g(x) = e^{-x} \ln x$.
Solution:
$$g'(x) = e^{-x}\left(\frac{1}{x}\right) + (-e^{-x})\ln x = e^{-x}\left(\frac{1}{x} - \ln x\right)$$

33. Find the derivative of $f(x) = \ln e^{x^2}$.
Solution:
$$f(x) = \ln e^{x^2} = x^2$$
$$f'(x) = 2x$$

35. Find dx/dp for $x = \ln(1000/p)$.
Solution:
$$x = \ln \frac{1000}{p} = \ln 1000 - \ln p$$
$$\frac{dx}{dp} = 0 - \frac{1}{p} = -\frac{1}{p}$$

37. Find dx/dp for $x = 500 \ln[p/(p^2 + 1)]$.
Solution:
$$x = 500 \ln \frac{p}{p^2 + 1} = 500[\ln p - \ln(p^2 + 1)]$$
$$\frac{dx}{dp} = 500\left(\frac{1}{p} - \frac{2p}{p^2 + 1}\right) = \frac{500(1 - p^2)}{p(p^2 + 1)}$$

39. For $f(x) = x \ln x - 4x$, find $f''(x)$.
Solution:
$$f'(x) = x\left(\frac{1}{x}\right) + \ln x - 4 = \ln x - 3$$
$$f''(x) = \frac{1}{x}$$

41. Find any relative extrema and inflection points and sketch the graph of $y = x - \ln x$.
Solution:
$$y' = 1 - \frac{1}{x} = \frac{x - 1}{x}$$
$y' = 0$ when $x = 1$.
$$y'' = \frac{1}{x^2}$$

Since $y''(1) = 1 > 0$, there is a relative minimum at $(1, 1)$. Moreover, since $y'' > 0$ on $(0, \infty)$ it follows that the graph is concave up in its domain and there are no inflection points.

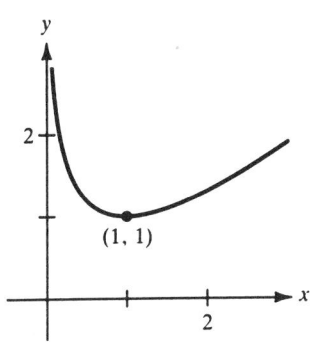

43. Find any relative extrema and inflection points and sketch the graph of $y = (\ln x)/x$.
Solution: The domain of the function is $(0, \infty)$.

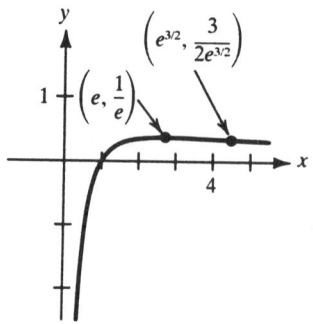

$$y' = \frac{1 - \ln x}{x^2}$$

$y' = 0$ when $x = e$.

$$y'' = \frac{2\ln x - 3}{x^3}$$

Since $y''(e) < 0$, it follows that $(e, 1/e)$ is a relative maximum. Since $y'' = 0$ when $2\ln x - 3 = 0$ and $x = e^{3/2}$, there is an inflection point at

$$(e^{3/2}, \frac{3}{2e^{3/2}})$$

45. Find any relative extrema and inflection points and sketch the graph of $y = x^2 \ln x$.
Solution:
$$y' = x(1 + 2\ln x)$$

$y' = 0$ when $x = e^{-1/2}$. ($x = 0$ is not in the domain.)

$$y'' = 3 + 2\ln x$$

Since $y''(e^{-1/2}) > 0$, it follows that there is a relative minimum at

$$(\frac{1}{\sqrt{e}}, -\frac{1}{2e})$$

Since $y'' = 0$ when $x = e^{-3/2}$, it follows that there is an inflection point at

$$(\frac{1}{e^{3/2}}, -\frac{3}{2e^3})$$

47. Evaluate $\int \frac{1}{x + 1} dx$.
Solution:

$$\int \frac{1}{x + 1} dx = \ln|x + 1| + C$$

49. Evaluate $\int \frac{1}{3 - 2x} dx$.
Solution:

$$-\frac{1}{2} \int \frac{1(-2)}{3 - 2x} dx = -\frac{1}{2}\ln|3 - 2x| + C$$

51. Evaluate $\int \frac{x}{x^2 + 1} dx$.
Solution:

$$\frac{1}{2} \int \frac{x(2)}{x^2 + 1} dx = \frac{1}{2}\ln(x^2 + 1) + C$$
$$= \ln\sqrt{x^2 + 1} + C$$

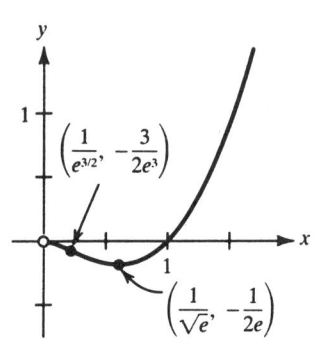

Section 5.4   237

53. Evaluate $\int_0^1 \frac{x^2}{x^3 + 1} dx$.

   Solution:

   $$\int_0^1 \frac{x^2}{x^3 + 1} dx = \frac{1}{3} \int_0^1 \frac{3x^2}{x^3 + 1} dx$$

   $$= \frac{1}{3} \ln |x^3 + 1| \Big]_0^1 = \frac{1}{3} \ln 2 - \frac{1}{3} \ln 1$$

   $$= \frac{1}{3} \ln 2 \approx 0.231$$

55. Evaluate $\int \frac{x^2 - 4}{x} dx$.

   Solution:

   $$\int \frac{x^2 - 4}{x} dx = \int [x - 4(\frac{1}{x})] dx$$

   $$= \frac{x^2}{2} - 4 \ln |x| + C$$

57. Evaluate $\int_1^e \frac{(1 + \ln x)^2}{x} dx$.

   Solution:

   $$\int_1^e \frac{(1 + \ln x)^2}{x} dx = \int_1^e (1 + \ln x)^2 (\frac{1}{x}) dx$$

   $$= \frac{(1 + \ln x)^3}{3} \Big]_1^e$$

   $$= \frac{8}{3} - \frac{1}{3} = \frac{7}{3}$$

59. Evaluate $\int_0^2 \frac{x^2 - 2}{x + 1} dx$.

   Solution:

   $$\int_0^2 \frac{x^2 - 2}{x + 1} dx = \int_0^2 (x - 1 - \frac{1}{x + 1}) dx$$

   $$= \left[\frac{x^2}{2} - x - \ln |x + 1| \right]_0^2$$

   $$= -\ln 3 \approx -1.099$$

61. Evaluate $\int \frac{1}{\sqrt{x + 1}} dx$.

   Solution:

   $$\int \frac{1}{\sqrt{x + 1}} dx = \int (x + 1)^{-1/2} dx = 2\sqrt{x + 1} + C$$

63. Evaluate $\displaystyle\int \frac{x^2 + 2x + 3}{x^3 + 3x^2 + 9x + 1}\, dx$.

Solution:

$$\int \frac{x^2 + 2x + 3}{x^3 + 3x^2 + 9x + 1}\, dx = \frac{1}{3}\int \frac{3(x^2 + 2x + 3)}{x^3 + 3x^2 + 9x + 1}\, dx$$

$$= \frac{1}{3}\int \frac{3x^2 + 6x + 9}{x^3 + 3x^2 + 9x + 1}\, dx$$

$$= \frac{1}{3}\ln |x^3 + 3x^2 + 9x + 1| + C$$

65. Evaluate $\displaystyle\int \frac{e^x + e^{-x}}{e^x - e^{-x}}\, dx$.

Solution:

$$\int \frac{e^x + e^{-x}}{e^x - e^{-x}}\, dx = \int \frac{u'}{u}\, dx$$

$$= \ln |u| + C = \ln |e^x - e^{-x}| + C$$

67. Find the area of the region bounded by the graphs of $y = (x^2 + 4)/x$, $y = 0$, $x = 1$ and $x = 4$.

Solution:

$$A = \int_1^4 \frac{x^2 + 4}{x}\, dx = \int_1^4 \left[x + 4\left(\frac{1}{x}\right)\right] dx$$

$$= \left[\frac{x^2}{2} + 4\ln |x|\right]_1^4 = (8 + 4\ln 4) - \left(\frac{1}{2} + 4\ln 1\right)$$

$$= \frac{15}{2} + 8\ln 2 \approx 13.045$$

69. A population of bacteria is growing at the rate of $dP/dt = 3000/(1 + 0.25t)$ where t is the time in days. Assuming that the initial population (when $t = 0$) is 1000, write an equation that gives the population at any time t, and then find the population when $t = 3$ days.

Solution:

$$P = \int \frac{3000}{1 + 0.25t}\, dt = \frac{3000}{0.25}\int \frac{0.25}{1 + 0.25t}\, dt$$

$$= 12{,}000 \ln |1 + 0.25t| + C$$

Since $P(0) = 12{,}000 \ln 1 + C = 1000$, it follows that $C = 1000$. Therefore,

$$P(t) = 12{,}000 \ln |1 + 0.25t| + 1000$$
$$= 1000[12 \ln |1 + 0.25t| + 1]$$
$$= 1000[1 + \ln (1 + 0.25t)^{12}]$$

and the population when $t = 3$ is

$$P(3) = 1000[1 + \ln (1 + 0.75)^{12}] \approx 7715 \text{ bacteria}$$

Section 5.5   239

● Section 5.5 Exponential Growth and Decay

1. Find the exponential function $y = Ce^{kt}$ that passes through the points (0, 2) and (4, 3).
   Solution: Since $y = 2$ when $t = 0$, it follows that $C = 2$. Moreover, since $y = 3$ when $t = 4$, we have $3 = 2e^{4k}$ and $k = [\ln(3/2)]/4 \approx 0.1014$. Thus,
   $$y = 2e^{0.1014t}$$

3. Find the exponential function $y = Ce^{kt}$ that passes through the points (0, 4) and (5, 1/2).
   Solution: Since $y = 4$ when $t = 0$, it follows that $C = 4$. Moreover, since $y = 1/2$ when $t = 5$, we have $1/2 = 4e^{5k}$ and $k = [\ln(1/8)]/5 \approx -0.4159$. Thus,
   $$y = 4e^{-0.4159t}$$

5. Find the exponential function $y = Ce^{kt}$ that passes through the points (1, 1) and (5, 5).
   Solution: Using the fact that $y = 1$ when $t = 1$ and $y = 5$ when $t = 5$, we have $1 = Ce^{k}$ and $5 = Ce^{5k}$. From these two equations, we have $Ce^{k} = (1/5)Ce^{5k}$. Thus, $k = (\ln 5)/4 \approx 0.4024$ and we have
   $$y = Ce^{0.4024t}$$
   Since $1 = Ce^{0.4024}$, it follows that $C \approx 0.6687$ and
   $$y = 0.6687e^{0.4024t}$$

7. The half-life of the isotope Ra^{226} is 1,620 years. If the initial amount is 10 grams, how much will remain after (a) 1,000 years and (b) 10,000 years?
   Solution: From Example 1, we have
   $$y = 10e^{[\ln(1/2)/1620]t}$$
   (a) When $t = 1,000$,
   $$y = 10e^{[\ln(1/2)/1620](1000)} \approx 6.519 \text{ grams}$$
   (b) When $t = 10,000$,
   $$y = 10e^{[\ln(1/2)/1620](10000)} \approx 0.139 \text{ gram}$$

9. The half-life of the isotope C^{14} is 5,730 years. If 2 grams remain after 10,000 years (a) what is the initial amount and (b) how much is present after 1,000 years?
   Solution: Since $y = Ce^{[\ln(1/2)/5730]t}$, we have
   $$2 = Ce^{[\ln(1/2)/5730](10000)} \implies C \approx 6.705$$
   Which implies that the initial quantity is 6.705 grams. When $t = 1000$, we have
   $$y = 6.705e^{[\ln(1/2)/5730](1000)} \approx 5.941 \text{ grams}$$

11. The half-life of the isotope Pu^{230} is 24,360 years. If 2.1 grams remain after 1,000 years, (a) what is the initial amount and (b) how much will remain after 10,000 years?
    Solution: Since $y = Ce^{[\ln(1/2)/24360]t}$, we have
    $$2.1 = Ce^{[\ln(1/2)/24360](1000)} \implies C \approx 2.161$$
    Thus, the initial quantity is 2.161 grams. When $t = 10,000$,
    $$y = 2.161e^{[\ln(1/2)/24360](10000)} \approx 1.626 \text{ grams}$$

13. What percentage of a present amount of radioactive radium (Ra^{226}) will remain after 100 years?
    Solution: When $t = 100$, we have
    $$y = Ce^{[\ln(1/2)/1620](100)} \approx 0.958C = 95.8\%C$$
    After 100 years, approximately 95.8% of the radioactive radium will remain.

15. The number of a certain type of bacteria increases continuously at a rate proportional to the number present. If there are 100 present at a given time and 300 present 5 hours later, how many will there be 10 hours after the initial time? How long does it take the number of bacteria to double?
    Solution: Since $y = Ce^{kt}$, we use the initial conditions (0, 100) and (5, 300) to determine that $C = 100$ and $k = (\ln 3)/5 \approx 0.2197$. Therefore,
    $$y = 100e^{0.2197t}$$
    When $t = 10$, we have
    $$y(10) \approx 900$$
    The time required to double is given by
    $$200 = 100e^{0.2197t}$$
    $$e^{0.2197t} = 2$$
    $$t = (\ln 2)/0.2197 \approx 3.15 \text{ hours}$$

17. $1,000 is deposited in a savings account at a rate of 12% compounded continuously. (a) Find the time for the amount to double. What is the amount after (b) 10 years, (c) after 25 years?
    Solution: Since $A = 1000e^{0.12t}$, the time to double is given by $2000 = 1000e^{0.12t}$ and we have
    $$t = (\ln 2)/0.12 \approx 5.776 \text{ years}$$
    Amount after 10 years: $A = 1000e^{1.2} \approx \$3320.12$
    Amount after 25 years: $A = 1000e^{0.12(25)} \approx \$20,085.54$

Section 5.5

19. $750 is deposited in an account with continuously compounded interest. If the amount doubles in 7.75 years, what is the rate? How much will be in the account after 10 years and after 25 years?
    Solution: Since $A = 750e^{rt}$ and $A = 1500$ when $t = 7.75$, we have

    $$1500 = 750e^{7.75r}$$
    $$r = (\ln 2)/7.75 \approx 0.0894 = 8.94\%$$

    Amount after 10 years: $A = 750e^{0.0894(10)} \approx \$1833.67$
    Amount after 25 years: $A = 750e^{0.0894(25)} \approx \$7009.86$

21. $500 is deposited in an account with continuously compounded interest. If the balance is $1,292.85 after 10 years, what is the rate? How long will it take the amount to double? What is the amount after 25 years?
    Solution: Since $A = 500e^{rt}$ and $A = 1292.85$ when $t = 10$, we have

    $$1292.85 = 500e^{10r}$$
    $$r = [\ln(1292.85/500)]/10 \approx 0.095 = 9.5\%$$

    The time to double is given by

    $$1000 = 500e^{0.095t}$$
    $$t = (\ln 2)/0.095 \approx 7.296 \text{ years}$$

    Amount after 25 years: $A = 500e^{0.095(25)} \approx \$5375.51$

23. During a slump in the economy, a company found that its annual revenues dropped from $742,000 in 1984 to $632,000 in 1986. If the revenue is following an exponential pattern of decline, what is the expected revenue for 1987? (Let $t = 0$ represent 1984.)
    Solution: Since $y = Ce^{kt}$ and $y = 742,000$ when $t = 0$ and $y = 632,000$ when $t = 2$, we have $742,000 = Ce^0$, which implies that $C = 742,000$. Therefore,

    $$632,000 = 742,000e^{2k}$$
    $$k = [\ln(632,000/742,000)]/2 \approx -0.0802$$
    $$y = 742,000e^{-0.0802t}$$

    When $t = 3$, $y = 742,000e^{-0.0802(3)} \approx \$583,275.41$

25. The sales S (in thousands of units) of a new product after it has been on the market t years are given by $S(t) = Ce^{k/t}$.
    (a) Find S as a function of t if 5000 units have been sold after 1 year and the saturation point for the market is 30,000. That is, $S(1) = 5$ and $\lim_{t \to \infty} S(t) = 30$.
    (b) How many units will be sold after 5 years?
    (c) Sketch a graph of this sales function.

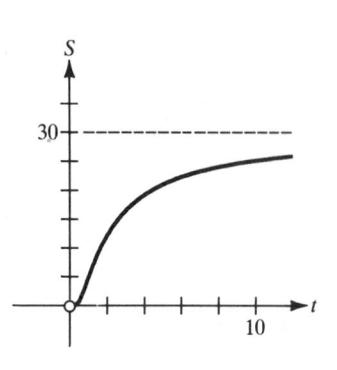

Solution:
(a) Since $S = 5$ when $t = 1$, we have $5 = Ce^k$ and

$$\lim_{t \to \infty} Ce^{k/t} = C = 30$$

Therefore, $5 = 30e^k$, $k = \ln(1/6) \approx -1.7918$, and

$$S = 30e^{-1.7918/t}$$

(b) $S(5) = 30e^{-1.7918/5} \approx 20.9646 \approx 20,965$ units
(c) See graph.

27. The management at a certain factory has found that the maximum number of units a worker can produce in a day is 30. The learning curve for the number of units N produced per day after a new employee has worked t days is given by $N = 30(1 - e^{kt})$. After 20 days on the job, a particular worker produced 19 units.
(a) Find the learning curve for this worker (that is, find the value of k).
(b) How many days should pass before this worker is producing 25 units per day?

Solution:
(a) Since $19 = 30(1 - e^{20k})$, it follows that

$$30e^{20k} = 11$$
$$k = [\ln(11/30)]/20 \approx -0.0502$$
$$N = 30(1 - e^{-0.0502t})$$

(b) $\quad 25 = 30(1 - e^{-0.0502t})$
$e^{-0.0502t} = 1/6$
$t = (\ln 6)/(0.0502) \approx 36$ days

29. A small business assumes that the demand function for one of its products is the exponential function $p = Ce^{kx}$.
(a) Find p as a function of x if the quantity demanded is 1000 units and 1200 units when the price is $45 and $40, respectively.
(b) Find the values of x and p that will maximize revenue for this product.

Solution:
(a) Since $p = Ce^{kx}$ where $p = 45$ when $x = 1000$ and $p = 40$ when $x = 1200$, we have

$$45 = Ce^{1000k} \text{ and } 40 = Ce^{1200k}$$
$$\ln 45 = \ln C + 1000k$$
$$\ln 40 = \ln C + 1200k$$
$$\ln 45 - \ln 40 = -200k$$
$$k = [\ln(45/40)]/(-200) \approx -0.0005889$$

Therefore, we have $45 = Ce^{1000(-0.0005889)}$ which implies that $C \approx 81.0915$ and

$$p = 81.0915e^{-0.0005889x}$$

Review Exercises for Chapter 5   243

(b) Since $R = xp = 81.0915xe^{-0.0005889x}$, we have

$$R' = 81.0915[-0.0005889xe^{-0.0005889x} + e^{-0.0005889x}]$$
$$= 81.0915e^{-0.0005889x}[1 - 0.0005889x] = 0$$

Since $R' = 0$ when $x = 1/0.0005889 \approx 1698$ units, we have

$$p = 81.0915e^{-0.0005889(1698)} \approx \$29.83$$

● Review Exercises for Chapter 5

1. Sketch the graph of $f(x) = e^{-2x}$.
   Solution:

x	-2	-1	0	1	2
f(x)	54.60	7.39	1	0.14	0.02

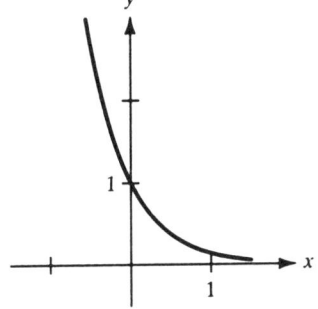

3. Sketch the graph of $g(x) = \ln(x+1)$.
   Solution:

x	-0.5	0	1	2	3
g(x)	-0.69	0	0.69	1.10	1.39

5. Solve $e^{\ln x} = 3$ for x.
   Solution:
   $x = 3$

7. Solve $\ln 4x - \ln(x+1) = 0$ for x.
   Solution:
   $\ln 4x = \ln(x+1)$
   $4x = x + 1$
   $3x = 1$
   $x = 1/3$

9. Differentiate $y = \ln e^{-x^2}$.
   Solution:
   $y = \ln e^{-x^2} = -x^2$
   $y' = -2x$

11. Differentiate $y = x^2 e^x$.
    Solution:
    $y' = x^2 e^x + 2xe^x = xe^x(x+2)$

13. Differentiate $y = \sqrt{e^{2x} + e^{-2x}}$.
    Solution:
    $$y' = \frac{1}{2}(e^{2x} + e^{-2x})^{-1/2}(2e^{2x} - 2e^{-2x})$$
    $$= \frac{e^{2x} - e^{-2x}}{\sqrt{e^{2x} + e^{-2x}}}$$

15. Differentiate $y = e^{2x}/x$.
    Solution:
    $$y' = \frac{x(2e^{2x}) - e^{2x}(1)}{x^2} = \frac{e^{2x}(2x - 1)}{x^2}$$

17. Find y' for $ye^x = xy + 1$.
    Solution:
    $$y(e^x - x) = 1$$
    $$y = \frac{1}{e^x - x} = (e^x - x)^{-1}$$
    $$y' = -(e^x - x)^{-2}(e^x - 1) = \frac{1 - e^x}{(e^x - x)^2}$$

19. Differentiate $y = \ln\sqrt{x}$.
    Solution:
    $$y = \ln\sqrt{x} = \frac{1}{2}\ln x$$
    $$y' = \frac{1}{2x}$$

21. Differentiate $y = x\sqrt{\ln x}$.
    Solution:
    $$y' = x[\frac{1}{2}(\ln x)^{-1/2}(\frac{1}{x})] + \sqrt{\ln x}$$
    $$= \frac{1}{2\sqrt{\ln x}} + \sqrt{\ln x} = \frac{1 + 2\ln x}{2\sqrt{\ln x}}$$

23. Find y' for $y\ln x + y^2 = 1$.
    Solution:
    $$y(\frac{1}{x}) + y'\ln x + 2yy' = 0$$
    $$y'(2y + \ln x) = -\frac{y}{x}$$
    $$y' = -\frac{y}{x(2y + \ln x)}$$

25. Find y' for $\ln y = x\ln x$.
    Solution:
    $$(\frac{1}{y})y' = x(\frac{1}{x}) + \ln x$$
    $$y' = y(1 + \ln x)$$

27. Differentiate $y = (1/b^2)[\ln(a + bx) + (a/(a + bx))]$.
    Solution:
    $$y' = \frac{1}{b^2}[\frac{b}{a + bx} - \frac{ab}{(a + bx)^2}]$$
    $$= \frac{1}{b^2}[\frac{ab + b^2x - ab}{(a + bx)^2}] = \frac{x}{(a + bx)^2}$$

Review Exercises for Chapter 5

29. When $y = -(1/a) \ln[(a + bx)/x]$, find dy/dx.

Solution:
$$y = -\frac{1}{a} \ln \frac{a + bx}{x} = -\frac{1}{a}[\ln(a + bx) - \ln x]$$

$$y' = -\frac{1}{a}\left[\frac{b}{a + bx} - \frac{1}{x}\right] = \frac{1}{x(a + bx)}$$

31. Evaluate $\int xe^{-3x^2}\, dx$.

Solution:
$$\int xe^{-3x^2}\, dx = -\frac{1}{6}\int e^{-3x^2}(-6x)\, dx = -\frac{1}{6}e^{-3x^2} + C$$

33. Evaluate $\int \dfrac{e^{4x} - e^{2x} + 1}{e^x}\, dx$.

Solution:
$$\int \frac{e^{4x} - e^{2x} + 1}{e^x}\, dx = \int (e^{3x} - e^x + e^{-x})\, dx$$

$$= \frac{1}{3}e^{3x} - e^x - e^{-x} + C$$

$$= \frac{e^{3x}}{3} - e^x - \frac{1}{e^x} + C$$

$$= \frac{e^{4x} - 3e^{2x} - 3}{3e^x} + C$$

35. Evaluate $\int \dfrac{e^x}{e^x - 1}\, dx$.

Solution:
$$\int \frac{e^x}{e^x - 1}\, dx = \ln|e^x - 1| + C$$

37. Evaluate $\int \dfrac{e^{-2x}}{1 + e^{-2x}}\, dx$.

Solution:
$$\int \frac{e^{-2x}}{1 + e^{-2x}}\, dx = -\frac{1}{2}\int \frac{-2e^{-2x}}{1 + e^{-2x}}\, dx$$

$$= -\frac{1}{2}\ln|1 + e^{-2x}| + C$$

$$= -\ln\sqrt{1 + e^{-2x}} + C$$

39. Evaluate $\int \dfrac{x}{1 - x^2}\, dx$.

Solution:
$$\int \frac{x}{1 - x^2}\, dx = -\frac{1}{2}\int \frac{-2x}{1 - x^2}\, dx$$

$$= -\frac{1}{2}\ln|1 - x^2| + C$$

41. Evaluate $\displaystyle\int \frac{x}{\sqrt{1-x^2}}\,dx$.

Solution:

$$\int \frac{x}{\sqrt{1-x^2}}\,dx = -\frac{1}{2}\int (1-x^2)^{-1/2}(-2x)\,dx$$

$$= -\frac{1}{2}(2)(1-x^2)^{1/2} + C$$

$$= -\sqrt{1-x^2} + C$$

43. Evaluate $\displaystyle\int \frac{1}{7x-2}\,dx$.

Solution:

$$\int \frac{1}{7x-2}\,dx = \frac{1}{7}\int \frac{7}{7x-2}\,dx$$

$$= \frac{1}{7}\ln|7x-2| + C$$

45. Evaluate $\displaystyle\int \frac{1}{(7x-2)^3}\,dx$.

Solution:

$$\int \frac{1}{(7x-2)^3}\,dx = \frac{1}{7}\int (7x-2)^{-3}(7)\,dx$$

$$= \frac{1}{7}\left(\frac{(7x-2)^{-2}}{-2}\right) + C$$

$$= -\frac{1}{14(7x-2)^2} + C$$

47. Evaluate $\displaystyle\int \frac{\ln x^2}{x}\,dx$.

Solution:

$$\int \frac{\ln x^2}{x}\,dx = 2\int (\ln x)\left(\frac{1}{x}\right)\,dx$$

$$= 2\frac{(\ln x)^2}{2} + C = (\ln x)^2 + C$$

49. Evaluate $\displaystyle\int \frac{x^2+3}{x}\,dx$.

Solution:

$$\int \frac{x^2+3}{x}\,dx = \int \left[x + 3\left(\frac{1}{x}\right)\right]dx = \frac{1}{2}x^2 + 3\ln|x| + C$$

51. Evaluate $\displaystyle\int \frac{x^2}{x^3-1}\,dx$.

Solution:

$$\int \frac{x^2}{x^3-1}\,dx = \frac{1}{3}\int \frac{3x^2}{x^3-1}\,dx = \frac{1}{3}\ln|x^3-1| + C$$

Review Exercises for Chapter 5

53. Evaluate $\int \frac{x}{x+1} dx$.

Solution:
$$\int \frac{x}{x+1} dx = \int \left(1 - \frac{1}{x+1}\right) dx$$
$$= x - \ln|x+1| + C$$

55. Evaluate $\int_1^2 \frac{e^{1/x}}{x^2} dx$.

Solution:
$$\int_1^2 \frac{e^{1/x}}{x^2} dx = -\int_1^2 e^{1/x}\left(-\frac{1}{x^2}\right) dx = -e^{1/x}\Big]_1^2$$
$$= -e^{1/2} + e = e - e^{1/2} \approx 1.07$$

57. Evaluate $\int_1^4 \frac{x+1}{x} dx$.

Solution:
$$\int_1^4 \frac{x+1}{x} dx = \int_1^4 \left(1 + \frac{1}{x}\right) dx = \left[x + \ln|x|\right]_1^4$$
$$= (4 + \ln 4) - (1 + \ln 1)$$
$$= 3 + \ln 4 \approx 4.386$$

59. Find the area of the region bounded by the graphs of $y = 1/(x+1)$, $y = 0$, $x = 0$, and $x = 4$.

Solution:
$$A = \int_0^4 \frac{1}{x+1} dx = \ln|x+1|\Big]_0^4$$
$$= \ln 5 - \ln 1 = \ln 5 \approx 1.609$$

61. A deposit of $500 earns interest at the rate of 5% compounded continuously. Find its value after (a) 1 year, (b) 10 years, and (c) 100 years.

Solution:
(a) $A = 500e^{0.05(1)} \approx \$525.64$
(b) $A = 500e^{0.05(10)} \approx \$824.36$
(c) $A = 500e^{0.05(100)} \approx \$74,206.58$

63. How large a deposit, at 7% interest compounded continuously, must be made to obtain a balance of $10,000 in 15 years?

Solution: Since $A = Pe^{rt}$, we have $10000 = Pe^{0.07(15)}$ and $P \approx \$3499.38$.

65. A population is growing continuously at the rate of $2\frac{1}{2}$% per year. Find the time necessary for the population to (a) double in size, and (b) triple in size.

Solution: Since $P = Ce^{0.025t}$, we have $P/C = e^{0.025t}$ and $t = [\ln(P/C)]/0.025$.

(a) When $P = 2C$, $t = \dfrac{\ln(2C/C)}{0.025} = \dfrac{\ln 2}{0.025} \approx 27.73$ years

(b) When $P = 3C$, $t = \dfrac{\ln(3C/C)}{0.025} = \dfrac{\ln 3}{0.025} \approx 43.94$ years

67. A solution of a certain drug contains 500 units per milliliter when it is prepared. After 40 days, it contains 300 units per milliliter. Assuming that the rate of decomposition is proportional to the amount present, find an equation giving the amount A after t days.

Solution: Since $A = Ce^{kt}$ where $A = 500$ when $t = 0$ and $A = 300$ when $t = 40$, we have $500 = Ce^{k(0)}$ which implies that $C = 500$. Therefore, we have

$$A = 500e^{kt}$$
$$300 = 500e^{k(40)}$$
$$(3/5) = e^{k(40)}$$
$$k = [\ln(3/5)]/40$$
$$A = 500e^{[t\ln(3/5)]/40} \approx 500e^{-0.013t}$$

69. The **exponential density function** is given by $f(t) = (1/\mu)e^{-t/\mu}$, where t is the time in minutes and μ is the average time between successive events. The definite integral

$$P(a \leq t \leq b) = \int_a^b \frac{1}{\mu} e^{-t/\mu}\, dt$$

gives the probability that the elapsed time before the next occurrence lies between a and b units of time. The average time between incoming calls at a switchboard is $\mu = 3$ minutes. Find the probabilities for the following time intervals: 0–2, 2–4, 4–6, 6–8, and 8–10.

Solution:

$$\int_a^b \frac{1}{\mu} e^{-t/\mu}\, dt = -\int_a^b e^{t(-1/\mu)}\left(-\frac{1}{\mu}\right) dt = -e^{-t/\mu}\Big]_a^b$$
$$= -e^{-b/\mu} + e^{-a/\mu}$$

Since $\mu = 3$, we have

$$\int_0^2 \frac{1}{3} e^{-t/3}\, dt = -e^{-2/3} + e^{-0/3} = 1 - e^{-2/3} \approx 0.4866$$

$$\int_2^4 \frac{1}{3} e^{-t/3}\, dt = -e^{-4/3} + e^{-2/3} \approx 0.2498$$

$$\int_4^6 \frac{1}{3} e^{-t/3}\, dt = -e^{-6/3} + e^{-4/3} \approx 0.1283$$

$$\int_6^8 \frac{1}{3} e^{-t/3}\, dt = -e^{-8/3} + e^{-6/3} \approx 0.0659$$

$$\int_8^{10} \frac{1}{3} e^{-t/3}\, dt = -e^{-10/3} + e^{-8/3} \approx 0.0338$$

PRACTICE TEST FOR CHAPTER 5

1. Evaluate each expression:
   (a) $27^{4/3}$
   (b) $4^{-5/2}$
   (c) $(8^{2/3})(64^{-1/3})$

2. Solve for x:
   (a) $4^{x+1} = 64$
   (b) $x^{6/5} = 64$
   (c) $(2x + 3)^{10} = 13^{10}$

3. Sketch the graph of
   (a) $f(x) = 3^x$
   (b) $g(x) = (4/9)^x$

4. Find the amount in an account in which $2,000 is invested for 7 years at 8.5% if the interest is compounded:
   (a) annually
   (b) monthly
   (c) continuously

5. Differentiate $y = e^{3x^2}$.

6. Differentiate $y = e^{\sqrt[3]{x}}$.

7. Differentiate $y = \sqrt{e^x + e^{-x}}$.

8. Differentiate $y = x^3 e^{2x}$.

9. Differentiate $y = \dfrac{e^x + 3}{4x}$.

10. Evaluate $\int e^{7x}\, dx$.

11. Evaluate $\int xe^{4x^2}\, dx$.

12. Evaluate $\int e^x(1 + 4e^x)^3\, dx$.

13. Evaluate $\int (e^x + 2)^2\, dx$.

14. Evaluate $\int \dfrac{e^{3x} - 4e^x + 1}{e^x}\, dx$.

15. Write $\ln 5 = 1.6094...$ as an exponential equation.

16. Sketch the graph of
    (a) $y = \ln(x + 2)$
    (b) $y = \ln x + 2$

17. Write the given expression as a single logarithm.
    (a) $\ln(3x+1) - \ln(2x-5)$
    (b) $4\ln x - 3\ln y - (1/2)\ln z$

18. Solve for x.
    (a) $\ln x = 17$
    (b) $5^{3x} = 2$

19. Differentiate $y = \ln(6x - 7)$.

20. Differentiate $y = \ln \dfrac{x^3}{4x+10}$.

21. Differentiate $y = \ln \sqrt[3]{\dfrac{x}{x+3}}$.

22. Differentiate $y = x^4 \ln x$.

23. Differentiate $y = \sqrt{\ln x + 1}$.

24. Evaluate $\displaystyle\int \dfrac{1}{x+6}\,dx$.

25. Evaluate $\displaystyle\int \dfrac{x^2}{8-x^3}\,dx$.

26. Evaluate $\displaystyle\int \dfrac{e^x}{1+3e^x}\,dx$.

27. Evaluate $\displaystyle\int \dfrac{(\ln x)^6}{x}\,dx$.

28. Evaluate $\displaystyle\int \dfrac{x^2+5}{x-1}\,dx$.

29. Find the exponential function $y = Ce^{kt}$ that passes through the two given points.
    (a) $(0, 7)$ and $(4, 1/3)$
    (b) $(3, 2/3)$ and $(8, 8)$

30. If $5,000 is invested in an account in which the interest rate of 12% is compounded continuously, find the time required for the investment to double.

# Chapter 6   Techniques of Integration

● Section 6.1   Integration by Substitution

1. Evaluate $\int (3x - 2)^4 \, dx$.

   Solution:
   $$\int (3x - 2)^4 \, dx = \frac{1}{3} \int (3x - 2)^4 (3) \, dx$$
   $$= \left(\frac{1}{3}\right) \frac{(3x - 2)^5}{5} + C$$
   $$= \frac{1}{15}(3x - 2)^5 + C$$

3. Evaluate $\int \frac{2}{(t - 9)^2} \, dt$.

   Solution:
   $$\int \frac{2}{(t - 9)^2} \, dt = 2 \int (t - 9)^{-2} \, dt$$
   $$= (2)\frac{(t - 9)^{-1}}{-1} + C$$
   $$= -\frac{2}{t - 9} + C$$
   $$= \frac{2}{9 - t} + C$$

5. Evaluate $\int 2x\sqrt{1 + x^2} \, dx$.

   Solution:
   $$\int 2x\sqrt{1 + x^2} \, dx = \int (1 + x^2)^{1/2}(2x) \, dx$$
   $$= \frac{2}{3}(1 + x^2)^{3/2} + C$$

7. Evaluate $\int \frac{12x + 2}{3x^2 + x} \, dx$.

   Solution:
   $$\int \frac{12x + 2}{3x^2 + x} \, dx = 2 \int \frac{6x + 1}{3x^2 + x} \, dx$$
   $$= 2 \ln |3x^2 + x| + C$$
   $$= \ln (3x^2 + x)^2 + C$$

Section 6.1

9. Evaluate $\int \frac{1}{(5x+1)^3} dx$.

    Solution:
    $$\int \frac{1}{(5x+1)^3} dx = \frac{1}{5}\int (5x+1)^{-3}(5)\, dx$$
    $$= \left(\frac{1}{5}\right)\frac{(5x+1)^{-2}}{-2} + C$$
    $$= -\frac{1}{10(5x+1)^2} + C$$

11. Evaluate $\int \frac{\ln 2x^2}{x} dx$.

    Solution:
    $$\int \frac{\ln 2x^2}{x} dx = \frac{1}{2}\int (\ln 2x^2)\left(\frac{2}{x}\right) dx$$
    $$= \left(\frac{1}{2}\right)\frac{(\ln 2x^2)^2}{2} + C = \frac{1}{4}(\ln 2x^2)^2 + C$$

13. Evaluate $\int \frac{e^{3x}}{1-e^{3x}} dx$.

    Solution:
    $$\int \frac{e^{3x}}{1-e^{3x}} dx = -\frac{1}{3}\int \frac{-3e^{3x}}{1-e^{3x}} dx$$
    $$= -\frac{1}{3}\ln|1-e^{3x}| + C$$

15. Evaluate $\int \frac{x^2}{x-1} dx$.

    Solution:
    $$\int \frac{x^2}{x-1} dx = \int \left[x + 1 + \frac{1}{x-1}\right] dx$$
    $$= \frac{x^2}{2} + x + \ln|x-1| + C$$

17. Evaluate $\int x\sqrt{4-2x^2}\, dx$.

    Solution:
    $$\int x\sqrt{4-2x^2}\, dx = -\frac{1}{4}\int (4-2x^2)^{1/2}(-4x)\, dx$$
    $$= -\frac{1}{4}\left(\frac{2}{3}\right)(4-2x^2)^{3/2} + C$$
    $$= -\frac{1}{6}(4-2x^2)^{3/2} + C$$

Section 6.1 253

19. Evaluate $\int e^{5x}\,dx$.

Solution:
$$\int e^{5x}\,dx = \frac{1}{5}\int e^{5x}(5)\,dx = \frac{1}{5}e^{5x} + C$$

21. Evaluate $\int \frac{x}{(x+1)^3}\,dx$.

Solution: Let $u = x + 1$, then $x = u - 1$ and $dx = du$.

$$\int \frac{x}{(x+1)^3}\,dx = \int \frac{u-1}{u^3}\,du$$

$$= \int (u^{-2} - u^{-3})\,du = \frac{u^{-1}}{-1} - \frac{u^{-2}}{-2} + C$$

$$= -\frac{1}{u} + \frac{1}{2u^2} + C = \frac{1 - 2u}{2u^2} + C$$

$$= \frac{1 - 2(x+1)}{2(x+1)^2} + C = -\frac{2x+1}{2(x+1)^2} + C$$

23. Evaluate $\int \frac{x}{(3x-1)^2}\,dx$.

Solution: Let $u = 3x - 1$, then $x = \frac{u+1}{3}$ and $dx = \frac{1}{3}du$.

$$\int \frac{x}{(3x-1)^2}\,dx = \int \frac{(u+1)/3}{u^2}\left(\frac{1}{3}\right)du$$

$$= \frac{1}{9}\int \left(\frac{1}{u} + \frac{1}{u^2}\right)du$$

$$= \frac{1}{9}\left[\ln|u| - \frac{1}{u}\right] + C$$

$$= \frac{1}{9}\left[\ln|3x-1| - \frac{1}{3x-1}\right] + C$$

25. Evaluate $\int x(1-x)^4\,dx$.

Solution: Let $u = 1 - x$, then $x = 1 - u$ and $dx = -du$.

$$\int x(1-x)^4\,dx = \int (1-u)u^4(-du)$$

$$= \int (u^5 - u^4)\,du = \frac{u^6}{6} - \frac{u^5}{5} + C$$

$$= \frac{u^5}{30}(5u - 6) + C$$

$$= -\frac{(1-x)^5}{30}(1 + 5x) + C$$

254   Section 6.1

27. Evaluate $\int \frac{x-1}{x^2-2x} dx$.

   Solution:
   $$\int \frac{x-1}{x^2-2x} dx = \frac{1}{2}\int \frac{2x-2}{x^2-2x} dx$$
   $$= \frac{1}{2}\ln|x^2 - 2x| + C$$

29. Evaluate $\int x\sqrt{x-3}\, dx$.

   Solution: Let $u = x - 3$, then $x = u + 3$ and $dx = du$.
   $$\int x\sqrt{x-3}\, dx = \int (u+3)u^{1/2}\, du$$
   $$= \int (u^{3/2} + 3u^{1/2})\, du$$
   $$= \frac{2}{5}u^{5/2} + 2u^{3/2} + C$$
   $$= \frac{2u^{3/2}}{5}(u + 5) + C$$
   $$= \frac{2}{5}(x-3)^{3/2}(x+2) + C$$

31. Evaluate $\int x^2\sqrt{1-x}\, dx$.

   Solution: Let $u = \sqrt{1-x}$, then $x = 1 - u^2$ and $dx = -2u\, du$.

   $$\int x^2\sqrt{1-x}\, dx$$
   $$= \int (1-u^2)^2 u(-2u)\, du$$
   $$= -2\int (u^6 - 2u^4 + u^2)\, du$$
   $$= -2\left(\frac{u^7}{7} - \frac{2u^5}{5} + \frac{u^3}{3}\right) + C$$
   $$= -\frac{2u^3}{105}(35 - 42u^2 + 15u^4) + C$$
   $$= -\frac{2}{105}(1-x)^{3/2}(15x^2 + 12x + 8) + C$$

Section 6.1   255

33. Evaluate $\int \dfrac{x^2 - 1}{\sqrt{2x - 1}}\, dx$.

Solution: Let $u = \sqrt{2x - 1}$, then $x = (u^2 + 1)/2$ and $dx = u\, du$.

$$\int \dfrac{x^2 - 1}{\sqrt{2x - 1}}\, dx$$

$$= \int \dfrac{[(u^2 + 1)/2]^2 - 1}{u}(u)\, du$$

$$= \dfrac{1}{4}\int (u^4 + 2u^2 - 3)\, du$$

$$= \dfrac{1}{4}\left(\dfrac{u^5}{5} + \dfrac{2u^3}{3} - 3u\right) + C$$

$$= \dfrac{u}{60}(3u^4 + 10u^2 - 45) + C$$

$$= \dfrac{\sqrt{2x - 1}}{60}[3(2x-1)^2 + 10(2x-1) - 45] + C$$

$$= \dfrac{\sqrt{2x - 1}}{60}[12x^2 + 8x - 52] + C$$

$$= \dfrac{\sqrt{2x - 1}}{15}(3x^2 + 2x - 13) + C$$

35. Evaluate $\int \dfrac{1}{1 + \sqrt{t}}\, dt$.

Solution: Let $u = 1 + \sqrt{t}$, then $t = (u - 1)^2$ and $dt = 2(u - 1)\, du$.

$$\int \dfrac{1}{1 + \sqrt{t}}\, dt = \int \left(\dfrac{1}{u}\right) 2(u - 1)\, du$$

$$= 2\int \left[1 - \dfrac{1}{u}\right] du$$

$$= 2[u - \ln|u|] + C_1$$

$$= 2[1 + \sqrt{t} - \ln|1 + \sqrt{t}|] + C_1$$

$$= 2\sqrt{t} - 2\ln|1 + \sqrt{t}| + C$$

37. Evaluate $\int \dfrac{2\sqrt{t} + 1}{t}\, dt$.

Solution:

$$\int \dfrac{2\sqrt{t} + 1}{t}\, dt = \int \left(2t^{-1/2} + \dfrac{1}{t}\right) dt$$

$$= 4t^{1/2} + \ln|t| + C$$

$$= 4\sqrt{t} + \ln|t| + C$$

39. Evaluate $\int \dfrac{1}{6x + \sqrt{2x}} \, dx$.

   Solution: Let $u = \sqrt{2x}$, then $x = \dfrac{u^2}{2}$ and $dx = u \, du$.

   $$\int \dfrac{1}{6x + \sqrt{2x}} \, dx = \int \dfrac{1}{6(u^2/2) + u}(u) \, du$$

   $$= \int \dfrac{u}{3u^2 + u} \, du = \int \dfrac{1}{3u + 1} \, du$$

   $$= \dfrac{1}{3} \int \dfrac{3}{3u + 1} \, du = \dfrac{1}{3} \ln |3u + 1| + C$$

   $$= \dfrac{1}{3} \ln |3\sqrt{2x} + 1| + C$$

41. Evaluate $\int_0^4 \sqrt{2x + 1} \, dx$.

   Solution:

   $$\int_0^4 \sqrt{2x + 1} \, dx = \dfrac{1}{2} \int_0^4 (2x + 1)^{1/2} (2) \, dx$$

   $$= \dfrac{1}{2} \left(\dfrac{2}{3}\right) (2x + 1)^{3/2} \Big]_0^4$$

   $$= \dfrac{1}{3}(9)^{3/2} - \dfrac{1}{3}(1)^{3/2} = \dfrac{26}{3}$$

43. Evaluate $\int_0^1 3x e^{x^2} \, dx$.

   Solution:

   $$\int_0^1 3x e^{x^2} \, dx = \dfrac{3}{2} \int_0^1 2x e^{x^2} \, dx$$

   $$= \dfrac{3}{2} e^{x^2} \Big]_0^1 = \dfrac{3}{2}(e - 1) \approx 2.577$$

45. Evaluate $\int_0^5 \dfrac{x}{(x + 5)^2} \, dx$.

   Solution: Let $u = x + 5$, then $x = u - 5$ and $dx = du$. $u = 10$ when $x = 5$ and $u = 5$ when $x = 0$.

   $$\int_0^5 \dfrac{x}{(x + 5)^2} \, dx = \int_5^{10} \dfrac{u - 5}{u^2} \, du = \int_5^{10} \left(\dfrac{1}{u} - \dfrac{5}{u^2}\right) du$$

   $$= \left[\ln |u| + \dfrac{5}{u}\right]_5^{10}$$

   $$= \left(\ln 10 + \dfrac{1}{2}\right) - (\ln 5 + 1)$$

   $$= (\ln 2) - \dfrac{1}{2} \approx 0.193$$

Section 6.1

47. Evaluate $\int_0^{0.5} x(1-x)^3 \, dx$.

Solution: Let $u = 1 - x$, then $x = 1 - u$ and $dx = -du$. $u = 1$ when $x = 0$ and $u = 0.5$ when $x = 0.5$.

$$\int_0^{0.5} x(1-x)^3 \, dx = \int_1^{0.5} (1-u) u^3 (-du)$$

$$= \int_1^{0.5} (u^4 - u^3) \, du = \left(\frac{u^5}{5} - \frac{u^4}{4} \right) \Big]_1^{0.5}$$

$$= \left(\frac{1}{160} - \frac{1}{64} \right) - \left(\frac{1}{5} - \frac{1}{4} \right) = \frac{13}{320}$$

49. Evaluate $\int_3^7 x\sqrt{x-3} \, dx$.

Solution: Let $u = \sqrt{x-3}$, then $x = u^2 + 3$ and $dx = 2u \, du$. $u = 0$ when $x = 3$ and $u = 2$ when $x = 7$.

$$\int_3^7 x\sqrt{x-3} \, dx = \int_0^2 (u^2 + 3) u (2u \, du)$$

$$= 2 \int_0^2 (u^4 + 3u^2) \, du = 2 \left(\frac{u^5}{5} + u^3 \right) \Big]_0^2$$

$$= 2 \left(\frac{32}{5} + 8 \right) = \frac{144}{5}$$

51. Evaluate $\int_0^7 x\sqrt[3]{x+1} \, dx$.

Solution: Let $u = \sqrt[3]{x+1}$, then $x = u^3 - 1$ and $dx = 3u^2 \, du$. $u = 1$ when $x = 0$ and $u = 2$ when $x = 7$.

$$\int_0^7 x\sqrt[3]{x+1} \, dx = 3 \int_1^2 (u^3 - 1) u^3 \, du$$

$$= 3 \int_1^2 (u^6 - u^3) \, du = 3 \left(\frac{u^7}{7} - \frac{u^4}{4} \right) \Big]_1^2$$

$$= 3 \left(\frac{128}{7} - 4 \right) - 3 \left(\frac{1}{7} - \frac{1}{4} \right) = \frac{1209}{28}$$

53. Find the area of the region bounded by the graphs of $y = x\sqrt{x+1}$ and $y = 0$.

Solution: Let $u = \sqrt{x+1}$, then $x = u^2 - 1$ and $dx = 2u \, du$. $u = 0$ when $x = -1$ and $u = 1$ when $x = 0$.

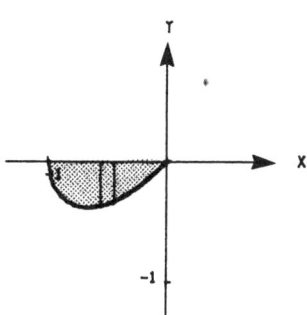

$$A = -\int_{-1}^0 x\sqrt{x+1} \, dx = -2 \int_0^1 (u^2 - 1) u^2 \, du$$

$$= -2 \left(\frac{u^5}{5} - \frac{u^3}{3} \right) \Big]_0^1$$

$$= \frac{4}{15} \text{ square units}$$

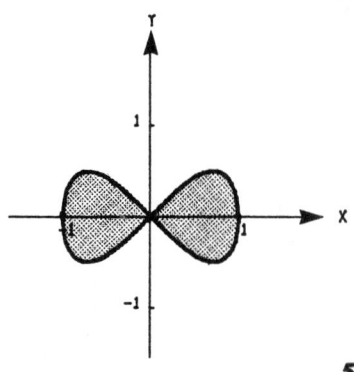

55. Find the area of the region bounded by the graph of $y^2 = x^2(1 - x^2)$.

Solution: Use $y = x\sqrt{1 - x^2}$ and $y = 0$ and multiply by 4.

$$A = 4 \int_0^1 x\sqrt{1 - x^2}\, dx = \frac{4}{-2} \int_0^1 (1 - x^2)^{1/2}(-2x)\, dx$$

$$= -2\left(\frac{2}{3}\right)(1 - x^2)^{3/2} \Big]_0^1$$

$$= -\frac{4}{3}[0 - 1] = \frac{4}{3} \text{ square units}$$

57. Find the volume of the solid generated by revolving the region bounded by the graph of $y = x\sqrt{1 - x^2}$ about the x-axis.

Solution:

$$V = 2\pi \int_0^1 [x\sqrt{1 - x^2}]^2\, dx = 2\pi \int_0^1 x^2(1 - x^2)\, dx$$

$$= 2\pi \left[\frac{x^3}{3} - \frac{x^5}{5}\right]_0^1$$

$$= 2\pi \left(\frac{1}{3} - \frac{1}{5}\right) = \frac{4\pi}{15} \text{ cubic units}$$

59. Find the average amount by which the function $f(x)$ exceeds the function $g(x)$ on the interval $[0, 1]$.

$$f(x) = \frac{1}{x + 1}, \quad g(x) = \frac{x}{(x + 1)^2}$$

Solution:

$$\frac{1}{1 - 0} \int_0^1 [f(x) - g(x)]\, dx = \int_0^1 \left[\frac{1}{x + 1} - \frac{x}{(x + 1)^2}\right] dx$$

$$= \int_0^1 \frac{1}{(x + 1)^2}\, dx = -\frac{1}{x + 1}\Big]_0^1$$

$$= -\frac{1}{2} + 1 = \frac{1}{2}$$

61. In the Introductory Example of this section, the drug-retention pattern was given by $y = x^2(1 - x)^{1/2}$. Find the portion of the population retaining between 0% and 50% of the drug after 24 hours.

Solution: The portion of population retaining between 0% and 50% is given by

$$\frac{\int_0^{0.5} x^2(1 - x)^{1/2}\, dx}{\int_0^1 x^2(1 - x)^{1/2}\, dx} = \frac{-\frac{2}{105}(1 - x)^{3/2}(15x^2 + 12x + 8)\Big]_0^{0.5}}{-\frac{2}{105}(1 - x)^{3/2}(15x^2 + 12x + 8)\Big]_0^1}$$

$$= \frac{(-2/105)[6.27557 - 8]}{(-2/105)[0 - 8]}$$

$$\approx 0.21555 \approx 21.56\%$$

63. The probability of recall in a certain experiment is found to be

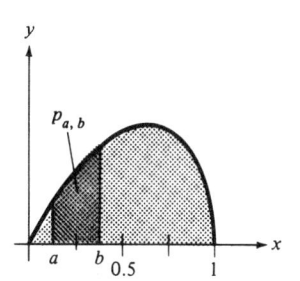

$$P(a \leq x \leq b) = \int_a^b \frac{15}{4} x\sqrt{1-x}\, dx$$

where x represents the percentage of recall.
(a) What is the probability that a randomly chosen individual will recall between 50% and 75% of the material?
(b) What is the median percentage recall? That is, for what value of b is it true that the probability from 0 to b is 0.5?

Solution: Let $u = \sqrt{1-x}$, then $x = 1 - u^2$ and $dx = -2u\, du$.

$$\int \frac{15}{4} x\sqrt{1-x}\, dx = \frac{1}{2}(1-x)^{3/2}(-3x-2) + C$$

(a) $P(0.50 \leq x \leq 0.75) = \frac{1}{2}(1-x)^{3/2}(-3x-2) \Big]_{0.50}^{0.75}$

$$= \frac{1}{2}[-0.53125 + 1.23744]$$

$$\approx 0.353$$

(b) $P(0 \leq x \leq b) = \frac{1}{2}(1-x)^{3/2}(-3x-2) \Big]_0^b$

$$= \frac{1}{2}[(1-b)^{3/2}(-3b-2) + 2] = 0.5$$

Solving this equation for b produces

$$(1-b)^{3/2}(-3b-2) + 2 = 1$$
$$(1-b)^{3/2}(-3b-2) = -1$$
$$(1-b)^{3/2}(3b+2) = 1$$
$$(1-b)^3(3b+2)^2 = 1$$
$$b \approx 0.586$$

65. A company sells a seasonal product that has a daily revenue approximated by

$$R = 0.06t^2(365-t)^{1/2} + 1250, \quad 0 \leq t \leq 365$$

Find the average daily revenue over the period of one year.

Solution: The average daily revenue is given by

$$\frac{1}{365-0} \int_0^{365} [0.06t^2(365-t)^{1/2} + 1250]\, dt$$

$$= \frac{1}{365} \int_0^{365} [0.06t^2(365-t)^{1/2}]\, dt + \frac{1}{365} \int_0^{365} 1250\, dt$$

$$= \frac{1}{365} \int_{\sqrt{365}}^0 0.06(365-u^2)^2 u(-2u)\, du + 1250$$

$$\approx \$24520.95$$

Section 6.2

● Section 6.2 Integration by Parts

1. Evaluate $\int e^{2x}\, dx$.

    Solution:
    $$\int e^{2x}\, dx = \frac{1}{2}\int e^{2x}(2)\, dx = \frac{1}{2}e^{2x} + C$$

3. Evaluate $\int xe^{2x}\, dx$.

    Solution: Let $u = x$ and $dv = e^{2x}\, dx$ then $du = dx$ and $v = (1/2)e^{2x}$.
    $$\int xe^{2x}\, dx = \frac{1}{2}xe^{2x} - \frac{1}{2}\int e^{2x}\, dx$$
    $$= \frac{1}{2}xe^{2x} - \frac{1}{4}e^{2x} + C = \frac{e^{2x}}{4}(2x - 1) + C$$

5. Evaluate $\int xe^{x^2}\, dx$.

    Solution:
    $$\int xe^{x^2}\, dx = \frac{1}{2}e^{x^2} + C$$

7. Evaluate $\int x^2 e^{-2x}\, dx$.

    Solution: Let $u = x^2$ and $dv = e^{-2x}\, dx$, then $du = 2x\, dx$ and $v = (-1/2)e^{-2x}$.
    $$\int x^2 e^{-2x}\, dx = -\frac{1}{2}x^2 e^{-2x} + \int xe^{-2x}\, dx$$
    Let $u = x$ and $dv = e^{-2x}\, dx$, then $du = dx$ and we have $v = (-1/2)e^{-2x}$.
    $$\int x^2 e^{-2x}\, dx = -\frac{1}{2}x^2 e^{-2x} - \frac{1}{2}xe^{-2x} + \frac{1}{2}\int e^{-2x}\, dx$$
    $$= -\frac{1}{2}x^2 e^{-2x} - \frac{1}{2}xe^{-2x} - \frac{1}{4}e^{-2x} + C$$
    $$= -\frac{e^{-2x}}{4}(2x^2 + 2x + 1) + C$$

9. Evaluate $\int x^3 e^x\, dx$.

    Solution: Let $u = x^3$ and $dv = e^x\, dx$, then $du = 3x^2\, dx$ and $v = e^x$.
    $$\int x^3 e^x\, dx = x^3 e^x - 3\int x^2 e^x\, dx$$

Section 6.2

Let $u = x^2$ and $dv = e^x\, dx$, then $du = 2x\, dx$ and $v = e^x$.

$$\int x^3 e^x\, dx = x^3 e^x - 3(x^2 e^x - 2\int x e^x\, dx)$$

$$= x^3 e^x - 3x^2 e^x + 6\int x e^x\, dx$$

Let $u = x$ and $dv = e^x\, dx$, then $du = dx$ and $v = e^x$.

$$\int x^3 e^x\, dx = x^3 e^x - 3x^2 e^x + 6(x e^x - \int e^x\, dx)$$

$$= x^3 e^x - 3x^2 e^x + 6x e^x - 6 e^x + C$$

$$= e^x(x^3 - 3x^2 + 6x - 6) + C$$

11. Evaluate $\int x^3 \ln x\, dx$.

   Solution: Let $u = \ln x$ and $dv = x^3\, dx$, then $du = 1/x\, dx$ and $v = x^4/4$.

   $$\int x^3 \ln x\, dx = \frac{x^4}{4}\ln x - \frac{1}{4}\int x^3\, dx$$

   $$= \frac{x^4}{4}\ln x - \frac{x^4}{16} + C = \frac{x^4}{16}(4\ln x - 1) + C$$

13. Evaluate $\int t \ln(t + 1)\, dt$.

   Solution: Let $u = \ln(t + 1)$ and $dv = t\, dt$, then $du = 1/(t+1)\, dt$ and $v = t^2/2$.

   $$\int t \ln(t+1)\, dt = \frac{t^2}{2}\ln(t+1) - \frac{1}{2}\int \frac{t^2}{t+1}\, dt$$

   $$= \frac{t^2}{2}\ln(t+1) - \frac{1}{2}\int (t - 1 + \frac{1}{t+1})\, dt$$

   $$= \frac{t^2}{2}\ln(t+1) - \frac{1}{2}[\frac{t^2}{2} - t + \ln(t+1)] + C$$

   $$= \frac{1}{4}[2(t^2 - 1)\ln(t+1) + t(2 - t)] + C$$

15. Evaluate $\int (\ln x)^2\, dx$.

   Solution: Let $u = (\ln x)^2$ and $dv = dx$, then $du = (2\ln x)/x\, dx$ and $v = x$.

   $$\int (\ln x)^2\, dx = x(\ln x)^2 - 2\int \ln x\, dx$$

   Let $u = \ln x$ and $dv = dx$, then $du = 1/x\, dx$ and $v = x$.

   $$\int (\ln x)^2\, dx = x(\ln x)^2 - 2(x\ln x - \int dx)$$

   $$= x(\ln x)^2 - 2x\ln x + 2x + C$$

   $$= x[(\ln x)^2 - 2\ln x + 2] + C$$

Section 6.2

17. Evaluate $\int \dfrac{(\ln x)^2}{x}\, dx$.

Solution:
$$\int \frac{(\ln x)^2}{x}\, dx = \int (\ln x)^2 \left(\frac{1}{x}\right) dx$$
$$= \frac{(\ln x)^3}{3} + C$$

19. Evaluate $\int x\sqrt{x-1}\, dx$.

Solution: Let $u = x$ and $dv = \sqrt{x-1}\, dx$, then $du = dx$ and $v = (2/3)(x-1)^{3/2}$.
$$\int x\sqrt{x-1}\, dx = \frac{2}{3} x(x-1)^{3/2} - \frac{2}{3}\int (x-1)^{3/2}\, dx$$
$$= \frac{2}{3} x(x-1)^{3/2} - \frac{4}{15}(x-1)^{5/2} + C$$
$$= \frac{2}{15}(x-1)^{3/2}(3x+2) + C$$

21. Evaluate $\int (x^2 - 1)e^x\, dx$.

Solution: Let $u = x^2 - 1$ and $dv = e^x\, dx$, then $du = 2x\, dx$ and $v = e^x$.
$$\int (x^2 - 1)e^x\, dx = (x^2 - 1)e^x - 2\int xe^x\, dx$$

Let $u = x$ and $dv = e^x\, dx$, then $du = dx$ and $v = e^x$.
$$\int (x^2 - 1)e^x\, dx = (x^2 - 1)e^x - 2\left[xe^x - \int e^x\, dx\right]$$
$$= (x^2 - 1)e^x - 2xe^x + 2e^x + C$$
$$= e^x(x^2 - 2x + 1) + C$$
$$= (x-1)^2 e^x + C$$

23. Evaluate $\int \dfrac{xe^{2x}}{(2x+1)^2}\, dx$.

Solution: Let $u = xe^{2x}$ and $dv = (2x+1)^{-2}\, dx$ then $du = e^{2x}(2x+1)\, dx$ and $v = -1/[2(2x+1)]$.
$$\int \frac{xe^{2x}}{(2x+1)^2}\, dx = -\frac{xe^{2x}}{2(2x+1)} + \frac{1}{2}\int e^{2x}\, dx$$
$$= -\frac{xe^{2x}}{2(2x+1)} + \frac{1}{4}e^{2x} + C$$
$$= \frac{e^{2x}}{4(2x+1)} + C$$

Section 6.2

25. Evaluate $\int_0^1 x^2 e^x \, dx$.

Solution: Let $u = x^2$ and $dv = e^x \, dx$, then $du = 2x \, dx$ and $v = e^x$.

$$\int_0^1 x^2 e^x \, dx = x^2 e^x \Big]_0^1 - 2\int_0^1 x e^x \, dx$$

Let $u = x$ and $dv = e^x \, dx$, then $du = dx$ and $v = e^x$.

$$\int_0^1 x^2 e^x \, dx = x^2 e^x \Big]_0^1 - 2[x e^x \Big]_0^1 - \int_0^1 e^x \, dx]$$

$$= e - 2[e - e^x \Big]_0^1]$$

$$= e - 2e + 2(e - 1) = e - 2 \approx 0.718$$

27. Evaluate $\int_1^e x^4 \ln x \, dx$.

Solution: Let $u = \ln x$ and $dv = x^4 \, dx$, then $du = 1/x \, dx$ and $v = x^5/5$.

$$\int_1^e x^4 \ln x \, dx = \frac{x^5}{5} \ln x \Big]_1^e - \int_1^e \frac{1}{x} \cdot \frac{x^5}{5} \, dx$$

$$= \frac{1}{5}(e^5 - 0) - \frac{1}{5}\int_1^e x^4 \, dx$$

$$= \frac{1}{5} e^5 - \frac{1}{25} x^5 \Big]_1^e$$

$$= \frac{1}{5} e^5 - \frac{1}{25} e^5 + \frac{1}{25}$$

$$= \frac{1}{25}(4e^5 + 1) \approx 23.786$$

29. Integrate $\int 2x\sqrt{2x - 3} \, dx$.

   (a) by parts, letting $dv = \sqrt{2x - 3} \, dx$
   (b) by substitution, letting $u = \sqrt{2x - 3}$

Solution:
(a) Let $u = 2x$ and $dv = \sqrt{2x - 3} \, dx$, then $du = 2 \, dx$ and $v = (1/3)(2x - 3)^{3/2}$.

$$\int 2x\sqrt{2x - 3} \, dx$$

$$= \frac{2}{3} x(2x - 3)^{3/2} - \frac{2}{3} \int (2x - 3)^{3/2} \, dx$$

$$= \frac{2}{3} x(2x - 3)^{3/2} - \frac{2}{15}(2x - 3)^{5/2} + C$$

$$= \frac{2}{15}(2x - 3)^{3/2}(3x + 3) + C$$

$$= \frac{2}{5}(2x - 3)^{3/2}(x + 1) + C$$

(b) Let $u = \sqrt{2x - 3}$, then $x = (u^2 + 3)/2$ and $dx = u\, du$.

$$\int 2x\sqrt{2x - 3}\, dx = \int (u^2 + 3)u^2\, du = \frac{u^5}{5} + u^3 + C$$

$$= \frac{u^3}{5}(u^2 + 5) + C$$

$$= \frac{2}{5}(2x - 3)^{3/2}(x + 1) + C$$

31. Integrate $\int \dfrac{x}{\sqrt{4 + 5x}}\, dx$.

    (a) by parts, letting $dv = 1/\sqrt{4 + 5x}\, dx$
    (b) by substitution, letting $u = \sqrt{4 + 5x}$

    Solution:
    (a) Let $u = x$ and $dv = 1/\sqrt{4 + 5x}\, dx$, then $du = dx$ and $v = (2/5)\sqrt{4 + 5x}$.

    $$\int \frac{x}{\sqrt{4 + 5x}}\, dx = \frac{2}{5}x\sqrt{4 + 5x} - \frac{2}{5}\int \sqrt{4 + 5x}\, dx$$

    $$= \frac{2}{5}x\sqrt{4 + 5x} - \frac{4}{75}(4 + 5x)^{3/2} + C$$

    $$= \frac{2}{75}\sqrt{4 + 5x}(5x - 8) + C$$

    (b) Let $u = \sqrt{4 + 5x}$, then $x = (u^2 - 4)/5$ and $dx = 2u/5\, du$.

    $$\int \frac{x}{\sqrt{4 + 5x}}\, dx = \int \frac{(u^2 - 4)/5}{u}\left(\frac{2u}{5}\right) du$$

    $$= \frac{2}{25}\int (u^2 - 4)\, du = \frac{2}{25}\left(\frac{u^3}{3} - 4u\right) + C$$

    $$= \frac{2}{75}\sqrt{4 + 5x}(5x - 8) + C$$

33. Use integration by parts to verify the formula

    $$\int x^n \ln x\, dx = \frac{x^{n+1}}{(n + 1)^2}[-1 + (n + 1)\ln x] + C, \quad n \ne -1$$

    Solution: Let $u = \ln x$ and $dv = x^n\, dx$, then $du = 1/x\, dx$ and $v = x^{n+1}/(n + 1)$.

    $$\int x^n \ln x\, dx = \frac{x^{n+1}}{n + 1}\ln x - \int \frac{1}{x}\cdot\frac{x^{n+1}}{n + 1}\, dx$$

    $$= \frac{x^{n+1}}{n + 1}\ln x - \frac{1}{n + 1}\int x^n\, dx$$

    $$= \frac{x^{n+1}}{n + 1}\ln x - \frac{1}{n + 1}\cdot\frac{x^{n+1}}{n + 1} + C$$

    $$= \frac{x^{n+1}}{(n + 1)^2}[-1 + (n + 1)\ln x] + C$$

Section 6.2

35. Evaluate $\int x^2 e^{5x} \, dx$. [Use Exercise 34.]

Solution: Using $n = 2$ and $a = 5$, we have

$$\int x^2 e^{5x} \, dx = \frac{x^2 e^{5x}}{5} - \frac{2}{5} \int x e^{5x} \, dx$$

Now, using $n = 1$ and $a = 5$, we have

$$\int x^2 e^{5x} \, dx = \frac{x^2 e^{5x}}{5} - \frac{2}{5}[\frac{x e^{5x}}{5} - \frac{1}{5} \int e^{5x} \, dx]$$

$$= \frac{x^2 e^{5x}}{5} - \frac{2 x e^{5x}}{25} + \frac{2 e^{5x}}{125} + C$$

$$= \frac{e^{5x}}{125}(25x^2 - 10x + 2) + C$$

37. Evaluate $\int x^5 \ln x \, dx$. [Use Exercise 33.]

Solution: Using $n = 5$, we have

$$\int x^5 \ln x \, dx = \frac{x^6}{36}[-1 + 6 \ln x] + C$$

39. Find the area of the region bounded by the graphs of $y = xe^{-x}$, $y = 0$, and $x = 4$.

Solution: Letting $u = x$ and $dv = e^{-x} \, dx$, we have $du = dx$ and $v = -e^{-x}$, and the area is

$$A = \int_0^4 x e^{-x} \, dx = -x e^{-x} \Big]_0^4 + \int_0^4 e^{-x} \, dx$$

$$= -4 e^{-4} - e^{-x} \Big]_0^4 = -4 e^{-4} - e^{-4} + 1$$

$$= 1 - 5 e^{-4} \approx 0.908$$

41. Given the region bounded by the graphs of $y = \ln x$, $y = 0$, and $x = e$, find (a) the area of the region and (b) the volume of the solid generated by revolving the region about the x-axis.

Solution:

(a) Letting $u = \ln x$ and $dv = dx$, we have $du = 1/x \, dx$ and $v = x$, and the area is

$$A = \int_1^e \ln x \, dx = x \ln x \Big]_1^e - \int_1^e \frac{1}{x}(x) \, dx = e - x \Big]_1^e$$

$$= e - e + 1 = 1$$

(b) Letting $u = (\ln x)^2$ and $dv = dx$, we have $du = 2(\ln x)(1/x) \, dx$ and $v = x$, and the volume is

$$V = \pi \int_1^e (\ln x)^2 \, dx = \pi [x(\ln x)^2 \Big]_1^e - 2 \int_1^e \ln x \, dx]$$

$$= \pi[e - 2\Big[x \ln x - x\Big]_1^e]$$

$$= \pi(e - 2[(e - e) - (0 - 1)])$$

$$= \pi(e - 2) \approx 2.257$$

Section 6.2

43. A model for the ability M of a child to memorize, measured on a scale from 0 to 10, is given by $M = 1 + 1.6t \ln t$, $0 < t \leq 4$, where t is the child's age in years. Find the average value of M
(a) between the child's first and second birthdays
(b) between the child's third and fourth birthdays
Solution:

(a) Average $= \int_1^2 (1.6t \ln t + 1) \, dt$

$= (0.8t^2 \ln t - 0.4t^2 + t) \Big]_1^2$

$= 3.2(\ln 2) - 0.2 \approx 2.0181$

(b) Average $= \int_3^4 (1.6t \ln t + 1) \, dt$

$= (0.8t^2 \ln t - 0.4t^2 + t) \Big]_3^4$

$= 12.8(\ln 4) - 7.2(\ln 3) - 1.8 \approx 8.0346$

45. Find the present value of the income given by $c(t) = 5000$, measured in dollars, over $t_1 = 4$ years at the annual interest rate $r = 11\%$.
Solution:

$V = \int_0^{t_1} c(t) e^{-rt} \, dt$

$= \int_0^4 5000 e^{-0.11t} \, dt = \frac{5000}{-0.11} e^{-0.11t} \Big]_0^4$

$= \frac{5000}{-0.11} [e^{-0.44} - 1] \approx \$16,180.16$

47. Find the present value of the income given by $c(t) = 100,000 + 4000t$, measured in dollars, over $t_1 = 10$ years at the annual interest rate $r = 10\%$.
Solution:

$V = \int_0^{t_1} c(t) e^{-rt} \, dt = \int_0^{10} (100,000 + 4000t) e^{-0.10t} \, dt$

$= 100,000 \int_0^{10} e^{-0.10t} \, dt + 4000 \int_0^{10} t e^{-0.10t} \, dt$

$\approx \$737,817.01$

49. Find the present value of the income given by $c(t) = 1000 + 50e^{t/2}$, measured in dollars, over $t_1 = 4$ years at the annual interest rate $r = 6\%$.
Solution:

$V = \int_0^{t_1} c(t) e^{-rt} \, dt = \int_0^4 (1000 + 50e^{t/2}) e^{-0.06t} \, dt$

$\approx \$4103.07$

Section 6.3

Section 6.3 Partial Fractions

1. Write $\dfrac{2(x + 20)}{x^2 - 25}$ as a sum of partial fractions.

   Solution:
   $$\frac{2x + 40}{(x - 5)(x + 5)} = \frac{A}{x - 5} + \frac{B}{x + 5}$$
   Basic Equation: $2x + 40 = A(x + 5) + B(x - 5)$
   When $x = 5$: $50 = 10A$, $A = 5$
   When $x = -5$: $30 = -10B$, $B = -3$
   $$\frac{2(x + 20)}{x^2 - 25} = \frac{5}{x - 5} - \frac{3}{x + 5}$$

3. Write $\dfrac{8x + 3}{x^2 - 3x}$ as a sum of partial fractions.

   Solution:
   $$\frac{8x + 3}{x(x - 3)} = \frac{A}{x} + \frac{B}{x - 3}$$
   Basic Equation: $8x + 3 = A(x - 3) + Bx$
   When $x = 0$: $3 = -3A$, $A = -1$
   When $x = 3$: $27 = 3B$, $B = 9$
   $$\frac{8x + 3}{x^2 - 3x} = \frac{9}{x - 3} - \frac{1}{x}$$

5. Write $\dfrac{4x - 13}{x^2 - 3x - 10}$ as a sum of partial fractions.

   Solution:
   $$\frac{4x - 13}{(x - 5)(x + 2)} = \frac{A}{x - 5} + \frac{B}{x + 2}$$
   Basic Equation: $4x - 13 = A(x + 2) + B(x - 5)$
   When $x = 5$: $7 = 7A$, $A = 1$
   When $x = -2$: $-21 = -7B$, $B = 3$
   $$\frac{4x - 13}{x^2 - 3x - 10} = \frac{1}{x - 5} + \frac{3}{x + 2}$$

7. Write $\dfrac{2x^2 - 2x - 3}{x^3 + x^2}$ as a sum of partial fractions.

   Solution:
   $$\frac{2x^2 - 2x - 3}{x^2(x + 1)} = \frac{A}{x} + \frac{B}{x^2} + \frac{C}{x + 1}$$
   Basic Equation:
   $2x^2 - 2x - 3 = Ax(x + 1) + B(x + 1) + Cx^2$
   When $x = 0$: $B = -3$
   When $x = -1$: $C = 1$
   When $x = 1$: $-3 = 2A + 2B + C = 2A - 5$, $A = 1$
   $$\frac{2x^2 - 2x - 3}{x^3 + x^2} = \frac{1}{x} - \frac{3}{x^2} + \frac{1}{x + 1}$$

Section 6.3

9. Write $\dfrac{x+1}{3(x-2)^2}$ as a sum of partial fractions.

   Solution:

   $$\frac{1}{3}\left[\frac{x+1}{(x-2)^2}\right] = \frac{1}{3}\left[\frac{A}{x-2} + \frac{B}{(x-2)^2}\right]$$

   Basic Equation: $x + 1 = A(x-2) + B$
   When $x = 2$: $\quad 3 = B$
   When $x = 3$: $\quad 4 = A + B \implies A = 1$

   $$\frac{x+1}{3(x-2)^2} = \frac{1}{3}\left[\frac{1}{x-2} + \frac{3}{(x-2)^2}\right]$$

   $$= \frac{1}{3(x-2)} + \frac{1}{(x-2)^2}$$

11. Find $\displaystyle\int \frac{1}{x^2-1}\,dx$.

    Solution:

    $$\frac{1}{(x+1)(x-1)} = \frac{A}{x+1} + \frac{B}{x-1}$$

    Basic Equation: $1 = A(x-1) + B(x+1)$
    When $x = -1$: $\quad 1 = -2A, \quad A = -1/2$
    When $x = 1$: $\quad 1 = 2B, \quad B = 1/2$

    $$\int \frac{1}{x^2-1}\,dx = \frac{-1}{2}\int \frac{1}{x+1}\,dx + \frac{1}{2}\int \frac{1}{x-1}\,dx$$

    $$= \frac{-1}{2}\ln|x+1| + \frac{1}{2}\ln|x-1| + C$$

    $$= \frac{1}{2}\ln\left|\frac{x-1}{x+1}\right| + C$$

13. Find $\displaystyle\int \frac{-2}{x^2-16}\,dx$.

    Solution:

    $$\frac{-2}{(x+4)(x-4)} = \frac{A}{x+4} + \frac{B}{x-4}$$

    Basic Equation: $-2 = A(x-4) + B(x+4)$
    When $x = -4$: $\quad -2 = -8A, \quad A = 1/4$
    When $x = 4$: $\quad -2 = 8B, \quad B = -1/4$

    $$\int \frac{-2}{x^2-16}\,dx = \frac{1}{4}\int \frac{1}{x+4}\,dx - \frac{1}{4}\int \frac{1}{x-4}\,dx$$

    $$= \frac{1}{4}\ln|x+4| - \frac{1}{4}\ln|x-4| + C$$

    $$= \frac{1}{4}\ln\left|\frac{x+4}{x-4}\right| + C$$

Section 6.3

15. Find $\int \dfrac{1}{x^2 + x}\, dx$.

Solution:

$$\dfrac{1}{x(x+1)} = \dfrac{A}{x} + \dfrac{B}{x+1}$$

Basic Equation: $1 = A(x+1) + Bx$
When $x = 0$: $A = 1$
When $x = -1$: $B = -1$

$$\int \dfrac{1}{x^2+x}\, dx = \int \dfrac{1}{x}\, dx - \int \dfrac{1}{x+1}\, dx$$
$$= \ln|x| - \ln|x+1| + C$$
$$= \ln\left|\dfrac{x}{x+1}\right| + C$$

17. Find $\int \dfrac{1}{2x^2 + x}\, dx$.

Solution:

$$\dfrac{1}{x(2x+1)} = \dfrac{A}{x} + \dfrac{B}{2x+1}$$

Basic Equation: $1 = A(2x+1) + Bx$
When $x = 0$: $A = 1$
When $x = -1/2$: $B = -2$

$$\int \dfrac{1}{2x^2+x}\, dx = \int \dfrac{1}{x}\, dx - \int \dfrac{2}{2x+1}\, dx$$
$$= \ln|x| - \ln|2x+1| + C$$
$$= \ln\left|\dfrac{x}{2x+1}\right| + C$$

19. Find $\int \dfrac{3}{x^2 + x - 2}\, dx$.

Solution:

$$\dfrac{3}{(x-1)(x+2)} = \dfrac{A}{x-1} + \dfrac{B}{x+2}$$

Basic Equation: $3 = A(x+2) + B(x-1)$
When $x = 1$: $3 = 3A$, $A = 1$
When $x = -2$: $3 = -3B$, $B = -1$

$$\int \dfrac{3}{x^2+x-2}\, dx = \int \dfrac{1}{x-1}\, dx - \int \dfrac{1}{x+2}\, dx$$
$$= \ln|x-1| - \ln|x+2| + C$$
$$= \ln\left|\dfrac{x-1}{x+2}\right| + C$$

Section 6.3

21. Find $\int \dfrac{5-x}{2x^2+x-1}\,dx$.

   Solution:

   $$\dfrac{5-x}{(2x-1)(x+1)} = \dfrac{A}{2x-1} + \dfrac{B}{x+1}$$

   Basic Equation: $5 - x = A(x+1) + B(2x-1)$
   When $x = 1/2$: $\quad 4.5 = 1.5A, \quad A = 3$
   When $x = -1$: $\quad\ \ 6 = -3B, \quad\ \ B = -2$

   $$\int \dfrac{5-x}{2x^2+x-1}\,dx = 3\int \dfrac{1}{2x-1}\,dx - 2\int \dfrac{1}{x+1}\,dx$$

   $$= \dfrac{3}{2}\ln|2x-1| - 2\ln|x+1| + C$$

23. Find $\int \dfrac{x^2+12x+12}{x^3-4x}\,dx$.

   Solution:

   $$\dfrac{x^2+12x+12}{x(x+2)(x-2)} = \dfrac{A}{x} + \dfrac{B}{x+2} + \dfrac{C}{x-2}$$

   Basic Equation:
   $x^2 + 12x + 12 = A(x+2)(x-2) + Bx(x-2) + Cx(x+2)$

   When $x = 0$: $\quad 12 = -4A, \quad A = -3$
   When $x = -2$: $\quad -8 = 8B, \quad\ \ B = -1$
   When $x = 2$: $\quad\ \ 40 = 8C, \quad\ \ C = 5$

   $$\int \dfrac{x^2+12x+12}{x^3-4x}\,dx$$

   $$= 5\int \dfrac{1}{x-2}\,dx - \int \dfrac{1}{x+2}\,dx - 3\int \dfrac{1}{x}\,dx$$

   $$= 5\ln|x-2| - \ln|x+2| - 3\ln|x| + C$$

25. Find $\int \dfrac{x+2}{x^2-4x}\,dx$.

   Solution:

   $$\dfrac{x+2}{x(x-4)} = \dfrac{A}{x-4} + \dfrac{B}{x}$$

   Basic Equation: $x + 2 = Ax + B(x-4)$
   When $x = 4$: $\quad 6 = 4A, \quad\ \ A = 3/2$
   When $x = 0$: $\quad 2 = -4B, \quad B = -1/2$

   $$\int \dfrac{x+2}{x^2-4x}\,dx = \dfrac{1}{2}\left[3\int \dfrac{1}{x-4}\,dx - \int \dfrac{1}{x}\,dx\right]$$

   $$= \dfrac{1}{2}[3\ln|x-4| - \ln|x|] + C$$

Section 6.3

27. Find $\int \dfrac{2x-3}{(x-1)^2}\,dx$.

Solution:

$$\dfrac{2x-3}{(x-1)^2} = \dfrac{A}{x-1} + \dfrac{B}{(x-1)^2}$$

Basic Equation: $2x - 3 = A(x-1) + B$
When $x = 1$: $B = -1$
When $x = 0$: $A = 2$

$$\int \dfrac{2x-3}{(x-1)^2}\,dx = 2\int \dfrac{1}{x-1}\,dx - \int \dfrac{1}{(x-1)^2}\,dx$$

$$= 2\ln|x-1| + \dfrac{1}{x-1} + C$$

29. Find $\int \dfrac{4x^2-1}{2x(x^2+2x+1)}\,dx$.

Solution:

$$\dfrac{4x^2-1}{2x(x+1)^2} = \dfrac{A}{2x} + \dfrac{B}{x+1} + \dfrac{C}{(x+1)^2}$$

Basic Equation:
$$4x^2 - 1 = A(x+1)^2 + B(2x)(x+1) + C(2x)$$
$$= (A + 2B)x^2 + (2A + 2B + 2C)x + A$$

Therefore, $A + 2B = 4$, $2A + 2B + 2C = 0$, and $A = -1$. Solving these equations we get $A = -1$, $B = 5/2$, and $C = -3/2$.

$$\int \dfrac{4x^2-1}{2x(x^2+2x+1)}\,dx$$

$$= \dfrac{-1}{2}\int \dfrac{1}{x}\,dx + \dfrac{5}{2}\int \dfrac{1}{x+1}\,dx - \dfrac{3}{2}\int \dfrac{1}{(x+1)^2}\,dx$$

$$= \dfrac{1}{2}\left[5\ln|x+1| - \ln|x| + \dfrac{3}{x+1}\right] + C$$

31. Evaluate $\displaystyle\int_3^4 \dfrac{1}{x^2-4}\,dx$.

Solution:

$$\dfrac{1}{(x-2)(x+2)} = \dfrac{A}{x-2} + \dfrac{B}{x+2}$$

Basic Equation: $1 = A(x+2) + B(x-2)$
When $x = 2$: $A = 1/4$
When $x = -2$: $B = -1/4$

$$\int_3^4 \dfrac{1}{x^2-4}\,dx = \dfrac{1}{4}\int_3^4 \dfrac{1}{x-2}\,dx - \dfrac{1}{4}\int_3^4 \dfrac{1}{x+2}\,dx$$

$$= \dfrac{1}{4}\ln\left|\dfrac{x-2}{x+2}\right|\Big]_3^4 = \dfrac{1}{4}\ln\left(\dfrac{5}{3}\right) \approx 0.128$$

33. Evaluate $\int_1^5 \dfrac{x-1}{x^2(x+1)}\,dx$.

Solution:

$$\dfrac{x-1}{x^2(x+1)} = \dfrac{A}{x} + \dfrac{B}{x^2} + \dfrac{C}{x+1}$$

Basic Equation: $x - 1 = Ax(x+1) + B(x+1) + Cx^2$
When $x = 0$: $B = -1$
When $x = -1$: $C = -2$
When $x = 1$: $0 = 2A + 2B + C$, $\quad 0 = 2A - 4$, $\quad A = 2$

$$\int_1^5 \dfrac{x-1}{x^2(x+1)}\,dx = 2\int_1^5 \dfrac{1}{x}\,dx - \int_1^5 \dfrac{1}{x^2}\,dx - 2\int_1^5 \dfrac{1}{x+1}\,dx$$

$$= \left[2\ln|x| + \dfrac{1}{x} - 2\ln|x+1|\right]_1^5$$

$$= \left[2\ln\left|\dfrac{x}{x+1}\right| + \dfrac{1}{x}\right]_1^5 = 2\ln\left(\dfrac{5}{3}\right) - \dfrac{4}{5}$$

$$\approx 0.222$$

35. Evaluate $\int \dfrac{e^x}{(e^x - 1)(e^x + 4)}\,dx$.

Solution: Let $u = e^x$, then $du = e^x\,dx$.

$$\int \dfrac{e^x}{(e^x - 1)(e^x + 4)}\,dx = \int \dfrac{1}{(u-1)(u+4)}\,du$$

$$= \dfrac{1}{5}\left[\int \dfrac{1}{u-1}\,du - \int \dfrac{1}{u+4}\,du\right]$$

$$= \dfrac{1}{5}\ln\left|\dfrac{u-1}{u+4}\right| + C$$

$$= \dfrac{1}{5}\ln\left|\dfrac{e^x - 1}{e^x + 4}\right| + C$$

37. Evaluate $\int \dfrac{1}{x\sqrt{4+x^2}}\,dx$.

Solution: Let $u = \sqrt{4+x^2}$, then $x^2 = u^2 - 4$ and $x\,dx = u\,du$.

$$\int \dfrac{1}{x\sqrt{4+x^2}}\,dx = \int \dfrac{x}{x^2\sqrt{4+x^2}}\,dx = \int \dfrac{1}{(u^2-4)u}(u)\,du$$

$$= \int \dfrac{1}{u^2-4}\,du = \dfrac{1}{4}\int\left(\dfrac{-1}{u+2} + \dfrac{1}{u-2}\right)du$$

$$= \dfrac{1}{4}(\ln|u-2| - \ln|u+2|) + C$$

$$= \dfrac{1}{4}\ln\left|\dfrac{u-2}{u+2}\right| + C$$

$$= \dfrac{1}{4}\ln\left|\dfrac{\sqrt{4+x^2} - 2}{\sqrt{4+x^2} + 2}\right| + C$$

Section 6.3

39. Evaluate $\int \dfrac{1}{\sqrt{x}(\sqrt{x} + 1)^2} \, dx$.

Solution: Let $u = \sqrt{x}$, then $x = u^2$ and $dx = 2u\,du$.

$$\int \frac{1}{\sqrt{x}(\sqrt{x}+1)^2}\,dx = 2\int \frac{1}{u(u+1)^2}(u)\,du$$

$$= 2\int \frac{1}{(u+1)^2}\,du$$

$$= -\frac{2}{u+1} + C = -\frac{2}{\sqrt{x}+1} + C$$

41. (a) Write the rational expression $1/(a^2 - x^2)$ as a sum of partial fractions.

(b) Use the result of part (a) to show that

$$\int \frac{1}{a^2 - x^2}\,dx = \frac{1}{2a}\ln\left|\frac{a+x}{a-x}\right| + C$$

Solution:

(a) $\dfrac{1}{a^2 - x^2} = \dfrac{1}{(a+x)(a-x)} = \dfrac{A}{a+x} + \dfrac{B}{a-x}$

$$= \frac{1}{2a}\left(\frac{1}{a+x} + \frac{1}{a-x}\right)$$

(b) $\displaystyle\int \frac{1}{a^2 - x^2}\,dx = \frac{1}{2a}\int\left(\frac{1}{a+x} + \frac{1}{a-x}\right)dx$

$$= \frac{1}{2a}(\ln|a+x| - \ln|a-x|) + C$$

$$= \frac{1}{2a}\ln\left|\frac{a+x}{a-x}\right| + C$$

43. Find the area of the region bounded by the graphs of $y = 7/(16 - x^2)$ and $y = 1$.

Solution: The points of intersection of the two graphs occur when $16 - x^2 = 7$, which implies that $x = \pm 3$. Therefore, the area is

$$A = \int_{-3}^{3}\left[1 - \frac{7}{16-x^2}\right]dx = \int_{-3}^{3}dx - 7\int_{-3}^{3}\frac{1}{16-x^2}\,dx$$

$$= 6 - 7\int_{-3}^{3}\frac{1}{16-x^2}\,dx$$

$$= 6 - \frac{7}{8}\int_{-3}^{3}\left(\frac{1}{4-x} + \frac{1}{4+x}\right)dx$$

$$= 6 - \frac{7}{8}\Big[-\ln|4-x| + \ln|4+x|\Big]_{-3}^{3}$$

$$= 6 - \frac{7}{8}\ln\left|\frac{4+x}{4-x}\right|\Big]_{-3}^{3} = 6 - \frac{7}{8}[\ln 7 - \ln\frac{1}{7}]$$

$$= 6 - \frac{7}{8}[\ln 7 - (\ln 1 - \ln 7)] = 6 - \frac{7}{4}\ln 7 \approx 2.595$$

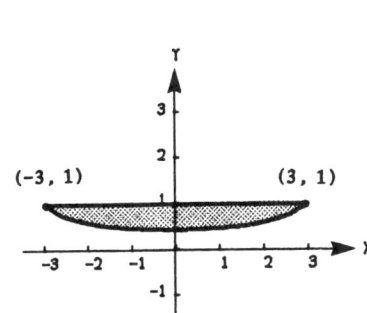

45. Show that $y = 10/(1 + 9e^{-kt})$ is a solution of

$$\int \frac{1}{y(10 - y)} \, dy = \int k \, dt$$

To solve for the constant of integration, use the condition that $y = 1$ when $t = 0$.
Solution:

$$\int \frac{1}{y(10 - y)} \, dy = \int k \, dt$$

$$\frac{1}{10} \int \left(\frac{1}{y} + \frac{1}{10 - y}\right) dy = kt + C_2$$

$$\frac{1}{10} \ln \left|\frac{y}{10 - y}\right| + C_1 = kt + C_2$$

$$\frac{y}{10 - y} = Ce^{10kt}$$

Since $y = 1$ when $t = 0$, we have $C = 1/9$

$$\frac{9y}{10 - y} = e^{10kt}$$

Solving for y, produces

$$y = \frac{10e^{10kt}}{e^{10kt} + 9} = \frac{10}{1 + 9e^{-10kt}}$$

47. A single infected individual enters a community of n individuals susceptible to the disease. Let x be the number of newly infected individuals after time t. Assume that the disease spreads at a rate proportional to the product of the total number infected and the number of susceptible individuals not yet infected. Thus, $dx/dt = k(x + 1)(n - x)$ and we obtain

$$\int \frac{1}{(x + 1)(n - x)} \, dx = \int k \, dt$$

Solve for x as a function of t.
Solution:

$$\int \frac{1}{(x + 1)(n - x)} \, dx = \int k \, dt$$

$$\frac{1}{n + 1} \int \left(\frac{1}{x + 1} + \frac{1}{n - x}\right) dx = kt + C$$

$$\frac{1}{n + 1} \ln \left|\frac{x + 1}{n - x}\right| = kt + C$$

When $t = 0$ and $x = 0$, we have $C = [1/(n + 1)] \ln (1/n)$.

$$\frac{1}{n + 1} \ln \left|\frac{x + 1}{n - x}\right| = kt + \frac{1}{n + 1} \ln \left(\frac{1}{n}\right)$$

Solving for x produces

$$x = \frac{n[e^{(n+1)kt} - 1]}{n + e^{(n+1)kt}}$$

Section 6.4 Integration by Tables and Completing the Square

1. Evaluate $\int \dfrac{x}{(2 + 3x)^2}\, dx$.

    Solution: Formula 4: $u = x$, $du = dx$, $a = 2$, $b = 3$

    $$\int \dfrac{x}{(2 + 3x)^2}\, dx = \dfrac{1}{9}\left[\dfrac{2}{2 + 3x} + \ln|2 + 3x|\right] + C$$

3. Evaluate $\int \dfrac{x}{\sqrt{2 + 3x}}\, dx$.

    Solution: Formula 19: $u = x$, $du = dx$, $a = 2$, $b = 3$

    $$\int \dfrac{x}{\sqrt{2 + 3x}}\, dx = \dfrac{-2(4 - 3x)}{27}\sqrt{2 + 3x} + C$$

    $$= \dfrac{2(3x - 4)}{27}\sqrt{2 + 3x} + C$$

5. Evaluate $\int \dfrac{2x}{\sqrt{x^4 - 9}}\, dx$.

    Solution: Formula 27: $u = x^2$, $du = 2x\, dx$, $a = 3$

    $$\int \dfrac{2x}{\sqrt{x^4 - 9}}\, dx = \ln|x^2 + \sqrt{x^4 - 9}| + C$$

7. Evaluate $\int x^3 e^{x^2}\, dx$.

    Solution: Formula 37: $u = x^2$, $du = 2x\, dx$

    $$\int x^3 e^{x^2}\, dx = \dfrac{1}{2}\int x^2 e^{x^2}\, 2x\, dx = \dfrac{1}{2}(x^2 - 1)e^{x^2} + C$$

9. Evaluate $\int x \ln(x^2 + 1)\, dx$.

    Solution: Formula 41: $u = x^2 + 1$, $du = 2x\, dx$

    $$\int x \ln(x^2 + 1)\, dx = \dfrac{1}{2}\int \ln(x^2 + 1)(2x)\, dx$$

    $$= \dfrac{1}{2}(x^2 + 1)[-1 + \ln(x^2 + 1)] + C$$

11. Evaluate $\int \dfrac{1}{x(1 + x)}\, dx$.

    Solution: Formula 10: $u = x$, $du = dx$, $a = b = 1$

    $$\int \dfrac{1}{x(1 + x)}\, dx = \ln\left|\dfrac{x}{1 + x}\right| + C$$

13. Evaluate $\int \dfrac{1}{x\sqrt{x^2 + 1}}\, dx$.

    Solution: Formula 28: $u = x$, $du = dx$, $a = 1$

    $$\int \dfrac{1}{x\sqrt{x^2 + 1}}\, dx = -\ln\left|\dfrac{1 + \sqrt{x^2 + 1}}{x}\right| + C$$

Section 6.4

15. Evaluate $\displaystyle\int \frac{1}{x\sqrt{4-x^2}}\,dx$.

Solution: Formula 33: $u = x$, $du = dx$, $a = 2$

$$\int \frac{1}{x\sqrt{4-x^2}}\,dx = -\frac{1}{2}\ln\left|\frac{2+\sqrt{4-x^2}}{x}\right| + C$$

17. Evaluate $\displaystyle\int x\ln x\,dx$.

Solution: Formula 42: $u = x$, $du = dx$

$$\int x\ln x\,dx = \frac{x^2}{4}(-1 + 2\ln x) + C$$

19. Evaluate $\displaystyle\int \frac{e^x}{e^{2x}(1+e^x)}\,dx$.

Solution: Formula 12: $u = e^x$, $du = e^x\,dx$, $a = b = 1$

$$\int \frac{e^x}{e^{2x}(1+e^x)}\,dx = -\left[e^{-x} + \ln\left(\frac{e^x}{1+e^x}\right)\right] + C$$

21. Evaluate $\displaystyle\int x\sqrt{x^4-9}\,dx$.

Solution: Formula 23: $u = x^2$, $du = 2x\,dx$, $a = 3$

$$\int x\sqrt{x^4-9}\,dx = \frac{1}{2}\int \sqrt{(x^2)^2-3^2}\,(2x)\,dx$$

$$= \frac{1}{4}\left(x^2\sqrt{x^4-9} - 9\ln\left|x^2 + \sqrt{x^4-9}\right|\right) + C$$

23. Evaluate $\displaystyle\int \frac{t^2}{(2+3t)^3}\,dt$.

Solution: Formula 8: $u = t$, $du = dt$, $a = 2$, $b = 3$

$$\int \frac{t^2}{(2+3t)^3}\,dt$$

$$= \frac{1}{27}\left[\frac{4}{2+3t} - \frac{4}{2(2+3t)^2} + \ln|2+3t|\right] + C$$

25. Evaluate $\displaystyle\int \frac{s}{s^2\sqrt{3+s}}\,ds$.

Solution: Formula 15: $u = s$, $du = ds$, $a = 3$, $b = 1$

$$\int \frac{s}{s^2\sqrt{3+s}}\,ds = \int \frac{1}{s\sqrt{3+s}}\,ds$$

$$= \frac{1}{\sqrt{3}}\ln\left|\frac{\sqrt{3+s}-\sqrt{3}}{\sqrt{3+s}+\sqrt{3}}\right| + C$$

Section 6.4

27. Evaluate $\int \dfrac{x^2}{1+x}\,dx$.

    Solution: Formula 6: $u = x$, $du = dx$, $a = b = 1$

    $$\int \dfrac{x^2}{1+x}\,dx = -\dfrac{x}{2}(2 - x) + \ln|1 + x| + C$$

    $$= \dfrac{x}{2}(x - 2) + \ln|x + 1| + C$$

29. Evaluate $\int \dfrac{1}{x^2\sqrt{1-x^2}}\,dx$.

    Solution: Formula 34: $u = x$, $du = dx$, $a = 1$

    $$\int \dfrac{1}{x^2\sqrt{1-x^2}}\,dx = -\dfrac{\sqrt{1-x^2}}{x} + C$$

31. Evaluate $\int x^2 \ln x\,dx$.

    Solution: Formula 43: $u = x$, $du = dx$, $n = 2$

    $$\int x^2 \ln x\,dx = \dfrac{x^3}{9}(-1 + 3\ln x) + C$$

33. Evaluate $\int \dfrac{x^2}{(3x-5)^2}\,dx$.

    Solution: Formula 7: $u = x$, $du = dx$, $a = -5$, $b = 3$

    $$\int \dfrac{x^2}{(3x-5)^2}\,dx$$

    $$= \dfrac{1}{27}\left[3x - \dfrac{25}{3x-5} + 10\ln|3x-5|\right] + C$$

35. Evaluate $\int x^2\sqrt{x^2+4}\,dx$.

    Solution: Formula 24: $u = x$, $du = dx$, $a = 2$

    $$\int x^2\sqrt{x^2+4}\,dx$$

    $$= \dfrac{1}{8}\left[x(2x^2+4)\sqrt{x^2+4} - 16\ln|x + \sqrt{x^2+4}|\right] + C$$

    $$= \dfrac{1}{4}\left[x(x^2+2)\sqrt{x^2+4} - 8\ln|x + \sqrt{x^2+4}|\right] + C$$

37. Evaluate $\int \dfrac{1}{1+e^{2x}}\,dx$.

    Solution: Formula 39: $u = 2x$, $du = 2\,dx$

    $$\int \dfrac{1}{1+e^{2x}}\,dx = x - \dfrac{1}{2}\ln(1 + e^{2x}) + C$$

Section 6.4

39. Evaluate $\int \dfrac{\ln x}{x(3 + 2 \ln x)} \, dx$.

Solution: Formula 3: $u = \ln x$, $du = dx/x$, $a = 3$, $b = 2$

$$\int \dfrac{\ln x}{x(3 + 2 \ln x)} \, dx = \int \dfrac{\ln x}{3 + 2 \ln x} \left(\dfrac{1}{x}\right) dx$$

$$= \dfrac{1}{4}[2 \ln x - 3 \ln |3 + 2 \ln x|] + C$$

41. Complete the square to express each polynomial as the sum or difference of squares.
 (a) $x^2 + 6x$ (b) $x^2 - 8x + 9$
 (c) $x^4 + 2x^2 - 5$ (d) $3 - 2x - x^2$

Solution:
(a) $x^2 + 6x = x^2 + 6x + 9 - 9 = (x + 3)^2 - 9$

(b) $x^2 - 8x + 9 = x^2 - 8x + 16 - 16 + 9 = (x - 4)^2 - 7$

(c) $x^4 + 2x^2 - 5 = x^4 + 2x^2 + 1 - 1 - 5$
$= (x^2 + 1)^2 - 6$

(d) $3 - 2x - x^2 = -(x^2 + 2x - 3)$
$= -(x^2 + 2x + 1 - 1 - 3)$
$= -[(x + 1)^2 - 4] = 4 - (x + 1)^2$

43. Evaluate $\int \dfrac{1}{x^2 - 2x - 3} \, dx$.

Solution: Formula 21: $u = x - 1$, $du = dx$, $a = 2$

$$\int \dfrac{1}{x^2 - 2x - 3} \, dx = \int \dfrac{1}{(x - 1)^2 - 4} \, dx$$

$$= \dfrac{1}{4} \ln \left| \dfrac{(x - 1) - 2}{(x - 1) + 2} \right| + C$$

$$= \dfrac{1}{4} \ln \left| \dfrac{x - 3}{x + 1} \right| + C$$

45. Evaluate $\int \dfrac{1}{(x - 1)\sqrt{x^2 - 2x + 2}} \, dx$.

Solution: Formula 28: $u = x - 1$, $du = dx$, $a = 1$

$$\int \dfrac{1}{(x - 1)\sqrt{x^2 - 2x + 2}} \, dx = \int \dfrac{1}{(x - 1)\sqrt{(x - 1)^2 + 1}} \, dx$$

$$= -\ln \left| \dfrac{1 + \sqrt{x^2 - 2x + 2}}{x - 1} \right| + C$$

47. Evaluate $\int \dfrac{1}{2x^2 - 4x - 6} \, dx$.

Solution: Formula 21: $u = x - 1$, $du = dx$, $a = 2$

$$\int \dfrac{1}{2x^2 - 4x - 6} \, dx = \dfrac{1}{2} \int \dfrac{1}{(x - 1)^2 - 4} \, dx$$

$$= \dfrac{1}{8} \ln \left| \dfrac{x - 3}{x + 1} \right| + C$$

Section 6.4   279

49. Evaluate $\int \dfrac{x}{\sqrt{x^4 + 2x^2 + 2}} \, dx$.

Solution: Formula 27: $u = x^2 + 1$, $du = 2x \, dx$, $a = 1$

$$\int \dfrac{x}{\sqrt{x^4 + 2x^2 + 2}} \, dx = \int \dfrac{x}{\sqrt{(x^2 + 1)^2 + 1}} \, dx$$

$$= \dfrac{1}{2} \int \dfrac{2x}{\sqrt{(x^2 + 1)^2 + 1}} \, dx$$

$$= \dfrac{1}{2} \ln |x^2 + 1 + \sqrt{x^4 + 2x^2 + 2}| + C$$

51. Find the area of the region bounded by the graphs of $y = x/\sqrt{x + 1}$, $y = 0$, and $x = 8$.

Solution: Formula 19: $u = x$, $du = dx$, $a = b = 1$

$$A = \int_0^8 \dfrac{x}{\sqrt{x + 1}} \, dx = -\dfrac{2(2 - x)}{3} \cdot \sqrt{x + 1} \,\Big]_0^8$$

$$= 40/3 \text{ square units}$$

53. Find the average value of $N = 50/(1 + e^{4.8 - 1.9t})$ over the interval [3, 4], where N is the size of a population and t is the time in days.

Solution: Formula 40: $u = 4.8 - 1.9t$, $du = -1.9 \, dt$, $n = 1$

$$\text{Average} = \int_3^4 \dfrac{50}{1 + e^{4.8 - 1.9t}} \, dt$$

$$= -\dfrac{50}{1.9} \int_3^4 \dfrac{-1.9}{1 + e^{4.8 - 1.9t}} \, dt$$

$$= -\dfrac{50}{1.9} \Big[4.8 - 1.9t - \ln(1 + e^{4.8 - 1.9t}) \Big]_3^4$$

$$\approx 42.58 \approx 43$$

55. Find the consumer surplus and producer surplus for a given product if the demand and supply functions are Demand: $p = 60/\sqrt{x^2 + 81}$ and Supply: $p = x/3$.

Solution: Equating the supply and demand functions produces $x = 12$, $p = 4$ as the point of equilibrium.

Consumer Surplus: Formula 27: $u = x$, $du = dx$, $a = 9$

$$\int_0^{12} \left(\dfrac{60}{\sqrt{x^2 + 81}} - 4 \right) dx$$

$$= \Big[60 \ln |x + \sqrt{x^2 + 81}| - 4x \Big]_0^{12} \approx 17.92$$

Producer Surplus:

$$\int_0^{12} \left(4 - \dfrac{x}{3} \right) dx = \left(4x - \dfrac{x^2}{6} \right) \Big]_0^{12} = 24$$

Section 6.5 Numerical Integration

1. Use the Trapezoidal Rule and Simpson's Rule to approximate to four decimal places the value of

$$\int_0^2 x^2 \, dx, \quad n = 4$$

Compare these results with the exact value.
Solution:
Exact:

$$\int_0^2 x^2 \, dx = \frac{1}{3}x^3 \Big]_0^2 = \frac{8}{3} \approx 2.6667$$

Trapezoidal Rule:

$$\int_0^2 x^2 \, dx \approx \frac{1}{4}[0 + 2(\frac{1}{2})^2 + 2(1)^2 + 2(\frac{3}{2})^2 + (2)^2]$$

$$= \frac{11}{4} = 2.7500$$

Simpson's Rule:

$$\int_0^2 x^2 \, dx \approx \frac{1}{6}[0 + 4(\frac{1}{2})^2 + 2(1)^2 + 4(\frac{3}{2})^2 + (2)^2]$$

$$= \frac{8}{3} \approx 2.6667$$

3. Use the Trapezoidal Rule and Simpson's Rule to approximate to four decimal places the value of

$$\int_0^2 x^3 \, dx, \quad n = 4$$

Compare these results with the exact value.
Solution:
Exact:

$$\int_0^2 x^3 \, dx = \frac{x^4}{4} \Big]_0^2 = 4.0000$$

Trapezoidal Rule:

$$\int_0^2 x^3 \, dx \approx \frac{1}{4}[0 + 2(\frac{1}{2})^3 + 2(1)^3 + 2(\frac{3}{2})^3 + (2)^3]$$

$$= \frac{17}{4} = 4.2500$$

Simpson's Rule:

$$\int_0^2 x^3 \, dx \approx \frac{1}{6}[0 + 4(\frac{1}{2})^3 + 2(1)^3 + 4(\frac{3}{2})^3 + (2)^3]$$

$$= \frac{24}{6} = 4.0000$$

Section 6.5   281

5. Use the Trapezoidal Rule and Simpson's Rule to approximate to four decimal places the value of

$$\int_0^2 x^3 \, dx, \quad n = 8$$

Compare these results with the exact value.
Solution:
Exact:

$$\int_0^2 x^3 \, dx = \frac{1}{4}x^4 \Big]_0^2 = 4.0000$$

Trapezoidal Rule:

$$\int_0^2 x^3 \, dx \approx \frac{1}{8}[0 + 2(\frac{1}{4})^3 + 2(\frac{2}{4})^3 + 2(\frac{3}{4})^3 + 2(1)^3$$
$$+ 2(\frac{5}{4})^3 + 2(\frac{6}{4})^3 + 2(\frac{7}{4})^3 + 8]$$

$$= 4.0625$$

Simpson's Rule:

$$\int_0^2 x^3 \, dx \approx \frac{1}{12}[0 + 4(\frac{1}{4})^3 + 2(\frac{2}{4})^3 + 4(\frac{3}{4})^3 + 2(1)^3$$
$$+ 4(\frac{5}{4})^3 + 2(\frac{6}{4})^3 + 4(\frac{7}{4})^3 + 8]$$

$$= 4.0000$$

7. Use the Trapezoidal Rule and Simpson's Rule to approximate to four decimal places the value of

$$\int_1^2 \frac{1}{x^2} \, dx, \quad n = 4$$

Compare these results with the exact value.
Solution:
Exact:

$$\int_1^2 \frac{1}{x^2} \, dx = \frac{-1}{x} \Big]_1^2 = 0.5000$$

Trapezoidal Rule:

$$\int_1^2 \frac{1}{x^2} \, dx \approx \frac{1}{8}[1 + 2(\frac{4}{5})^2 + 2(\frac{4}{6})^2 + 2(\frac{4}{7})^2 + \frac{1}{4}]$$

$$\approx 0.5090$$

Simpson's Rule:

$$\int_1^2 \frac{1}{x^2} \, dx \approx \frac{1}{12}[1 + 4(\frac{4}{5})^2 + 2(\frac{4}{6})^2 + 4(\frac{4}{7})^2 + \frac{1}{4}]$$

$$\approx 0.5004$$

Section 6.5

9. Use the Trapezoidal Rule and Simpson's Rule to approximate to four decimal places the value of

$$\int_0^1 \frac{1}{1+x}\, dx, \quad n = 4$$

Compare these results with the exact value.

Solution:

Exact:

$$\int_0^1 \frac{1}{1+x}\, dx = \ln|1+x|\Big]_0^1 \approx 0.6931$$

Trapezoidal Rule:

$$\int_0^1 \frac{1}{1+x}\, dx \approx \frac{1}{8}[1 + 2(\tfrac{4}{5}) + 2(\tfrac{2}{3}) + 2(\tfrac{4}{7}) + \tfrac{1}{2}]$$
$$\approx 0.6970$$

Simpson's Rule:

$$\int_0^1 \frac{1}{1+x}\, dx \approx \frac{1}{12}[1 + 4(\tfrac{4}{5}) + 2(\tfrac{2}{3}) + 4(\tfrac{4}{7}) + \tfrac{1}{2}]$$
$$\approx 0.6933$$

11. Using (a) the Trapezoidal Rule and (b) Simpson's Rule, approximate to three significant digits

$$\int_0^2 \sqrt{1+x^3}\, dx, \quad n = 2$$

Solution:

(a) Trapezoidal Rule:

$$\frac{1}{2}(1 + 2\sqrt{2} + 3) = 2 + \sqrt{2} \approx 3.41$$

(b) Simpson's Rule:

$$\frac{1}{3}(1 + 4\sqrt{2} + 3) = \frac{4}{3}(1 + \sqrt{2}) \approx 3.22$$

13. Using (a) the Trapezoidal Rule and (b) Simpson's Rule, approximate to three significant digits

$$\int_0^1 \sqrt{x}\, \sqrt{1-x}\, dx, \quad n = 4$$

Solution:

$$\int_0^1 \sqrt{x}\, \sqrt{1-x}\, dx = \int_0^1 \sqrt{x(1-x)}\, dx$$

(a) Trapezoidal Rule:

$$\frac{1}{8}[0 + 2\sqrt{\tfrac{1}{4}(1-\tfrac{1}{4})} + 2\sqrt{\tfrac{1}{2}(1-\tfrac{1}{2})} + 2\sqrt{\tfrac{3}{4}(1-\tfrac{3}{4})} + 0]$$
$$\approx 0.342$$

(b) Simpson's Rule:

$$\frac{1}{12}[0 + 4\sqrt{\tfrac{1}{4}(1-\tfrac{1}{4})} + 2\sqrt{\tfrac{1}{2}(1-\tfrac{1}{2})} + 4\sqrt{\tfrac{3}{4}(1-\tfrac{3}{4})} + 0]$$
$$\approx 0.372$$

15. Using (a) the Trapezoidal Rule and (b) Simpson's Rule, approximate to three significant digits

$$\int_0^1 \sqrt{1 - x^2}\, dx, \quad n = 4$$

Solution:
(a) Trapezoidal Rule:

$$\frac{1}{8}[1 + 2\sqrt{\frac{15}{16}} + 2\sqrt{\frac{3}{4}} + 2\sqrt{\frac{7}{16}} + 0] \approx 0.749$$

(b) Simpson's Rule:

$$\frac{1}{12}[1 + 4\sqrt{\frac{15}{16}} + 2\sqrt{\frac{3}{4}} + 4\sqrt{\frac{7}{16}} + 0] \approx 0.771$$

17. Using (a) the Trapezoidal Rule and (b) Simpson's Rule, approximate to three significant digits

$$\int_0^1 \sqrt{1 - x^2}\, dx, \quad n = 8$$

Solution:
(a) Trapezoidal Rule:

$$\frac{1}{16}[1 + 2\sqrt{\frac{63}{64}} + 2\sqrt{\frac{60}{64}} + 2\sqrt{\frac{55}{64}} + 2\sqrt{\frac{48}{64}}$$
$$+ 2\sqrt{\frac{39}{64}} + 2\sqrt{\frac{28}{64}} + 2\sqrt{\frac{15}{64}} + 0]$$

$$\approx 0.772$$

(b) Simpson's Rule:

$$\frac{1}{24}[1 + 2\sqrt{\frac{63}{64}} + 4\sqrt{\frac{60}{64}} + 2\sqrt{\frac{55}{64}} + 4\sqrt{\frac{48}{64}}$$
$$+ 2\sqrt{\frac{39}{64}} + 4\sqrt{\frac{28}{64}} + 2\sqrt{\frac{15}{64}} + 0]$$

$$\approx 0.780$$

19. Using (a) the Trapezoidal Rule and (b) Simpson's Rule, approximate to three significant digits

$$\int_1^3 \frac{1}{2 + x + x^2}\, dx, \quad n = 2$$

Solution:
(a) Trapezoidal Rule:

$$\frac{1}{2}[\frac{1}{4} + 2(\frac{1}{8}) + \frac{1}{14}] \approx 0.286$$

(b) Simpson's Rule:

$$\frac{1}{3}[\frac{1}{4} + 4(\frac{1}{8}) + \frac{1}{14}] \approx 0.274$$

21. Use Simpson's Rule with n = 8 to approximate the present value of the income $c(t) = 6000 + 200\sqrt{t}$ over $t_1 = 4$ years at the annual interest r = 7%.

Solution:

$$V = \int_0^{t_1} c(t)e^{-rt}\,dt = \int_0^4 (6000 + 200\sqrt{t})e^{-0.07t}\,dt$$

$$\approx \frac{4-0}{24}[(6000 + 200\sqrt{0})e^0 + 4(6000 + 200\sqrt{1/2})e^{-0.07(1/2)}$$

$$+ 2(6000 + 200\sqrt{1})e^{-0.07} + 4(6000 + 200\sqrt{3/2})e^{-0.07(3/2)}$$

$$+ 2(6000 + 200\sqrt{2})e^{-0.14} + 4(6000 + 200\sqrt{5/2})e^{-0.07(5/2)}$$

$$+ 2(6000 + 200\sqrt{3})e^{-0.21} + 4(6000 + 200\sqrt{7/2})e^{-0.07(7/2)}$$

$$+ (6000 + 200\sqrt{4})e^{-0.28}]$$

$$\approx \$21,831.20$$

23. Use Simpson's Rule with n = 6 to approximate the indicated normal probability given by $P(0 \leq x \leq 1)$.

Solution:

$$P(a \leq x \leq b) = \int_a^b \frac{1}{\sqrt{2\pi}} e^{-x^2/2}\,dx$$

$$P(0 \leq x \leq 1) = \frac{1}{\sqrt{2\pi}} \int_0^1 e^{-x^2/2}\,dx$$

$$\approx \frac{1}{\sqrt{2\pi}}(\frac{1}{18})[e^0 + 4e^{-(1/6)^2/2} + 2e^{-(1/3)^2/2} + 4e^{-(1/2)^2/2}$$

$$+ 2e^{-(2/3)^2/2} + 4e^{-(5/6)^2/2} + e^{-1/2}]$$

$$= \frac{1}{18\sqrt{2\pi}}[1 + 4e^{-1/72} + 2e^{-1/18} + 4e^{-1/8} + 2e^{-2/9}$$

$$+ 4e^{-25/72} + e^{-1/2}]$$

$$\approx 0.3413 = 34.13\%$$

25. Use Simpson's Rule to estimate the number of square feet of land in the lot shown in the figure where x and y are measured in feet. The land is bounded by a stream and two straight roads.

x	0	100	200	300	400	500	600	700	800	900	1000
y	125	125	120	112	90	90	95	88	75	35	0

Solution:

$$A \approx \frac{1000}{3(10)}[125 + 4(125) + 2(120) + 4(112) + 2(90) + 4(90)$$

$$+ 2(95) + 4(88) + 2(75) + 4(35) + 0]$$

$$= 89,500 \text{ square feet}$$

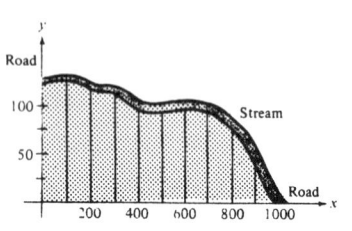

Section 6.5   285

27. Find the maximum possible error in approximating

$$\int_0^2 x^3 \, dx, \quad n = 4$$

using (a) the Trapezoidal Rule and (b) Simpson's Rule.
Solution:

$$\begin{aligned} f(x) &= x^3 \\ f'(x) &= 3x^2 \\ f''(x) &= 6x \\ f'''(x) &= 6 \\ f^{(4)}(x) &= 0 \end{aligned}$$

(a) Trapezoidal Rule: Since $f''(x)$ is maximum in [0, 2] when $x = 2$, we have

$$|\text{Error}| \leq \frac{(2-0)^3}{12(4^2)}(12) = 0.5$$

(b) Simpson's Rule: Since $f^{(4)}(x) = 0$, we have

$$|\text{Error}| \leq \frac{(2-0)^5}{180(4^4)}(0) = 0$$

29. Find the maximum possible error in approximating

$$\int_0^1 e^{x^3} \, dx, \quad n = 4$$

using (a) the Trapezoidal Rule and (b) Simpson's Rule.
Solution:

$$\begin{aligned} f(x) &= e^{x^3} \\ f'(x) &= 3x^2 e^{x^3} \\ f''(x) &= 3(3x^4 + 2x)e^{x^3} \\ f'''(x) &= 3(9x^6 + 18x^3 + 2)e^{x^3} \\ f^{(4)}(x) &= 3(27x^8 + 108x^5 + 60x^2)e^{x^3} \end{aligned}$$

(a) Trapezoidal Rule: Since $|f''(x)|$ is maximum in [0, 1] when $x = 1$, we have

$$|\text{Error}| \leq \frac{(1-0)^3}{12(4^2)}(15e) = \frac{5e}{64} \approx 0.212$$

(b) Simpson's Rule: Since $|f^{(4)}(x)|$ is maximum in [0, 1] when $x = 1$, we have

$$|\text{Error}| \leq \frac{(1-0)^5}{180(4^4)}(585e) = \frac{13e}{1024} \approx 0.035$$

31. Find n so the error in the approximation of

$$\int_0^1 e^{-x^2} \, dx$$

is less than 0.00001 using (a) the Trapezoidal Rule and (b) Simpson's Rule.

Solution:

$$f(x) = e^{-x^2}$$
$$f'(x) = -2xe^{-x^2}$$
$$f''(x) = (4x^2 - 2)e^{-x^2}$$
$$f'''(x) = 4(-2x^3 + 3x)e^{-x^2}$$
$$f^{(4)}(x) = 4(4x^4 - 12x^2 + 3)e^{-x^2}$$

(a) Trapezoidal Rule: Since $|f''(x)|$ is maximum in $[0, 1]$ when $x = 0$ and $|f''(0)| = 2$, we have

$$|\text{Error}| \leq \frac{(1-0)^3}{12n^2}(2) < 0.00001,$$

$$\frac{1}{6n^2} < 0.00001, \quad n^2 > \frac{1}{6(0.00001)}, \quad n > 130$$

(b) Simpson's Rule: Since $|f^{(4)}(x)|$ is maximum in $[0, 1]$ when $x = 0$ and $|f^{(4)}(0)| = 12$ we have

$$|\text{Error}| \leq \frac{12}{180n^4} < 0.00001, \quad n^4 > 6666.67,$$

$n > 9.04$
and we let $n = 10$.

33. Prove that Simpson's Rule is exact when used to approximate the integral of a cubic polynomial function, and demonstrate the result for

$$\int_0^1 x^3 \, dx, \quad n = 2$$

Solution:

$$P_3(x) = ax^3 + bx^2 + cx + d$$
$$P_3'(x) = 3ax^2 + 2bx + c$$
$$P_3''(x) = 6ax + 2b$$
$$P_3'''(x) = 6a$$
$$P_3^{(4)}(x) = 0$$

$$|\text{Error}| \leq \frac{(b-a)^5}{180n^4}[\max |P_3^{(4)}(x)|]$$

$$= \frac{(b-a)^5}{180n^4}(0) = 0$$

Therefore Simpson's Rule is exact when used to approximate a cubic polynomial.

$$\int_0^1 x^3 \, dx = \frac{1-0}{3(2)}[0^3 + 4(\frac{1}{2})^3 + (1)^3]$$

$$= \frac{1}{6}[0 + \frac{1}{2} + 1] = \frac{1}{4}$$

The exact value of this integral is

$$\int_0^1 x^3 \, dx = \frac{x^4}{4}\Big]_0^1 = \frac{1}{4}$$

which is the same as the Simpson approximation.

Section 6.6 Improper integrals

1. Evaluate $\int_0^4 \frac{1}{\sqrt{x}}\, dx$.

   Solution: This integral converges since
   $$\int_0^4 \frac{1}{\sqrt{x}}\, dx = \lim_{b \to 0} 2\sqrt{x}\,\Big]_b^4 = 4$$

3. Evaluate $\int_0^2 \frac{1}{(x-1)^{2/3}}\, dx$.

   Solution: This integral converges since
   $$\int_0^2 \frac{1}{(x-1)^{2/3}}\, dx = \int_0^1 \frac{1}{(x-1)^{2/3}}\, dx + \int_1^2 \frac{1}{(x-1)^{2/3}}\, dx$$
   $$= \lim_{b \to 1} [3(x-1)^{1/3}\,\Big]_0^b + 3(x-1)^{1/3}\,\Big]_b^2\,]$$
   $$= 3 + 3 = 6$$

5. Evaluate $\int_0^\infty e^{-x}\, dx$.

   Solution: This integral converges since
   $$\int_0^\infty e^{-x}\, dx = \lim_{b \to \infty} -e^{-x}\,\Big]_0^b = 0 + 1 = 1$$

7. Evaluate $\int_0^1 \frac{1}{1-x}\, dx$.

   Solution: This integral diverges since
   $$\int_0^1 \frac{1}{1-x}\, dx = \lim_{b \to 1} -\ln|1-x|\,\Big]_0^b = \infty$$

9. Evaluate $\int_0^8 \frac{1}{\sqrt[3]{8-x}}\, dx$.

   Solution: This integral converges since
   $$\int_0^8 \frac{1}{\sqrt[3]{8-x}}\, dx = \lim_{b \to 8} -\frac{3}{2}(8-x)^{2/3}\,\Big]_0^b = 6$$

11. Evaluate $\int_0^1 \frac{1}{x^2}\, dx$.

    Solution: This integral diverges since
    $$\int_0^1 \frac{1}{x^2}\, dx = \lim_{b \to 0} -\frac{1}{x}\,\Big]_b^1 = \infty$$

13. Evaluate $\int_0^2 \frac{1}{\sqrt[3]{x-1}} \, dx$.

    Solution: This integral converges since

    $$\int_0^2 \frac{1}{\sqrt[3]{x-1}} \, dx = \int_0^1 \frac{1}{\sqrt[3]{x-1}} \, dx + \int_1^2 \frac{1}{\sqrt[3]{x-1}} \, dx$$

    $$= \lim_{b \to 1} \left[\frac{3}{2}(x-1)^{2/3} \Big]_0^b + \frac{3}{2}(x-1)^{2/3} \Big]_b^2 \right]$$

    $$= -\frac{3}{2} + \frac{3}{2} = 0$$

15. Evaluate $\int_2^4 \frac{1}{\sqrt{x^2-4}} \, dx$.

    Solution: This integral converges since

    $$\int_2^4 \frac{1}{\sqrt{x^2-4}} \, dx = \lim_{b \to 2} \ln|x + \sqrt{x^2-4}| \Big]_b^4$$

    $$= \ln(4 + 2\sqrt{3}) - \ln 2$$

    $$= \ln(2 + \sqrt{3}) \approx 1.317$$

17. Evaluate $\int_1^\infty \frac{1}{x^2} \, dx$.

    Solution: This integral converges since

    $$\int_1^\infty \frac{1}{x^2} \, dx = \lim_{b \to \infty} -\frac{1}{x} \Big]_1^b = 0 + 1 = 1$$

19. Evaluate $\int_0^\infty e^{x/2} \, dx$.

    Solution: This integral diverges since

    $$\int_0^\infty e^{x/2} \, dx = \lim_{b \to \infty} 2e^{x/2} \Big]_0^b = \infty$$

21. Evaluate $\int_{-\infty}^\infty x^2 e^{-x^3} \, dx$.

    Solution: This integral diverges since

    $$\int_{-\infty}^\infty x^2 e^{-x^3} \, dx = \lim_{b \to \infty} -\frac{1}{3} e^{-x^3} \Big]_{-b}^b = \infty$$

23. Evaluate $\int_5^\infty \frac{x}{\sqrt{x^2-16}} \, dx$.

    Solution: This integral diverges since

    $$\int_5^\infty \frac{x}{\sqrt{x^2-16}} \, dx = \lim_{b \to \infty} \sqrt{x^2-16} \Big]_5^b = \infty$$

Section 6.6   289

25. Evaluate the convergent improper integral using the fact that $\lim_{x \to \infty} x^n e^{-ax} = 0$, $a > 0$, $n > 0$.

$$\int_0^\infty x^2 e^{-x} \, dx$$

Solution:

$$\int_0^\infty x^2 e^{-x} \, dx = \lim_{b \to \infty} \left[-x^2 e^{-x} - 2x e^{-x} - 2e^{-x} \right]_0^b = 2$$

27. Evaluate the convergent improper integral using the fact that $\lim_{x \to \infty} x^n e^{-ax} = 0$, $a > 0$, $n > 0$.

$$\int_0^\infty x e^{-2x} \, dx$$

Solution:

$$\int_0^\infty x e^{-2x} \, dx = \lim_{b \to \infty} \left[-\frac{1}{2} x e^{-2x} - \frac{1}{4} e^{-2x} \right]_0^b$$

$$= (0 - 0) - \left(0 - \frac{1}{4}\right) = \frac{1}{4}$$

29. Find (a) the area of the region bounded by the graphs of $y = 1/x^2$, $y = 0$, $x \geq 1$ and (b) the volume of the solid generated by revolving the region about the x-axis.

Solution:

(a) $\quad A = \int_1^\infty \frac{1}{x^2} \, dx = \lim_{b \to \infty} \left. -\frac{1}{x} \right]_1^b = 0 - (-1)$

$\quad = 1$ square unit

(b) $\quad V = \pi \int_1^\infty \frac{1}{x^4} \, dx = \lim_{b \to \infty} \left. -\frac{\pi}{3x^3} \right]_1^b = 0 - \left(-\frac{\pi}{3}\right)$

$\quad = \frac{\pi}{3}$ cubic units

31. The capitalized cost is given by

$$C = C_0 + \int_0^n c(t) e^{-rt} \, dt$$

Find the capitalized cost C of an asset

(a) for n = 5 years
(b) for n = 10 years
(c) forever

where $C_0 = \$650,000$ is the original investment, t is the time in years, r = 10% is the annual rate compounded continuously, and c(t) = \$25,000 is the annual cost of maintenance.

Solution:

$$C = 650{,}000 + \int_0^n 25{,}000 e^{-0.10t}\, dt$$

$$= 650{,}000 + \frac{25000}{-0.10} e^{-0.10t} \Big]_0^n$$

$$= 650{,}000 - 250{,}000(e^{-0.10n} - 1)$$

(a) For n = 5:

$$C = 650{,}000 - 250{,}000(e^{-0.50} - 1)$$
$$\approx \$748{,}367.34$$

(b) For n = 10:

$$C = 650{,}000 - 250{,}000(e^{-1} - 1)$$
$$\approx \$808{,}030.14$$

(c) As $n \longrightarrow \infty$:

$$C = 650{,}000 - 250{,}000(0 - 1)$$
$$= \$900{,}000.00$$

33. Demonstrate that

$$\lim_{x \longrightarrow \infty} x^n e^{-x} = 0$$

by completing the following table for (a) n = 1, (b) n = 2, and (c) n = 5.

x	1	10	25	50
$x^n e^{-x}$				

Solution:

(a)

x	1	10	25	50
xe^{-x}	0.3679	0.0005	0.0000	0.0000

(b)

x	1	10	25	50
$x^2 e^{-x}$	0.3679	0.0045	0.0000	0.0000

(c)

x	1	10	25	50
$x^5 e^{-x}$	0.3679	4.5400	0.0001	0.0000

Section 6.7 Random Variables and Probability

1. A couple has 4 children. Assume that it is equally likely that each child will be a girl or a boy.
   (a) What is the sample space?
   (b) Complete a table to form the probability distribution if the random variable x is the number of girls in the family.
   (c) Use the table of part (b) to graph the probability distribution.
   (d) Use the table of part (b) to find the probability of having at least 1 boy.

   Solution: 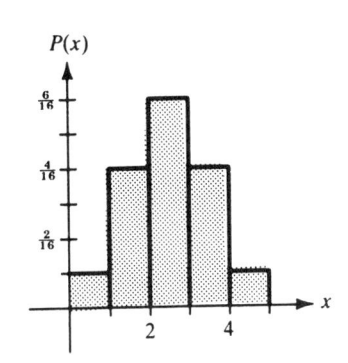
   (a) S = {gggg, gggb, ggbg, gbgg, bggg, ggbb, gbbg, gbgb, bgbg, bbgg, bggb, gbbb, bgbb, bbgb, bbbg, bbbb}
   (b)

Random Variable, x	0	1	2	3	4
Probability of x, P(x)	1/16	4/16	6/16	4/16	1/16

   (c) See accompanying graph.
   (d) Probability of at least one boy = 1 − probability of all girls. P = 1 − (1/16) = 15/16

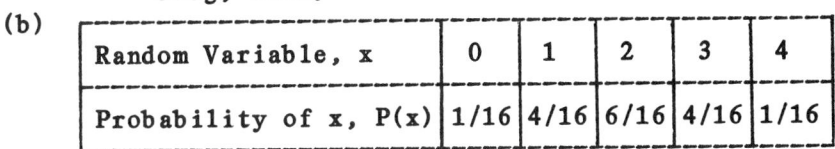

3. Verify that $f(x) = 6x(1-x)$ is a probability density function over the interval [0, 1].
   Solution:
   $$\int_0^1 6x(1-x)\,dx = \int_0^1 (6x - 6x^2)\,dx = \left[3x^2 - 2x^3\right]_0^1 = 1$$

5. Verify that $f(x) = (4/27)x^2(3-x)$ is a probability density function over the interval [0, 3].
   Solution:
   $$\int_0^3 \frac{4}{27} x^2(3-x)\,dx = \frac{4}{27}\int_0^3 (3x^2 - x^3)\,dx$$
   $$= \frac{4}{27}\left[x^3 - \frac{x^4}{4}\right]_0^3 = \frac{4}{27}\left(27 - \frac{81}{4}\right) = 1$$

7. Verify that $f(x) = (1/3)e^{-x/3}$ is a probability density function over the interval [0, ∞).
   Solution:
   $$\int_0^\infty \frac{1}{3} e^{-x/3}\,dx = \lim_{b \to \infty} \left. -e^{-x/3} \right]_0^b = 0 - (-1) = 1$$

9. Find the constant k so that $f(x) = kx$ is a probability density function over the interval [1, 5].
   Solution:
   $$\int_1^5 kx\,dx = \left.\frac{kx^2}{2}\right]_1^5 = 12k = 1 \implies k = \frac{1}{12}$$

11. Find the constant k so that $f(x) = k(4 - x^2)$ is a probability density function over the interval $[-2, 2]$.
    Solution:
    $$\int_{-2}^{2} k(4 - x^2)\,dx = k\left(4x - \frac{x^3}{3}\right)\Big]_{-2}^{2} = \frac{32k}{3} = 1$$
    $$k = \frac{3}{32}$$

13. Find the constant k so that $f(x) = ke^{-x/2}$ is a probability density function over the interval $[0, \infty)$.
    Solution:
    $$\int_{0}^{\infty} ke^{-x/2}\,dx = \lim_{b \to \infty} -2ke^{-x/2}\Big]_{0}^{b} = 2k = 1 \implies k = \frac{1}{2}$$

15. Sketch the graph of $f(x) = 1/10$ over the interval $[0, 10]$ and find the indicated probabilities.
    (a) $P(0 < x < 6)$ (b) $P(4 < x < 6)$
    (c) $P(8 < x < 10)$ (d) $P(x \geq 2)$
    Solution: 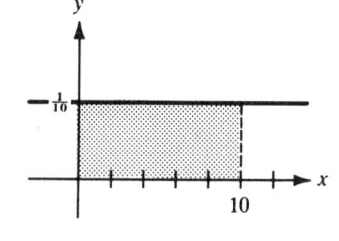
    $$\int_{a}^{b} \frac{1}{10}\,dx = \frac{x}{10}\Big]_{a}^{b} = \frac{b-a}{10}$$

    (a) $P(0 < x < 6) = (6 - 0)/10 = 3/5$
    (b) $P(4 < x < 6) = (6 - 4)/10 = 1/5$
    (c) $P(8 < x < 10) = (10 - 8)/10 = 1/5$
    (d) $P(x \geq 2) = P(2 < x < 10) = (10 - 2)/10 = 4/5$

17. Sketch the graph of $f(x) = (3/16)\sqrt{x}$ over the interval $[0, 4]$ and find the indicated probabilities.
    (a) $P(0 < x < 2)$ (b) $P(2 < x < 4)$
    (c) $P(1 < x < 3)$ (d) $P(x \leq 3)$
    Solution: 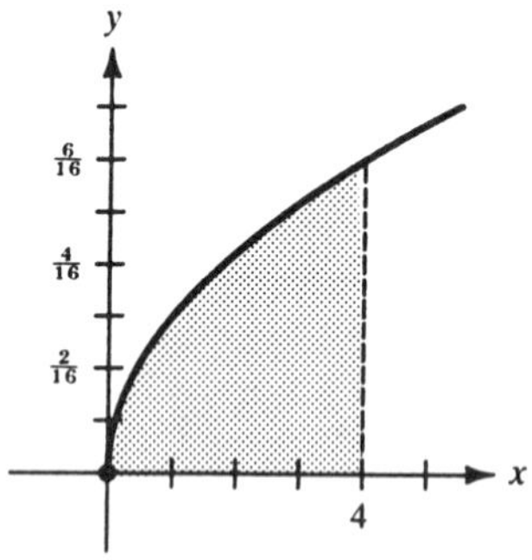
    $$\int_{a}^{b} \frac{3}{16}\sqrt{x}\,dx = \left(\frac{3}{16}\right)\frac{2}{3}x^{3/2}\Big]_{a}^{b} = \frac{1}{8}[b\sqrt{b} - a\sqrt{a}]$$

    (a) $P(0 < x < 2) = \sqrt{2}/4 \approx 0.354$
    (b) $P(2 < x < 4) = 1 - (\sqrt{2}/4) \approx 0.646$
    (c) $P(1 < x < 3) = (1/8)(3\sqrt{3} - 1) \approx 0.525$
    (d) $P(x \leq 3) = 3\sqrt{3}/8 \approx 0.650$

19. Sketch the graph of $f(x) = (1/2)e^{-t/2}$ over the interval $[0, \infty)$ and find the indicated probabilities.
    (a) $P(t < 2)$ (b) $P(t \geq 2)$
    (c) $P(1 < t < 4)$ (d) $P(t = 3)$
    Solution: 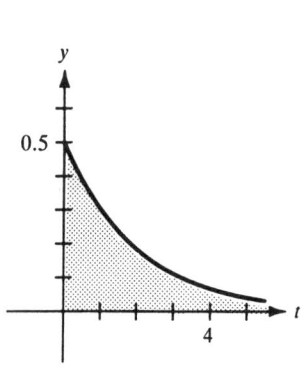
    $$\int_{a}^{b} \frac{1}{2}e^{-t/2}\,dt = -e^{-t/2}\Big]_{a}^{b} = e^{-a/2} - e^{-b/2}$$

    (a) $P(t < 2) = e^{-0/2} - e^{-2/2} = 1 - e^{-1} \approx 0.632$
    (b) $P(t > 2) = e^{-2/2} - 0 = e^{-1} \approx 0.368$
    (c) $P(1 < t < 4) = e^{-1/2} - e^{-4/2} = e^{-1/2} - e^{-2} \approx 0.471$
    (d) $P(t = 3) = e^{-3/2} - e^{-3/2} = 0$

Section 6.7

21. Buses arrive and depart from a college every 20 minutes. The probability density function for the waiting time t (in minutes) for a person arriving at random at the bus stop is

$$f(t) = \frac{1}{20}, \quad [0, 20]$$

Find the probability that the person will wait (a) no more than 5 minutes and (b) at least 12 minutes.
Solution:

$$P(a < x < b) = \int_a^b \frac{1}{20} dx = \frac{b-a}{20}$$

(a) $P(0 \le x \le 5) = \frac{5-0}{20} = \frac{1}{4}$

(b) $P(12 < x < 20) = \frac{20-12}{20} = \frac{2}{5}$

23. The waiting time (in minutes) for service at the checkout at a certain grocery store is exponentially distributed with $\lambda = 3$. Using the exponential density function

$$f(t) = (1/\lambda)e^{-t/\lambda}, \quad [0, \infty)$$

find the probability of waiting (a) less than 2 minutes (b) more than 2 minutes but less than 4 minutes and (c) at least 2 minutes.
Solution:

$$\int_a^b \frac{1}{3} e^{-t/3} dt = -e^{-t/3} \Big]_a^b = e^{-a/3} - e^{-b/3}$$

(a) $P(0 < x < 2) = e^{-0/3} - e^{-2/3} = 1 - e^{-2/3} \approx 0.487$

(b) $P(2 < x < 4) = e^{-2/3} - e^{-4/3} \approx 0.250$

(c) $P(x > 2) = 1 - P(0 < x < 2) = e^{-2/3} \approx 0.513$

25. The length of time (in hours) required to unload trucks at a depot is exponentially distributed with $\lambda = 3/4$. Using the exponential density function

$$f(t) = (1/\lambda)e^{-t/\lambda}, \quad [0, \infty)$$

what proportion of the trucks can be unloaded in less than 1 hour?
Solution:

$$P(0 < x < 1) = \int_0^1 \frac{4}{3} e^{-4t/3} dt = -e^{-4t/3} \Big]_0^1$$

$$= 1 - e^{-4/3} \approx 0.736$$

27. The weekly demand x (in tons) for a certain product is a continuous random variable with the density function

$$f(x) = \frac{1}{25} xe^{-x/5}, \quad [0, \infty)$$

Find the following.
(a) $P(x < 5)$
(b) $P(5 < x < 10)$
(c) $P(x > 10) = 1 - P(x \leq 10)$

Solution:

$$\int_a^b \frac{1}{25} xe^{-x/5} \, dx = -\frac{1}{5} e^{-x/5} (x + 5) \Big]_a^b$$

(a) $P(x < 5) = -\frac{1}{5} e^{-x/5} (x + 5) \Big]_0^5 = -\frac{1}{5}[10e^{-1} - 5]$

$\qquad = 1 - 2e^{-1} \approx 0.264$

(b) $P(5 < x < 10) = -\frac{1}{5} e^{-x/5} (x + 5) \Big]_5^{10}$

$\qquad = -\frac{1}{5}[15e^{-2} - 10e^{-1}]$

$\qquad = 2e^{-1} - 3e^{-2} \approx 0.330$

(c) $P(x > 10) = 1 - P(x \leq 10)$
$\qquad = 1 - [P(x < 5) + P(5 < x < 10)]$
$\qquad = 1 - 1 + 2e^{-1} - 2e^{-1} + 3e^{-2}$
$\qquad = 3e^{-2} \approx 0.406$

29. The probability density function for the percentage of recall in a certain learning experiment is found to be

$$f(x) = \frac{15}{4} x\sqrt{1-x}, \quad [0, 1]$$

What is the probability that a randomly chosen individual in the experiment will recall between
(a) 0% and 25% of the material
(b) 50% and 75% of the material

Solution:

$$\int_a^b \frac{15}{4} x\sqrt{1-x} \, dx = -\frac{1}{2}(1-x)^{3/2}(3x+2) \Big]_a^b$$

(a) $P(0 < x < 0.25) = -\frac{1}{2}(1-x)^{3/2}(3x+2) \Big]_0^{0.25}$

$\qquad = -\frac{1}{2}[(0.75)^{3/2}(2.75) - 2]$

$\qquad \approx 0.1069$

(b) $P(0.50 < x < 0.75) = -\frac{1}{2}(1-x)^{3/2}(3x+2) \Big]_{0.50}^{0.75}$

$\qquad = -\frac{1}{2}[(0.25)^{3/2}(4.25) - (0.50)^{3/2}(3.50)]$

$\qquad \approx 0.3531$

Section 6.8

● **Section 6.8 Expected Value, Standard Deviation, and Median**

1. Use the probability density function f(x) = 1/8 over the interval [0, 8] to find (a) the mean, (b) the variance, and (c) the median of the random variable. Locate the mean and median on the graph of the density function.
   Solution:

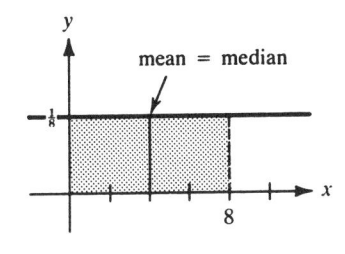

   (a) $\mu = \int_a^b xf(x)\,dx = \int_0^8 x\left(\dfrac{1}{8}\right)\,dx = \dfrac{x^2}{16}\Big]_0^8 = 4$

   (b) $\sigma^2 = \int_a^b x^2 f(x)\,dx - \mu^2 = \int_0^8 x^2\left(\dfrac{1}{8}\right)\,dx - (4)^2$

   $= \dfrac{x^3}{24}\Big]_0^8 - 16 = \dfrac{64}{3} - 16 = \dfrac{16}{3}$

   (c) $P(0 < x < m) = \int_0^m \dfrac{1}{8}\,dx = \dfrac{x}{8}\Big]_0^m = \dfrac{m}{8} = \dfrac{1}{2} \implies m = 4$

3. Use the probability density function f(x) = t/32 over the interval [0, 8] to find (a) the mean, (b) the variance, and (c) the median of the random variable. Locate the mean and median on the graph of the density function.
   Solution:

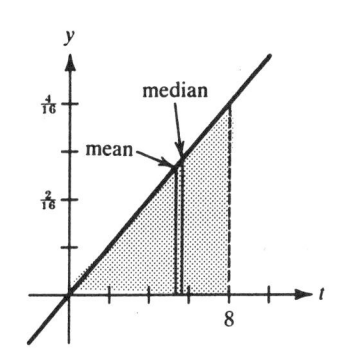

   (a) $\mu = \int_a^b tf(t)\,dt = \int_0^8 t\left(\dfrac{t}{32}\right)\,dt = \dfrac{t^3}{96}\Big]_0^8 = \dfrac{16}{3}$

   (b) $\sigma^2 = \int_a^b t^2 f(t)\,dt - \mu^2 = \int_0^8 t^2\left(\dfrac{t}{32}\right)\,dt - \left(\dfrac{16}{3}\right)^2$

   $= \dfrac{t^4}{128}\Big]_0^8 - \dfrac{256}{9} = 32 - \dfrac{256}{9} = \dfrac{32}{9}$

   (c) $P(0 < t < m) = \int_0^m \dfrac{t}{32}\,dt = \dfrac{t^2}{64}\Big]_0^m = \dfrac{m^2}{64} = \dfrac{1}{2}$

   $m^2 = 32$
   $m = 4\sqrt{2} \approx 5.657$

5. Use the probability density function f(x) = 6x(1 − x) over the interval [0, 1] to find (a) the mean, (b) the variance, and (c) the median of the random variable. Locate the mean and median on the graph of the density function.
   Solution:

   (a) $\mu = \int_a^b xf(x)\,dx = \int_0^1 x[6x(1-x)]\,dx$

   $= \int_0^1 (6x^2 - 6x^3)\,dx = \left[2x^3 - \dfrac{3}{2}x^4\right]_0^1 = \dfrac{1}{2}$

(b) $\sigma^2 = \int_a^b x^2 f(x)\, dx - \mu^2 = \int_0^1 x^2[6x(1-x)]\, dx - (\frac{1}{2})^2$

$= \int_0^1 (6x^3 - 6x^4)\, dx - \frac{1}{4} = \left[\frac{3}{2}x^4 - \frac{6}{5}x^5\right]_0^1 - \frac{1}{4}$

$= \frac{3}{2} - \frac{6}{5} - \frac{1}{4} = \frac{1}{20}$

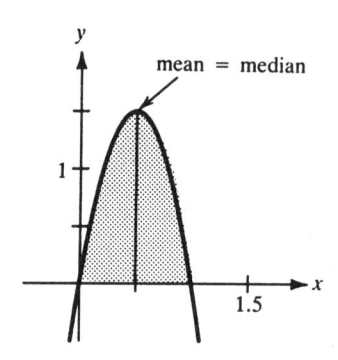

(c) $P(0 < x < m) = \int_0^m 6x(1-x)\, dx = \int_0^m (6x - 6x^2)\, dx$

$= \left[3x^2 - 2x^3\right]_0^m = 3m^2 - 2m^3 = \frac{1}{2}$

$6m^2 - 4m^3 = 1$
$0 = 4m^3 - 6m^2 + 1$
$0 = (2m - 1)(2m^2 - 2m - 1)$

$m = \frac{1}{2}$ or $m = \frac{2 \pm \sqrt{12}}{4} = \frac{1 \pm \sqrt{3}}{2}$

In the interval [0, 1], $m = 1/2$.

7. Use the probability density function $f(x) = (5/2)x^{3/2}$ over the interval [0, 1] to find (a) the mean, (b) the variance, and (c) the median of the random variable. Locate the mean and median on the graph of the density function.
Solution:

(a) $\mu = \int_a^b x f(x)\, dx = \int_0^1 x(\frac{5}{2}x^{3/2})\, dx = \frac{5}{2}\int_0^1 x^{5/2}\, dx$

$= (\frac{5}{2})\frac{2}{7}x^{7/2}\Big]_0^1 = \frac{5}{7}$

(b) $\sigma^2 = \int_a^b x^2 f(x)\, dx - \mu^2 = \int_0^1 x^2(\frac{5}{2}x^{3/2})\, dx - (\frac{5}{7})^2$

$= \frac{5}{2}\int_0^1 x^{7/2}\, dx - \frac{25}{49}$

$= (\frac{5}{2})\frac{2}{9}x^{9/2}\Big]_0^1 - \frac{25}{49}$

$= \frac{5}{9} - \frac{25}{49} = \frac{20}{441}$

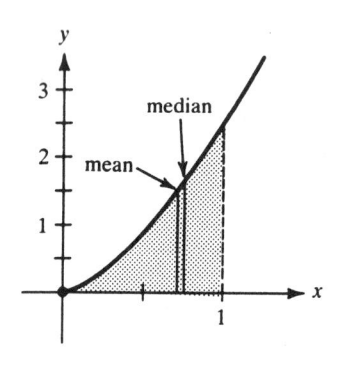

(c) $P(0 < x < m) = \int_0^m \frac{5}{2}x^{3/2}\, dx = (\frac{5}{2})\frac{2}{5}x^{5/2}\Big]_0^m = \frac{1}{2}$

$m^{5/2} = \frac{1}{2}$

$m = (\frac{1}{2})^{2/5} \approx 0.758$

Section 6.8

9. Use the probability density function
$$f(x) = \frac{4}{3(x+1)^2}$$
over the interval [0, 3] to find (a) the mean, (b) the variance, and (c) the median of the random variable. Locate the mean and median on the graph of the density function.

Solution:

(a) $\mu = \int_a^b x f(x) \, dx = \int_0^3 \frac{4x}{3(x+1)^2} \, dx$

$= \frac{4}{3}\left[\frac{1}{x+1} + \ln|x+1|\right]_0^3$ (Formula 4)

$= \frac{4}{3}\left[\left(\frac{1}{4} + \ln 4\right) - (1 + 0)\right]$

$= \frac{4}{3}(\ln 4) - 1 \approx 0.848$

(b) $\sigma^2 = \int_a^b x^2 f(x) \, dx - \mu^2$

$= \int_0^3 \frac{4x^2}{3(x+1)^2} \, dx - \left[\frac{4}{3}(\ln 4) - 1\right]^2$

$= \frac{4}{3}\left[x - \frac{1}{x+1} - 2\ln|x+1|\right]_0^3 - \left[\frac{4}{3}(\ln 4) - 1\right]^2$

(Formula 7)

$= \frac{4}{3}\left[\left(3 - \frac{1}{4} - 2\ln 4\right) - (0 - 1 - 0)\right]$
$\qquad\qquad - \left[\frac{4}{3}(\ln 4) - 1\right]^2$

$= \left[5 - \frac{8}{3}\ln 4\right] - \left[\frac{4}{3}(\ln 4) - 1\right]^2$

$= 4 - \left(\frac{4}{3}\ln 4\right)^2 \approx 0.583$

(c) $P(0 < x < m) = \int_0^m \frac{4}{3(x+1)^2} \, dx = -\frac{4}{3(x+1)}\Big]_0^m$

$= -\frac{4}{3(m+1)} + \frac{4}{3} = \frac{4}{3}\left[1 - \frac{1}{m+1}\right] = \frac{1}{2}$

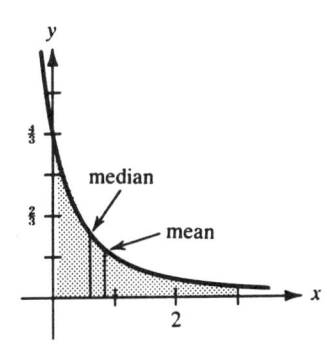

$1 - \frac{1}{m+1} = \frac{3}{8}$

$\frac{5}{8} = \frac{1}{m+1}$

$m + 1 = \frac{8}{5}$

$m = 3/5$

Section 6.8

11. Find the median of the exponential probability density function

$$f(t) = \frac{1}{7}e^{-t/7}, \quad [0, \infty)$$

Solution:

$$\text{Median} = \int_0^m \frac{1}{7}e^{-t/7}\,dt = -e^{-t/7}\Big]_0^m = 1 - e^{-m/7} = \frac{1}{2}$$

$$e^{-m/7} = \frac{1}{2} \implies -\frac{m}{7} = \ln\frac{1}{2}$$

$$m = -7\ln\frac{1}{2} = 7\ln 2 \approx 4.852$$

13. Find the mean and standard deviation of the exponential probability density function

$$f(x) = \frac{1}{\lambda}e^{-x/\lambda}, \quad [0, \infty)$$

using the limit $\lim_{x \to \infty} x^n e^{-x} = 0$.

Solution:

$$\mu = \int_0^\infty \frac{x}{\lambda}e^{-x/\lambda}\,dx = \lambda\int_0^\infty \left(-\frac{x}{\lambda}\right)e^{-x/\lambda}\left(-\frac{1}{\lambda}\right)dx$$

$$= \lim_{b \to \infty} \lambda\left(-\frac{x}{\lambda} - 1\right)e^{-x/\lambda}\Big]_0^b \quad \text{(Formula 37)}$$

$$= \lim_{b \to \infty}\left[-xe^{-x/\lambda} - \lambda e^{-x/\lambda}\right]_0^b = (0 - 0) - (0 - \lambda) = \lambda$$

$$\sigma^2 = \int_0^\infty \frac{x^2}{\lambda}e^{-x/\lambda}\,dx - \mu^2 = -\lambda^2\int_0^\infty \frac{x^2}{\lambda^2}e^{-x/\lambda}\left(-\frac{1}{\lambda}\right)dx - \mu^2$$

$$= \lim_{b \to \infty}\left[-\lambda^2\left[\frac{x^2}{\lambda^2}e^{-x/\lambda} - 2\left(-\frac{x}{\lambda} - 1\right)e^{-x/\lambda}\right]\right]_0^b - \lambda^2$$

(Formulas 38 and 37)

$$= \lim_{b \to \infty}\left[-x^2 e^{-x/\lambda} - 2x\lambda e^{-x/\lambda} - 2\lambda^2 e^{-x/\lambda}\right]_0^b - \lambda^2$$

$$= 2\lambda^2 - \lambda^2 = \lambda^2 \implies \sigma = \sqrt{\lambda^2} = \lambda$$

15. Determine the mean, variance, and standard deviation of the probability density function

$$f(x) = \frac{1}{8}e^{-x/8}, \quad [0, \infty)$$

Solution: From Exercise 13 we have the following for $f(x) = (1/\lambda)e^{-x/\lambda}$ on the interval $[0, \infty)$.

Mean: $\mu = \lambda$
Standard deviation: $\sigma^2 = \lambda^2$
Variance: $\sigma = \lambda$

Therefore, $\lambda = 8$, $\mu = 8$, $\sigma^2 = 64$, and $\sigma = 8$.

Section 6.8   299

17. The time t until failure of a machine component is exponentially distributed, with a mean of 4 years.
  (a) Find the probability density function for the random variable t.
  (b) Find the probability that a given component will fail in less than 3 years.

Solution: From Exercise 13 we have the following for $f(x) = (1/\lambda)e^{-x/\lambda}$ on the interval $[0, \infty)$.

  Mean: $\mu = \lambda$
  Standard deviation: $\sigma^2 = \lambda^2$
  Variance: $\sigma = \lambda$

  (a) Since $\mu = \lambda = 4$, we have $f(t) = (1/4)e^{-t/4}$.

  (b) $P(0 < t < 3) = \int_0^3 \frac{1}{4} e^{-t/4} \, dt = -e^{-t/4} \Big]_0^3$
   $= 1 - e^{-3/4} \approx 0.528 = 52.8\%$

19. The waiting time t for service at a customer service desk in a department store is exponentially distributed, with a mean of 5 minutes.
  (a) Find the probability density function for the random variable t.
  (b) Find $P(\mu - \sigma < t < \mu + \sigma)$.

Solution: From Exercise 13 we have the following for $f(x) = (1/\lambda)e^{-x/\lambda}$ on the interval $[0, \infty)$.

  Mean: $\mu = \lambda$
  Standard deviation: $\sigma^2 = \lambda^2$
  Variance: $\sigma = \lambda$

  (a) Since $\mu = \lambda = 5$, we have $f(t) = (1/5)e^{-t/5}$.
  (b) $P(\mu - \sigma < t < \mu + \sigma) = P(0 < t < 10)$
   Since $\mu = \sigma = \lambda = 5$, we have

   $P(0 < t < 10) = \int_0^{10} \frac{1}{5} e^{-t/5} \, dt = -e^{-t/5} \Big]_0^{10}$
   $= 1 - e^{-2} \approx 0.865 = 86.5\%$

21. The daily demand x for a certain product (in tons) is a random variable with the probability density function

$$f(x) = \frac{1}{36} x e^{-x/6}, \quad [0, \infty)$$

  (a) Determine the expected daily demand.
  (b) Find $P(x \leq 5)$.

Solution: From Exercise 14 we have the following for $f(x) = (1/\lambda^2) x e^{-x/\lambda}$ on the interval $[0, \infty)$.

$$\mu = \int_0^\infty \frac{x^2}{\lambda^2} e^{-x/\lambda} \, dx = -\lambda \int_0^\infty (-\frac{x}{\lambda})^2 e^{-x/\lambda}(-\frac{1}{\lambda}) \, dx$$

$$= \lim_{b \to \infty} -\lambda \left[\frac{x^2}{\lambda^2} e^{-x/\lambda} - 2(-\frac{x}{\lambda} - 1) e^{-x/\lambda} \right]_0^b$$

$$= 2\lambda \qquad\qquad\qquad\qquad \text{(Formulas 38 and 37)}$$

(a) Since $\lambda = 6$, we have $\mu = 2\lambda = 12$.

(b) $P(x \leq 5) = \int_0^5 \dfrac{x}{36} e^{-x/6} \, dx = \int_0^5 \left(-\dfrac{x}{6}\right) e^{-x/6} \left(-\dfrac{1}{6}\right) dx$

$= \left(-\dfrac{x}{6} - 1\right) e^{-x/6} \Big]_0^5 = -\dfrac{11}{6} e^{-5/6} + 1 \approx 20.3\%$

23. The daily demand x for a certain product (in hundreds of pounds) is a random variable with the probability density function

$$f(x) = \dfrac{1}{36} x(6-x), \quad [0, 6]$$

(a) Determine the expected value and the standard deviation of demand.
(b) Determine the median of the random variable.
(c) Find $P(\mu - \sigma < x < \mu + \sigma)$.

Solution:

(a) $\mu = \int_0^6 \dfrac{1}{36} x^2 (6-x) \, dx = \dfrac{1}{36} \int_0^6 (6x^2 - x^3) \, dx$

$= \dfrac{1}{36}\left[2x^3 - \dfrac{x^4}{4}\right]_0^6 = 3$

$\sigma^2 = \int_0^6 \dfrac{1}{36} x^3 (6-x) \, dx - (3)^2$

$= \dfrac{1}{36}\int_0^6 (6x^3 - x^4) \, dx - 9 = \dfrac{1}{36}\left[\dfrac{3x^4}{2} - \dfrac{x^5}{5}\right]_0^6 - 9$

$= \dfrac{54}{5} - 9 = \dfrac{9}{5}$

$\sigma = \sqrt{\dfrac{9}{5}} = \dfrac{3\sqrt{5}}{5} \approx 1.342$

(b) $\int_0^m \dfrac{1}{36} x(6-x) \, dx = \dfrac{1}{36} \int_0^m (6x - x^2) \, dx$

$= \dfrac{1}{36}\left[3x^2 - \dfrac{x^3}{3}\right]_0^m = \dfrac{1}{36}\left[3m^2 - \dfrac{m^3}{3}\right] = \dfrac{1}{2}$

$3m^2 - \dfrac{m^3}{3} = 18 \quad \Longrightarrow \quad 0 = (m-3)(m^2 - 6m - 18)$

$m = 3 \quad \text{or} \quad m = \dfrac{6 \pm \sqrt{108}}{2} = \dfrac{6 \pm 6\sqrt{3}}{2} = 3 \pm 3\sqrt{3}$

In the interval [0, 6], $m = 3$.

(c) $P(\mu - \sigma < x < \mu + \sigma) = P\left(3 - \dfrac{3\sqrt{5}}{5} < x < 3 + \dfrac{3\sqrt{5}}{5}\right)$

$\approx P(1.6584 < x < 4.3416) = \int_{1.6584}^{4.3416} \dfrac{1}{36} x(6-x) \, dx$

$= \dfrac{1}{36}\left[3x^2 - \dfrac{x^3}{3}\right]_{1.6584}^{4.3416} \approx 0.626 = 62.6\%$

25. The percentage recall x in a learning experiment is a random variable with the probability density function

$$f(x) = \frac{15}{4} x\sqrt{1-x}, \quad [0, 1]$$

Determine the mean and variance of the random variable x.
Solution:

$$\mu = \int_0^1 \frac{15}{4} x^2 \sqrt{1-x}\, dx \qquad \text{(Formula 14)}$$

$$= \frac{15}{4}(\frac{2}{-7})\left[x^2(1-x)^{3/2} - 2(\frac{2}{-5})[x(1-x)^{3/2} + \frac{2}{3}(1-x)^{3/2}]\right]_0^1$$

$$= -\frac{15}{14}[0 - \frac{4}{5}(\frac{2}{3})] = \frac{4}{7}$$

$$\sigma^2 = \int_0^1 \frac{15}{4} x^3 \sqrt{1-x}\, dx - (\frac{4}{7})^2$$

$$= \frac{15}{4}(\frac{2}{-9})\left[\left[x^3(1-x)^{3/2}\right]_0^1 - 3\int_0^1 x^2\sqrt{1-x}\, dx\right] - \frac{16}{49}$$

$$= -\frac{5}{6}[0 - 3(\frac{4}{7})(\frac{4}{15})] - \frac{16}{49}$$

$$= \frac{8}{21} - \frac{16}{49} = \frac{8}{147}$$

27. Use the normal probability density function

$$f(x) = \frac{1}{\sigma\sqrt{2\pi}} e^{-(x-\mu)^2/2\sigma^2}, \quad -\infty < x < \infty$$

to sketch the graph of the normal density function if $\mu = 0$ and $\sigma = 1$. Show the maximum and the points of inflection on the graph. 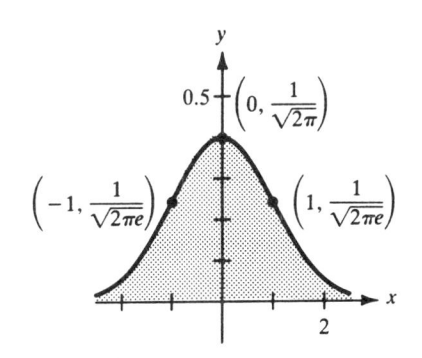
Solution: For $\mu = 0$ and $\sigma = 1$ we have

$$f(x) = (1/\sqrt{2\pi}) e^{-x^2/2}$$

$$f'(x) = (1/\sqrt{2\pi})(-x) e^{-x^2/2}$$

$$f''(x) = (1/\sqrt{2\pi})[x^2 e^{-x^2/2} - e^{-x^2/2}]$$

$$= (1/\sqrt{2\pi}) e^{-x^2/2}(x^2 - 1)$$

Since $f'(x) = 0$ when $x = 0$ and $f''(0) < 0$, it follows that $(0, 1/\sqrt{2\pi})$ is a relative maximum.
Since $f''(x) = 0$ when $x = \pm 1$, it follows that

$$(\pm 1, \frac{1}{\sqrt{2\pi}} e^{-1/2}) = (\pm 1, \frac{1}{\sqrt{2\pi e}})$$

are inflection points.

29. Use the normal probability density function

$$f(x) = \frac{1}{\sigma\sqrt{2\pi}} e^{-(x-\mu)^2/2\sigma^2}, \quad -\infty < x < \infty$$

to determine the weekly profit x (in thousands of dollars) of a store that is normally distributed with $\mu = 8$ and $\sigma = 3$. Use Simpson's Rule with n = 6 to approximate $P(8 < x < 11)$.

Solution: For $\mu = 8$ and $\sigma = 3$ we have

$$f(x) = \frac{1}{3\sqrt{2\pi}} e^{-(x-8)^2/18}$$

$$P(8 < x < 11) = \int_8^{11} \frac{1}{3\sqrt{2\pi}} e^{-(x-8)^2/18} dx$$

$$\approx \frac{1}{3\sqrt{2\pi}} \left(\frac{11-8}{18}\right) [e^{-(8-8)^2/18} + 4e^{-(8.5-8)^2/18}$$

$$+ 2e^{-(9-8)^2/18} + 4e^{-(9.5-8)^2/18}$$

$$+ 2e^{-(10-8)^2/18} + 4e^{-(10.5-8)^2/18}$$

$$+ e^{-(11-8)^2/18}]$$

$$= \frac{1}{18\sqrt{2\pi}} [1 + 4e^{-0.25/18} + 2e^{-1/18} + 4e^{-2.25/18}$$

$$+ 2e^{-2/9} + 4e^{-6.25/18} + e^{-1/2}]$$

$$\approx 0.3413 = 34.13\%$$

● Review Exercises for Chapter 6

1. Evaluate $\int x(2-x)^3 dx$.

Solution:

$$\int x(2-x)^3 dx = \int x(8 - 12x + 6x^2 - x^3) dx$$

$$= \int (8x - 12x^2 + 6x^3 - x^4) dx$$

$$= 4x^2 - 4x^3 + \frac{3}{2}x^4 - \frac{1}{5}x^5 + C$$

$$= \frac{x^2}{10}(40 - 40x + 15x^2 - 2x^3) + C$$

Review Exercises for Chapter 6

3. Evaluate $\int x(x+1)^{3/2}\, dx$.

    Solution: Use substitution and let $u = x + 1$, then $x = u - 1$ and $dx = du$.

    $$\int x(x+1)^{3/2}\, dx = \int (u-1)u^{3/2}\, du = \int (u^{5/2} - u^{3/2})\, du$$

    $$= \frac{2}{7}u^{7/2} - \frac{2}{5}u^{5/2} + C = \frac{2}{35}u^{5/2}(5u - 7) + C$$

    $$= \frac{2}{35}(x+1)^{5/2}[5(x+1) - 7] + C$$

    $$= \frac{2}{35}(x+1)^{5/2}(5x - 2) + C$$

5. Evaluate $\int x\sqrt{x+3}\, dx$.

    Solution: Use substitution and let $u = x + 3$, then $x = u - 3$ and $dx = du$.

    $$\int x\sqrt{x+3}\, dx = \int (u-3)u^{1/2}\, du = \int (u^{3/2} - 3u^{1/2})\, du$$

    $$= \frac{2}{5}u^{5/2} - 2u^{3/2} + C = \frac{2}{5}u^{3/2}(u - 5) + C$$

    $$= \frac{2}{5}(x+3)^{3/2}[(x+3) - 5] + C$$

    $$= \frac{2}{5}(x+3)^{3/2}(x - 2) + C$$

7. Evaluate $\int x^2\sqrt{1-x}\, dx$.

    Solution: Use substitution and let $u = 1 - x$, then $x = 1 - u$ and $dx = -du$.

    $$\int x^2\sqrt{1-x}\, dx = -\int (1-u)^2 u^{1/2}\, du$$

    $$= -\int (u^{1/2} - 2u^{3/2} + u^{5/2})\, du$$

    $$= -[\frac{2}{3}u^{3/2} - \frac{4}{5}u^{5/2} + \frac{2}{7}u^{7/2}] + C$$

    $$= -\frac{2}{105}u^{3/2}(35 - 42u + 15u^2) + C$$

    $$= -\frac{2}{105}(1-x)^{3/2}[35 - 42(1-x) + 15(1-x)^2] + C$$

    $$= -\frac{2}{105}(1-x)^{3/2}(15x^2 + 12x + 8) + C$$

9. Evaluate $\int \dfrac{\sqrt{x}}{1+\sqrt{x}}\, dx$.

Solution: Use substitution and let $u = \sqrt{x}$, then $x = u^2$ and $dx = 2u\, du$.

$$\int \frac{\sqrt{x}}{1+\sqrt{x}}\, dx = \int \frac{u}{1+u}(2u)\, du = 2\int \frac{u^2}{1+u}\, du$$

$$= 2\int \left[u - 1 + \frac{1}{1+u}\right] du$$

$$= 2\left[\frac{u^2}{2} - u + \ln|1+u|\right] + C$$

$$= x - 2\sqrt{x} + \ln|1+\sqrt{x}| + C$$

11. Evaluate $\int \dfrac{\ln 2x}{\sqrt{x}}\, dx$.

Solution: Use integration by parts and let $u = \ln 2x$ and $dv = (1/\sqrt{x})\, dx$, then $du = 1/x\, dx$ and $v = 2\sqrt{x}$.

$$\int \frac{\ln 2x}{\sqrt{x}}\, dx = 2\sqrt{x}\ln 2x - \int 2\sqrt{x}\left(\frac{1}{x}\right) dx$$

$$= 2\sqrt{x}\ln 2x - 2\int \frac{1}{\sqrt{x}}\, dx$$

$$= 2\sqrt{x}\ln 2x - 4\sqrt{x} + C$$

$$= 2\sqrt{x}[(\ln 2x) - 2] + C$$

13. Evaluate $\int (x-1)e^x\, dx$.

Solution: Use integration by parts and let $u = x - 1$ and $dv = e^x\, dx$, then $du = dx$ and $v = e^x$.

$$\int (x-1)e^x\, dx = (x-1)e^x - \int e^x\, dx$$

$$= (x-1)e^x - e^x + C$$

$$= (x-2)e^x + C$$

15. Evaluate $\int \sqrt{x}\ln x\, dx$.

Solution: Use integration by parts and let $u = \ln x$ and $dv = \sqrt{x}\, dx$, then $du = 1/x\, dx$ and $v = (2/3)x^{3/2}$.

$$\int \sqrt{x}\ln x\, dx = \frac{2}{3}x^{3/2}\ln x - \int \frac{2}{3}x^{3/2}\left(\frac{1}{x}\right) dx$$

$$= \frac{2}{3}x^{3/2}\ln x - \frac{2}{3}\int x^{1/2}\, dx$$

$$= \frac{2}{3}x^{3/2}\ln x - \frac{4}{9}x^{3/2} + C$$

$$= \frac{2}{9}x^{3/2}[(3\ln x) - 2] + C$$

Review Exercises for Chapter 6

17. Evaluate $\int \dfrac{9}{x^2 - 9}\, dx$.

Solution: Use partial fractions

$$\dfrac{9}{x^2 - 9} = \dfrac{9}{(x-3)(x+3)} = \dfrac{A}{x-3} + \dfrac{B}{x+3}$$

Basic Equation: $9 = A(x+3) + B(x-3)$
When $x = 3$: $\quad 9 = 6A, \quad A = 3/2$
When $x = -3$: $\quad 9 = -6B, \quad B = -3/2$

$$\int \dfrac{9}{x^2 - 9}\, dx = \dfrac{3}{2} \int \left[\dfrac{1}{x-3} - \dfrac{1}{x+3}\right] dx$$

$$= \dfrac{3}{2}[\ln|x-3| - \ln|x+3|] + C$$

$$= \dfrac{3}{2} \ln \left|\dfrac{x-3}{x+3}\right| + C$$

19. Evaluate $\int \dfrac{1}{x(x+5)}\, dx$.

Solution: Use partial fractions

$$\dfrac{1}{x(x+5)} = \dfrac{A}{x} + \dfrac{B}{x+5}$$

Basic Equation: $1 = A(x+5) + Bx$
When $x = 0$: $\quad 1 = 5A, \quad A = 1/5$
When $x = -5$: $\quad 1 = -5B, \quad B = -1/5$

$$\int \dfrac{1}{x(x+5)}\, dx = \dfrac{1}{5} \int \left[\dfrac{1}{x} - \dfrac{1}{x+5}\right] dx$$

$$= \dfrac{1}{5}[\ln|x| - \ln|x+5|] + C$$

$$= \dfrac{1}{5} \ln \left|\dfrac{x}{x+5}\right| + C$$

21. Evaluate $\int \dfrac{4x - 2}{3(x-1)^2}\, dx$.

Solution: Use partial fractions

$$\dfrac{2x - 1}{(x-1)^2} = \dfrac{A}{x-1} + \dfrac{B}{(x-1)^2}$$

Basic Equation: $2x - 1 = A(x-1) + B$
When $x = 1$: $1 = B$
When $x = 2$: $3 = A + B, \quad A = 2$

$$\int \dfrac{4x-2}{3(x-1)^2}\, dx = \dfrac{2}{3} \int \dfrac{2x-1}{(x-1)^2}\, dx$$

$$= \dfrac{2}{3} \int \left[\dfrac{2}{x-1} + \dfrac{1}{(x-1)^2}\right] dx$$

$$= \dfrac{2}{3}\left[2\ln|x-1| - \dfrac{1}{x-1}\right] + C$$

23. Evaluate $\int \dfrac{\sqrt{x^2 + 9}}{x}\, dx$.

Solution: Use Formula 25 from the tables and let $u = x$, $a = 3$ and $du = dx$.

$$\int \dfrac{\sqrt{x^2 + 9}}{x}\, dx = \sqrt{x^2 + 9} - 3 \ln \left| \dfrac{3 + \sqrt{x^2 + 9}}{x} \right| + C$$

25. Approximate $\int_0^1 \dfrac{x^{3/2}}{2 - x^2}\, dx$, using $n = 4$.

Solution:
Trapezoidal Rule:

$$\dfrac{1}{8}[0 + 2[\dfrac{1/8}{2-(1/16)}] + 2[\dfrac{1/(2\sqrt{2})}{2-(1/4)}] + 2[\dfrac{3\sqrt{3}/8}{2-(9/16)}] + 1]$$

$$= \dfrac{1}{8}[0 + \dfrac{1}{4}(\dfrac{16}{31}) + \dfrac{1}{\sqrt{2}}(\dfrac{4}{7}) + \dfrac{3\sqrt{3}}{4}(\dfrac{16}{23}) + 1]$$

$$= \dfrac{1}{8}[0 + \dfrac{4}{31} + \dfrac{2\sqrt{2}}{7} + \dfrac{12\sqrt{3}}{23} + 1] \approx 0.305$$

Simpson's Rule:

$$\dfrac{1}{12}[0 + \dfrac{8}{31} + \dfrac{2\sqrt{2}}{7} + \dfrac{24\sqrt{3}}{23} + 1] \approx 0.289$$

27. Approximate $\int_1^2 \dfrac{1}{1 + \ln x}\, dx$ using $n = 4$.

Solution:
Trapezoidal Rule:

$$\dfrac{1}{8}[1 + 2[\dfrac{1}{1 + \ln(5/4)}] + 2[\dfrac{1}{1 + \ln(3/2)}]$$

$$+ 2[\dfrac{1}{1 + \ln(7/4)}] + \dfrac{1}{1 + \ln 2}] \approx 0.741$$

Simpson's Rule:

$$\dfrac{1}{12}[1 + 4[\dfrac{1}{1 + \ln(5/4)}] + 2[\dfrac{1}{1 + \ln(3/2)}]$$

$$+ 4[\dfrac{1}{1 + \ln(7/4)}] + \dfrac{1}{1 + \ln 2}] \approx 0.737$$

29. Evaluate $\int_0^{16} \dfrac{1}{\sqrt[4]{x}}\, dx$.

Solution: This integral converges since

$$\int_0^{16} \dfrac{1}{\sqrt[4]{x}}\, dx = \lim_{b \to 0} \dfrac{4}{3} x^{3/4} \Big]_b^{16}$$

$$= \dfrac{4}{3}(8 - 0) = \dfrac{32}{3}$$

31. Evaluate $\int_1^\infty x^2 \ln x \, dx$.

   Solution: This integral diverges since (by Formula 43), we have

   $$\int_1^\infty x^2 \ln x \, dx = \lim_{b \to \infty} \frac{x^3}{9}(-1 + 3\ln x)\Big]_1^b$$

   $$= \lim_{b \to \infty} \left[-\frac{x^3}{9} + \frac{x^3}{3}\ln x\right]_1^b = \infty$$

33. Evaluate $\int_0^4 \frac{x}{\sqrt{16 - x^2}} \, dx$.

   Solution: This integral converges since

   $$\int_0^4 \frac{x}{\sqrt{16 - x^2}} \, dx = \lim_{b \to 4} -\sqrt{16 - x^2}\Big]_0^b = 0 - (-4) = 4$$

35. Find the area of the region enclosed by the graphs of $y = x\sqrt{4 - x}$ and $y = 0$.

   Solution: Let $u = 4 - x$, then $x = 4 - u$ and $dx = -du$. When $x = 0$, $u = 4$ and when $x = 4$, $u = 0$.

   $$A = \int_0^4 x\sqrt{4 - x} \, dx = \int_4^0 (4 - u)u^{1/2}(-du)$$

   $$= -\int_4^0 (4u^{1/2} - u^{3/2}) \, du$$

   $$= -\left[\frac{8}{3}u^{3/2} - \frac{2}{5}u^{5/2}\right]_4^0$$

   $$= \frac{2}{15}u^{3/2}(3u - 20)\Big]_4^0$$

   $$= 0 - \frac{2}{15}(8)(-8) = \frac{128}{15}$$

37. Find the area of the region enclosed by the graphs of $y = (1/8)e^{-x/8}$, $y = 0$, and $x \geq 0$.
   Solution:

   $$A = \int_0^\infty \frac{1}{8} e^{-x/8} \, dx = \lim_{b \to \infty} -e^{-x/8}\Big]_0^\infty = 0 - (-1) = 1$$

39. Find the volume of the solid generated by revolving the region bounded by the graphs of $y = \sqrt{x}(1 - x)^2$ and $y = 0$ about the x-axis.
   Solution:

   $$V = \pi \int_0^1 [\sqrt{x}(1 - x)^2]^2 \, dx = \pi \int_0^1 x(1 - x)^4 \, dx$$

   $$= \pi \int_0^1 (x - 4x^2 + 6x^3 - 4x^4 + x^5) \, dx$$

   $$= \pi \left[\frac{x^2}{2} - \frac{4x^3}{3} + \frac{3x^4}{2} - \frac{4x^5}{5} + \frac{x^6}{6}\right]_0^1 = \frac{\pi}{30}$$

41. Find the constant k so that $f(x) = k(4 - x)$ is a probability density function over the interval $[0, 4]$.
Solution:
$$\int_0^4 k(4 - x)\, dx = k\left[4x - \frac{x^2}{2}\right]_0^4 = 8k = 1 \implies k = \frac{1}{8}$$

43. Find the constant k so that $f(x) = k/\sqrt{x}$ is a probability density function over the interval $[1, 9]$.
Solution:
$$\int_1^9 \frac{k}{\sqrt{x}}\, dx = 2k\sqrt{x}\,\Big]_1^9 = 4k = 1 \implies k = \frac{1}{4}$$

45. Find the mean, median, standard deviation, and required probability for the probability density function
$$f(x) = \frac{1}{50}(10 - x), \quad [0, 10]$$

(a) $P(0 < x < 2)$ (b) $P(8 < x < 10)$

Solution:

Mean: $\mu = \int_0^{10} \frac{1}{50} x(10 - x)\, dx = \frac{1}{50}\left[5x^2 - \frac{x^3}{3}\right]_0^{10} = \frac{10}{3}$

Median: $\int_0^m \frac{1}{50}(10 - x)\, dx = \frac{1}{50}\left[10x - \frac{x^2}{2}\right]_0^m = \frac{1}{2}$

$$\frac{1}{50}\left[10m - \frac{m^2}{2}\right] = \frac{1}{2}$$
$$10m - \frac{m^2}{2} = 25$$
$$0 = m^2 - 20m + 50$$
$$m = \frac{20 \pm \sqrt{200}}{2} = 10 \pm 5\sqrt{2}$$

In the interval $[0, 10]$, $m = 10 - 5\sqrt{2} \approx 2.929$.

Variance: $\sigma^2 = \int_0^{10} \frac{1}{50} x^2(10 - x)\, dx - \left(\frac{10}{3}\right)^2$

$$= \frac{1}{50}\left[\frac{10x^3}{3} - \frac{x^4}{4}\right]_0^{10} - \frac{100}{9} = \frac{50}{3} - \frac{100}{9} = \frac{50}{9}$$

Standard deviation: $\sigma = \sqrt{\frac{50}{9}} = \frac{5\sqrt{2}}{3} \approx 2.357$

(a) $P(0 < x < 2) = \int_0^2 \frac{1}{50}(10 - x)\, dx = \frac{1}{50}\left[10x - \frac{x^2}{2}\right]_0^2$
$$= \frac{9}{25}$$

(b) $P(8 < x < 10) = \int_8^{10} \frac{1}{50}(10 - x)\, dx = \frac{1}{50}\left[10x - \frac{x^2}{2}\right]_8^{10}$
$$= 1 - \frac{48}{50} = \frac{1}{25}$$

47. Find the mean, median, standard deviation, and required probability for the probability density function $f(x) = 2/(x + 1)^2$ on the interval $[0, 1]$.
(a) $P(0 < x < 1/2)$ (b) $P(1/2 < x < 1)$

Solution: Let $u = x + 1$, then $x = u - 1$ and $dx = du$. When $x = 0$, $u = 1$ and when $x = 1$, $u = 2$.

Mean: $\mu = \int_0^1 \dfrac{2x}{(x+1)^2}\, dx = \int_1^2 \dfrac{2(u-1)}{u^2}\, du$

$= 2\int_1^2 [\dfrac{1}{u} - \dfrac{1}{u^2}]\, du = 2\left[\ln|u| + \dfrac{1}{u}\right]_1^2$

$= 2(\ln 2 + \dfrac{1}{2}) - 2(0 + 1)$

$= 2(\ln 2) - 1 \approx 0.386$

Median: $\int_0^m \dfrac{2}{(x+1)^2}\, dx = -\dfrac{2}{x+1}\Big]_0^m = -\dfrac{2}{m+1} + 2 = \dfrac{1}{2}$

Solving the equation $3/2 = 2/(m+1)$ produces $m = 1/3$.

Variance: $\sigma^2 = \int_0^1 \dfrac{2x^2}{(x+1)^2}\, dx - [2(\ln 2) - 1]^2$

Let $u = x + 1$, then $x = u - 1$ and $dx = du$. When $x = 0$, $u = 1$ and when $x = 1$, $u = 2$.

$\int_0^1 \dfrac{2x^2}{(x+1)^2}\, dx = \int_1^2 \dfrac{2(u-1)^2}{u^2}\, du$

$= 2\int_1^2 [1 - \dfrac{2}{u} + \dfrac{1}{u^2}]\, du$

$= 2\left[u - 2\ln|u| - \dfrac{1}{u}\right]_1^2$

$= 2(2 - 2\ln 2 - \dfrac{1}{2}) - 2(1 - 0 - 1)$

$= 3 - 4\ln 2$

$\sigma^2 = 3 - 4(\ln 2) - [2(\ln 2) - 1]^2$
$= 2 - (2\ln 2)^2 \approx 0.078$
$\sigma = \sqrt{\sigma^2} \approx 0.280$

(a) $P(0 < x < \dfrac{1}{2}) = \int_0^{1/2} \dfrac{2}{(x+1)^2}\, dx = -\dfrac{2}{x+1}\Big]_0^{1/2}$

$= -\dfrac{4}{3} + 2 = \dfrac{2}{3}$

(b) $P(\dfrac{1}{2} < x < 1) = \int_{1/2}^1 \dfrac{2}{(x+1)^2}\, dx = -\dfrac{2}{x+1}\Big]_{1/2}^1$

$= -1 + \dfrac{4}{3} = \dfrac{1}{3}$

49. Find the median of the exponential probability density function

$$f(x) = \frac{1}{12} e^{-x/12}, \quad [0, \infty)$$

Solution:

Median: $\int_0^m \frac{1}{12} e^{-x/12} \, dx = -e^{-x/12} \Big]_0^m = \frac{1}{2}$

$$1 - e^{-m/12} = \frac{1}{2}$$

$$e^{-m/12} = \frac{1}{2}$$

$$-\frac{m}{12} = \ln\left(\frac{1}{2}\right) = -\ln 2$$

$$m = 12 \ln 2 \approx 8.318$$

51. The time t (in days) until recovery after a certain medical procedure is a random variable with the probability density function

$$f(t) = \frac{1}{2\sqrt{t-2}}, \quad [3, 6]$$

(a) Find the probability that a patient selected at random will take more than 4 days for recovery.
(b) Determine the expected time for recovery.

Solution:

(a) $P(4 < t < 6) = \int_4^6 \frac{1}{2\sqrt{t-2}} \, dt = \sqrt{t-2} \Big]_4^6$

$$= 2 - \sqrt{2} \approx 0.586$$

(b) Let $u = t - 2$, then $t = u + 2$ and $dt = du$. When $t = 3$, $u = 1$ and when $t = 6$, $u = 4$.

$$\mu = \int_3^6 \frac{t}{2\sqrt{t-2}} \, dt = \int_1^4 \frac{u+2}{2\sqrt{u}} \, du$$

$$= \frac{1}{2} \int_1^4 \left(\sqrt{u} + \frac{2}{\sqrt{u}}\right) du = \frac{1}{2}\left[\frac{2}{3} u^{3/2} + 4u^{1/2}\right]_1^4$$

$$= \frac{1}{2}\left(\frac{16}{3} + 8\right) - \frac{1}{2}\left(\frac{2}{3} + 4\right)$$

$$= \frac{13}{3} \approx 4.33 \text{ days}$$

PRACTICE TEST FOR CHAPTER 6

1. Evaluate $\int x\sqrt{x+3}\,dx$.

2. Evaluate $\int \dfrac{x}{(x-2)^3}\,dx$.

3. Evaluate $\int \dfrac{1}{3x+\sqrt{x}}\,dx$.

4. Evaluate $\int xe^{2x}\,dx$.

5. Evaluate $\int x^3 \ln x\,dx$.

6. Evaluate $\int x^2\sqrt{x-6}\,dx$.

7. Evaluate $\int \dfrac{-5}{x^2+x-6}\,dx$.

8. Evaluate $\int \dfrac{x+12}{x^2+4x}\,dx$.

9. Evaluate $\int \dfrac{5x+3}{(x+2)^2}\,dx$.

10. Evaluate $\int \dfrac{1}{x^2\sqrt{16-x^2}}\,dx$. (Use tables.)

11. Evaluate $\int (\ln x)^3\,dx$. (Use tables.)

12. Approximate the integral using (a) the Trapezoidal Rule and (b) Simpson's Rule.

    $\int_0^4 \sqrt{3+x^3}\,dx, \quad n=8$

13. Approximate the integral using (a) the Trapezoidal Rule and (b) Simpson's Rule.

    $\int_0^2 e^{-x^2/2}\,dx, \quad n=4$

14. Determine the divergence or convergence of the integral. Evaluate the integral if it converges.

    $\int_0^9 \dfrac{1}{\sqrt{x}}\,dx$

15. Determine the divergence or convergence of the integral. Evaluate the integral if it converges.

$$\int_1^\infty \frac{1}{x-3}\, dx$$

16. Determine the divergence or convergence of the integral. Evaluate the integral if it converges.

$$\int_{-\infty}^0 e^{-3x}\, dx$$

17. Find the constant k so that $f(x) = ke^{-x/4}$ is a probability density function over the interval $[0, \infty)$.

18. Find (a) $P(0 < x < 5)$ and (b) $P(x > 1)$ for the probability density function

$$f(x) = \frac{x}{32}, \quad [0, 8].$$

19. Find (a) the mean, (b) the standard deviation, and (c) the median for the probability density function

$$f(x) = \frac{3}{256} x(8 - x), \quad [0, 8].$$

20. Find (a) the mean, (b) the standard deviation, and (c) the median for the probability density function

$$f(x) = \frac{6}{x^2}, \quad [2, 3].$$

Chapter 7 Differential Equations

Section 7.1 Solutions of Differential Equations

1. Verify that $y = (1/x) + C$ is a solution of the differential equation $dy/dx = -1/x^2$.
   Solution: By differentiating, we have
   $$\frac{dy}{dx} = -\frac{1}{x^2}$$

3. Verify that $y = Ce^{4x}$ is a solution of the differential equation $dy/dx = 4y$.
   Solution: By differentiating, we have
   $$\frac{dy}{dx} = 4Ce^{4x} = 4y$$

5. Verify that $y = Ce^{-t/2} + 5$ is a solution of the differential equation $2(dy/dt) + y - 5 = 0$.
   Solution: Since $dy/dt = -(1/2)Ce^{-t/2}$, we have
   $$2\frac{dy}{dt} + y - 5 = 2(-\frac{1}{2}Ce^{-t/2}) + (Ce^{-t/2} + 5) - 5$$
   $$= 0$$

7. Verify that $y = Cx^2 - 3x$ is a solution of the differential equation $xy' - 3x - 2y = 0$.
   Solution: Since $y' = 2Cx - 3$, we have
   $$xy' - 3x - 2y = x(2Cx - 3) - 3x - 2(Cx^2 - 3x) = 0$$

9. Verify that $y = x\ln x + Cx - 2$ is a solution of the differential equation $x(y' - 1) - (y + 2) = 0$.
   Solution: Since $y' = (\ln x) + 1 + C$, we have
   $$x(y' - 1) - (y + 2) = x(\ln x + C) - (x\ln x + Cx)$$
   $$= 0$$

11. Verify that $y = x^2 + 2x + (C/x)$ is a solution of the differential equation $xy' + y = x(3x + 4)$.
    Solution: Since $y' = 2x + 2 - (C/x^2)$, we have
    $$xy' + y = (2x^2 + 2x - \frac{C}{x}) + (x^2 + 2x + \frac{C}{x})$$
    $$= 3x^2 + 4x = x(3x + 4)$$

13. Verify that $y = C_1 e^{x/2} + C_2 e^{-2x}$ is a solution of the differential equation $2y'' + 3y' - 2y = 0$.
    Solution: Since $y' = (1/2)C_1 e^{x/2} - 2C_2 e^{-2x}$, we have $y'' = (1/4)C_1 e^{x/2} + 4C_2 e^{-2x}$, and it follows that
    $$2y'' + 3y' - 2y = \frac{1}{2}C_1 e^{x/2} + 8C_2 e^{-2x} + \frac{3}{2}C_1 e^{x/2}$$
    $$- 6C_2 e^{-2x} - 2C_1 e^{x/2} - 2C_2 e^{-2x}$$
    $$= 0$$

15. Verify that $y = (bx^4)/(4 - a) + Cx^a$ is a solution of the differential equation $y' - (ay/x) = bx^3$.
Solution: Since $y' = 4bx^3/(4 - a) + aCx^{a-1}$, we have

$$y' - \frac{ay}{x} = \left[\frac{4bx^3}{4 - a} + aCx^{a-1}\right] - \frac{a}{x}\left[\frac{bx^4}{4 - a} + Cx^a\right]$$

$$= \frac{4bx^3}{4 - a} + aCx^{a-1} - \frac{abx^3}{4 - a} - aCx^{a-1}$$

$$= \frac{bx^3(4 - a)}{4 - a} = bx^3$$

17. Verify that $x^2 + y^2 = Cy$ is a solution of the differential equation $y' = 2xy/(x^2 - y^2)$.
Solution: By implicit differentiation, we have $2x + 2yy' = Cy'$, which implies that $2x = y'(C - 2y)$ and

$$y' = \frac{2x}{C - 2y} = \frac{2xy}{Cy - 2y^2}$$

$$= \frac{2xy}{(x^2 + y^2) - 2y^2}$$

$$= \frac{2xy}{x^2 - y^2}$$

19. Verify that $y = 2/(1 + Ce^{x^2})$ is a solution of the differential equation $y' + 2xy = xy^2$.
Solution: Since

$$y' = -2(1 + Ce^{x^2})^{-2} 2xCe^{x^2} = -\frac{4xCe^{x^2}}{(1 + Ce^{x^2})^2}$$

we have

$$y' + 2xy = -\frac{4xCe^{x^2}}{(1 + Ce^{x^2})^2} + \frac{4x}{1 + Ce^{x^2}}$$

$$= \frac{-4xCe^{x^2} + 4x + 4xCe^{x^2}}{(1 + Ce^{x^2})^2}$$

$$= x\left(\frac{2}{1 + Ce^{x^2}}\right)^2 = xy^2$$

21. Determine whether $y = e^{-2x}$ is a solution of the differential equation $y^{(4)} - 16y = 0$.
Solution:
$$y' = -2e^{-2x}$$
$$y'' = 4e^{-2x}$$
$$y''' = -8e^{-2x}$$
$$y^{(4)} = 16e^{-2x}$$

Therefore, we have

$$y^{(4)} - 16y = 16e^{-2x} - 16(e^{-2x}) = 0$$

Section 7.1   315

23. Determine whether $y = 4/x$ is a solution of the differential equation $y^{(4)} - 16y = 0$.
Solution:
$$y = 4x^{-1}$$
$$y' = -4x^{-2}$$
$$y'' = 8x^{-3}$$
$$y''' = -24x^{-4}$$
$$y^{(4)} = 96ex^{-5}$$

Therefore, we have $y^{(4)} - 16y = 96x^{-5} - 16(4x^{-1}) \neq 0$ and y is not a solution of the given differential equation.

25. Verify that $y = Ce^{-2x}$ satisfies the differential equation $y' + 2y = 0$. Then find the particular solution satisfying the initial condition of $y = 3$ when $x = 0$.
Solution: Since $y' = -2Ce^{-2x} = -2y$, it follows that $y' + 2y = 0$. To find the particular solution, we use the fact that $y = 3$ when $x = 0$. That is,

$$3 = Ce^0 = C$$

Thus, $C = 3$ and the particular solution is $y = 3e^{-2x}$.

27. Verify that $y = C_1 + C_2 \ln |x|$ satisfies the differential equation $xy'' + y' = 0$. Then find the particular solution satisfying the initial conditions of $y = 5$ and $y' = 1/2$ when $x = 1$.
Solution: Since $y' = C_2(1/x)$ and $y'' = -C_2(1/x^2)$, it follows that $xy'' + y' = 0$. To find the particular solution, we use the fact that $y = 5$ and $y' = 1/2$ when $x = 1$. That is,

$$1/2 = C_2(1/1) \quad \Rightarrow \quad C_2 = 1/2$$
$$5 = C_1 + (1/2)(0) \quad \Rightarrow \quad C_1 = 5$$

Thus, the particular solution is

$$y = 5 + \frac{1}{2}\ln |x| = 5 + \ln \sqrt{|x|}$$

29. Verify that $y = C_1 e^{6x} + C_2 e^{-5x}$ satifies the differential equation $y'' - y' - 30y = 0$. Then find the particular solution satisfying the initial conditions of $y = 0$ and $y' = -4$ when $x = 0$.
Solution: Since $y' = 6C_1 e^{6x} - 5C_2 e^{-5x}$, it follows that $y'' - y' - 30y = 0$. To find the particular solution, we use the fact that $y = 0$ and $y' = -4$ when $x = 0$. That is,

$$6C_1 - 5C_2 = -4$$
$$C_1 + C_2 = 0$$

which implies that $C_1 = -4/11$ and $C_2 = 4/11$ and the particular solution is

$$y = -\frac{4}{11}e^{6x} + \frac{4}{11}e^{-5x}$$

31. Verify that $y = e^{2x/3}(C_1 + C_2 x)$ satifies the differential equation $9y'' - 12y' + 4y = 0$. Then find the particular solution satisfying the initial conditions of $y = 4$ when $x = 0$ and $y = 0$ when $x = 3$.
Solution: Since

$$y' = e^{2x/3}[(2/3)C_1 + (2/3)C_2 x + C_2]$$
$$y'' = e^{2x/3}[(4/9)C_1 + (4/9)C_2 x + (4/3)C_2]$$

it follows that $9y'' - 12y' + 4y = 0$. To find the particular solution, we use the fact that $y = 4$ when $x = 0$ and $y = 0$ when $x = 3$. That is,

$$4 = e^0[C_1 + C_2(0)] \implies C_1 = 4$$
$$0 = e^2[4 + C_2(3)] \implies C_2 = -4/3$$

Therefore the particular solution is

$$y = e^{2x/3}(4 - \frac{4}{3}x) = \frac{4}{3}e^{2x/3}(3 - x)$$

33. The general solution of $2xy' - 3y = 0$ is $y^2 = Cx^3$. Find the particular solution that passes through the point $(4, 4)$.
Solution: Since $y = 4$ when $x = 4$, we have

$$4^2 = C4^3$$

which implies that $C = 1/4$ and the particular solution is

$$y^2 = \frac{x^3}{4}$$

35. The general solution of $y' - y = 0$ is $y = Ce^x$. Find the particular solution that passes through the point $(0, 3)$.
Solution: Since $y = 3$ when $x = 0$, we have

$$3 = Ce^0$$

which implies that $C = 3$ and the particular solution is

$$y = 3e^x$$

37. The general solution of $yy' + x = 0$ is $x^2 + y^2 = C$. Sketch the particular solutions corresponding to $C = 0$, $C = 1$, and $C = 4$.
Solution:
When $C = 0$, the graph is a single point: $(0, 0)$

When $C = 1$, the graph is a circle of radius 1, centered at the origin.

When $C = 4$, the graph is a circle of radius 2, centered at the origin.

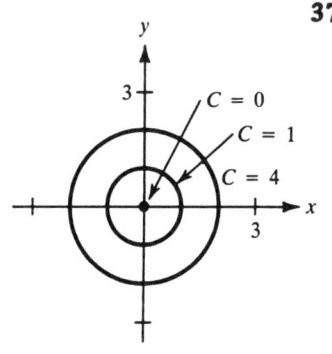

Section 7.1

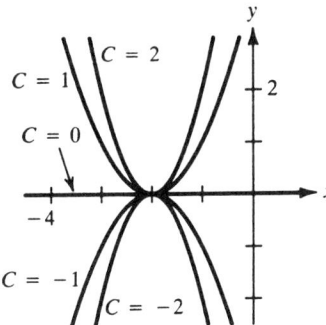

39. The general solution of $(x + 2)y' - 2y = 0$ is $y = C(x + 2)^2$. Sketch the particular solutions corresponding to $C = 0$, $C = \pm 1$, and $C = \pm 2$.
Solution:
When $C = 0$, the graph is a straight line.
When $C = 1$, the graph is a parabola opening up with a vertex at $(-2, 0)$.
When $C = -1$, the graph is a parabola opening down with a vertex at $(-2, 0)$.
When $C = 2$, the graph is a parabola opening up with a vertex at $(-2, 0)$.
When $C = -2$, the graph is a parabola opening down with a vertex at $(-2, 0)$.

41. Use integration to find the general solution of the differential equation $dy/dx = 3x^2$.
Solution:
$$y = \int 3x^2 \, dx = x^3 + C$$

43. Use integration to find the general solution of the differential equation $dy/dx = (x - 2)/x$.
Solution:
$$y = \int \frac{x - 2}{x} \, dx = \int \left(1 - \frac{2}{x}\right) dx$$
$$= x - 2 \ln |x| + C$$

45. Use integration to find the general solution of the differential equation $dy/dx = x\sqrt{x - 3}$.
Solution: Letting $u = x - 3$, we have
$$y = \int x\sqrt{x - 3} \, dx = \int (u + 3) u^{1/2} \, du$$
$$= \int (u^{3/2} + 3u^{1/2}) \, du = \frac{2}{5} u^{5/2} + 2u^{3/2} + C$$
$$= \frac{2}{5} u^{3/2}(u + 5) + C = \frac{2}{5}(x - 3)^{3/2}(x + 2) + C$$

47. The limiting capacity of the habitat for a particular wildlife herd is L. The growth rate dN/dt of the herd is proportional to the unutilized opportunity for growth, as described by the differential equation $dN/dt = k(L - N)$. The general solution to this differential equation is

$$N = L - Ce^{-kt}.$$

Suppose that 100 animals are released into a tract of land that can support 750 of these animals. After 2 years the herd as grown to 160 animals. (a) Find the population function in terms of the time t in years. (b) Sketch the graph of this population function.

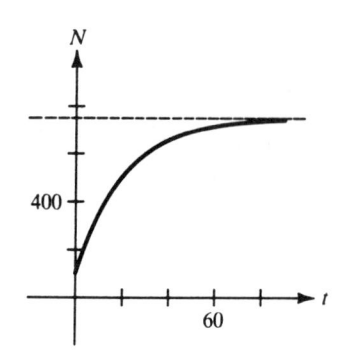

Solution:

(a) Since N = 100 when t = 0, it follows that C = 650. Therefore the population function is

$$N = 750 - 650e^{-kt}$$

Moreover, since N = 160 when t = 2, it follows that

$$160 = 750 - 650e^{-2k}$$
$$e^{-2k} = 59/65$$
$$k = -(1/2) \ln(59/65) = -0.0484$$

Thus, the population function is

$$N = 750 - 650e^{-0.0484t}$$

(b) See accompanying graph.

● Section 7.2 Separation of Variables

1. Find the general solution of $y' = 2x$.
   Solution:
   $$\frac{dy}{dx} = 2x$$
   $$\int dy = \int 2x\, dx$$
   $$y = x^2 + C$$

3. Find the general solution of $3y^2 y' = 1$.
   Solution:
   $$3y^2 \frac{dy}{dx} = 1$$
   $$\int 3y^2\, dy = \int dx$$
   $$y^3 = x + C$$
   $$y = \sqrt[3]{x + C}$$

5. Find the general solution of $y' - xy = 0$.
   Solution: Since $y' - xy = 0$, we have
   $$\frac{dy}{dx} = xy$$
   $$\int \frac{1}{y}\, dy = \int x\, dx$$
   $$\ln|y| = \frac{1}{2}x^2 + C_1$$
   $$y = e^{(x^2/2) + C_1} = e^{C_1} e^{x^2/2} = Ce^{x^2/2}$$

Section 7.2

7. Find the general solution of $(1 + x)y' - 2y = 0$.
   Solution:

   $$(1 + x)\frac{dy}{dx} = 2y$$

   $$\int \frac{1}{2y}\, dy = \int \frac{1}{1 + x}\, dx$$

   $$\frac{1}{2}\ln|y| = \ln|1 + x| + C_1$$

   $$\ln \sqrt{y} = \ln|C_2(1 + x)|$$
   $$\sqrt{y} = C_2(1 + x)$$
   $$y = C(1 + x)^2$$

9. Find the general solution of $e^y y' = 3t^2 + 1$.
   Solution:

   $$e^y \frac{dy}{dt} = 3t^2 + 1$$

   $$\int e^y\, dy = \int (3t^2 + 1)\, dt$$

   $$e^y = t^3 + t + C$$
   $$y = \ln|t^3 + t + C|$$

11. Find the general solution of $y' = x/y$.
    Solution:

    $$\frac{dy}{dx} = \frac{x}{y}$$

    $$\int y\, dy = \int x\, dx$$

    $$\frac{1}{2}y^2 = \frac{1}{2}x^2 + C_1$$
    $$y^2 = x^2 + 2C_1$$
    $$x^2 - y^2 = C$$

13. Find the general solution of $(2 + x)y' = 2y$.
    Solution:

    $$(2 + x)\frac{dy}{dx} = 2y$$

    $$\int \frac{1}{2y}\, dy = \int \frac{1}{2 + x}\, dx$$

    $$\frac{1}{2}\ln|y| = \ln|C_1(2 + x)|$$

    $$\sqrt{y} = C_1(2 + x)$$
    $$y = C(2 + x)^2$$

15. Find the general solution of $y \ln x - xy' = 0$.
    Solution:
    $$y \ln x = x \frac{dy}{dx}$$
    $$\frac{dy}{y} = \frac{\ln x}{x} dx$$
    $$\ln |y| = \frac{1}{2}(\ln x)^2 + C_1$$
    $$y = Ce^{(\ln x)^2/2}$$

17. Find the general solution of $y' - (x+1)(y+1) = 0$.
    Solution:
    $$\frac{dy}{dx} = (x+1)(y+1)$$
    $$\int \frac{1}{y+1} dy = \int (x+1) dx$$
    $$\ln |y+1| = \frac{x^2}{2} + x + C_1$$
    $$y + 1 = Ce^{(x^2/2)+x}$$
    $$y = Ce^{x(x+2)/2} - 1$$

19. Find the general solution of $xyy' + y^2 = 1$.
    Solution:
    $$xyy' = 1 - y^2$$
    $$\frac{dy}{dx} = \frac{1-y^2}{xy}$$
    $$\int \frac{y}{1-y^2} dy = \int \frac{1}{x} dx$$
    $$-\frac{1}{2} \ln |1-y^2| = \ln |x| + C_1$$
    $$\ln |1-y^2| = \ln \frac{C}{x^2}$$
    $$1 - y^2 = \frac{C}{x^2}$$
    $$x^2(1-y^2) = C$$

21. Find the particular solution of $yy' - e^x = 0$ that satisfies the initial condition $y(0) = 4$.

Section 7.2

Solution:

$$y \frac{dy}{dx} = e^x$$

$$\int y \, dy = \int e^x \, dx$$

$$\frac{y^2}{2} = e^x + C$$

When $x = 0$, $y = 4$. Therefore, $C = 7$ and the particular solution is $y^2 = 2e^x + 14$.

23. Find the particular solution of $y(x + 1) + y' = 0$ that satisfies the initial condition $y(-2) = 1$.
    Solution:

$$\frac{dy}{dx} = -y(x + 1)$$

$$\int \frac{1}{y} \, dy = \int (-x - 1) \, dx$$

$$\ln |y| = -\frac{x^2}{2} - x + C_1$$

$$y = Ce^{-x(x+2)/2}$$

When $x = -2$, $y = 1$. Therefore, $C = 1$ and the particular solution is $y = e^{-x(x+2)/2}$.

25. Find the particular solution of $dP - kP \, dt = 0$ that satisfies the initial condition $P(0) = P_0$.
    Solution:

$$dP = kP \, dt$$

$$\int \frac{1}{P} \, dP = \int k \, dt$$

$$\ln P = kt + C_1$$
$$P = Ce^{kt}$$

When $t = 0$, $P = P_0$. Therefore, $C = P_0$ and the particular solution is $P = P_0 e^{kt}$.

27. Find an equation for the curve passing through the point (1, 1) with a slope of $y' = -9x/16y$.
    Solution:

$$\frac{dy}{dx} = -\frac{9x}{16y}$$

$$\int 16y \, dy = \int -9x \, dx$$

$$8y^2 = -(9/2)x^2 + C_1$$
$$9x^2 + 16y^2 = C$$

When $x = 1$, $y = 1$. Therefore, $C = 25$ and the equation for the curve is $9x^2 + 16y^2 = 25$.

29. Solve 12.5(dv/dt) = 43.2 − 1.25v to find velocity v as a function of time if v = 0 when t = 0. The differential equation was derived to describe the motion of two people on a toboggan after considering the force of gravity, friction and air resistance.
Solution:

$$\frac{dv}{dt} = 3.456 - 0.1v$$

$$\int \frac{dv}{3.456 - 0.1v} = \int dt$$

$$-10 \ln |3.456 - 0.1v| = t + C_1$$
$$(3.456 - 0.1v)^{-10} = C_2 e^t$$
$$3.456 - 0.1v = Ce^{-0.1t}$$
$$v = -10Ce^{-0.1t} + 34.56$$

When t = 0, v = 0. Therefore, C = 3.456 and the solution is

$$v = 34.56(1 - e^{-0.1t})$$

31. Use Newton's Law of Cooling given in the Introductory Example to solve the following problem. A room is kept at a constant temperature of 70° and an object placed in the room cooled from 350° to 150° in 45 minutes. At what time will the object cool to a temperature of 80°?
Solution:

$$\frac{dT}{dt} = k(T - 70)$$

$$\int \frac{dT}{T - 70} = \int k \, dt$$

$$\ln |T - 70| = kt + C_1$$
$$T - 70 = Ce^{kt}$$
$$T = Ce^{kt} + 70$$

When t = 0, T = 350. Therefore, C = 280 and the equation for the temperature is $T = 280e^{kt} + 70$. When t = 3/4, T = 150. Therefore, we have

$$150 = 280e^{3k/4} + 70$$

which implies that k = (4/3) ln (2/7) ≈ −1.6704. Thus, the equation for the temperature is

$$T = 280e^{-1.6704t} + 70$$

When T = 80°, we can solve for the time to obtain

$$t = \frac{\ln (1/28)}{-1.6704} \approx 1.9949 \text{ hr.}$$

$$\approx 119.69 \text{ min.}$$

Section 7.3 First-Order Linear Differential Equations

1. Solve the differential equation $y' + 3y = 6$.
   Solution: For this linear differential equation, we have $P(x) = 3$ and $Q(x) = 6$. Therefore, the integrating factor is
   $$u(x) = e^{\int 3\,dx} = e^{3x}$$
   and the general solution is
   $$y = \frac{1}{u(x)} \int Q(x)u(x)\,dx = e^{-3x} \int 6e^{3x}\,dx$$
   $$= e^{-3x}(2e^{3x} + C) = 2 + Ce^{-3x}$$

3. Solve the differential equation $y' + y = e^{-x}$.
   Solution: For this linear differential equation, we have $P(x) = 1$ and $Q(x) = e^{-x}$. Therefore, the integrating factor is
   $$u(x) = e^{\int dx} = e^{x}$$
   and the general solution is
   $$y = \frac{1}{u(x)} \int Q(x)u(x)\,dx = e^{-x} \int e^{-x}e^{x}\,dx$$
   $$= e^{-x}(x + C)$$

5. Solve the differential equation $y' + (y/x) = 3x + 4$.
   Solution: For this linear differential equation, we have $P(x) = 1/x$ and $Q(x) = 3x + 4$. Therefore, the integrating factor is
   $$u(x) = e^{\int 1/x\,dx} = e^{\ln x} = x$$
   and the general solution is
   $$y = \frac{1}{u(x)} \int Q(x)u(x)\,dx = \frac{1}{x} \int (3x + 4)x\,dx$$
   $$= \frac{1}{x}(x^3 + 2x^2 + C) = x^2 + 2x + \frac{C}{x}$$

7. Solve the differential equation $y' + 2xy = 2x$.
   Solution: For this linear differential equation, we have $P(x) = 2x$ and $Q(x) = 2x$. Therefore, the integrating factor is
   $$u(x) = e^{\int 2x\,dx} = e^{x^2}$$
   and the general solution is
   $$y = \frac{1}{u(x)} \int Q(x)u(x)\,dx = e^{-x^2} \int 2xe^{x^2}\,dx$$
   $$= e^{-x^2}(e^{x^2} + C) = 1 + Ce^{-x^2}$$

Section 7.3

9. Solve the differential equation $(x - 1)y' + y = x^2 - 1$.
   Solution: For this linear differential equation

   $$y' + y\left(\frac{1}{x-1}\right) = x + 1$$

   we have $P(x) = 1/(x - 1)$ and $Q(x) = x + 1$. Therefore, the integrating factor is

   $$u(x) = e^{\int 1/(x-1)\,dx} = e^{\ln(x-1)} = x - 1$$

   and the general solution is

   $$y = \frac{1}{u(x)} \int Q(x)u(x)\,dx = \frac{1}{x-1} \int (x+1)(x-1)\,dx$$

   $$= \frac{1}{x-1}\left(\frac{x^3}{3} - x + C_1\right) = \frac{x^3 - 3x + C}{3(x-1)}$$

11. Solve the differential equation $xy' + y = x\ln x$.
    Solution: For this linear differential equation

    $$y' + \frac{y}{x} = \ln x$$

    we have $P(x) = 1/x$ and $Q(x) = \ln x$. Therefore, the integrating factor is

    $$u(x) = e^{\int 1/x\,dx} = e^{\ln x} = x$$

    and the general solution is

    $$y = \frac{1}{u(x)} \int Q(x)u(x)\,dx = \frac{1}{x} \int x\ln x\,dx$$

    $$= \frac{1}{x}\left(\frac{x^2}{2}\ln x - \frac{x^2}{4} + C\right) = \frac{x}{2}\ln x - \frac{x}{4} + \frac{C}{x}$$

13. Find the particular solution of $y' + y = e^x$ that satisfies the initial condition of $y(0) = 2$.
    Solution: Since $P(x) = 1$ and $Q(x) = e^x$, the integrating factor is

    $$u(x) = e^{\int dx} = e^x$$

    and the general solution is

    $$y = e^{-x} \int e^x e^x\,dx = e^{-x}\left(\frac{1}{2}e^{2x} + C\right)$$

    $$= \frac{1}{2}e^x + Ce^{-x}$$

    Since $y = 2$ when $x = 0$, it follows that $C = 3/2$ and the particular solution is

    $$y = \frac{1}{2}e^x + \frac{3}{2}e^{-x} = \frac{1}{2}(e^x + 3e^{-x})$$

15. Find the particular solution of $xy' + y = 0$ that satisfies the initial condition of $y(2) = 2$.
Solution: Since $P(x) = 1/x$ and $Q(x) = 0$, the integrating factor is

$$u(x) = e^{\int 1/x \, dx} = e^{\ln x} = x$$

and the general solution is

$$y = \frac{1}{x} \int 0 \, dx = \frac{C}{x}$$

Since $y = 2$ when $x = 2$, it follows that $C = 4$ and the particular solution is $y = 4/x$ or $xy = 4$.

17. Find the particular solution of $y' + y = x$ that satisfies the initial condition of $y(0) = 4$.
Solution: Since $P(x) = 1$ and $Q(x) = x$, the integrating factor is

$$u(x) = e^{\int dx} = e^x$$

and the general solution is

$$y = e^{-x} \int xe^x \, dx = e^{-x}(xe^x - e^x + C)$$

$$= x - 1 + Ce^{-x}$$

Since $y = 4$ when $x = 0$, it follows that $C = 5$ and the particular solution is $y = x - 1 + 5e^{-x}$.

19. Find the particular solution of $xy' - 2y = -x^2$ that satisfies the initial condition of $y(1) = 5$.
Solution: Since $P(x) = -2/x$ and $Q(x) = -x$, the integrating factor is

$$u(x) = e^{\int -2/x \, dx} = e^{-2\ln x} = \frac{1}{x^2}$$

and the general solution is

$$y = x^2 \int (-x)\left(\frac{1}{x^2}\right) dx = x^2(-\ln|x| + C)$$

Since $y = 5$ when $x = 1$, it follows that $C = 5$ and the particular solution is $y = x^2(5 - \ln|x|)$.

21. A brokerage firm opens a new real estate investment plan for which the earnings are equivalent to continuous compounding at the rate of r. The firm estimates that deposits from investors will create a net cash flow of Pt dollars, where t is the time in years. The rate of increase in the amount A in the real estate investment is

$$\frac{dA}{dt} = rA + Pt$$

Solve the differential equation and find the amount A as a function of t. Assume that $A = 0$ when $t = 0$.

Solution: Since $P(t) = -r$ and $Q(t) = Pt$, the integrating factor is

$$u(x) = e^{\int -r\, dt} = e^{-rt}$$

and the general solution is

$$A = e^{rt} \int Pte^{-rt}\, dt = Pe^{rt}\left(-\frac{t}{r}e^{-rt} - \frac{1}{r^2}e^{-rt} + C_1\right)$$
$$= \frac{P}{r^2}(-rt - 1 + Ce^{rt})$$

Since $A = 0$ when $t = 0$, it follows that $C = 1$ and the particular solution is

$$A = \frac{P}{r^2}(e^{rt} - rt - 1)$$

23. The rate of increase (in thousands of units) in sales S is estimated to be

$$\frac{dS}{dt} = 0.2(100 - S) + 0.2t$$

where t is the time in years. Solve the differential equation and complete a table to estimate the sales of a new product for the first 10 years. (Assume that $S = 0$ when $t = 0$.)

Solution: Since $P(t) = 0.2$ and $Q(t) = 20 + 0.2t$, the integrating factor is

$$u(t) = e^{\int 0.2\, dt} = e^{t/5}$$

and the general solution is

$$S = e^{-t/5} \int e^{t/5}\left(20 + \frac{t}{5}\right) dt$$

Using integration by parts, the integral is

$$S = e^{-t/5}(100e^{t/5} + te^{t/5} - 5e^{t/5} + C)$$
$$= 100 + t - 5 + Ce^{-t/5}$$
$$= 95 + t + Ce^{-t/5}$$

Since $S = 0$ when $t = 0$, it follows that $C = -95$ and the particular solution is

$$S = t + 95(1 - e^{-t/5})$$

During the first ten years, the sales are as follows.

t	1	2	3	4	5
S	18.22	33.32	45.86	56.31	65.05

t	6	7	8	9	10
S	72.39	78.57	83.82	88.30	92.14

Section 7.4 Applications of Differential Equations

1. Solve the differential equation $dy/dx = ky$ and find the particular solution that passes through the points $(0, 1)$ and $(3, 2)$.
   Solution: The general solution is $y = Ce^{kx}$. Since $y = 1$ when $x = 0$, it follows that $C = 1$. Thus,
   $$y = e^{kx}$$
   Since $y = 2$ when $x = 3$, it follows that $2 = e^{3k}$ which implies that
   $$k = \frac{\ln 2}{3} \approx 0.2310$$
   Thus, the particular solution is
   $$y = e^{0.2310x}$$

3. Solve the differential equation $dy/dx = ky$ and find the particular solution that passes through the points $(0, 4)$ and $(4, 1)$.
   Solution: The general solution is $y = Ce^{kx}$. Since $y = 4$ when $x = 0$, it follows that $C = 4$. Thus,
   $$y = 4e^{kx}$$
   Since $y = 1$ when $x = 4$, it follows that $1/4 = e^{4k}$ which implies that
   $$k = \frac{1}{4}\ln\frac{1}{4} \approx -0.3466$$
   Thus, the particular solution is
   $$y = 4e^{-0.3466x}$$

5. Solve the differential equation $dy/dx = ky$ and find the particular solution that passes through the points $(2, 2)$ and $(3, 4)$.
   Solution: The general solution is $y = Ce^{kx}$. Since $y = 2$ when $x = 2$ and $y = 4$ when $x = 3$ it follows that
   $$2 = Ce^{2k} \quad \text{and} \quad 4 = Ce^{3k}$$
   By equating C-values from these two equations, we have
   $$2e^{-2k} = 4e^{-3k}$$
   $$1/2 = e^{-k} \quad \Longrightarrow \quad k = \ln 2 \approx 0.6931$$
   This implies that
   $$C = 2e^{-2\ln 2} = 2e^{\ln(1/4)} = 2\left(\frac{1}{4}\right) = \frac{1}{2}$$
   Thus, the particular solution is
   $$y = \frac{1}{2}e^{x\ln 2} \approx \frac{1}{2}e^{0.6931x}$$

7. Solve the differential equation $dy/dx = ky(20 - y)$ and find the particular solution that passes through the points $(0, 1)$ and $(5, 10)$.
Solution: The general solution is

$$y = Ce^{20kx}(20 - y)$$

Since $y = 1$ when $x = 0$, it follows that $C = 1/19$. Thus,

$$y = \frac{1}{19}e^{20kx}(20 - y)$$

Since $y = 10$ when $x = 5$, it follows that

$$19 = e^{100k}$$

$$20k = \frac{\ln 19}{5} \approx 0.5889$$

Thus, the particular solution is

$$y = \frac{1}{19}e^{0.5889x}(20 - y)$$

$$y(19 + e^{0.5889x}) = 20e^{0.5889x}$$

$$y = \frac{20e^{0.5889x}}{19 + e^{0.5889x}} = \frac{20}{1 + 19e^{-0.5889x}}$$

9. Solve the differential equation $dy/dx = ky(5000 - y)$ and find the particular solution that passes through the points $(0, 250)$ and $(25, 2000)$.
Solution: The general solution is

$$y = Ce^{5000kx}(5000 - y)$$

Since $y = 250$ when $x = 0$, it follows that $C = 1/19$. Thus,

$$y = \frac{1}{19}e^{5000kx}(5000 - y)$$

Since $y = 2000$ when $x = 25$, it follows that

$$\frac{38}{3} = e^{125000k}$$

$$5000k = \frac{\ln(38/3)}{25} \approx 0.10156$$

Thus, the particular solution is

$$y = \frac{1}{19}e^{0.10156x}(5000 - y)$$

$$y(19 + e^{0.10156x}) = 5000e^{0.10156x}$$

$$y = \frac{5000e^{0.10156x}}{19 + e^{0.10156x}}$$

$$= \frac{5000}{1 + 19e^{-0.10156x}}$$

Section 7.4   329

11. Use the chemical reaction model $y = -1/(kt + C)$ to find the amount y as a function of t and sketch the graph of the function. Use the initial conditions $y = 100$ when $t = 0$ and $y = 37$ when $t = 2$. 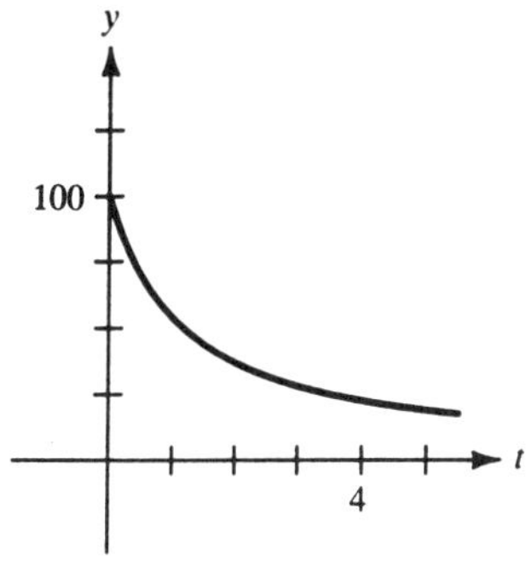
Solution: Since $y = 100$ when $t = 0$, it follows that $100 = -1/C$ which implies that $C = -0.01$. Therefore, we have $y = -1/(kt - 0.01)$. Since $y = 37$ when $t = 2$, it follows that $37 = -1/(2k - 0.01)$ and

$$k = -\frac{63}{30(200)} \approx -0.0085$$

Therefore, y is given by

$$y = \frac{-1}{-0.0085 - 0.01} = \frac{2000}{17t + 20}$$

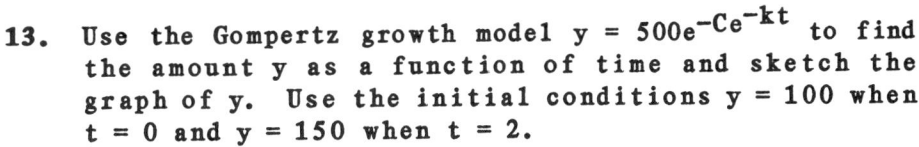

13. Use the Gompertz growth model $y = 500e^{-Ce^{-kt}}$ to find the amount y as a function of time and sketch the graph of y. Use the initial conditions $y = 100$ when $t = 0$ and $y = 150$ when $t = 2$. 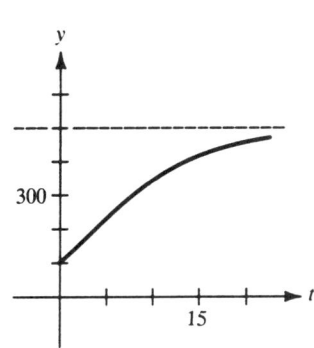
Solution: Since $y = 100$ when $t = 0$, it follows that $100 = 500e^{-C}$, which implies that $C = \ln 5$. Therefore we have

$$y = 500e^{(-\ln 5)e^{-kt}}$$

Since $y = 150$ when $t = 2$, it follows that

$$150 = 500e^{(-\ln 5)e^{-2k}}$$

$$e^{-2k} = \frac{\ln 0.3}{\ln 0.2}$$

$$k = \ln \frac{\ln 0.3}{\ln 0.2} \approx 0.1452$$

Therefore, y is given by $y = 500e^{-1.6904e^{-0.1451t}}$

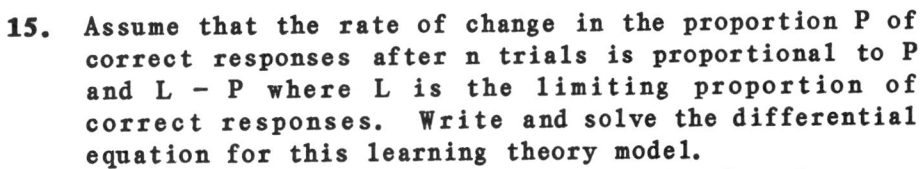

15. Assume that the rate of change in the proportion P of correct responses after n trials is proportional to P and L - P where L is the limiting proportion of correct responses. Write and solve the differential equation for this learning theory model.
Solution: The differential equation is given by

$$\frac{dP}{dn} = kP(L - P)$$

$$\int \frac{1}{P(L - P)} \, dP = \int k \, dn$$

$$\frac{1}{L}[\ln|P| - \ln|L - P|] = kn + C_1$$

$$\frac{P}{L - P} = Ce^{Lkn}$$

$$P = \frac{CLe^{Lkn}}{1 + Ce^{Lkn}} = \frac{CL}{e^{-Lkn} + C}$$

17. Assume that the rate of change in the number of miles s of road cleared of snow per hour by a snowplow is inversely proportional to the height h of snow. That is ds/dh = k/h. Find s as a function of h if s = 25 miles when h = 2 inches and s = 12 miles when h = 6 inches ($2 \le h \le 15$).
Solution:

$$\frac{ds}{dh} = \frac{k}{h}$$

$$\int ds = \int \frac{k}{h}\, dh$$

$$s = k \ln h + C_1 = k \ln Ch$$

Since s = 25 when h = 2 and s = 12 when h = 6, it follows that $25 = k \ln 2C$ and $12 = k \ln 6C$, which implies that

$$C = \frac{1}{2} e^{-(25/13) \ln 3} \approx 0.0605$$

and

$$k = \frac{25}{\ln 2C} = \frac{-13}{\ln 3} \approx -11.8331$$

Therefore, s is given by

$$s = -\frac{13}{\ln 3} \ln \left[\frac{h}{2} e^{-(25/13) \ln 3}\right]$$

$$= -\frac{13}{\ln 3} \left[\ln \frac{h}{2} - \frac{25}{13} \ln 3\right]$$

$$= -\frac{1}{\ln 3} \left[13 \ln \frac{h}{2} - 25 \ln 3\right]$$

$$= 25 - \frac{13 \ln (h/2)}{\ln 3}, \quad 2 \le h \le 15$$

19. Let x and y be the sizes of two organs of a particular mammal at time t. Empirical data indicate that the relative growth rates of these two organs are equal, and hence we have

$$\frac{1}{x} \frac{dx}{dt} = \frac{1}{y} \frac{dy}{dt}$$

Solve this differential equation, writing y as a function of x.
Solution:

$$\int \left(\frac{1}{y} \frac{dy}{dt}\right) dt = \int \left(\frac{1}{x} \frac{dx}{dt}\right) dt$$

$$\int \frac{1}{y}\, dy = \int \frac{1}{x}\, dx$$

$$\ln |y| = \ln |x| + C_1 = \ln |Cx|$$

$$y = Cx$$

21. A large corporation starts at time $t = 0$ to invest part of its receipts at a rate of P dollars per year in a fund for future corporate expansion. Assume that the fund earns r percent per year compounded continuously. Thus, the rate of growth of the amount A in the fund is given by $dA/dt = rA + P$, where $A = 0$ when $t = 0$. Solve this differential equation for A as a function of t.

Solution:

$$\int \frac{1}{rA + P} dA = \int dt$$

$$\frac{1}{r} \ln |rA + P| = t + C_1$$

$$rA + P = Ce^{rt}$$

$$A = \frac{1}{r}(Ce^{rt} - P)$$

Since $A = 0$ when $t = 0$, it follows that $C = P$. Therefore, we have

$$A = \frac{P}{r}(e^{rt} - 1)$$

23. Use the result of Exercise 21 to find P if the corporation needs \$120,000,000 in 8 years and the fund earns $16\frac{1}{4}\%$ compounded continuously.

Solution: Since $A = 120{,}000{,}000$ when $t = 8$ and $r = 0.1625$, we have

$$P = \frac{(0.1625)(120{,}000{,}000)}{e^{(0.1625)(8)} - 1} \approx \$7{,}305{,}295.15$$

25. A medical researcher wants to determine the concentration C (in moles per liter) of a tracer drug injected into a moving fluid. We can start solving this problem by considering a single-compartment dilution model. If the tracer is injected instantaneously at time $t = 0$, then the concentration of the fluid in the compartment begins diluting according to the differential equation $dC/dt = -(R/V)C$, $C = C_0$ when $t = 0$. (a) Solve this differential equation to find the concentration as a function of time. (b) Find the limit of C as $t \longrightarrow \infty$.

Solution:

$$\int \frac{dC}{C} = \int -\frac{R}{V} dt$$

$$\ln |C| = -\frac{R}{V} t + K_1$$

$$C = Ke^{-Rt/V}$$

Since $C = C_0$ when $t = 0$, it follows that $K = C_0$ and the function is $C = C_0 e^{-Rt/V}$. Finally, as $t \longrightarrow \infty$, we have

$$\lim_{t \to \infty} C = \lim_{t \to \infty} C_0 e^{-Rt/V} = 0$$

27. In Exercise 25, we assumed that there was a single initial injection of the tracer drug into the compartment. Now let us consider the case in which the tracer is continuously injected (beginning at t = 0) at the rate of Q moles per minute. By considering Q to be negligible compared with R, we have the differential equation $dC/dt = (Q/V) - (R/V)C$, C = 0 when t = 0. (a) Solve this differential equation to find the concentration as a function of time. (b) Find the limit of C as $t \longrightarrow \infty$.

Solution:

$$\int \frac{1}{Q - RC} \, dC = \int \frac{1}{V} \, dt$$

$$-\frac{1}{R} \ln |Q - RC| = \frac{t}{V} + K_1$$

$$Q - RC = e^{-R[(t/V)+K_1]}$$

$$C = \frac{1}{R}(Q - e^{-R[(t/V)+K_1]})$$

$$= \frac{1}{R}(Q - Ke^{-Rt/V})$$

Since C = 0 when t = 0, it follows that K = Q and we have

$$C = \frac{Q}{R}(1 - e^{-Rt/V})$$

As $t \longrightarrow \infty$, the limit of C is Q/R.

29. Glucose is added intraveneously to the bloodstream at the rate of q units per minute and the body is using or removing the glucose from the bloodstream at a rate proportional to the amount present. Assume Q(t) is the amount of glucose in the bloodstream at time t. (a) Determine the differential equation describing the rate of change as a function of time. (b) Solve the differential equation letting $Q = Q_0$ when t = 0. (c) Find the limit of C as $t \longrightarrow \infty$

Solution:

(a) $\dfrac{dQ}{dt} = q - kQ$, q is a constant

(b) To solve the first-order linear differential equation $Q' + kQ = q$, we let P(t) = k, Q(t) = q and obtain the integrating factor $u(t) = e^{kt}$. Thus, the solution is

$$Q = e^{-kt} \int qe^{kt} \, dt = e^{-kt}(\frac{q}{k}e^{kt} + C) = \frac{q}{k} + Ce^{-kt}$$

Since $Q = Q_0$ when t = 0, we have $C = Q_0 - (q/k)$ and the solution is

$$Q = \frac{q}{k} + (Q_0 - \frac{q}{k})e^{-kt}$$

(c) The limit of Q as $t \longrightarrow \infty$ is q/k.

Review Exercises for Chapter 7

1. Verify that $y = x \ln x^2 + 2x^{3/2} + Cx$ is a solution of the differential equation $xy' - y = x(2 + \sqrt{x})$.
   Solution: Since $y' = 2 + \ln x^2 + 3x^{1/2} + C$, we have
   $$xy' - y = x(2 + \ln x^2 + 3x^{1/2} + C)$$
   $$- (x \ln x^2 + 2x^{3/2} + Cx)$$
   $$= 2x + x \ln x^2 + 3x^{3/2} + Cx$$
   $$- x \ln x^2 - 2x^{3/2} - Cx$$
   $$= 2x + x^{3/2} = x(2 + \sqrt{x})$$

3. Verify that $y = C(x - 1)^2$ is a solution of the differential equation $y' - (2y/x) = (1/x)y'$.
   Solution: Since $y' = 2C(x - 1)$, we have
   $$y' - \frac{2y}{x} = 2C(x - 1) - \frac{2C(x - 1)^2}{x}$$
   $$= 2C(x - 1)(1 - \frac{x - 1}{x})$$
   $$= 2C(x - 1)(\frac{1}{x}) = y'(\frac{1}{x})$$

5. Verify that $y = -(1/3) + Ce^{-3/x}$ is a solution of the differential equation $y' - (3y/x^2) = 1/x^2$.
   Solution: Since $y' = (3/x^2)Ce^{-3x}$, we have
   $$y' - \frac{3y}{x^2} = \frac{3}{x^2}Ce^{-3/x} - \frac{3[-(1/3) + Ce^{-3/x}]}{x^2}$$
   $$= \frac{3}{x^2}Ce^{-3/x} + \frac{1}{x^2} - \frac{3}{x^2}Ce^{-3/x}$$
   $$= \frac{1}{x^2}$$

7. Verify that $y = [C_1 + C_2 x + (1/3)x^3]e^x$ is a solution of the differential equation $y'' - 2y' + y = 2xe^x$.
   Solution: Since
   $$y' = (C_1 + C_2 + C_2 x + x^2 + \frac{1}{3}x^3)e^x$$
   and
   $$y'' = (C_1 + 2C_2 + C_2 x + 2x + 2x^2 + \frac{1}{3}x^3)e^x$$

   it follows that
   $$y'' - 2y' + y = [C_1 + 2C_2 + C_2 x + 2x + 2x^2 + (1/3)x^3]e^x$$
   $$- 2[C_1 + C_2 + C_2 x + x^2 + (1/3)x^3]e^x$$
   $$+ [C_1 + C_2 x + (1/3)x^3]e^x$$
   $$= 2xe^x$$

9. Verify that $y = (1/5)x^3 - x + C\sqrt{x}$ is a solution of the differential equation $2xy' - y = x^3 - x$.
   Solution: Since
   $$y' = \frac{3}{5}x^2 - 1 + \frac{C}{2\sqrt{x}}$$
   it follows that
   $$2xy' - y = 2x(\frac{3}{5}x^2 - 1 + \frac{C}{2\sqrt{x}}) - (\frac{1}{5}x^3 - x + C\sqrt{x})$$
   $$= \frac{6}{5}x^3 - 2x + C\sqrt{x} - \frac{1}{5}x^3 + x - C\sqrt{x}$$
   $$= x^3 - x$$

11. Verify that $y = x^2 + 2x + (C/x)$ satisfies the differential equation $y' + (y/x) = 3x + 4$. Then find the particular solution satisfying the initial condition of $y = 3$ when $x = 1$.
    Solution: Since $y' = 2x + 2 - (C/x^2)$, it follows that $y' + (y/x) = 3x + 4$. To find the particular solution, we use the fact that $y = 3$ when $x = 1$. That is,
    $$3 = 1^2 + 2(1) + \frac{C}{1}$$
    which implies $C = 0$. Thus, the particular solution is
    $$y = x^2 + 2x$$

13. Verify that $y = 1 + Ce^{-x^2}$, satisfies the differential equation $y' + 2xy = 2x$. Then find the particular solution satisfying the initial condition of $y = -1$ when $x = 0$.
    Solution: Since $y' = -2xCe^{-x^2}$, it follows that $y' + 2xy = 2x$. To find the particular solution, we use the fact that $y = -1$ when $x = 0$. That is,
    $$-1 = 1 + C$$
    which implies $C = -2$. Thus, the particular solution is
    $$y = 1 - 2e^{-x^2}$$

15. Solve the differential equation $yy' - 3x^2 = 0$.
    Solution:
    $$y(\frac{dy}{dx}) - 3x^2 = 0$$
    $$\int y \, dy = \int 3x^2 \, dx$$
    $$\frac{y^2}{2} = x^3 + C_1$$
    $$y^2 = 2x^3 + C$$

Review Exercises for Chapter 7   335

17. Solve the differential equation $y' = x^2y^2 - 9x^2$.
Solution:
$$\frac{dy}{dx} = x^2(y^2 - 9)$$

$$\int \frac{1}{y^2 - 9} \, dy = \int x^2 \, dx$$

$$\int \left(\frac{1}{y-3} - \frac{1}{y+3}\right) dy = \int 6x^2 \, dx$$

$$\ln|y - 3| - \ln|y + 3| = 2x^3 + C_1$$

$$\frac{y - 3}{y + 3} = Ce^{2x^3}$$

Solving for y produces

$$y = \frac{3(1 + Ce^{2x^3})}{1 - Ce^{2x^3}}$$

19. Solve the differential equation $(1 + x)y' = 1 + y$.
Solution:
$$(1 + x)\frac{dy}{dx} = 1 + y$$

$$\int \frac{1}{1 + y} \, dy = \int \frac{1}{1 + x} \, dx$$

$$\ln|1 + y| = \ln|1 + x| + C_1 = \ln|C(1 + x)|$$

$$1 + y = C(1 + x)$$

$$y = C(1 + x) - 1$$

21. Solve the differential equation $y' - (y/x) = 2 + \sqrt{x}$.
Solution: For this linear differential equation, we have $P(x) = -1/x$ and $Q(x) = 2 + \sqrt{x}$. Therefore, the integrating factor is

$$u(x) = e^{\int -1/x \, dx} = \frac{1}{x}$$

and the solution is

$$y = x \int \left(\frac{2}{x} + x^{-1/2}\right) dx$$

$$= x(2 \ln x + 2\sqrt{x} + C) = 2x \ln x + 2x^{3/2} + Cx$$

23. Solve the differential equation $y' - (3y/x) = 1/x^2$.
Solution: For this linear differential equation, we have $P(x) = -3/x$ and $Q(x) = 1/x^2$. Therefore, the integrating factor is

$$u(x) = e^{\int -3/x \, dx} = e^{3/x}$$

and the solution is

$$y = e^{-3/x} \int \frac{1}{x^2} e^{3/x} \, dx$$

$$= e^{-3/x}\left(-\frac{1}{3}e^{3/x} + C\right) = -\frac{1}{3} + Ce^{-3/x}$$

25. Solve the differential equation $2xy' - y = x^3 - x$.
    Solution: This linear differential equation
    $$y' - \frac{y}{2x} = \frac{x^2}{2} - \frac{1}{2}$$
    has $P(x) = -1/2x$ and $Q(x) = (x^2/2) - (1/2)$. Therefore, the integrating factor is
    $$u(x) = e^{\int -1/2x \, dx} = \frac{1}{\sqrt{x}}$$
    and the solution is
    $$y = \sqrt{x} \int \frac{1}{2}(x^2 - 1)\frac{1}{\sqrt{x}} \, dx$$
    $$= \sqrt{x} \int \frac{1}{2}(x^{3/2} - x^{-1/2}) \, dx$$
    $$= \frac{1}{2}\sqrt{x}(\frac{2}{5}x^{5/2} - 2x^{1/2} + C_1)$$
    $$= \frac{1}{5}x^3 - x + C\sqrt{x}$$

27. In a chemical reaction a certain compound changes into another compound at a rate proportional to the unchanged amount. If initially there were 20 grams of the original compound and 16 grams after 1 hour, when will 75% of the compound be changed?
    Solution: Let x = amount of compound. Then dx/dt is the rate the compound changes. Since $dx/dt = kx$, we have
    $$\int \frac{1}{x} \, dx = \int k \, dt$$
    $$\ln |x| = kt + C_1$$
    $$x = Ce^{kt}$$
    Since $x = 20$ when $t = 0$, it follows that $C = 20$. Thus, we have
    $$x = 20e^{kt}$$
    Since $x = 16$ when $t = 1$, it follows that $k = \ln(4/5)$. Thus, the solution is
    $$x = 20e^{[\ln(4/5)]/t}$$
    After 75% has been changed, we have $x = 5$ which implies that
    $$5 = 20e^{[\ln(4/5)]/t}$$
    and solving for t produces
    $$t = \frac{\ln(1/4)}{\ln(4/5)} \approx 6.213 \text{ hours}$$

Review Exercises for Chapter 7   337

29. Let $A(t)$ be the amount in a fund earning interest at the annual rate of r percent compounded continuously. If a continuous cash flow of P dollars per year is withdrawn from the fund, then the rate of decrease of A is given by the differential equation $dA/dt = rA - P$ where $A = A_0$ when $t = 0$. Solve this differential equation for A as a function of t.
Solution: For this linear differential equation, we have $P(t) = -r$ and $Q(t) = -P$. Therefore, the integrating factor is

$$u(x) = e^{\int -r\, dt} = e^{-rt}$$

and the solution is

$$A = e^{rt} \int -Pe^{-rt}\, dt$$

$$= e^{rt}(\frac{P}{r} e^{-rt} + C)$$

$$= \frac{P}{r} + Ce^{rt}$$

Since $A = A_0$ when $t = 0$, we have $C = A_0 - (P/r)$ which implies that

$$A = \frac{P}{r} + (A_0 - \frac{P}{r})e^{rt}$$

31. Use the result of Exercise 29 to find A_0 if a retired couple want a continuous cash flow of $40,000 per year for 20 years. (That is, $A = 0$ when $t = 20$.) Assume that their investment will earn 13% compounded continuously.
Solution: Since $A = 0$ when $t = 20$, we have

$$0 = \frac{40,000}{0.13} + (A_0 - \frac{40,000}{0.13})e^{0.13(20)}$$

$$A_0 e^{2.6} = \frac{40,000}{0.13} e^{2.6} - \frac{40,000}{0.13}$$

$$A_0 = \frac{40,000(e^{2.6} - 1)}{0.13 e^{2.6}}$$

$$\approx \$284,838.90$$

33. A company introduces a new product for which it is believed that the limit on yearly sales is 500,000 units. Sales are expected to increase at a rate proportional to sales and the difference between sales and the upper limit. Initial and second year sales are 10,000 and 100,000 units, respectively.
(a) Estimate third year sales.
(b) During what year will sales surpass 400,000 units?

Solution: Let S = sales. The dS/dt = rate of change of sales and we have $dS/dt = kS(500{,}000 - S)$. The solution of this differential equation is

$$\int \frac{1}{S(500{,}000 - S)}\, dS = \int k\, dt$$

$$\int \left(\frac{1}{S} + \frac{1}{500{,}000 - S}\right) dS = 500{,}000 \int k\, dt$$

$$\ln|S| - \ln|500{,}000 - S| = 500{,}000 kt + C_1$$

$$\ln\left|\frac{S}{500{,}000 - S}\right| = 500{,}000 kt + C_1$$

$$\frac{S}{500{,}000 - S} = Ce^{500{,}000 kt}$$

Since $S = 10{,}000$ when $t = 0$, we have $C = 1/49$. Therefore, the solution is

$$\frac{49S}{500{,}000 - S} = e^{500{,}000 kt}$$

Since $S = 100{,}000$ when $t = 2$, it follows that

$$\frac{4{,}900{,}000}{500{,}000 - 100{,}000} = e^{1{,}000{,}000 k}$$

$$\frac{49}{4} = e^{1{,}000{,}000 k}$$

$$k = \frac{1}{1{,}000{,}000} \ln \frac{49}{4}$$

Therefore, the solution is

$$\frac{49S}{500{,}000 - S} = \left(\frac{7}{2}\right)^t$$

(a) When $t = 3$, we have

$$\frac{49S}{500{,}000 - S} = \left(\frac{7}{2}\right)^3$$

$$49S = \frac{343}{8}(500{,}000 - S)$$

$$392S = 343(500{,}000 - S)$$
$$735S = 171{,}500{,}000$$
$$S = 233{,}333 \text{ units}$$

(b) When $S = 400{,}000$, we have

$$\frac{49(400{,}000)}{500{,}000 - 400{,}000} = \left(\frac{7}{2}\right)^t$$

$$196 = (7/2)^t$$
$$\ln 196 = t \ln(7/2)$$

$$t = \frac{\ln 196}{\ln(7/2)} \approx 4.213 \text{ years}$$

During the 5th year, sales will surpass 400,000.

PRACTICE TEST FOR CHAPTER 7

1. Verify that
$$y = x^3 - 4x + \frac{C}{x}$$
is a solution of $xy' + y = 4x(x^2 - 2)$.

2. Verify that
$$y = Ce^{-5x}$$
is a solution of $y''' + 125y = 0$.

3. Find the general solution of $y^3 y' = x + 2$.

4. Find the general solution of $y' = \dfrac{y + 4}{x - 1}$.

5. Find the general solution of $y' \ln y = xe^x$.

6. Find the general solution of $y' + 4y = e^{-2x}$.

7. Find the general solution of $x^3 y' + 2y = e^{1/x^2}$.

8. Find the general solution of $xy' - 4y = 6x^2 - 1$.

9. Assume that the rate of change of y (with respect to time t) is proportional to (30 - y). Find the particular solution that passes through the points (0, 4) and (6, 11).

10. Use the Gompertz growth model
$$\frac{dy}{dt} = ky \ln \frac{1000}{y}$$
to find the growth function given y = 50 when t = 0 and y = 200 when t = 4.

Chapter 8 Functions of Several Variables

● Section 8.1 The Three-Dimensional Coordinate System

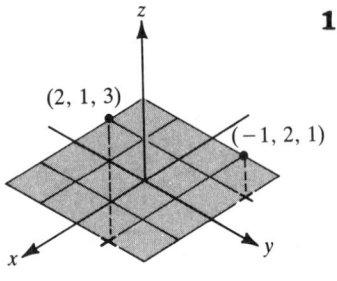

1. Plot the points on the same 3-dimensional coordinate system.
   (a) (2, 1, 3)   (b) (-1, 2, 1)
   Solution:
   (a) See graph
   (b) See graph

3. Plot the points on the same 3-dimensional coordinate system.
   (a) (5, -2, 2)   (b) (5, -2, -2)
   Solution:
   (a) See graph
   (b) See graph

5. Find the coordinates of the midpoint of the line segment joining the points (5, -9, 7), and (-2, 3, 3).
   Solution:
   Midpoint $= \left(\dfrac{5 + (-2)}{2}, \dfrac{-9 + 3}{2}, \dfrac{7 + 3}{2}\right) = \left(\dfrac{3}{2}, -3, 5\right)$

7. Find the coordinates of the midpoint of the line segment joining the points (-5, -2, 5.5) and (6.3, 4.2, -7.1).
   Solution:
   Midpoint $= \left(\dfrac{-5 + 6.3}{2}, \dfrac{-2 + 4.2}{2}, \dfrac{5.5 - 7.1}{2}\right)$
   $= (0.65, 1.1, -0.8)$

9. Find the coordinates of the other endpoint of a line segment when one endpoint is (-2, 1, 1) and the midpoint is (0, 2, 5).
   Solution: Let (x, y, z) be the coordinates of the other endpoint. Then,

   $\left(\dfrac{-2 + x}{2}, \dfrac{1 + y}{2}, \dfrac{1 + z}{2}\right) = (0, 2, 5)$

   $\dfrac{-2 + x}{2} = 0 \implies x = 2$

   $\dfrac{1 + y}{2} = 2 \implies y = 3$

   $\dfrac{1 + z}{2} = 5 \implies z = 9$

   Other endpoint: (2, 3, 9)

11. Find the distance between the points (4, 1, 5) and (8, 2, 6).
    Solution:
    $$d = \sqrt{(8-4)^2 + (2-1)^2 + (6-5)^2} = \sqrt{18} = 3\sqrt{2}$$

13. Find the distance between the points (-1, -5, 7) and (-3, 4, -4).
    Solution:
    $$d = \sqrt{(-3+1)^2 + (4+5)^2 + (-4-7)^2} = \sqrt{206}$$

15. Find the lengths of the sides of a triangle with vertices (0, 0, 0), (2, 2, 1), and (2, -4, 4). Determine whether the triangle is a right triangle, an isosceles triangle, or neither of these.
    Solution: Let A = (0, 0, 0), B = (2, 2, 1), and C = (2, -4, 4). Then we have
    $$d(AB) = \sqrt{(2-0)^2 + (2-0)^2 + (1-0)^2} = 3$$
    $$d(AC) = \sqrt{(2-0)^2 + (-4-0)^2 + (4-0)^2} = 6$$
    $$d(BC) = \sqrt{(2-2)^2 + (-4-2)^2 + (4-1)^2} = 3\sqrt{5}$$

    The triangle is a right triangle since

    $$d^2(AB) + d^2(AC) = (3)^2 + (6)^2 = 9 + 36 = 45 = d^2(BC)$$

17. Find the lengths of the sides of a triangle with vertices (1, -3, -2), (5, -1, 2), and (-1, 1, 2). Determine whether the triangle is a right triangle, and isosceles triangle, or neither of these.
    Solution: Let A = (1, -3, -2), B = (5, -1, 2), and C = (-1, 1, 2). Then we have
    $$d(AB) = \sqrt{(5-1)^2 + (-1+3)^2 + (2+2)^2} = 6$$
    $$d(AC) = \sqrt{(-1-1)^2 + (1+3)^2 + (2+2)^2} = 6$$
    $$d(BC) = \sqrt{(-1-5)^2 + (1+1)^2 + (2-2)^2} = 2\sqrt{10}$$

    Since d(AB) = d(AC) the triangle is isosceles.

19. Find the standard form of the equation of a sphere with center (0, 2, 5) and radius 2.
    Solution:
    $$(x-0)^2 + (y-2)^2 + (z-5)^2 = 2^2$$
    $$x^2 + (y-2)^2 + (z-5)^2 = 4$$

21. Find the standard form of the equation of a sphere with endpoints of a diameter (2, 0, 0) and (0, 6, 0).
    Solution: The midpoint of the diameter is the center.
    $$\text{Center} = \left(\frac{2+0}{2}, \frac{0+6}{2}, \frac{0+0}{2}\right) = (1, 3, 0)$$

    The radius is the distance between the center and either endpoint.

    $$\text{Radius} = \sqrt{(1-2)^2 + (3-0)^2 + (0-0)^2} = \sqrt{10}$$

    $$(x-1)^2 + (y-3)^2 + (z-0)^2 = (\sqrt{10})^2$$
    $$(x-1)^2 + (y-3)^2 + z^2 = 10$$

Section 8.1

23. Find the center and radius of the sphere
$x^2 + y^2 + z^2 - 2x + 6y + 8z + 1 = 0$.
Solution:
$$(x - 1)^2 + (y + 3)^2 + (z + 4)^2 = 25$$

Center: (1, -3, -4)   Radius: 5

25. Find the center and radius of the sphere
$x^2 + y^2 + z^2 - 8y = 0$.
Solution:
$$(x - 0)^2 + (y - 4)^2 + (z - 0)^2 = 16$$

Center: (0, 4, 0)   Radius: 4

27. Find the center and radius of the sphere
$2x^2 + 2y^2 + 2z^2 - 2x + 2y - 4z + 1 = 0$.
Solution:
$$(x - \tfrac{1}{2})^2 + (y + \tfrac{1}{2})^2 + (z - 1)^2 = 1$$

Center: (1/2, -1/2, 1)   Radius: 1

29. Find the center and radius of the sphere
$9x^2 + 9y^2 + 9z^2 - 6x + 18y + 1 = 0$.
Solution:
$$(x - \tfrac{1}{3})^2 + (y + 1)^2 + z^2 = 1$$

Center: (1/3, -1, 0)   Radius: 1

31. Sketch the xy-trace of the sphere
$(x - 2)^2 + (y - 2)^2 + (z - 3)^2 = 25$.
Solution: To find the xy-trace, we let $z = 0$.

$$(x - 2)^2 + (y - 2)^2 + (0 - 3)^2 = 25$$
$$(x - 2)^2 + (y - 2)^2 + 9 = 25$$
$$(x - 2)^2 + (y - 2)^2 = 16$$

The xy-trace is a circle centered at (2, 2) with a radius of 4 in the xy-plane.

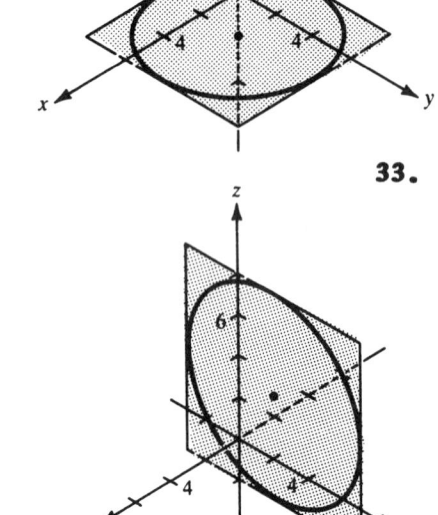

33. Sketch the yz-trace of the sphere
$x^2 + y^2 + z^2 - 4x - 4y - 6z - 12 = 0$.
Solution:
$$(x - 2)^2 + (y - 2)^2 + (z - 3)^2 = 29$$

To find the yz-trace, we let $x = 0$.

$$(0 - 2)^2 + (y - 2)^2 + (z - 3)^2 = 29$$
$$4 + (y - 2)^2 + (z - 3)^2 = 29$$
$$(y - 2)^2 + (z - 3)^2 = 25$$

The yz-trace is a circle centered at (2, 3) with a radius of 5 in the yz-plane.

Section 8.2 Surfaces in Space

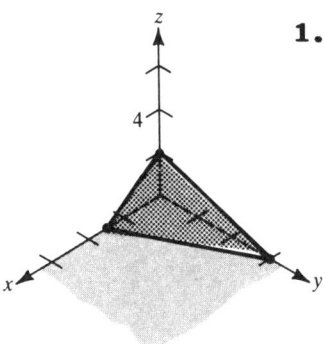

1. Find the intercepts and sketch the graph of $4x + 2y + 6z = 12$.
Solution:
To find the x-intercept, let $y = 0$ and $z = 0$.
$\quad 4x = 12 \implies x = 3$
To find the y-intercept, let $x = 0$ and $z = 0$.
$\quad 2y = 12 \implies y = 6$
To find the z-intercept, let $x = 0$ and $y = 0$.
$\quad 6z = 12 \implies z = 2$

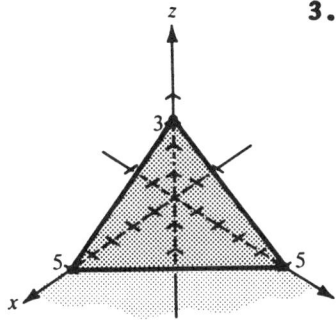

3. Find the intercepts and sketch the graph of $3x + 3y + 5z = 15$.
Solution:
To find the x-intercept, let $y = 0$ and $z = 0$.
$\quad 3x = 15 \implies x = 5$
To find the y-intercept, let $x = 0$ and $z = 0$.
$\quad 3y = 15 \implies y = 5$
To find the z-intercept, let $x = 0$ and $y = 0$.
$\quad 5z = 15 \implies z = 3$

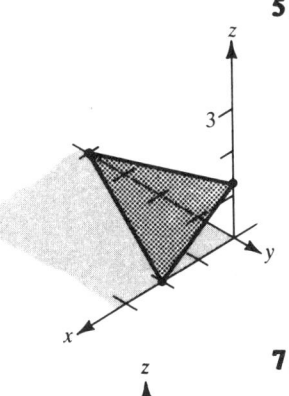

5. Find the intercepts and sketch the graph of $2x - y + 3z = 4$.
Solution:
To find the x-intercept, let $y = 0$ and $z = 0$.
$\quad 2x = 4 \implies x = 2$
To find the y-intercept, let $x = 0$ and $z = 0$.
$\quad -y = 4 \implies y = -4$
To find the z-intercept, let $x = 0$ and $y = 0$.
$\quad 3z = 4 \implies z = 4/3$

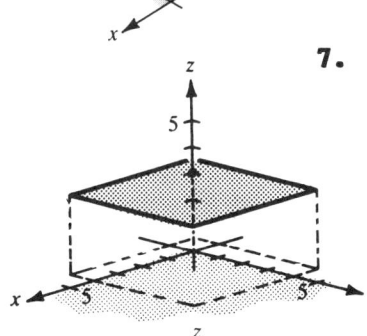

7. Find the intercepts and sketch the graph of $z = 3$.
Solution: Since the coefficients of x and y are zero, the only intercept is the z-intercept of 3. The plane is parallel to the xy-plane.

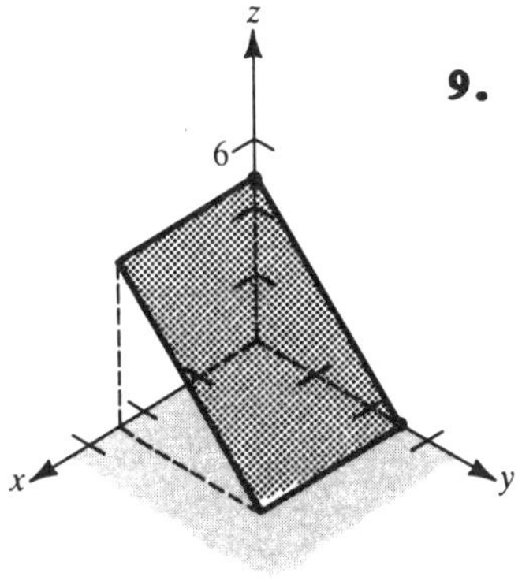

9. Find the intercepts and sketch the graph of $y + z = 5$.
Solution: The y and z intercepts are both 5, and the plane is parallel to the x-axis.

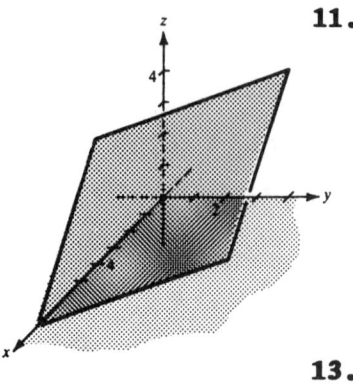

11. Find the intercepts and sketch the graph of $x + y - z = 0$.
    Solution:
    The only intercept is the origin.
    The xy-trace is the line $x + y = 0$.
    The xz-trace is the line $x - z = 0$.
    The yz-trace is the line $y - z = 0$.

13. Determine whether the planes $5x - 3y + z = 4$ and $x + 4y + 7z = 1$ are parallel, perpendicular, or neither.
    Solution: For the first plane, $a_1 = 5$, $b_1 = -3$, and $c_1 = 1$, and for the second plane, $a_2 = 1$, $b_2 = 4$, and $c_2 = 7$. Therefore, we have

    $$a_1a_2 + b_1b_2 + c_1c_2 = (5)(1) + (-3)(4) + (1)(7)$$
    $$= 5 - 12 + 7 = 0$$

    The planes are perpendicular.

15. Determine whether the planes $x - 5y - z = 1$ and $5x - 25y - 5z = -3$ are parallel, perpendicular, or neither.
    Solution: For the first plane, $a_1 = 1$, $b_1 = -5$, and $c_1 = -1$, and for the second plane, $a_2 = 5$, $b_2 = -25$, and $c_2 = -5$. Therefore, we have

    $$a_2 = 5a_1, \quad b_2 = 5b_1, \quad \text{and} \quad c_2 = 5c_1$$

    The planes are parallel.

17. Determine whether the planes $x + 2y = 3$ and $4x + 8y = 5$ are parallel, perpendicular, or neither.
    Solution: For the first plane, $a_1 = 1$, $b_1 = 2$, and $c_1 = 0$, and for the second plane, $a_2 = 4$, $b_2 = 8$, and $c_2 = 0$. Therefore, we have

    $$a_2 = 4a_1, \quad b_2 = 4b_1, \quad \text{and} \quad c_2 = 4c_1$$

    The planes are parallel.

19. Determine whether the planes $2x + y = 3$ and $x - 5z = 0$ are parallel, perpendicular, or neither.
    Solution: For the first plane, $a_1 = 2$, $b_1 = 1$, and $c_1 = 0$, and for the second plane, $a_2 = 1$, $b_2 = 0$, and $c_2 = -5$. The planes are not parallel since

    $$a_1 = 2a_2 \quad \text{and} \quad b_1 \neq 2b_2$$

    and the planes are not perpendicular since

    $$a_1a_2 + b_1b_2 + c_1c_2 = (2)(1) + (1)(0) + (0)(-5)$$
    $$= 2 \neq 0$$

21. Find the distance between the point (0, 0, 0) and the plane $2x + 3y + z = 12$.
Solution:
$$D = \frac{|ax_0 + by_0 + cz_0 + d|}{\sqrt{a^2 + b^2 + c^2}}$$
$$= \frac{|2(0) + 3(0) + 1(0) - 12|}{\sqrt{(2)^2 + (3)^2 + (1)^2}}$$
$$= \frac{12}{\sqrt{4 + 9 + 1}} = \frac{12}{\sqrt{14}} = \frac{6\sqrt{14}}{7}$$

23. Find the distance between the point (1, 2, 3) and the plane $2x - y + z = 4$.
Solution:
$$D = \frac{|ax_0 + by_0 + cz_0 + d|}{\sqrt{a^2 + b^2 + c^2}}$$
$$= \frac{|2(1) - 1(2) + 1(3) - 4|}{\sqrt{(2)^2 + (-1)^2 + (1)^2}} = \frac{1}{\sqrt{6}} = \frac{\sqrt{6}}{6}$$

25. Match $(x^2/9) + (y^2/16) + (z^2/9) = 1$ with the correct graph.
Solution: The graph is an ellipsoid that matches graph (c).

27. Match $4x^2 - y^2 + 4z^2 = 4$ with the correct graph.
Solution: The graph of
$$\frac{x^2}{1} - \frac{y^2}{4} + \frac{z^2}{1} = 1$$
is a hyperboloid of one sheet that matches graph (f).

29. Match $4x^2 - 4y + z^2 = 0$ with the correct graph.
Solution: The graph of
$$y = \frac{x^2}{1} + \frac{z^2}{4}$$
is an elliptic paraboloid that matches graph (d).

31. Match $4x^2 - y^2 + 4z = 0$ with the correct graph.
Solution: The graph of
$$z = \frac{y^2}{4} - \frac{x^2}{1}$$
is a hyperbolic paraboloid that matches graph (a).

33. Identify $x^2 + (y^2/4) + z^2 = 1$.
Solution: The graph is an ellipsoid.

35. Identify $16x^2 - y^2 + 16z^2 = 4$.
Solution:
$$\frac{x^2}{1/4} - \frac{y^2}{4} + \frac{z^2}{1/4} = 1$$
The graph is a hyperboloid of one sheet.

346   Section 8.3

37. Identify $x^2 - y + z^2 = 0$.
    Solution: The graph of $y = x^2 + z^2$ is an elliptic paraboloid.

39. Identify $x^2 - y^2 + z = 0$.
    Solution: The graph of $z = y^2 - x^2$ is a hyperbolic paraboloid.

41. Identify $4x^2 - y^2 + 4z^2 = -16$.
    Solution: The graph of
    $$\frac{y^2}{16} - \frac{x^2}{4} - \frac{z^2}{4} = 1$$
    is a hyperboloid of two sheets.

43. Identify $z^2 = x^2 + 4y^2$.
    Solution: The graph of
    $$x^2 + \frac{y^2}{1/4} - z^2 = 0$$
    is an elliptic cone.

45. Identify $3z = -y^2 + x^2$.
    Solution: The graph of
    $$z = \frac{x^2}{3} - \frac{y^2}{3}$$
    is a hyperbolic paraboloid.

47. Identify $2x^2 + 2y^2 + 2z^2 - 3x + 4z = 10$.
    Solution: By completing the square, we find that the graph of
    $$(x - \frac{3}{4})^2 + y^2 + (z + 1)^2 = \frac{105}{16}$$
    is a sphere.

● Section 8.3   Functions of Several Variables

1. When $f(x, y) = x/y$, find
   (a) $f(3, 2)$ (b) $f(-1, 4)$
   (c) $f(30, 5)$ (d) $f(5, y)$
   (e) $f(x, 2)$ (f) $f(5, t)$
   Solution:
   (a) $f(3, 2) = 3/2$ (b) $f(-1, 4) = -1/4$
   (c) $f(30, 5) = 6$ (d) $f(5, y) = 5/y$
   (e) $f(x, 2) = x/2$ (f) $f(5, t) = 5/t$

3. When $f(x, y) = xe^y$, find
   (a) $f(5, 0)$ (b) $f(3, 2)$
   (c) $f(2, -1)$ (d) $f(5, y)$
   (e) $f(x, 2)$ (f) $f(t, t)$
   Solution:
   (a) $f(5, 0) = 5$ (b) $f(3, 2) = 3e^2$
   (c) $f(2, -1) = 2/e$ (d) $f(5, y) = 5e^y$
   (e) $f(x, 2) = xe^2$ (f) $f(t, t) = te^t$

Section 8.3	347

5. When $h(x, y, z) = xy/z$, find
   (a) $h(2, 3, 9)$ (b) $h(1, 0, 1)$
   Solution:
   (a) $h(2, 3, 9) = 2/3$ (b) $h(1, 0, 1) = 0$

7. When $V(r, h) = \pi r^2 h$, find
   (a) $V(3, 10)$ (b) $V(5, 2)$
   Solution:
   (a) $V(3, 10) = \pi(3)^2(10) = 90\pi$
   (b) $V(5, 2) = \pi(5)^2(2) = 50\pi$

9. When $A(P, r, t) = P[(1 + (r/12))^{12t} - 1][1 + (12/r)]$, find
   (a) $A(100, 0.10, 10)$ (b) $A(275, 0.0925, 40)$
   Solution:
   (a) $A(100, 0.10, 10)$
   $$= 100\left[\left(1 + \frac{0.10}{12}\right)^{120} - 1\right]\left(1 + \frac{12}{0.10}\right)$$
   $$\approx \$20,655.20$$

   (b) $A(275, 0.0925, 40)$
   $$= 275\left[\left(1 + \frac{0.0925}{12}\right)^{480} - 1\right]\left(1 + \frac{12}{0.0925}\right)$$
   $$\approx \$1,397,672.67$$

11. When $f(x, y) = \int_x^y (2t - 3)\, dt$, find
    (a) $f(0, 4)$ (b) $f(1, 4)$
    Solution:
    (a) $f(0, 4) = \int_0^4 (2t - 3)\, dt = (t^2 - 3t)\Big]_0^4 = 4$
    (b) $f(1, 4) = \int_1^4 (2t - 3)\, dt = (t^2 - 3t)\Big]_1^4 = 6$

13. When $f(x, y) = x^2 - 2y$, find
    (a) $\dfrac{f(x + \Delta x, y) - f(x, y)}{\Delta x}$

    (b) $\dfrac{f(x, y + \Delta y) - f(x, y)}{\Delta y}$

    Solution:
    (a) $\dfrac{f(x + \Delta x, y) - f(x, y)}{\Delta x}$

    $$= \frac{[(x + \Delta x)^2 - 2y] - (x^2 - 2y)}{\Delta x}$$

    $$= \frac{x^2 + 2x\Delta x + (\Delta x)^2 - 2y - x^2 + 2y}{\Delta x}$$

    $$= \frac{2x\Delta x + (\Delta x)^2}{\Delta x} = 2x + \Delta x$$

(b) $\dfrac{f(x, y + \Delta y) - f(x, y)}{\Delta y}$

$= \dfrac{[x^2 - 2(y + \Delta y)] - (x^2 - 2y)}{\Delta y}$

$= \dfrac{x^2 - 2y - 2\Delta y - x^2 + 2y}{\Delta y} = -\dfrac{2\Delta y}{\Delta y} = -2$

15. Describe the region R in the xy-coordinate plane that corresponds to the domain of $f(x, y) = \sqrt{4 - x^2 - 4y^2}$, and find the range of the function.
    Solution: The domain is the set of all points inside and on the ellipse $x^2 + 4y^2 = 4$ since $4 - x^2 - 4y^2 \geq 0$, and the range is [0, 2].

17. Describe the region R in the xy-coordinate plane that corresponds to the domain of $f(x, y) = 4 - x^2 - y^2$ and find the range of the function.
    Solution: The domain is the set of all points in the xy-plane, and the range is $(-\infty, 4]$.

19. Describe the region R in the xy-coordinate plane that corresponds to the domain of $f(x, y) = \ln(4 - x - y)$ and find the range of the function.
    Solution: The domain is the half plane below the line $y = -x + 4$ since $4 - x - y > 0$, and the range is $(-\infty, \infty)$.

21. Describe the region R in the xy-coordinate plane that corresponds to the domain of

    $h(x, y) = \dfrac{1}{xy}$

    and find the range of the function.
    Solution: The domain is the set of all points in the xy-plane except those on the x- and y-axes, and the range is $(-\infty, 0)$ and $(0, \infty)$.

23. Describe the region R in the xy-coordinate plane that corresponds to the domain of $g(x, y) = x\sqrt{y}$ and find the range of the function.
    Solution: The domain is the set of all points in the xy-plane such that $y \geq 0$, and the range is $(-\infty, \infty)$.

25. Sketch a contour map for

    $f(x, y) = \sqrt{25 - x^2 - y^2}$

    using the c-values, c = 0, 1, 2, 3, 4, 5.
    Solution:

c = 0,	$0 = \sqrt{25 - x^2 - y^2}$,	$x^2 + y^2 = 25$
c = 1,	$1 = \sqrt{25 - x^2 - y^2}$,	$x^2 + y^2 = 24$
c = 2,	$2 = \sqrt{25 - x^2 - y^2}$,	$x^2 + y^2 = 21$
c = 3,	$3 = \sqrt{25 - x^2 - y^2}$,	$x^2 + y^2 = 16$
c = 4,	$4 = \sqrt{25 - x^2 - y^2}$,	$x^2 + y^2 = 9$
c = 5,	$5 = \sqrt{25 - x^2 - y^2}$,	$x^2 + y^2 = 0$

Section 8.3   349

27. Sketch a contour map for

$$f(x, y) = xy$$

using the c-values, c = 1, -1, 3, -3.
Solution:

c = 1,	xy = 1
c = -1,	xy = -1
c = 3,	xy = 3
c = -3,	xy = -3

(contour map shown with $c = 6, 5, 4, 3, 2, 1, -1, -2, -3, -4, -5, -6$ and $c = 0$ (x & y-axes))

29. Sketch a contour map for $f(x, y) = \ln(x - y)$ using the c-values, c = 0, ±1/2, ±1, ±3/2, ±2.
Solution:

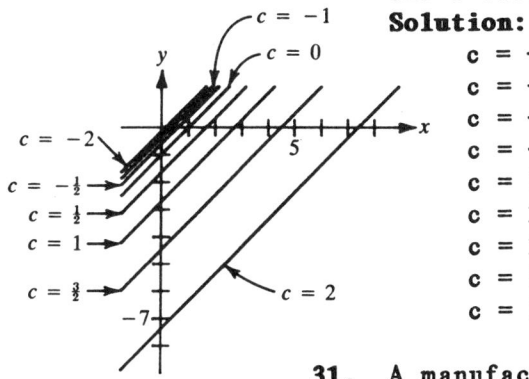

c = -2,	$\ln(x - y) = -2$	⇒	$x - y = e^{-2}$
c = -3/2,	$\ln(x - y) = -3/2$	⇒	$x - y = e^{-3/2}$
c = -1,	$\ln(x - y) = -1$	⇒	$x - y = e^{-1}$
c = -1/2,	$\ln(x - y) = -1/2$	⇒	$x - y = e^{-1/2}$
c = 0,	$\ln(x - y) = 0$	⇒	$x - y = 1$
c = 1/2,	$\ln(x - y) = 1/2$	⇒	$x - y = e^{1/2}$
c = 1,	$\ln(x - y) = 1$	⇒	$x - y = e$
c = 3/2,	$\ln(x - y) = 3/2$	⇒	$x - y = e^{3/2}$
c = 2,	$\ln(x - y) = 2$	⇒	$x - y = e^2$

31. A manufacturer estimates the Cobb-Douglas production function to be $f(x, y) = 100x^{0.75}y^{0.25}$. Estimate the production level when x = 1500 and y = 1000.
Solution:
$$f(1500, 1000) = 100(1500)^{0.75}(1000)^{0.25}$$
$$\approx 135{,}540 \text{ units}$$

33. A company manufactures two types of woodburning stoves: a free-standing model and a fireplace-insert model. The cost function for producing x free-standing stoves and y fireplace-insert stoves is $C(x, y) = 32\sqrt{xy} + 175x + 205y + 1050$. Find the cost when x = 80 and y = 20.
Solution:
$$C(80, 20) = 32\sqrt{(80)(20)} + 175(80) + 205(20) + 1050$$
$$= \$20{,}430.00$$

35. The **Doyle Log Rule** is one of several methods used to determine the lumber yield of a log in board feet in terms of its diameter d in inches and its length L in feet. The number of board feet is given by $N(d, L) = [(d - 4)/4]^2 L$. Find
(a) The number of board feet of lumber in a log with a diameter of 22 inches and a length of 12 feet.
(b) N(30, 12).
Solution:

(a) $N(22, 12) = \left(\dfrac{22 - 4}{4}\right)^2 (12) = 243$ board feet

(b) $N(30, 12) = \left(\dfrac{30 - 4}{4}\right)^2 (12) = 507$ board feet

37. (Queuing Model) The average length of time that a customer waits in line for service is given by $W(x, y) = 1/(x - y)$, $y < x$, where y is the average arrival rate expressed in the number of customers per unit time (hours) and x is the average service rate expressed in the same units. Evaluate W at the following points:
   (a) (15, 10)   (b) (12, 9)
   (c) (12, 6)   (d) (4, 2)
   Solution:
   (a) $W(15, 10) = 1/(15 - 10) = 1/5$ hr. = 12 min.
   (b) $W(12, 9) = 1/(12 - 9) = 1/3$ hr. = 20 min.
   (c) $W(12, 6) = 1/(12 - 6) = 1/6$ hr. = 10 min.
   (d) $W(4, 2) = 1/(4 - 2) = 1/2$ hr. = 30 min.

● Section 8.4   Partial Derivatives

1. For $f(x, y) = 2x - 3y + 5$, find the first partial derivatives with respect to x and with respect to y.
   Solution:
   $$f_x(x, y) = 2, \qquad f_y(x, y) = -3$$

3. For $f(x, y) = 5\sqrt{x} - 6y^2 + 4$, find the first partial derivatives with respect to x and with respect to y.
   Solution:
   $$f_x(x, y) = \frac{5}{2\sqrt{x}}, \qquad f_y(x, y) = -12y$$

5. For $f(x, y) = xy$, find the first partial derivatives with respect to x and with respect to y.
   Solution:
   $$f_x(x, y) = y, \qquad f_y(x, y) = x$$

7. For $z = x\sqrt{y}$, find the first partial derivatives with respect to x and with respect to y.
   Solution:
   $$\frac{\partial z}{\partial x} = \sqrt{y}, \qquad \frac{\partial z}{\partial y} = \frac{x}{2\sqrt{y}}$$

9. For $f(x, y) = \sqrt{x^2 + y^2}$, find the first partial derivatives with respect to x and with respect to y.
   Solution:
   $$f_x(x, y) = \frac{1}{2}(x^2 + y^2)^{-1/2}(2x) = \frac{x}{\sqrt{x^2 + y^2}}$$
   $$f_y(x, y) = \frac{1}{2}(x^2 + y^2)^{-1/2}(2y) = \frac{y}{\sqrt{x^2 + y^2}}$$

11. For $z = x^2 e^{2y}$, find the first partial derivatives with respect to x and with respect to y.
    Solution:
    $$\frac{\partial z}{\partial x} = 2xe^{2y}, \qquad \frac{\partial z}{\partial y} = 2x^2 e^{2y}$$

Section 8.4

13. For $h(x, y) = e^{-(x^2+y^2)}$, find the first partial derivatives with respect to x and with respect to y.
    Solution:
    $$h_x(x, y) = -2xe^{-(x^2+y^2)}$$
    $$h_y(x, y) = -2ye^{-(x^2+y^2)}$$

15. For $z = \ln(x^2 + y^2)$, find the first partial derivatives with respect to x and with respect to y.
    Solution:
    $$\frac{\partial z}{\partial x} = \frac{2x}{x^2 + y^2}, \qquad \frac{\partial z}{\partial y} = \frac{2y}{x^2 + y^2}$$

17. For $z = \ln[(x + y)/(x - y)]$, find the first partial derivatives with respect to x and with respect to y.
    Solution:
    $$z = \ln\left(\frac{x + y}{x - y}\right) = \ln(x + y) - \ln(x - y)$$
    $$\frac{\partial z}{\partial x} = \frac{1}{x + y} - \frac{1}{x - y} = \frac{-2y}{x^2 - y^2}$$
    $$\frac{\partial z}{\partial y} = \frac{1}{x + y} - \frac{-1}{x - y} = \frac{2x}{x^2 - y^2}$$

19. For $f(x, y) = \int_x^y (t^2 - 1)\, dt$

    find the first partial derivatives with respect to x and with respect to y.
    Solution:
    $$f(x, y) = \int_x^y (t^2 - 1)\, dt = \left[\frac{1}{3}t^3 - t\right]_x^y$$
    $$= \left(\frac{1}{3}y^3 - y\right) - \left(\frac{1}{3}x^3 - x\right)$$
    $$= \frac{1}{3}y^3 - y - \frac{1}{3}x^3 + x$$
    $$f_x(x, y) = -x^2 + 1 = 1 - x^2$$
    $$f_y(x, y) = y^2 - 1$$

21. When $f(x, y) = 3x^2 + xy - y^2$, evaluate f_x and f_y at point $(2, 1)$.
    Solution:
    $$f_x(x, y) = 6x + y, \qquad f_x(2, 1) = 13$$
    $$f_y(x, y) = x - 2y, \qquad f_y(2, 1) = 0$$

23. When $f(x, y) = xy/(x - y)$, evaluate f_x and f_y at the point $(2, -2)$.
    Solution:
    $$f_x(x, y) = \frac{(x - y)y - xy}{(x - y)^2} = -\frac{y^2}{(x - y)^2}$$
    $$f_x(2, -2) = -4/16 = -1/4$$
    $$f_y(x, y) = \frac{(x - y)x - xy(-1)}{(x - y)^2} = \frac{x^2}{(x - y)^2}$$
    $$f_y(2, -2) = 4/16 = 1/4$$

25. If $w = \sqrt{x^2 + y^2 + z^2}$, find the first partial derivatives with respect to x, y, and z, and evaluate these partial derivatives at point (2, -1, 2).
    Solution:

    $$\frac{\partial w}{\partial x} = \frac{x}{\sqrt{x^2 + y^2 + z^2}}, \quad \text{at } (2, -1, 2), \frac{\partial w}{\partial x} = \frac{2}{3}$$

    $$\frac{\partial w}{\partial y} = \frac{y}{\sqrt{x^2 + y^2 + z^2}}, \quad \text{at } (2, -1, 2), \frac{\partial w}{\partial y} = -\frac{1}{3}$$

    $$\frac{\partial w}{\partial z} = \frac{z}{\sqrt{x^2 + y^2 + z^2}}, \quad \text{at } (2, -1, 2), \frac{\partial w}{\partial z} = \frac{2}{3}$$

27. If $F(x, y, z) = \ln \sqrt{x^2 + y^2 + z^2}$, find the first partial derivatives with respect to x, y, and z, and evaluate these partial derivatives at point (3, 0, 4).
    Solution:

    $$F(x, y, z) = \ln \sqrt{x^2 + y^2 + z^2} = \frac{1}{2} \ln(x^2 + y^2 + z^2)$$

    $$F_x(x, y, z) = \frac{x}{x^2 + y^2 + z^2}, \quad F_x(3, 0, 4) = \frac{3}{25}$$

    $$F_y(x, y, z) = \frac{y}{x^2 + y^2 + z^2}, \quad F_y(3, 0, 4) = 0$$

    $$F_z(x, y, z) = \frac{z}{x^2 + y^2 + z^2}, \quad F_z(3, 0, 4) = \frac{4}{25}$$

29. If $f(x, y, z) = 3x^2y - 5xyz + 10yz^2$, find the first partial derivatives with respect to x, y, and z, and evaluate these partial derivatives at point (1, -1, 2).
    Solution:
    $f_x(x, y, z) = 6xy - 5yz, \quad f_x(1, -1, 2) = 4$
    $f_y(x, y, z) = 3x^2 - 5xz + 10z^2, \quad f_y(1, -1, 2) = 33$
    $f_z(x, y, z) = -5xy + 20yz, \quad f_z(1, -1, 2) = -35$

31. Find the second partial derivatives of $z = x^3 - 4y^2$.
    Solution: The first partial derivatives are

    $$\frac{\partial z}{\partial x} = 3x^2 \quad \text{and} \quad \frac{\partial z}{\partial y} = -8y$$

    and the second partial derivatives are

    $$\frac{\partial^2 z}{\partial x^2} = 6x, \quad \frac{\partial^2 z}{\partial y \, \partial x} = 0 = \frac{\partial^2 z}{\partial x \, \partial y}, \quad \frac{\partial^2 z}{\partial y^2} = -8$$

33. Find the second partial derivatives of

    $$z = x^3 + 3x^2y - 5y^2$$

    Solution: The first partial derivatives are

    $$\frac{\partial z}{\partial x} = 3x^2 + 6xy \quad \text{and} \quad \frac{\partial z}{\partial y} = 3x^2 - 10y$$

    and the second partial derivatives are

    $$\frac{\partial^2 z}{\partial x^2} = 6x + 6y, \quad \frac{\partial^2 z}{\partial y \, \partial x} = 6x = \frac{\partial^2 z}{\partial x \, \partial y}, \quad \frac{\partial^2 z}{\partial y^2} = -10$$

Section 8.4 353

35. Find the second partial derivatives of
$$z = 9 + 4x - 6y - x^2 - y^2$$
Solution: The first partial derivatives are
$$\frac{\partial z}{\partial x} = 4 - 2x \quad \text{and} \quad \frac{\partial z}{\partial y} = -6 - 2y$$
and the second partial derivatives are
$$\frac{\partial^2 z}{\partial x^2} = -2, \quad \frac{\partial^2 z}{\partial y\, \partial x} = 0 = \frac{\partial^2 z}{\partial x\, \partial y}, \quad \frac{\partial^2 z}{\partial y^2} = -2$$

37. Find the second partial derivatives of
$$z = \frac{xy}{x - y}$$
Solution: From Exercise 23, the first partial derivatives are
$$\frac{\partial z}{\partial x} = \frac{-y^2}{(x - y)^2} \quad \text{and} \quad \frac{\partial z}{\partial y} = \frac{x^2}{(x - y)^2}$$
and the second partial derivatives are
$$\frac{\partial^2 z}{\partial x^2} = \frac{2y^2}{(x - y)^3}, \quad \frac{\partial^2 z}{\partial y\, \partial x} = \frac{-2xy}{(x - y)^3} = \frac{\partial^2 z}{\partial x\, \partial y}$$
$$\frac{\partial^2 z}{\partial y^2} = \frac{2x^2}{(x - y)^3}$$

39. Find the second partial derivatives of $z = \sqrt{x^2 + y^2}$.
Solution: The first partial derivatives are
$$\frac{\partial z}{\partial x} = \frac{x}{\sqrt{x^2 + y^2}} \quad \text{and} \quad \frac{\partial z}{\partial y} = \frac{y}{\sqrt{x^2 + y^2}}$$
and the second partial derivatives are
$$\frac{\partial^2 z}{\partial x^2} = \frac{\sqrt{x^2 + y^2} - x(x/\sqrt{x^2 + y^2})}{x^2 + y^2} = \frac{y^2}{(x^2 + y^2)^{3/2}}$$
$$\frac{\partial^2 z}{\partial y\, \partial x} = -\frac{1}{2} x(x^2 + y^2)^{-3/2}(2y)$$
$$= -\frac{xy}{(x^2 + y^2)^{3/2}} = \frac{\partial^2 z}{\partial x\, \partial y}$$
$$\frac{\partial^2 z}{\partial y^2} = \frac{\sqrt{x^2 + y^2} - y(y/\sqrt{x^2 + y^2})}{x^2 + y^2} = \frac{x^2}{(x^2 + y^2)^{3/2}}$$

41. Find the second partial derivatives of $z = xe^{-y^2}$.
Solution: The first partial derivatives are
$$\frac{\partial z}{\partial x} = e^{-y^2} \quad \text{and} \quad \frac{\partial z}{\partial y} = -2xye^{-y^2}$$
and the second partial derivatives are
$$\frac{\partial^2 z}{\partial x^2} = 0, \quad \frac{\partial^2 z}{\partial y\, \partial x} = -2ye^{-y^2} = \frac{\partial^2 z}{\partial x\, \partial y}$$
$$\frac{\partial^2 z}{\partial y^2} = -2xe^{-y^2} + 4xy^2 e^{-y^2} = 2xe^{-y^2}(2y^2 - 1)$$

43. Show that $z = 5xy$ satisfies **Laplace's equation** $(\partial^2 z/\partial x^2) + (\partial^2 z/\partial y^2) = 0$.
Solution: The first partial derivatives are

$$\frac{\partial z}{\partial x} = 5y \quad \text{and} \quad \frac{\partial z}{\partial y} = 5x$$

and the second partial derivatives are

$$\frac{\partial^2 z}{\partial x^2} = 0 \quad \text{and} \quad \frac{\partial^2 z}{\partial y^2} = 0$$

Therefore, $\frac{\partial^2 z}{\partial x^2} + \frac{\partial^2 z}{\partial y^2} = 0$.

45. Show that $z = x^3 - 3xy^2$ satisfies **Laplace's equation** $(\partial^2 z/\partial x^2) + (\partial^2 z/\partial y^2) = 0$.
Solution: The first partial derivatives are

$$\frac{\partial z}{\partial x} = 3x^2 - 3y^2 \quad \text{and} \quad \frac{\partial z}{\partial y} = -6xy$$

and the second partial derivatives are

$$\frac{\partial^2 z}{\partial x^2} = 6x \quad \text{and} \quad \frac{\partial^2 z}{\partial y^2} = -6x$$

Therefore, $\frac{\partial^2 z}{\partial x^2} + \frac{\partial^2 z}{\partial y^2} = 0$.

47. For $z = 2x - 3y + 5$, find the slope of the surface at the point $(2, 1, 6)$ in (a) the x-direction and (b) the y-direction.
Solution:

(a) $\frac{\partial z}{\partial x} = 2,$ at $(2, 1, 6)$, $\frac{\partial z}{\partial x} = 2$

(b) $\frac{\partial z}{\partial y} = -3,$ at $(2, 1, 6)$, $\frac{\partial z}{\partial y} = -3$

49. For $z = 9x^2 - y^2$, find the slope of the surface at the point $(1, 3, 0)$ in (a) the x-direction and (b) the y-direction.
Solution:

(a) $\frac{\partial z}{\partial x} = 18x,$ at $(1, 3, 0)$, $\frac{\partial z}{\partial x} = 18$

(b) $\frac{\partial z}{\partial y} = -2y,$ at $(1, 3, 0)$, $\frac{\partial z}{\partial y} = -6$

51. For $z = \sqrt{25 - x^2 - y^2}$, find the slope of the surface at the point $(3, 0, 4)$ in (a) the x-direction and (b) the y-direction.
Solution:

(a) $\frac{\partial z}{\partial x} = -\frac{x}{\sqrt{25 - x^2 - y^2}},$ at $(3, 0, 4)$, $\frac{\partial z}{\partial x} = -\frac{3}{4}$

(b) $\frac{\partial z}{\partial y} = -\frac{y}{\sqrt{25 - x^2 - y^2}},$ at $(3, 0, 4)$, $\frac{\partial z}{\partial y} = 0$

Section 8.4   355

53. A company manufactures two types of woodburning stoves: a free-standing model and a fireplace-insert model. The cost function for producing x free-standing and y fireplace-insert stoves is $C = 32\sqrt{xy} + 175x + 205y + 1050$. Find the marginal costs ($\partial C/\partial x$ and $\partial C/\partial y$) when x = 80 and y = 20.
Solution:

$$\frac{\partial C}{\partial x} = \frac{16y}{\sqrt{xy}} + 175, \qquad \text{at } (80, 20), \frac{\partial C}{\partial x} = 183$$

$$\frac{\partial C}{\partial y} = \frac{16x}{\sqrt{xy}} + 205, \qquad \text{at } (80, 20), \frac{\partial C}{\partial y} = 237$$

55. Let x = 1000 and y = 500 in the Cobb-Douglas production function $f(x, y) = 100x^{0.6}y^{0.4}$.
(a) Find the marginal productivity of labor, $\partial f/\partial x$.
(b) Find the marginal productivity of capital, $\partial f/\partial y$.
Solution:

(a) $\partial f/\partial x = 60x^{-0.4}y^{0.4} = 60(y/x)^{0.4}$,
at (1000, 500), $\partial f/\partial x \approx 45.47$
(b) $\partial f/\partial y = 40x^{0.6}y^{-0.6} = 40(x/y)^{0.6}$,
at (1000, 500), $\partial f/\partial y \approx 60.63$

57. Let N be the number of applicants to a university, p the charge for food and housing at the university, and t the tuition. Suppose that N is a function of p and t such that $\partial N/\partial p < 0$ and $\partial N/\partial t < 0$. How would you interpret the fact that both partials are negative?
Solution: Since both first partials are negative, an increase in the charge for food and housing or tuition will cause a decrease in the number of applicants.

59. Let x_1 and x_2 be the demands for products 1 and 2, respectively, and p_1 and p_2 the prices of products 1 and 2, respectively. Determine if the following demand functions describe complementary or substitute product relationships:

(a) $x_1 = 150 - 2p_1 - (\frac{5}{2})p_2$, $x_2 = 350 - (\frac{3}{2})p_1 - 3p_2$

(b) $x_1 = 150 - 2p_1 + 1.8p_2$, $x_2 = 350 + 0.75p_1 - 1.9p_2$

(c) $x_1 = \dfrac{1000}{\sqrt{p_1 p_2}}$, $\qquad x_2 = \dfrac{750}{p_2\sqrt{p_1}}$

Solution:
(a) Complementary since

$$\frac{\partial x_1}{\partial p_2} = -\frac{5}{2} < 0 \quad \text{and} \quad \frac{\partial x_2}{\partial p_1} = -\frac{3}{2} < 0$$

(b) Substitute since

$$\frac{\partial x_1}{\partial p_2} = 1.8 > 0 \quad \text{and} \quad \frac{\partial x_2}{\partial p_1} = 0.75 > 0$$

(c) Complementary since

$$\frac{\partial x_1}{\partial p_2} = \frac{-500}{p_1^2 p_2 \sqrt{p_1 p_2}} < 0 \quad \text{and} \quad \frac{\partial x_2}{\partial p_1} = \frac{-375}{p_1 p_2 \sqrt{p_1}} < 0$$

Section 8.5 Extrema of Functions of Two Variables

1. Examine $f(x, y) = (x - 1)^2 + (y - 3)^2$ for relative extrema and saddle points.
   Solution: The first partial derivatives of f
   $$f_x(x, y) = 2(x - 1)$$
   $$f_y(x, y) = 2(y - 3)$$
   are zero at the point (1, 3). Moreover, since
   $$f_{xx}(x, y) = 2, \quad f_{yy}(x, y) = 2, \quad f_{xy}(x, y) = 0$$
   it follows that $f_{xx}(1, 3) > 0$ and
   $$f_{xx}(1, 3)f_{yy}(1, 3) - [f_{xy}(1, 3)]^2 = 4 > 0$$
   Thus, (1, 3, 0) is a relative minimum.

3. Examine $f(x, y) = 2x^2 + 2xy + y^2 + 2x - 3$ for relative extrema and saddle points.
   Solution: The first partial derivatives of f
   $$f_x(x, y) = 4x + 2y + 2$$
   $$f_y(x, y) = 2x + 2y$$
   are zero at the point (-1, 1). Moreover, since
   $$f_{xx}(x, y) = 4, \quad f_{yy}(x, y) = 2, \quad f_{xy}(x, y) = 2$$
   it follows that $f_{xx}(-1, 1) > 0$ and
   $$f_{xx}(-1, 1)f_{yy}(-1, 1) - [f_{xy}(-1, 1)]^2 = 4 > 0$$
   Thus, (-1, 1, -4) is a relative minimum.

5. Examine $f(x, y) = -5x^2 + 4xy - y^2 + 16x + 10$ for relative extrema and saddle points.
   Solution: The first partial derivatives of f
   $$f_x(x, y) = -10x + 4y + 16$$
   $$f_y(x, y) = 4x - 2y$$
   are zero at the point (8, 16). Moreover, since
   $$f_{xx}(x, y) = -10, \quad f_{yy}(x, y) = -2, \quad f_{xy}(x, y) = 4$$
   it follows that $f_{xx}(8, 16) < 0$ and
   $$f_{xx}(8, 16)f_{yy}(8, 16) - [f_{xy}(8, 16)]^2 = 4 > 0$$
   Thus, (8, 16, 74) is a relative maximum.

Section 8.5

7. Examine $f(x, y) = 2x^2 + 3y^2 - 4x - 12y + 13$ for relative extrema and saddle points.
   Solution: The first partial derivatives of f
   $$f_x(x, y) = 4x - 4 = 4(x - 1)$$
   $$f_y(x, y) = 6y - 12 = 6(y - 2)$$
   are zero at the point (1, 2). Moreover, since
   $$f_{xx}(x, y) = 4, \quad f_{yy}(x, y) = 6, \quad f_{xy}(x, y) = 0$$
   it follows that $f_{xx}(1, 2) > 0$ and
   $$f_{xx}(1, 2)f_{yy}(1, 2) - [f_{xy}(1, 2)]^2 = 24 > 0$$
   Thus, (1, 2, -1) is a relative minimum.

9. Examine $f(x, y) = x^2 - y^2 - 2x - 4y - 4$ for relative extrema and saddle points.
   Solution: The first partial derivatives of f
   $$f_x(x, y) = 2x - 2 = 2(x - 1)$$
   $$f_y(x, y) = -2y - 4 = -2(y + 2)$$
   are zero at the point (1, -2). Moreover, since
   $$f_{xx}(x, y) = 2, \quad f_{yy}(x, y) = -2, \quad f_{xy}(x, y) = 0$$
   it follows that
   $$f_{xx}(1, -2)f_{yy}(1, -2) - [f_{xy}(1, -2)]^2 = -4 < 0$$
   Thus, (1, -2, -1) is a saddle point.

11. Examine $f(x, y) = xy$ for relative extrema and saddle points.
    Solution: The first partial derivatives of f
    $$f_x(x, y) = y$$
    $$f_y(x, y) = x$$
    are zero at the point (0, 0). Moreover, since
    $$f_{xx}(x, y) = 0, \quad f_{yy}(x, y) = 0, \quad f_{xy}(x, y) = 1$$
    it follows that
    $$f_{xx}(0, 0)f_{yy}(0, 0) - [f_{xy}(0, 0)]^2 = -1 < 0$$
    Thus, (0, 0, 0) is a saddle point.

13. Examine $f(x, y) = x^3 - 3xy + y^3$ for relative extrema and saddle points.
    Solution: The first partial derivatives of f

$$f_x(x, y) = 3(x^2 - y) \quad \text{and} \quad f_y(x, y) = 3(-x + y^2)$$

are zero at (0, 0) and (1, 1). Moreover, since

$$f_{xx}(x, y) = 6x, \quad f_{yy}(x, y) = 6y, \quad f_{xy}(x, y) = -3$$

it follows that $f_{xx}(1, 1) = 6 > 0$ and

$$f_{xx}(1, 1)f_{yy}(1, 1) - [f_{xy}(1, 1)]^2 = 27 > 0$$
$$f_{xx}(0, 0)f_{yy}(0, 0) - [f_{xy}(0, 0)]^2 = -9 < 0$$

Thus, (1, 1, -1) is a relative minimum and (0, 0, 0) is a saddle point.

15. Examine $f(x, y) = (1/2)(3x^2 + 1 - 2x^3 - 2xy^2)$ for relative extrema and saddle points.
    Solution: The first partial derivatives of f

$$f_x(x, y) = 3x - 3x^2 - y^2 \quad \text{and} \quad f_y(x, y) = -2xy$$

are zero at (0, 0) and (1, 0). Moreover, since

$$f_{xx}(x, y) = 3 - 6x$$
$$f_{yy}(x, y) = -2x$$
$$f_{xy}(x, y) = -2y$$

it follows that $f_{xx}(0, 0) > 0$, $f_{xx}(1, 0) < 0$, and

$$f_{xx}(0, 0)f_{yy}(0, 0) - [f_{xy}(0, 0)]^2 = 0$$
$$f_{xx}(1, 0)f_{yy}(1, 0) - [f_{xy}(1, 0)]^2 = 6 > 0$$

The point (1, 0, 1) is a relative maximum. The second-partials test gives no information at (0, 0, 1/2), but by examining the graph we conclude that it is a saddle point.

17. Examine $f(x, y) = (x^2 + 4y^2)e^{1-x^2-y^2}$ for relative extrema and saddle points.
    Solution: The first partial derivatives of f

$$f_x(x, y) = 2xe^{1-x^2-y^2}(1 - x^2 - 4y^2)$$
$$f_y(x, y) = 2ye^{1-x^2-y^2}(4 - x^2 - 4y^2)$$

are zero at (0, 0), (0, ±1), and (±1, 0). Moreover, since

$$f_{xx}(x, y) = 2e^{1-x^2-y^2}(1 - 5x^2 + 2x^4 - 4y^2 + 8x^2y^2)$$
$$f_{yy}(x, y) = 2e^{1-x^2-y^2}(4 - x^2 - 20y^2 + 8y^4 + 2x^2y^2)$$
$$f_{xy}(x, y) = -4xye^{1-x^2-y^2}(5 - x^2 - 4y^2)$$

we can determine that (0, 0, 0) is a relative minimum, (0, ±1, 4) are relative maxima, and (±1, 0, 1) are saddle points.

19. Examine $f(x, y) = e^{xy}$ for relative extrema and saddle points.
   Solution: The first partial derivatives of f

   $$f_x(x, y) = ye^{xy} \quad \text{and} \quad f_y(x, y) = xe^{xy}$$

   are zero at the point (0, 0). Moreover, since

   $$f_{xx}(x, y) = y^2 e^{xy}, \qquad f_{yy}(x, y) = x^2 e^{xy}$$
   $$f_{xy}(x, y) = e^{xy}(1 + xy)$$

   it follows that $f_{xx}(0, 0) = 0$ and

   $$f_{xx}(0, 0)f_{yy}(0, 0) - [f_{xy}(0, 0)]^2 = -1 < 0$$

   Thus, (0, 0, 1) is a saddle point.

21. A company manufactures two products. The total revenue from x_1 units of product 1 and x_2 units of product 2 is $R = -5x_1^2 - 8x_2^2 - 2x_1x_2 + 42x_1 + 102x_2$. Find x_1 and x_2 so as to maximize the revenue.
   Solution: The first partial derivatives of R are

   $$R_{x_1} = -10x_1 - 2x_2 + 42$$
   $$R_{x_2} = -16x_2 - 2x_1 + 102$$

   Setting these equal to zero produces the system

   $$5x_1 + x_2 = 21$$
   $$x_1 + 8x_2 = 51$$

   which yields $x_1 = 3$ and $x_2 = 6$. By the second-partials test, it follows that the revenue is maximized when $x_1 = 3$ and $x_2 = 6$.

23. Find p_1 and p_2 so as to maximize the total revenue $R = x_1p_1 + x_2p_2$ for a retail outlet that sells two competitive products with the demand functions $x_1 = 1000 - 2p_1 + p_2$ and $x_2 = 1500 + 2p_1 - 1.5p_2$.
   Solution: The revenue function is given by

   $$R = x_1p_1 + x_2p_2$$
   $$= 1000p_1 + 1500p_2 + 3p_1p_2 - 2p_1^2 - 1.5p_2^2$$

   and the first partials of R are

   $$R_{p_1} = 1000 + 3p_2 - 4p_1$$
   $$R_{p_2} = 1500 + 3p_1 - 3p_2$$

   Setting these equal to zero produces the system

   $$4p_1 - 3p_2 = 1000$$
   $$-3p_1 + 3p_2 = 1500$$

   Solving this system yields $p_1 = 2500$ and $p_2 = 3000$, and by the second-partials test, we conclude that the revenue is maximized when $p_1 = 2500$ and $p_2 = 3000$.

Section 8.5

25. A corporation manufactures a product at two locations. The costs of producing x_1 units at location 1 and x_2 units at location 2 are $C_1 = 0.02x_1^2 + 4x_1 + 500$ and $C_2 = 0.05x_2^2 + 4x_2 + 275$ respectively. If the product sells for \$15 per unit, find the quantity that must be produced at each location to maximize the profit, $P = 15(x_1 + x_2) - C_1 - C_2$.
Solution: The profit function is given by

$$P = 15(x_1 + x_2) - C_1 - C_2$$
$$= -0.02x_1^2 - 0.05x_2^2 + 11x_1 + 11x_2 - 775$$

and the first partial derivatives of P

$$P_{x_1} = -0.04x_1 + 11 \quad \text{and} \quad P_{x_2} = -0.10x_2 + 11$$

are zero when $x_1 = 275$ and $x_2 = 110$. By the second-partials test, it follows that the profit is a maximum when $x_1 = 275$ and $x_2 = 110$.

27. A corporation manufactures a product at two locations. The cost functions for producing x_1 units at location 1 and x_2 units at location 2 are given by

$$C_1 = 0.05x_1^2 + 15x_1 + 5400$$
$$C_2 = 0.03x_2^2 + 15x_2 + 6100$$

respectively. The demand function for the product is given by $p = 225 - 0.4(x_1 + x_2)$ and therefore the total revenue function is

$$R = [225 - 0.4(x_1 + x_2)](x_1 + x_2)$$

Find the production levels at the two locations that will maximize the profit $P = R - C_1 - C_2$.
Solution: The profit is given by

$$P = R - C_1 - C_2$$
$$= [225 - 0.4(x_1 + x_2)](x_1 + x_2)$$
$$\quad - (0.05x_1^2 + 15x_1 + 5400) - (0.03x_2^2 + 15x_2 + 6100)$$
$$= -0.45x_1^2 - 0.43x_2^2 - 0.8x_1x_2 + 210x_1 + 210x_2 - 11500$$

and the first partial derivatives of P are

$$P_{x_1} = -0.9x_1 - 0.8x_2 + 210$$
$$P_{x_2} = -0.86x_2 - 0.8x_1 + 210$$

By setting these equal to zero, we obtain the system

$$0.9x_1 + 0.8x_2 = 210$$
$$0.8x_1 + 0.86x_2 = 210$$

Solving this system yields $x_1 \approx 94$ and $x_2 \approx 157$, and by the second-partials test, we conclude that the profit is maximum when $x_1 \approx 94$ and $x_2 \approx 157$.

Section 8.5	361

29. Find the dimensions of a rectangular package of largest volume that may be sent by parcel post assuming that the sum of the length and the girth (perimeter of a cross section) cannot exceed 108 inches.
Solution: Let x = length, y = width, and z = height. The sum of length and girth is given by

$$x + (2y + 2z) = 108$$
$$x = 108 - 2y - 2z$$

and the volume of the package is given by

$$V = xyz = 108yz - 2zy^2 - 2yz^2$$

The first partial derivatives of V are

$$V_y = 108z - 4yz - 2z^2 = z(108 - 4y - 2z)$$
$$V_z = 108y - 2y^2 - 4yz = y(108 - 2y - 4z)$$

Setting these equal to zero produces the system

$$4y + 2z = 108$$
$$2y + 4z = 108$$

which yields the solution

$$x = 36, \quad y = 18, \quad \text{and} \quad z = 18$$

31. When the sum is 30 and the product is maximum, find three positive numbers x, y, and z that satisfy these conditions.
Solution: Let x, y, and z be the numbers. The sum is given by

$$x + y + z = 30$$
$$z = 30 - x - y$$

and the product is given by

$$P = xyz = 30xy - x^2y - xy^2$$

The first partial derivatives of P are

$$P_x = 30y - 2xy - y^2 = y(30 - 2x - y)$$
$$P_y = 30x - x^2 - 2xy = x(30 - x - 2y)$$

Setting these equal to zero produces the system

$$2x + y = 30$$
$$x + 2y = 30$$

Solving the system, we have

$$x = 10, \quad y = 10, \quad \text{and} \quad z = 10$$

33. When the sum is 30 and the sum of the squares is minimum, find three positive numbers x, y, and z that satisfy these conditions.
Solution: The sum is given by

$$x + y + z = 30$$
$$z = 30 - x - y$$

and the sum of the squares is given by

$$S = x^2 + y^2 + z^2 = x^2 + y^2 + (30 - x - y)^2$$

The first partial derivatives of S are

$$S_x = 2x - 2(30 - x - y) = 4x + 2y - 60$$
$$S_y = 2y - 2(30 - x - y) = 2x + 4y - 60$$

Setting these equal to zero produces the system

$$2x + y = 30$$
$$x + 2y = 30$$

Solving this system yields $x = 10$ and $y = 10$. Thus, the sum of squares is a minimum when $x = y = z = 10$.

Section 8.6 Lagrange Multipliers and Constrained Optimization

1. Assuming that x, y, and z are positive, use Lagrange multipliers to maximize $f(x, y) = xy$ subject to the constraint $x + y = 10$.
Solution:
$$F(x, y, \lambda) = xy + \lambda(x + y - 10)$$

$$F_x = y + \lambda = 0, \qquad y = -\lambda$$
$$F_y = x + \lambda = 0, \qquad x = -\lambda$$
$$F_\lambda = x + y - 10 = 0, \qquad -2\lambda = 10$$

Thus, $\lambda = -5$, $x = 5$, and $y = 5$, and $f(x, y)$ is maximum at $(5, 5)$. The maximum is $f(5, 5) = 25$.

3. Assuming that x, y, and z are positive, use Lagrange multipliers to minimize $f(x, y) = x^2 + y^2$ subject to the constraint $x + y - 4 = 0$.
Solution:
$$F(x, y, \lambda) = x^2 + y^2 + \lambda(x + y - 4)$$

$$F_x = 2x + \lambda = 0, \qquad x = -(1/2)\lambda$$
$$F_y = 2y + \lambda = 0, \qquad y = -(1/2)\lambda$$
$$F_\lambda = x + y - 4 = 0, \qquad -\lambda = 4$$

Thus, $\lambda = -4$, $x = 2$, and $y = 2$, and $f(x, y)$ is minimum at $(2, 2)$. The minimum is $f(2, 2) = 8$.

Section 8.5
363

5. Assuming that x, y, and z are positive, use Lagrange multipliers to maximize $f(x, y) = x^2 - y^2$ subject to the constraint $y - x^2 = 0$.
Solution:
$$F(x, y, \lambda) = x^2 - y^2 + \lambda(y - x^2)$$

$F_x = 2x - 2x\lambda = 0,$ $2x(1 - \lambda) = 0$
$F_y = -2y + \lambda = 0,$ $y = (1/2)\lambda$
$F_\lambda = y - x^2 = 0,$ $x = \sqrt{y}$

Thus, $\lambda = 1$, $x = \sqrt{2}/2$, and $y = 1/2$, and $f(x, y)$ is maximum at $(\sqrt{2}/2, 1/2)$. The maximum is

$$f(\sqrt{2}/2, 1/2) = 1/4$$

7. Assuming that x, y, and z are positive, use Lagrange multipliers to maximize $f(x, y) = 2x + 2xy + y$ subject to the constraint $2x + y = 100$.
Solution:
$$F(x, y, \lambda) = 2x + 2xy + y + \lambda(2x + y - 100)$$

$F_x = 2 + 2y + 2\lambda = 0,$ $y = -\lambda - 1$
$F_y = 2x + 1 + \lambda = 0,$ $x = (1/2)(-\lambda - 1)$
$F_\lambda = 2x + y - 100 = 0,$ $2(-\lambda - 1) = 100$

Thus, $\lambda = -51$, $x = 25$, and $y = 50$, and $f(x, y)$ is maximum at $(25, 50)$. The maximum is $f(25, 50) = 2600$.

9. Assuming that x, y, and z are positive, use Lagrange multipliers to maximize $f(x, y) = \sqrt{6 - x^2 - y^2}$ subject to the constraint $x + y - 2 = 0$.
Solution: Note: $f(x, y)$ has a maximum value when $g(x, y) = 6 - x^2 - y^2$ is maximum.

$$F(x, y, \lambda) = 6 - x^2 - y^2 + \lambda(x + y - 2)$$

$F_x = -2x + \lambda = 0,$ $2x = \lambda$
$F_y = -2y + \lambda = 0,$ $2y = \lambda$
$F_\lambda = x + y - 2 = 0,$ $2x = 2$

Thus, $x = y = 1$, and $f(x, y)$ is maximum at $(1, 1)$. The maximum is $f(1, 1) = 2$.

11. Assuming that x, y, and z are positive, use Lagrange multipliers to maximize $f(x, y) = e^{xy}$ subject to the constraint $x^2 + y^2 - 8 = 0$.
Solution:
$$F(x, y, \lambda) = e^{xy} + \lambda(x^2 + y^2 - 8)$$

$F_x = ye^{xy} + 2x\lambda = 0,$ $e^{xy} = -2x\lambda/y$ $\Big\}$ $x = y$
$F_y = xe^{xy} + 2y\lambda = 0,$ $e^{xy} = -2y\lambda/x$
$F_\lambda = x^2 + y^2 - 8 = 0,$ $2x^2 = 8$

Thus, $x = y = 2$, and $f(x, y)$ is maximum at $(2, 2)$. The maximum is $f(2, 2) = e^4$.

13. Assuming that x, y, and z are positive, use Lagrange multipliers to minimize $f(x, y, z) = x^2 + y^2 + z^2$ subject to the constraint $x + y + z - 6 = 0$.
Solution:
$$F(x, y, z, \lambda) = x^2 + y^2 + z^2 + \lambda(x + y + z - 6)$$

$$\left.\begin{array}{l} F_x = 2x + \lambda = 0 \\ F_y = 2y + \lambda = 0 \\ F_z = 2z + \lambda = 0 \end{array}\right\} \quad x = y = z$$
$$F_\lambda = x + y + z - 6 = 0, \quad 3x = 6$$

Thus, $x = y = z = 2$, and $f(x, y, z)$ is minimum at $(2, 2, 2)$. The minimum is $f(2, 2, 2) = 12$.

15. Assuming that x, y, and z are positive, use Lagrange multipliers to maximize $f(x, y, z) = xyz$ subject to the constraints $x + y + z = 32$ and $x - y + z = 0$.
Solution:

$$F(x, y, z, \lambda, \eta) = xyz + \lambda(x + y + z - 32) + \eta(x - y + z)$$

$$\left.\begin{array}{l} F_x = yz + \lambda + \eta = 0 \\ F_y = xz + \lambda - \eta = 0 \\ F_z = xy + \lambda + \eta = 0 \end{array}\right\} \quad x = z$$
$$F_\lambda = x + y + z - 32 = 0, \quad x + 2x + x = 32$$
$$F_\eta = x - y + z = 0, \quad\quad\quad\quad y = 2x$$

Thus, $x = 8$, $y = 16$, and $z = 8$. The maximum is $f(8, 16, 8) = 1024$

17. Assuming that x, y, and z are positive, use Lagrange multipliers to maximize $f(x, y, z) = xyz$ subject to the constraints $x^2 + z^2 = 5$ and $x - 2y = 0$.
Solution:
$$F(x, y, z, \lambda, \eta) = xyz + \lambda(x^2 + z^2 - 5) + \eta(x - 2y)$$

$$F_x = yz + 2x\lambda + \eta = 0$$
$$F_y = xz - 2\eta = 0, \quad\quad \eta = xz/2$$
$$F_z = xy + 2z\lambda = 0, \quad\quad \lambda = -xy/2z$$
$$F_\lambda = x^2 + z^2 - 5 = 0, \quad z = \sqrt{5 - x^2}$$
$$F_\eta = x - 2y = 0, \quad\quad\quad y = x/2$$

From F_x, we can write

$$\frac{x\sqrt{5 - x^2}}{2} - \frac{x^3}{2\sqrt{5 - x^2}} + \frac{x\sqrt{5 - x^2}}{2} = 0$$

$$x\sqrt{5 - x^2} = \frac{x^3}{2\sqrt{5 - x^2}}$$
$$2x(5 - x^2) = x^3$$
$$3x^3 - 10x = 0$$
$$x(3x^2 - 10) = 0$$

Since x, y, and z are positive, we have $x = \sqrt{10/3}$, $y = (1/2)(\sqrt{10/3})$, and $z = \sqrt{5/3}$.

$$f(\sqrt{10/3}, \frac{1}{2}\sqrt{10/3}, \sqrt{5/3}) = \frac{5\sqrt{15}}{9}$$

19. Minimize $f(x, y, z) = x^2 + y^2 + z^2$ subject to the constraint $x + y + z = 1$. (Assume x, y, z positive.)
Solution:
$$F(x, y, z, \lambda) = x^2 + y^2 + z^2 + \lambda(x + y + z - 1)$$

$$\left.\begin{array}{l} F_x = 2x + \lambda = 0 \\ F_y = 2y + \lambda = 0 \\ F_z = 2z + \lambda = 0 \end{array}\right\} \quad x = y = z$$
$$F_\lambda = x + y + z - 1 = 0, \quad 3x = 1$$

Thus, $x = y = z = 1/3$, and $f(x, y, z)$ is minimum at at $f(1/3, 1/3, 1/3) = 1/3$.

21. Find the dimensions of the rectangular package of largest volume. Assume that the sum of the length and the girth cannot exceed 108 inches. (Maximize $V = xyz$ subject to the constraint $x + 2y + 2z = 108$.)
Solution:
$$F(x, y, z, \lambda) = xyz + \lambda(x + 2y + 2z - 108)$$

$$\left.\begin{array}{l} F_x = yz + \lambda = 0 \\ F_y = xz + 2\lambda = 0 \\ F_z = xy + 2\lambda = 0, \end{array}\right\} \quad \begin{array}{l} x = 2y \\ y = z \end{array}$$
$$F_\lambda = x + 2y + 2z - 108 = 0, \quad 6y = 108$$

Thus $x = 36$, $y = 18$, and $z = 18$. The volume is maximum when the dimensions are $36 \times 18 \times 18$ inches.

23. A manufacturer has an order for 1000 units that can be produced at two locations. Let x_1 and x_2 be the number of units produced at the two plants. Find the number of units that should be produced at each plant to minimize the cost if the cost function is given by $C = 0.25x_1^2 + 10x_1 + 0.15x_2^2 + 12x^2$.
Solution:
$$F(x_1, x_2, \lambda) = 0.25x_1^2 + 10x_1 + 0.15x_2^2 + 12x_2 + \lambda(x_1 + x_2 - 1000)$$

$$F_{x_1} = 0.50x_1 + 10 + \lambda = 0$$
$$F_{x_2} = 0.30x_2 + 12 + \lambda = 0$$
$$F_\lambda = x_1 + x_2 - 1000 = 0$$

Solving these equations produces $x_1 = 377.5$ units and $x_2 = 1000 - x_1 = 622.5$ units.

25. Find the minimum distance from the line $2x + 3y = -1$ to the point $(0, 0)$. Start by minimizing $d^2 = x^2 + y^2$.
Solution:
$$F(x, y, \lambda) = x^2 + y^2 + \lambda(2x + 3y + 1)$$

$$\begin{array}{ll} F_x = 2x + 2\lambda = 0, & x = -\lambda \\ F_y = 2y + 3\lambda = 0, & y = -3\lambda/2 \\ F_\lambda = 2x + 3y + 1 = 0 \end{array}$$

Thus, $\lambda = 2/13$, $x = -\lambda = -2/13$, $y = -3\lambda/2 = -3/13$, and

$$d = \sqrt{x^2 + y^2} = \sqrt{(-2/13)^2 + (-3/13)^2} = \sqrt{13}/13$$

27. Find the minimum distance from the plane $x + y + z = 1$ to the point $(2, 1, 1)$. Start by minimizing $d^2 = (x - 2)^2 + (y - 1)^2 + (z - 1)^2$.
Solution:
$$F(x, y, z, \lambda) = (x - 2)^2 + (y - 1)^2 + (z - 1)^2 + \lambda(x + y + z - 1)$$

$$\left. \begin{array}{l} F_x = 2(x - 2) + \lambda = 0 \\ F_y = 2(y - 1) + \lambda = 0 \\ F_z = 2(z - 1) + \lambda = 0 \end{array} \right\} \quad \begin{array}{l} x - 2 = y - 1 = z - 1 \\ x - 1 = y = z \end{array}$$
$$F_\lambda = x + y + z - 1 = 0$$

Thus, $x = 1$, $y = z = x - 1 = 0$, and
$$d = \sqrt{(1 - 2)^2 + (0 - 1)^2 + (0 - 1)^2} = \sqrt{3}$$

29. The production function for a company is $f(x, y) = 100x^{0.25}y^{0.75}$ where x is the number of units of labor and y is the number of units of capital. Suppose that labor costs \$48 per unit, capital costs \$36 per unit, and management sets a production goal of 20,000 units. Find the number of units of labor and capital needed to meet the production goal while minimizing the cost.
Solution:
$$F(x, y, \lambda) = 48x + 36y + \lambda(x^{0.25}y^{0.75} - 200)$$

$$F_x = 48 + 0.25\lambda x^{-0.75}y^{0.75} = 0$$
$$F_y = 36 + 0.75\lambda x^{0.25}y^{-0.25} = 0$$
$$F_\lambda = x^{0.25}y^{0.75} - 200 = 0$$

This produces
$$(y/x)^{0.75} = -48/0.25\lambda \quad \text{and} \quad (y/x)^{0.25} = -0.75\lambda/36$$

Thus, $y/x = (-48/0.25\lambda)(-0.75\lambda/36) = 4$, and
$$x = 200/(4^{0.75}) = 200/(2\sqrt{2}) = 50\sqrt{2} \approx 71$$
$$y = 4x = 200\sqrt{2} \approx 283$$

31. The production function for a company is $f(x, y) = 100x^{0.25}y^{0.75}$ where x is the number of units of labor and y is the number of units of capital. Suppose that labor costs \$48 per unit and capital costs \$36 per unit. The total cost of labor and capital is limited to \$100,000. Find the maximum production level for this manufacturer.
Solution: From Exercise 29, we have $y = 4x$.
$$F(x, y, \lambda) = 100x^{0.25}y^{0.75} + \lambda(100{,}000 - 48x - 36y)$$

$$F_\lambda = 48x + 36y - 100{,}000 = 0$$

Thus, $x = 3125/6$ and $y = 6250/3$ and
$$f\left(\frac{3125}{6}, \frac{6250}{3}\right) \approx 147{,}313.91 \approx 147{,}314$$

Section 8.7 The Method of Least Squares

1. (a) Use the method of least squares to find the least squares regression line, and (b) calculate the sum of the squared errors.
   Solution:

   (a) $\sum x_i = 0$, $\sum y_i = 4$, $\sum x_i y_i = 6$, $\sum x_i^2 = 8$

   $a = \dfrac{3(6) - 0(4)}{3(8) - 0^2} = \dfrac{3}{4}$, $b = \dfrac{1}{3}(4 - \dfrac{3}{4}(0)) = \dfrac{4}{3}$

   The regression line is $y = (3/4)x + (4/3)$.

   (b) $(-\dfrac{3}{2} + \dfrac{4}{3} - 0)^2 + (\dfrac{4}{3} - 1)^2 + (\dfrac{3}{2} + \dfrac{4}{3} - 3)^2 = \dfrac{1}{6}$

3. (a) Use the method of least squares to find the least squares regression line, and (b) calculate the sum of the squared errors.
   Solution:

   (a) $\sum x_i = 4$, $\sum y_i = 8$, $\sum x_i y_i = 4$, $\sum x_i^2 = 6$

   $a = \dfrac{4(4) - 4(8)}{4(6) - 4^2} = -2$, $b = \dfrac{1}{4}(8 + 2(4)) = 4$

   The regression line is $y = -2x + 4$.

   (b) $(4 - 4)^2 + (2 - 3)^2 + (2 - 1)^2 + (0 - 0)^2 = 2$

5. Find the least squares regression line for $(-2, 0)$, $(-1, 1)$, $(0, 1)$, $(1, 2)$ and $(2, 3)$.
   Solution:

   $\sum x_i = 0$, $\sum y_i = 7$, $\sum x_i y_i = 7$, $\sum x_i^2 = 10$

   Since $a = 7/10$ and $b = 7/5$, the regression line is $y = (7/10)x + (7/5)$.

7. Find the least squares regression line for $(-3, 0)$, $(1, 4)$, and $(2, 6)$.
   Solution:

   $\sum x_i = 0$, $\sum y_i = 10$, $\sum x_i y_i = 16$, $\sum x_i^2 = 14$

   Since $a = 8/7$ and $b = 10/3$, the regression line is $y = (8/7)x + (10/3)$.

9. Find the least squares regression line for $(-3, 4)$, $(-1, 2)$, $(1, 1)$, and $(3, 0)$.
   Solution:

   $\sum x_i = 0$, $\sum y_i = 7$, $\sum x_i y_i = -13$, $\sum x_i^2 = 20$

   Since $a = -13/20$ and $b = 7/4$, the regression line is $y = -(13/20)x + (7/4)$.

11. Find the least squares regression line for (0, 0), (1, 1), (3, 4), (4, 2), and (5, 5).
Solution:

$$\sum x_i = 13, \quad \sum y_i = 12, \quad \sum x_i y_i = 46, \quad \sum x_i^2 = 51$$

Since $a = 37/43$ and $b = 7/43$, the regression line is $y = (37/43)x + (7/43)$.

13. Find the least squares regression line for (0, 6), (4, 3), (5, 0), (8, -4), and (10, -5).
Solution:

$$\sum x_i = 27, \quad \sum y_i = 0, \quad \sum x_i y_i = -70, \quad \sum x_i^2 = 205$$

Since $a = -175/148$ and $b = 945/148$, the regression line is $y = -(175/148)x + (945/148)$.

15. Find the values of a and b such that the linear model $f(x) = ax + b$ has a minimum sum of the squared errors for the points (-2, 0), (0, 1), and (2, 3).
Solution: The sum of the squared errors is

$$S = (-2a + b)^2 + (b - 1)^2 + (2a + b - 3)^2$$
$$\partial S/\partial a = -4(-2a + b) + 4(2a + b - 3)$$
$$\partial S/\partial b = 2(-2a + b) + 2(b - 1) + 2(2a + b - 3)$$

Setting the first partial derivatives to equal zero produces $16a - 12 = 0$ and $6b - 8 = 0$. Thus, $a = 3/4$ and $b = 4/3$, and we have is $y = (3/4)x + (4/3)$.

17. Find the values of a and b such that the linear model $f(x) = ax + b$ has a minimum sum of the squared errors for the points (0, 4), (1, 1), (1, 3), and (2, 0).
Solution: The sum of the squared errors is

$$S = (b - 4)^2 + (a + b - 1)^2 + (a + b - 3)^2 + (2a + b)^2$$
$$\partial S/\partial a = 2(a + b - 1) + 2(a + b - 3) + 4(2a + b)$$
$$\partial S/\partial b = 2(b - 4) + 2(a + b - 1) + 2(a + b - 3) + 2(2a + b)$$

Setting the first partial derivatives to equal zero produces $12a + 8b = 8$ and $8a + 8b = 16$. Thus, $a = -2$ and $b = 4$, and we have $y = -2x + 4$.

19. Find the least squares regression quadratic for the points (-2, 0), (-1, 0), (0, 1), (1, 2), and (2, 5). Then plot these points and sketch the graph of the least squares quadratic.
Solution:

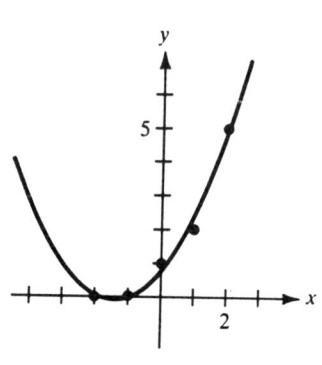

$$\sum x_i = 0, \quad \sum y_i = 8, \quad \sum x_i^2 = 10, \quad \sum x_i^3 = 0$$
$$\sum x_i^4 = 34, \quad \sum x_i y_i = 12, \quad \sum x_i^2 y_i = 22$$

This produces the system $34a + 10c = 22$, $10b = 12$, and $10a + 5c = 8$, which yields $a = 3/7$, $b = 6/5$, and $c = 26/35$, and we have $y = (3/7)x^2 + (6/5)x + (26/35)$.

Section 8.7   369

21. Find the least squares regression quadratic for the points (0, 0), (2, 2), (3, 6), and (4, 12). Then plot these points and sketch the graph of the least squares quadratic.
Solution:

$$\sum x_i = 9, \quad \sum y_i = 20, \quad \sum x_i^2 = 29, \quad \sum x_i^3 = 99$$

$$\sum x_i^4 = 353, \quad \sum x_i y_i = 70, \quad \sum x_i^2 y_i = 254$$

This produces the system

$$353a + 99b + 29c = 254$$
$$99a + 29b + 9c = 70$$
$$29a + 9b + 4c = 20$$

which yields a = 1, b = -1, and c = 0, and we have $y = x^2 - x$.

23. A store manager wants to know the demand for a certain product as a function of price. The daily sales for three prices(x) of the product are $1.00, $1.25, and $1.50 for demands(y) of 450, 375, and 330.
   (a) Find the least squares regression line for these data.
   (b) Estimate the demand when the price is $1.40.
   Solution: (1, 450), (1.25, 375), (1.5, 330)

   (a) $\sum x_i = 3.75, \quad \sum y_i = 1155$

   $\sum x_i y_i = 1413.75, \quad \sum x_i^2 = 4.8125$

   Thus, a = -240, b = 685, and y = -240x + 685.

   (b) When x = 1.4, y = 349.

25. A farmer used four test plots to determine the relationship between wheat yield(y) in bushels per acre and the amount of fertilizer(x) in hundreds of pounds per acre. The results are x-values of 1.0, 1.5, 2.0, and 2.5 for y-values of 32, 41, 48, and 53.
   (a) Find the least squares regression line for these data.
   (b) Estimate the yield for a fertilizer application of 160 pounds per acre.
   Solution: (1, 32), (1.5, 41), (2, 48), (2.5, 53)

   (a) $\sum x_i = 7, \quad \sum y_i = 174$

   $\sum x_i y_i = 322, \quad \sum x_i^2 = 13.5$

   Thus, a = 14, b = 19, and y = 14x + 19.

   (b) When x = 1.6, y = 41.4 bushels per acre.

27. The number of imported cars sold in the United States has increased dramatically in the past several years, as indicated in the table (sales in millions).

Year (x)	1960	1965	1970	1975	1980	1983
Sales (y)	0.50	0.57	1.28	1.59	2.40	2.37

Let $x = 0$ represent the year 1960. (a) Find the least squares regression line for these data. (b) Estimate the sales of imports for the year 1990.

Solution: (0, 0.50), (5, 0.57), (10, 1.28), (15, 1.59), (20, 2.40), (23, 2.37)

(a) $\sum x_i = 73$, $\sum y_i = 8.71$

$\sum x_i y_i = 142.01$ $\sum x_i^2 = 1279$

Thus $a = 3089/33500$, $b = 11048/33500$, and

$$y = \frac{1}{33500}(3089x + 11048)$$

(b) In 1990, $x = 30$ and

$$y = \frac{1}{33500}[3089(30) + 11048] \approx 3.10 \text{ million}$$

29. The following table gives the world population in billions for five different years.

Year (x)	1960	1970	1975	1980	1985
Population (y)	3.0	3.7	4.1	4.5	4.8

Let $x = 0$ represent the year 1975. (a) Find the least squares regression quadratic for these data. (b) Use this quadratic to estimate the world population for the year 1990.

Solution: (-15, 3), (-5, 3.7), (0, 4.1), (5, 4.5), (10, 4.8)

(a) $\sum x_i = -5$, $\sum y_i = 20.1$, $\sum x_i^2 = 375$, $\sum x_i^3 = -2375$

$\sum x_i^4 = 61,875$, $\sum x_i y_i = 7$, $\sum x_i^2 y_i = 1360$

$61875a - 2375b + 375c = 1360$
$-2375a + 375b - 5c = 7$
$375a - 5b + 5c = 20.1$

This yields $a = -3/67900$, $b = 4957/67900$, and $c = 278140/67900$ which implies that quadratic is $y = (1/67900)(-3x^2 + 4957x + 278,140)$.

(b) In 1990, $x = 15$ and

$$y = \frac{1}{67900}[-3(15)^2 + 4957(15) + 278140]$$

≈ 5.2 billion

Section 8.8 Double Integrals and Area in the Plane

1. Evaluate $\int_0^x (2x - y) \, dy$.

Solution:
$$\int_0^x (2x - y) \, dy = \left(2xy - \frac{y^2}{2}\right)\Big]_0^x = \frac{3x^2}{2}$$

3. Evaluate $\int_1^{2y} \frac{y}{x} \, dx$.

Solution:
$$\int_1^{2y} \frac{y}{x} \, dx = y \ln |x| \Big]_1^{2y} = y \ln |2y|$$

5. Evaluate $\int_0^{\sqrt{4-x^2}} x^2 y \, dy$.

Solution:
$$\int_0^{\sqrt{4-x^2}} x^2 y \, dy = \frac{x^2 y^2}{2}\Big]_0^{\sqrt{4-x^2}} = \frac{x^2(4 - x^2)}{2} = \frac{4x^2 - x^4}{2}$$

7. Evaluate $\int_{e^y}^{y} \frac{y \ln x}{x} \, dx$.

Solution:
$$\int_{e^y}^{y} \frac{y \ln x}{x} \, dx = \frac{y(\ln x)^2}{2}\Big]_{e^y}^{y} = \frac{y}{2}[(\ln y)^2 - y^2]$$

9. Evaluate $\int_0^{x^3} y e^{-y/x} \, dy$.

Solution: Using integration by parts, we have
$$\int_0^{x^3} y e^{-y/x} \, dy = -xy e^{-y/x}\Big]_0^{x^3} + x\int_0^{x^3} e^{-y/x} \, dx$$
$$= -x^4 e^{-x^2} - \left[x^2 e^{-y/x}\right]_0^{x^3}$$
$$= -x^4 e^{-x^2} - x^2 e^{-x^2} + x^2$$
$$= x^2(1 - e^{-x^2} - x^2 e^{-x^2})$$

11. Evaluate $\int_0^1 \int_0^2 (x + y) \, dy \, dx$.

Solution:
$$\int_0^1 \int_0^2 (x + y) \, dy \, dx = \int_0^1 \left[xy + \frac{y^2}{2}\right]_0^2 dx$$
$$= \int_0^1 (2x + 2) \, dx$$
$$= (x^2 + 2x)\Big]_0^1 = 3$$

Section 8.8

13. Evaluate $\int_0^4 \int_0^3 xy \, dy \, dx$.

Solution:

$$\int_0^4 \int_0^3 xy \, dy \, dx = \int_0^4 \left[\frac{xy^2}{2}\right]_0^3 dx$$

$$= \frac{9}{2} \int_0^4 x \, dx = \frac{9}{2}\left[\frac{x^2}{2}\right]_0^4 = 36$$

15. Evaluate $\int_1^2 \int_0^4 (x^2 - 2y^2 + 1) \, dx \, dy$.

Solution:

$$\int_1^2 \int_0^4 (x^2 - 2y^2 + 1) \, dx \, dy = \int_1^2 \left[\frac{x^3}{3} - 2xy^2 + x\right]_0^4 dy$$

$$= \int_1^2 \left(\frac{64}{3} - 8y^2 + 4\right) dy$$

$$= \left[\frac{76}{3}y - \frac{8y^3}{3}\right]_1^2$$

$$= \frac{4}{3}(19y - 2y^3)\Big]_1^2 = \frac{20}{3}$$

17. Evaluate $\int_0^1 \int_0^{\sqrt{1-y^2}} (x + y) \, dx \, dy$.

Solution:

$$\int_0^1 \int_0^{\sqrt{1-y^2}} (x + y) \, dx \, dy = \int_0^1 \left[\frac{x^2}{2} + xy\right]_0^{\sqrt{1-y^2}} dy$$

$$= \int_0^1 \left[\frac{1}{2}(1 - y^2) + y\sqrt{1 - y^2}\right] dy$$

$$= \left[\frac{1}{2}\left(y - \frac{y^3}{3}\right) - \frac{1}{2}\left(\frac{2}{3}\right)(1 - y^2)^{2/3}\right]_0^1$$

$$= \frac{1}{2}\left[y - \frac{y^3}{3} - \frac{2}{3}(1 - y^2)^{3/2}\right]_0^1 = \frac{2}{3}$$

19. Evaluate $\int_0^2 \int_0^{\sqrt{4-y^2}} \frac{2}{\sqrt{4 - y^2}} \, dx \, dy$.

Solution:

$$\int_0^2 \int_0^{\sqrt{4-y^2}} \frac{2}{\sqrt{4 - y^2}} \, dx \, dy = \int_0^2 \frac{2x}{\sqrt{4 - y^2}}\Big]_0^{\sqrt{4-y^2}} dy$$

$$= \int_0^2 2 \, dy = 4$$

Section 8.8   373

21. Evaluate $\int_0^2 \int_0^{4-x^2} x^3 \, dy \, dx$.

Solution:
$$\int_0^2 \int_0^{4-x^2} x^3 \, dy \, dx = \int_0^2 x^3 y \Big]_0^{4-x^2} dx$$
$$= \int_0^2 (4x^3 - x^5) \, dx = (x^4 - \frac{x^6}{6})\Big]_0^2$$
$$= \frac{16}{3}$$

23. Evaluate $\int_0^\infty \int_0^\infty e^{-(x+y)/2} \, dy \, dx$.

Solution: Since (for fixed x)
$$\lim_{b \to \infty} \left[-2e^{-(x+y)/2} \right]_0^b = 2e^{-x/2}$$

we have
$$\int_0^\infty \int_0^\infty e^{-(x+y)/2} \, dy \, dx = \int_0^\infty 2e^{-x/2} \, dx$$
$$= \lim_{b \to \infty} -4e^{-x/2} \Big]_0^b = 4$$

25. Sketch the region R whose area is given by
$$\int_0^1 \int_0^2 dy \, dx$$
Then switch the order of integration and show that both orders yield the same area.
Solution:
$$\int_0^1 \int_0^2 dy \, dx = \int_0^1 2 \, dx = 2$$
$$\int_0^2 \int_0^1 dx \, dy = \int_0^2 dy = 2$$

27. Sketch the region R whose area is given by
$$\int_0^1 \int_{2y}^2 dx \, dy$$
Then switch the order of integration and show that both orders yield the same area.
Solution:
$$\int_0^1 \int_{2y}^2 dx \, dy = \int_0^1 (2 - 2y) \, dy = (2y - y^2)\Big]_0^1 = 1$$
$$\int_0^2 \int_0^{x/2} dy \, dx = \int_0^2 \frac{x}{2} \, dx = \frac{x^2}{4}\Big]_0^2 = 1$$

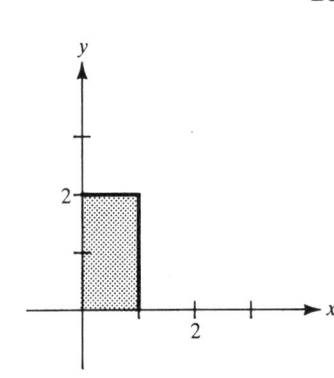

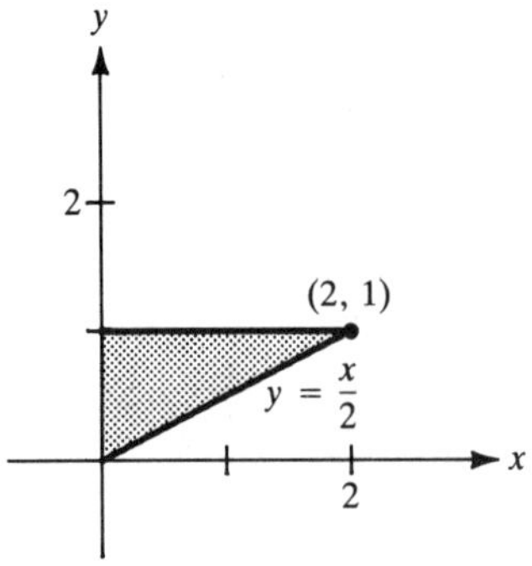

29. Sketch the region R whose area is given by
$$\int_0^2 \int_{x/2}^1 dy\, dx$$
Then switch the order of integration and show that both orders yield the same area.
Solution:
$$\int_0^2 \int_{x/2}^1 dy\, dx = \int_0^2 \left(1 - \frac{x}{2}\right) dx = \left(x - \frac{x^2}{4}\right)\Big]_0^2 = 1$$
$$\int_0^1 \int_0^{2y} dx\, dy = \int_0^1 2y\, dy = y^2 \Big]_0^1 = 1$$

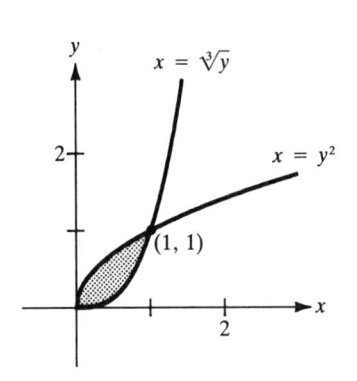

31. Sketch the region R whose area is given by
$$\int_0^1 \int_{y^2}^{\sqrt[3]{y}} dx\, dy$$
Then switch the order of integration and show that both orders yield the same area.
Solution:
$$\int_0^1 \int_{y^2}^{\sqrt[3]{y}} dx\, dy = \int_0^1 (\sqrt[3]{y} - y^2)\, dy$$
$$= \left(\frac{3}{4} y^{4/3} - \frac{y^3}{3}\right)\Big]_0^1 = \frac{5}{12}$$
$$\int_0^1 \int_{x^3}^{\sqrt{x}} dy\, dx = \int_0^1 (\sqrt{x} - x^3)\, dx$$
$$= \left(\frac{2}{3} x^{3/2} - \frac{x^4}{4}\right)\Big]_0^1 = \frac{5}{12}$$

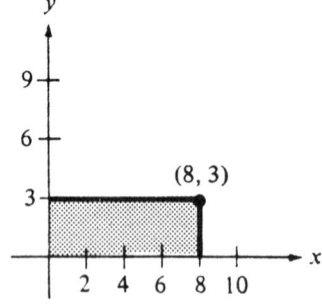

33. Use a double integral to find the area of the specified region.
Solution:
$$A = \int_0^8 \int_0^3 dy\, dx = \int_0^8 3\, dx$$
$$= 3x \Big]_0^8 = 24$$

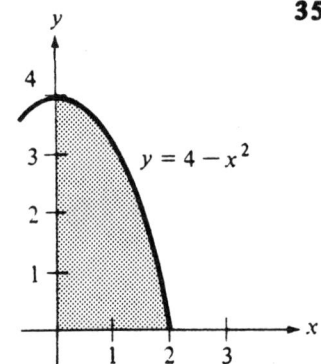

35. Use a double integral to find the area of the specified region.
Solution:
$$A = \int_0^2 \int_0^{4-x^2} dy\, dx = \int_0^2 (4 - x^2)\, dx$$
$$= \left[4x - \frac{x^3}{3}\right]_0^2 = \frac{16}{3}$$

Section 8.8   375

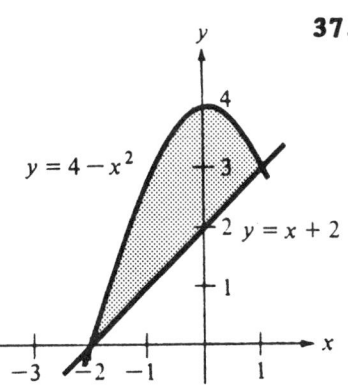

37. Use a double integral to find the area of the specified region.
Solution:

$$A = \int_{-2}^{1} \int_{x+2}^{4-x^2} dy\, dx = \int_{-2}^{1} [(4 - x^2) - (x + 2)]\, dx$$

$$= \int_{-2}^{1} (2 - x - x^2)\, dx$$

$$= \left[2x - \frac{x^2}{2} - \frac{x^3}{3}\right]_{-2}^{1}$$

$$= (2 - \frac{1}{2} - \frac{1}{3}) - (-4 - 2 + \frac{8}{3})$$

$$= \frac{9}{2}$$

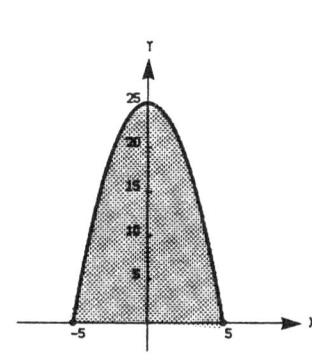

39. Use a double integral to find the area of the region bounded by the graphs of $y = 25 - x^2$ and $y = 0$.
Solution:

$$A = \int_{-5}^{5} \int_{0}^{25-x^2} dy\, dx = \int_{-5}^{5} (25 - x^2)\, dx$$

$$= \left[25x - \frac{x^3}{3}\right]_{-5}^{5}$$

$$= (125 - \frac{125}{3}) - (-125 + \frac{125}{3}) = \frac{500}{3}$$

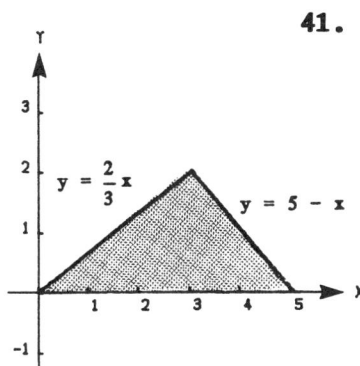

41. Use a double integral to find the area of the region bounded by the graphs of $2x - 3y = 0$, $x + y = 5$, and $y = 0$.
Solution: The point of intersection of the two graphs is given by equating $y = (2/3)x$ and $y = 5 - x$ which yields $x = 3$ and $y = 2$.

$$A = \int_{0}^{2} \int_{3y/2}^{5-y} dx\, dy = \int_{0}^{2} \left[5 - y - \frac{3y}{2}\right] dy$$

$$= \int_{0}^{2} (5 - \frac{5y}{2})\, dy = \left[5y - \frac{5y^2}{4}\right]_{0}^{2} = 5$$

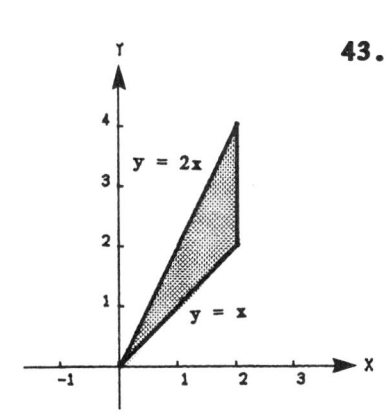

43. Use a double integral to find the area of the region bounded by the graphs of $y = x$, $y = 2x$, and $x = 2$.
Solution:

$$A = \int_{0}^{2} \int_{x}^{2x} dy\, dx = \int_{0}^{2} (2x - x)\, dx$$

$$= \int_{0}^{2} x\, dx$$

$$= \frac{x^2}{2}\Big]_{0}^{2} = 2$$

Section 8.9 Applications of Double Integrals

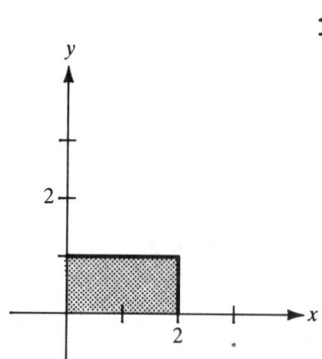

1. Evaluate $\int_0^2 \int_0^1 (1 + 2x + 2y) \, dy \, dx$.

    Solution:
    $$\int_0^2 \int_0^1 (1 + 2x + 2y) \, dy \, dx = \int_0^2 (y + 2xy + y^2)\Big]_0^1 dx$$
    $$= \int_0^2 (2 + 2x) \, dx$$
    $$= (2x + x^2)\Big]_0^2 = 8$$

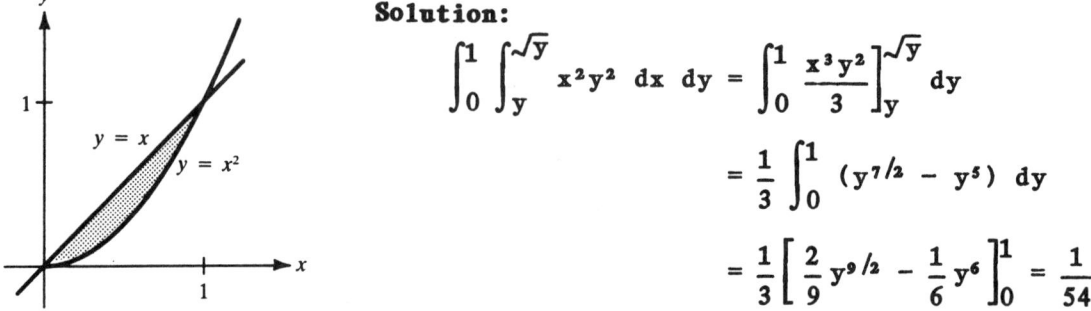

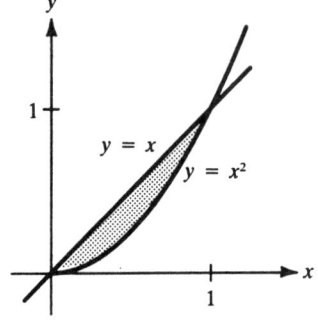

3. Evaluate $\int_0^1 \int_y^{\sqrt{y}} x^2 y^2 \, dx \, dy$.

    Solution:
    $$\int_0^1 \int_y^{\sqrt{y}} x^2 y^2 \, dx \, dy = \int_0^1 \frac{x^3 y^2}{3}\Big]_y^{\sqrt{y}} dy$$
    $$= \frac{1}{3} \int_0^1 (y^{7/2} - y^5) \, dy$$
    $$= \frac{1}{3}\left[\frac{2}{9} y^{9/2} - \frac{1}{6} y^6\right]_0^1 = \frac{1}{54}$$

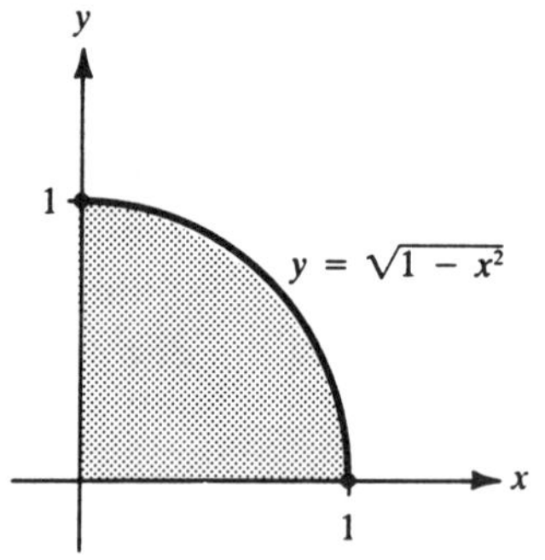

5. Evaluate $\int_0^1 \int_0^{\sqrt{1-x^2}} y \, dy \, dx$.

    Solution:
    $$\int_0^1 \int_0^{\sqrt{1-x^2}} y \, dy \, dx = \int_0^1 \frac{y^2}{2}\Big]_0^{\sqrt{1-x^2}} dx$$
    $$= \frac{1}{2} \int_0^1 (1 - x^2) \, dx$$
    $$= \frac{1}{2}\left(x - \frac{x^3}{3}\right)\Big]_0^1 = \frac{1}{3}$$

7. Set up a double integral for both orders of integration and use the more convenient order to integrate $f(x, y) = xy$ over rectangular region with vertices at (0, 0), (0, 5), (3, 5), and (3, 0).

    Solution:
    $$\int_0^3 \int_0^5 xy \, dy \, dx = \int_0^5 \int_0^3 xy \, dx \, dy$$
    $$\int_0^3 \int_0^5 xy \, dy \, dx = \int_0^3 \frac{xy^2}{2}\Big]_0^5 dx = \int_0^3 \frac{25}{2} x \, dx$$
    $$= \frac{25}{4} x^2\Big]_0^3 = \frac{225}{4}$$

Section 8.9   377

9. Set up a double integral for both orders of integration and use the more convenient order to integrate $f(x, y) = y/(1 + x^2)$ over the region bounded by the graphs of $y = 0$, $y = \sqrt{x}$, and $x = 4$.
Solution:
$$\int_0^4 \int_0^{\sqrt{x}} \frac{y}{1 + x^2} \, dy \, dx = \int_0^2 \int_{y^2}^4 \frac{y}{1 + x^2} \, dx \, dy$$

$$\int_0^4 \int_0^{\sqrt{x}} \frac{y}{1 + x^2} \, dy \, dx = \int_0^4 \frac{y^2}{2(1 + x^2)} \Big]_0^{\sqrt{x}} dx$$

$$= \int_0^4 \frac{x}{2(1 + x^2)} \, dx$$

$$= \frac{1}{4} \ln(1 + x^2) \Big]_0^4$$

$$= \frac{1}{4} \ln 17 \approx 0.708$$

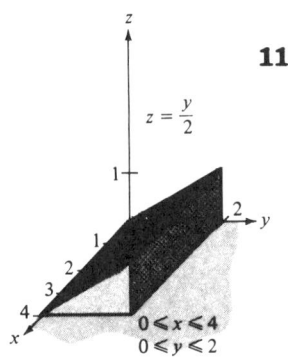

11. Use a double integral to find the volume of the specified solid.
Solution:
$$V = \int_0^2 \int_0^4 \frac{y}{2} \, dx \, dy = \int_0^2 \frac{xy}{2} \Big]_0^4 dy$$

$$= \int_0^2 2y \, dy = y^2 \Big]_0^2 = 4$$

13. Use a double integral to find the volume of the specified solid.
Solution:
$$V = \int_0^2 \int_x^2 (6 - x - y) \, dy \, dx = \int_0^2 \left(6y - xy - \frac{y^2}{2}\right) \Big]_x^2 dx$$

$$= \int_0^2 \left(10 - 2x - 6x + x^2 + \frac{x^2}{2}\right) dx$$

$$= \int_0^2 \left(10 - 8x + \frac{3x^2}{2}\right) dx = \left[10x - 4x^2 + \frac{x^3}{2}\right]_0^2 = 8$$

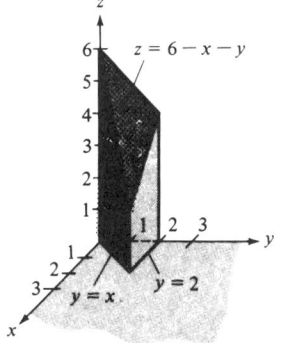

15. Use a double integral to find the volume of the specified solid.
Solution:
$$V = \int_0^6 \int_0^{4-(2x/3)} \left(3 - \frac{x}{2} - \frac{3y}{4}\right) dy \, dx$$

$$= \int_0^6 \left[3y - \frac{xy}{2} - \frac{3y^2}{8}\right]_0^{4-(2x/3)} dx$$

$$= \int_0^6 \left(6 - 2x + \frac{x^2}{6}\right) dx = \left[6x - x^2 + \frac{x^3}{18}\right]_0^6 = 12$$

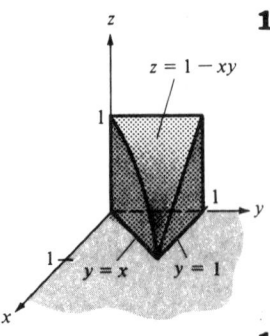

Section 8.9

17. Use a double integral to find the volume of the specified solid.
Solution:

$$V = \int_0^1 \int_0^y (1 - xy) \, dx \, dy = \int_0^1 \left[x - \frac{x^2 y}{2} \right]_0^y dy$$

$$= \int_0^1 \left(y - \frac{y^3}{2} \right) dy = \left[\frac{y^2}{2} - \frac{y^4}{8} \right]_0^1 = \frac{3}{8}$$

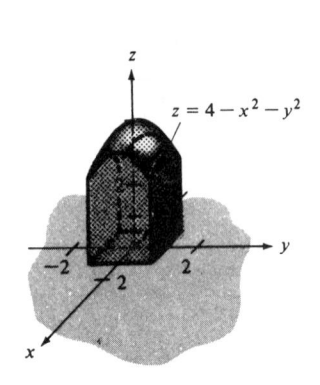

19. Use a double integral to find the volume of the specified solid.
Solution:

$$V = 4 \int_0^1 \int_0^1 (4 - x^2 - y^2) \, dy \, dx$$

$$= 4 \int_0^1 \left[4y - x^2 y - \frac{y^3}{3} \right]_0^1 dx$$

$$= 4 \int_0^1 \left[4 - x^2 - \frac{1}{3} \right] dx$$

$$= 4 \int_0^1 \left(\frac{11}{3} - x^2 \right) dx = 4 \left[\frac{11x}{3} - \frac{x^3}{3} \right]_0^1$$

$$= 4 \left(\frac{11}{3} - \frac{1}{3} \right) = 4 \left(\frac{10}{3} \right) = \frac{40}{3}$$

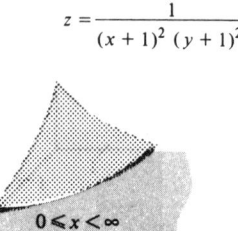

21. Use a double integral to find the volume of the specified solid.
Solution:

$$\int_0^\infty \int_0^\infty \frac{1}{(x+1)^2 (y+1)^2} \, dx \, dy$$

$$= \int_0^\infty \lim_{b \to \infty} \left. \frac{-1}{(x+1)(y+1)^2} \right]_0^b dy$$

$$= \int_0^\infty \frac{1}{(y+1)^2} \, dy = \lim_{b \to \infty} \left. -\frac{1}{y+1} \right]_0^b = 1$$

23. Use a double integral to find the volume of the solid bounded by the graphs of $z = xy$, $z = 0$, $y = 0$, $y = 4$, $x = 0$, and $x = 1$.
Solution:

$$V = \int_0^4 \int_0^1 xy \, dx \, dy = \int_0^4 \left. \frac{x^2 y}{2} \right]_0^1 dy$$

$$= \frac{1}{2} \int_0^4 y \, dy$$

$$= \left. \frac{y^2}{4} \right]_0^4 = 4$$

25. Use a double integral to find the volume of the solid bounded by the graphs of $z = x^2$, $z = 0$, $x = 0$, $x = 2$, $y = 0$, and $y = 4$.
Solution:
$$V = \int_0^2 \int_0^4 x^2 \, dy \, dx = \int_0^2 x^2 y \Big]_0^4 dx$$
$$= \int_0^2 4x^2 \, dx = \frac{4x^3}{3}\Big]_0^2 = \frac{32}{3}$$

27. Find the average value of $f(x, y) = x$ over the rectangular region with vertices $(0, 0)$, $(4, 0)$, $(4, 2)$, and $(0, 2)$.
Solution:
$$\text{Average} = \frac{1}{8} \int_0^4 \int_0^2 x \, dy \, dx = \frac{1}{8} \int_0^4 2x \, dx = \frac{x^2}{8}\Big]_0^4 = 2$$

29. Find the average value of $f(x, y) = x^2 + y^2$ over the rectangular region with vertices $(0, 0)$, $(2, 0)$, $(2, 2)$, and $(0, 2)$.
Solution:
$$\text{Average} = \frac{1}{4} \int_0^2 \int_0^2 (x^2 + y^2) \, dx \, dy$$
$$= \frac{1}{4} \int_0^2 \left[\frac{x^3}{3} + xy^2 \right]_0^2 dy$$
$$= \frac{1}{4} \int_0^2 \left(\frac{8}{3} + 2y^2\right) dy = \frac{1}{4}\left(\frac{8}{3}y + \frac{2}{3}y^3\right)\Big]_0^2 = \frac{8}{3}$$

31. A company sells two products whose demand functions are $x_1 = 500 - 3p_1$ and $x_2 = 750 - 2.4p_2$. Therefore, the total revenue is given by $R = x_1 p_1 + x_2 p_2$. Estimate the average revenue if the price p_1 varies between $50 and $75 and the price p_2 varies between $100 and $150.
Solution:
$$\text{Average:} = \frac{1}{1250} \int_{100}^{150} \int_{50}^{75} [(500 - 3p_1)p_1 + (750 - 2.4p_2)p_2] \, dp_1 \, dp_2$$
$$= \frac{1}{1250} \int_{100}^{150} \int_{50}^{75} [-3p_1^2 + 500p_1 - 2.4p_2^2 + 750p_2] \, dp_1 \, dp_2$$
$$= \frac{1}{1250} \int_{100}^{150} \left[-p_1^3 + 250p_1^2 - 2.4p_1 p_2^2 + 750 p_1 p_2 \right]_{50}^{75} dp_2$$
$$= \frac{1}{1250} \int_{100}^{150} [484,375 - 60p_2^2 + 18750p_2] \, dp_2$$
$$= \frac{1}{1250} \left[484,375p_2 - 20p_2^3 + 9375p_2^2 \right]_{100}^{150}$$

$$= \$75,125$$

33. For a particular company, the Cobb-Douglas annual production function is $f(x, y) = 100x^{0.6}y^{0.4}$. Estimate the average annual production level if the number of units of labor varies between 200 and 250 units and the number of units of capital varies between 300 and 325 units.
Solution:

$$\text{Average} = \frac{1}{1250} \int_{300}^{325} \int_{200}^{250} 100x^{0.6}y^{0.4} \, dx \, dy$$

$$= \frac{1}{1250} \int_{300}^{325} (100y^{0.4}) \frac{x^{1.6}}{1.6} \Big]_{200}^{250} dy$$

$$= \frac{128,844.1}{1250} \int_{300}^{325} y^{0.4} \, dy$$

$$= 103.0753 \left[\frac{y^{1.4}}{1.4} \right]_{300}^{325} \approx 25,645.24$$

● Review Exercises for Chapter 8

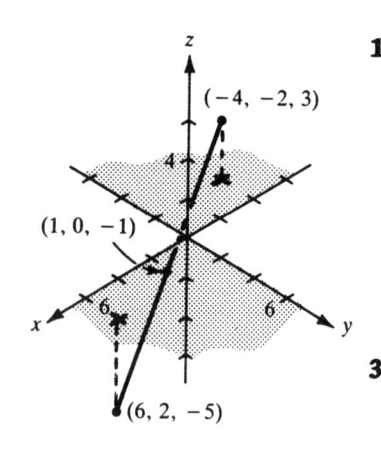

1. (a) Plot the points $(-4, -2, 3)$ and $(6, 2, -5)$, (b) find the distance between these points, and (c) find the midpoint of the line segment joining them.
Solution:
(a) See graph.
(b) $d = \sqrt{(6 + 4)^2 + (2 + 2)^2 + (-5 - 3)^2}$
$= \sqrt{100 + 16 + 64} = \sqrt{180} = 6\sqrt{5}$

(c) Midpoint $= (\frac{6 - 4}{2}, \frac{2 - 2}{2}, \frac{-5 + 3}{2}) = (1, 0, -1)$

3. Find the standard form of the equation of a sphere with center $(1, 0, -3)$ and radius 3.
Solution:
$(x - 1)^2 + (y - 0)^2 + (z - (-3))^2 = 3^2$
$(x - 1)^2 + y^2 + (z + 3)^2 = 9$

5. Find the center and radius of the sphere
$x^2 + y^2 + z^2 + 8x - 4y + 2z + 5 = 0$.
Solution:
$(x + 4)^2 + (y - 2)^2 + (z + 1)^2 = 16$

Center: $(-4, 2, -1)$ Radius: 4

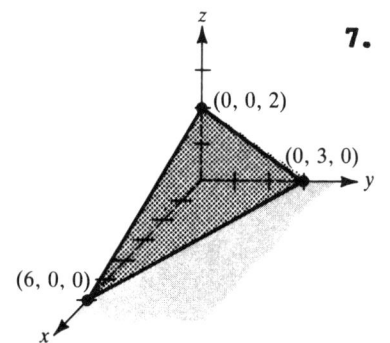

7. Sketch the graph of the plane $x + 2y + 3z = 6$.
Solution:
x-intercept: $(6, 0, 0)$
y-intercept: $(0, 3, 0)$
z-intercept: $(0, 0, 2)$

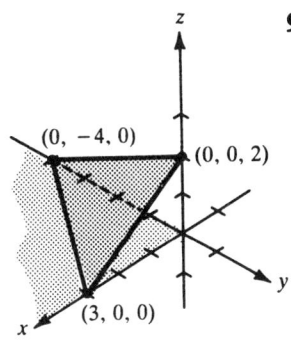

Review Exercises for Chapter 8   381

9. Sketch the graph of the plane $4x - 3y + 6z = 12$.
   Solution:
   x-intercept: $(3, 0, 0)$
   y-intercept: $(0, -4, 0)$
   z-intercept: $(0, 0, 2)$

11. When $x^2 + y^2 + z^2 - 2x + 4y - 6z + 5 = 0$, identify the surface.
    Solution: The graph is a sphere whose standard equation is
    $$(x - 1)^2 + (y + 2)^2 + (z - 3)^2 = 9$$

13. When $(x^2/16) + (y^2/9) + z^2 = 1$, identify the surface.
    Solution: The graph is an ellipsoid.

15. When $(x^2/16) - (y^2/9) + z^2 = -1$, identify the surface.
    Solution: The graph is a hyperboloid of two sheets whose standard equation is
    $$\frac{y^2}{9} - \frac{x^2}{16} - z^2 = 1$$

17. When $-4x^2 + y^2 + z^2 = 4$, identify the surface.
    Solution: The graph is a hyperboloid of one sheet whose standard equation is
    $$\frac{y^2}{4} + \frac{z^2}{4} - x^2 = 1$$

19. When $z = \sqrt{x^2 + y^2}$, identify the surface.
    Solution: The graph is the top half of a circular cone whose standard equation is
    $$x^2 + y^2 - z^2 = 0$$

21. Describe the region R in the xy-plane that corresponds to the domain of $f(x, y) = x^2 + y^2$, and find the range of this function.
    Solution: The domain is the set of all points in the xy-plane and the range is the interval $[0, \infty)$.

23. Describe the region R in the xy-plane that corresponds to the domain of $f(x, y) = \ln(1 - x^2 - y^2)$ and find the range of this function.
    Solution: The domain is the set of all points inside the circle $x^2 + y^2 = 1$ and the range is $(-\infty, \infty)$.

25. Find the first partial derivatives of
    $$f(x, y) = x\sqrt{y} + 3x - 2y$$
    Solution:
    $f_x(x, y) = \sqrt{y} + 3,\qquad f_y(x, y) = \dfrac{x}{2\sqrt{y}} - 2$

27. Find the first partial derivatives of
$$g(x, y) = \ln\sqrt{2x + 3y} = \frac{1}{2}\ln(2x + 3y)$$

Solution:

$$g_x(x, y) = \frac{1}{2}\left(\frac{2}{2x + 3y}\right) = \frac{1}{2x + 3y}$$

$$g_y(x, y) = \frac{1}{2}\left(\frac{3}{2x + 3y}\right) = \frac{3}{2(2x + 3y)}$$

29. Find the first partial derivatives of $z = xe^y + ye^x$.
Solution:

$$\frac{\partial z}{\partial x} = e^y + ye^x, \qquad \frac{\partial z}{\partial y} = xe^y + e^x$$

31. Find the first partial derivatives of
$$g(x, y) = \frac{xy}{x^2 + y^2}$$

Solution:

$$g_x(x, y) = \frac{(x^2 + y^2)y - xy(2x)}{(x^2 + y^2)^2}$$

$$= \frac{y^3 - x^2 y}{(x^2 + y^2)^2} = \frac{y(y^2 - x^2)}{(x^2 + y^2)^2}$$

$$g_y(x, y) = \frac{(x^2 + y^2)x - xy(2y)}{(x^2 + y^2)^2}$$

$$= \frac{x^3 - xy^2}{(x^2 + y^2)^2} = \frac{x(x^2 - y^2)}{(x^2 + y^2)^2}$$

33. Find the first partial derivatives of $w = 2\sqrt{xyz}$.
Solution:

$$\frac{\partial w}{\partial x} = 2\left[\frac{1}{2}(xyz)^{-1/2} yz\right] = \frac{yz}{\sqrt{xyz}}$$

$$\frac{\partial w}{\partial y} = 2\left[\frac{1}{2}(xyz)^{-1/2} xz\right] = \frac{xz}{\sqrt{xyz}}$$

$$\frac{\partial w}{\partial z} = 2\left[\frac{1}{2}(xyz)^{-1/2} xy\right] = \frac{xy}{\sqrt{xyz}}$$

35. Find all second partial derivatives and verify that the second mixed partials are equal for $f(x, y) = 3x^2 - xy + 2y^3$.
Solution: The first partial derivatives are

$$f_x(x, y) = 6x - y \quad \text{and} \quad f_y(x, y) = -x + 6y^2$$

and the second partial derivatives are

$$f_{xx}(x, y) = 6, \qquad f_{yx}(x, y) = -1$$
$$f_{xy}(x, y) = -1, \qquad f_{yy}(x, y) = 12y$$

Therefore, $f_{xy}(x, y) = f_{yx}(x, y) = -1$.

37. Show that $z = x^2 - y^2$ satisfies the Laplace equation

$$\frac{\partial^2 z}{\partial x^2} + \frac{\partial^2 z}{\partial y^2} = 0.$$

Solution:

$$\frac{\partial z}{\partial x} = 2x, \qquad \frac{\partial z}{\partial y} = -2y$$

$$\frac{\partial^2 z}{\partial x^2} = 2, \qquad \frac{\partial^2 z}{\partial y^2} = -2$$

Therefore, $\frac{\partial^2 z}{\partial x^2} + \frac{\partial^2 z}{\partial y^2} = 2 - 2 = 0.$

39. Locate and classify any extrema of

$$f(x, y) = x^3 - 3xy + y^2$$

Solution: The first partial derivatives of f

$$f_x(x, y) = 3x^2 - 3y$$
$$f_y(x, y) = -3x + 2y$$

are zero at (0, 0) and (3/2, 9/4). The second partial derivatives are

$$f_{xx}(x, y) = 6x, \quad f_{yy}(x, y) = 2, \quad f_{xy}(x, y) = -3$$

The point (0, 0, 0) is a saddle point since

$$f_{xx}(0, 0)f_{yy}(0, 0) - [f_{xy}(0, 0)]^2 = -9 < 0$$

The point (3/2, 9/4, -27/16) is relative minimum since $f_{xx}(3/2, 9/4) = 9 > 0$ and

$$f_{xx}(\tfrac{3}{2}, \tfrac{9}{4}) f_{yy}(\tfrac{3}{2}, \tfrac{9}{4}) - [f_{xy}(\tfrac{3}{2}, \tfrac{9}{4})]^2 = 9 > 0$$

41. Locate and classify any extrema of

$$f(x, y) = 2x^2 + 6xy + 9y^2 + 8x + 14$$

Solution: The first partial derivatives of f

$$f_x(x, y) = 4x + 6y + 8$$
$$f_y(x, y) = 6x + 18y$$

are zero at (-4, 4/3). The second partial derivatives are

$$f_{xx}(x, y) = 4, \quad f_{yy}(x, y) = 18, \quad f_{xy}(x, y) = 6$$

Since $f_{xx}(-4, 4/3) = 4 > 0$ and

$$f_{xx}(-4, \tfrac{4}{3}) f_{yy}(-4, \tfrac{4}{3}) - [f_{xy}(-4, \tfrac{4}{3})]^2 = 36 > 0$$

the point (-4, 4/3, -2) is a relative minimum.

43. Using Lagrange multipliers, locate any extrema of $z = x^2y$ subject to the constraint $x + 2y = 2$.
    Solution:
    $$F(x, y, \lambda) = x^2y + \lambda(x + 2y - 2)$$

    $$\left. \begin{array}{l} F_x(x, y, \lambda) = 2xy + \lambda = 0 \\ F_y(x, y, \lambda) = x^2 + 2\lambda = 0 \\ F_\lambda(x, y, \lambda) = x + 2y - 2 = 0, \end{array} \right\} \begin{array}{l} 4xy = x^2 \\ \\ y = (2 - x)/2 \end{array}$$

    Thus, $x = 0$ or $x = 4/3$ and the corresponding y-values are $y = 1$ or $y = 1/3$. This implies that the extrema occur at $(0, 1, 0)$ and $(4/3, 1/3, 16/27)$.

45. The production function for a manufacturer is
    $$f(x, y) = 4x + xy + 2y$$

    Assume that the total amount available for labor x and capital y is $2000 and that units of labor and capital cost $20 and $4, respectively. Find the maximum production level for this manufacturer.
    Solution: Maximize $f(x, y) = 4x + xy + 2y$, subject to the constraint $20x + 4y = 2000$.

    $$F(x, y, \lambda) = 4x + xy + 2y + \lambda(20x + 4y - 2000)$$

    $$\left. \begin{array}{l} F_x(x, y, \lambda) = 4 + y + 20\lambda = 0 \\ F_y(x, y, \lambda) = x + 2 + 4\lambda = 0 \\ F_\lambda(x, y, \lambda) = 20x + 4y - 2000, \end{array} \right\} \begin{array}{l} 4 + y = 5(x + 2) \\ y = 5x + 6 \\ y = 500 - 5x \end{array}$$

    Thus, $x = 49.4$ and $y = 5(49.4) + 6 = 253$, which implies that the maximum production level is

    $$f(49.4, 253) \approx 13202$$

47. Find the least squares regression line for the points $(1, 5)$, $(2, 4)$, $(3, 2)$ and $(5, 1)$. Plot the points and sketch the least squares regression line on the same coordinate axes.
    Solution:

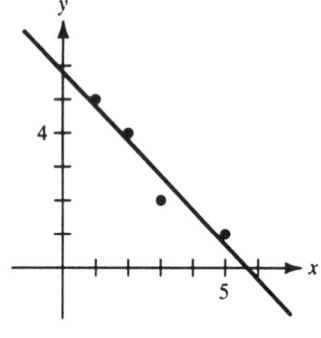

    $$\sum x_i = 11, \quad \sum y_i = 12, \quad \sum x_i^2 = 39, \quad \sum x_iy_i = 24$$

    Therefore, we have

    $$a = \frac{4(24) - 11(12)}{4(39) - (11)^2} = -\frac{36}{35}$$

    $$b = \frac{1}{4}\left[12 - \left(-\frac{36}{35}\right)(11)\right] = \frac{204}{35}$$

    and the least squares regression line is

    $$y = \frac{1}{35}(-36x + 204) = \frac{12}{35}(-3x + 17)$$

Review Exercises for Chapter 8

49. Find the least squares regression quadratic for the points (1, 1), (3, 2), (4, 4), and (5, 7). Plot the points and sketch the least squares quadratic on the same coordinate axes.

Solution:

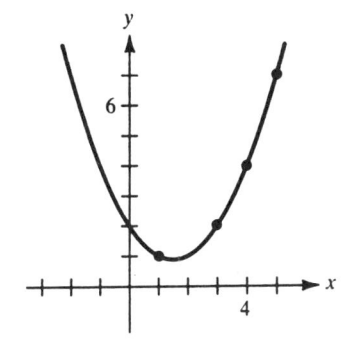

$$\sum x_i = 13, \qquad \sum y_i = 14, \qquad \sum x_i^2 = 51$$

$$\sum x_i y_i = 58, \qquad \sum x_i^3 = 217, \qquad \sum x_i^2 y_i = 258$$

$$\sum x_i^4 = 963$$

The system

$$963a + 217b + 51c = 258$$
$$217a + 51b + 13c = 58$$
$$51a + 13b + 4c = 14$$

has solutions $a = 1/2$, $b = -3/2$, and $c = 2$. Therefore, the least squares quadratic is

$$y = \frac{1}{2}x^2 - \frac{3}{2}x + 2$$

51. Evaluate $\int_0^1 \int_0^{1+x} (3x + 2y)\, dy\, dx$.

Solution:

$$\int_0^1 \int_0^{1+x} (3x + 2y)\, dy\, dx = \int_0^1 (3xy + y^2)\Big]_0^{1+x} dx$$

$$= \int_0^1 [3x(1 + x) + (1 + x)^2]\, dx$$

$$= \int_0^1 (4x^2 + 5x + 1)\, dx$$

$$= \left[\frac{4x^3}{3} + \frac{5x^2}{2} + x\right]_0^1$$

$$= \frac{4}{3} + \frac{5}{2} + 1 = \frac{29}{6}$$

53. Evaluate $\int_1^2 \int_1^{2y} \frac{x}{y^2}\, dx\, dy$.

Solution:

$$\int_1^2 \int_1^{2y} \frac{x}{y^2}\, dx\, dy = \int_1^2 \frac{x^2}{2y^2}\Big]_1^{2y} dy = \int_1^2 \left[\frac{4y^2}{2y^2} - \frac{1}{2y^2}\right] dy$$

$$= \int_1^2 \left(2 - \frac{1}{2}y^{-2}\right) dy = \left[2y + \frac{1}{2y}\right]_1^2$$

$$= \left(4 + \frac{1}{4}\right) - \left(2 + \frac{1}{2}\right) = \frac{7}{4}$$

55. Evaluate $\int_0^3 \int_0^{\sqrt{9-x^2}} 4x \, dy \, dx$.

Solution:

$$\int_0^3 \int_0^{\sqrt{9-x^2}} 4x \, dy \, dx = \int_0^3 4xy \Big]_0^{\sqrt{9-x^2}} dx$$

$$= \int_0^3 4x\sqrt{9-x^2} \, dx$$

$$= \frac{4}{-2} \int_0^3 (9-x^2)^{1/2}(-2x) \, dx$$

$$= -2\left(\frac{2}{3}\right)(9-x^2)^{3/2} \Big]_0^3$$

$$= -\frac{4}{3}(0^{3/2} - 9^{3/2}) = -\frac{4}{3}(0 - 27) = 36$$

57. Evaluate $\int_{-2}^{4} \int_{y^2/4}^{(4+y)/2} (x-y) \, dx \, dy$.

Solution:

$$\int_{-2}^{4} \int_{y^2/4}^{(4+y)/2} (x-y) \, dx \, dy$$

$$= \int_{-2}^{4} \left(\frac{x^2}{2} - xy\right)\Big]_{y^2/4}^{(4+y)/2} dy$$

$$= \int_{-2}^{4} \left\{\left[\frac{(4+y)^2}{8} - \frac{(4+y)y}{2}\right] - \left[\frac{y^4}{32} - \frac{y^3}{4}\right]\right\} dy$$

$$= \left[\frac{(4+y)^3}{24} - y^2 - \frac{y^3}{6} - \frac{y^5}{160} + \frac{y^4}{16}\right]_{-2}^{4}$$

$$= \left(\frac{64}{3} - 16 - \frac{32}{3} - \frac{32}{5} + 16\right) - \left(\frac{1}{3} - 4 + \frac{4}{3} + \frac{1}{5} + 1\right)$$

$$= \frac{27}{5}$$

59. Given a triangle with vertices (0, 0), (3, 0) and (0, 1), write the limits to the double integral

$$\iint_R f(x, y) \, dA$$

for both orders of integration. Compute the area of R by letting $f(x, y) = 1$ and integrating.

Solution:

$$A = \int_0^3 \int_0^{(3-x)/3} dy \, dx = \int_0^1 \int_0^{3-3y} dx \, dy$$

$$\int_0^1 \int_0^{3-3y} dx \, dy = \int_0^1 (3 - 3y) \, dy = \left[3y - \frac{3y^2}{2}\right]_0^1$$

$$= 3 - \frac{3}{2} = \frac{3}{2}$$

Review Exercises for Chapter 8 387

61. Given a region bounded by the graphs of $y = 9 - x^2$ and $y = 5$, write the limits to the double integral

$$\iint_R f(x, y) \, dA$$

for both orders of integration. Compute the area of R by letting $f(x, y) = 1$ and integrating.
Solution:

$$A = \int_{-2}^{2} \int_{5}^{9-x^2} dy \, dx = \int_{5}^{9} \int_{-\sqrt{9-y}}^{\sqrt{9-y}} dx \, dy$$

$$\int_{-2}^{2} \int_{5}^{9-x^2} dy \, dx = \int_{-2}^{2} [(9 - x^2) - 5] \, dx$$

$$= \int_{-2}^{2} (4 - x^2) \, dx = \left(4x - \frac{x^3}{3}\right)\Big]_{-2}^{2}$$

$$= \left(8 - \frac{8}{3}\right) - \left(-8 + \frac{8}{3}\right) = \frac{32}{3}$$

63. Given a region bounded by the graphs of $x = y + 3$ and $x = y^2 + 1$, write the limits to the double integral

$$\iint_R f(x, y) \, dA$$

for both orders of integration. Compute the area of R by letting $f(x, y) = 1$ and integrating.
Solution: The points of intersection of $x = y + 3$ and $x = y^2 + 1$ occur at $(2, -1)$ and $(5, 2)$.

$$A = \int_{1}^{2} \int_{-\sqrt{x-1}}^{\sqrt{x-1}} dy \, dx + \int_{2}^{5} \int_{x-3}^{\sqrt{x-1}} dy \, dx = \int_{-1}^{2} \int_{y^2+1}^{y+3} dx \, dy$$

$$\int_{-1}^{2} \int_{y^2+1}^{y+3} dx \, dy = \int_{-1}^{2} [(y + 3) - (y^2 + 1)] \, dy$$

$$= \int_{-1}^{2} (-y^2 + y + 2) \, dy$$

$$= \left(-\frac{y^3}{3} + \frac{y^2}{2} + 2y\right)\Big]_{-1}^{2}$$

$$= \left(-\frac{8}{3} + 2 + 4\right) - \left(\frac{1}{3} + \frac{1}{2} - 2\right) = \frac{9}{2}$$

65. Use an appropriate double integral to find the volume of a solid bounded by the graphs of $z = (xy)^2$, $z = 0$, $y = 0$, $y = 4$, $x = 0$, and $x = 4$.
Solution:

$$V = \int_{0}^{4} \int_{0}^{4} (xy)^2 \, dy \, dx = \int_{0}^{4} \int_{0}^{4} x^2 y^2 \, dy \, dx$$

$$= \int_{0}^{4} \frac{x^2 y^3}{3}\Big]_{0}^{4} dx = \int_{0}^{4} \frac{64x^2}{3} \, dx = \frac{64x^3}{9}\Big]_{0}^{4} = \frac{4096}{9}$$

PRACTICE TEST FOR CHAPTER 8

1. Find the distance between the points $(3, -7, 2)$ and $(5, 11, -6)$ and find the midpoint of the line segment joining the two points.

2. Find the standard form of the equation of the sphere whose center is $(1, -3, 0)$ and radius is $\sqrt{5}$.

3. Find the center and radius of the sphere whose equation is

$$x^2 + y^2 + z^2 - 4x + 2y + 8z = 0$$

4. Sketch the graph of the plane:

   (a) $3x + 8y + 6z = 24$
   (b) $y = 2$

5. Identify the surface:

   (a) $\dfrac{x^2}{16} + \dfrac{y^2}{4} - \dfrac{z^2}{9} = 1$

   (b) $z = \dfrac{x^2}{25} + y^2$

6. Find the domain of the function:

   (a) $f(x, y) = \ln(3 - x - y)$

   (b) $f(x, y) = \dfrac{1}{x^2 + y^2}$

7. Find the first partial derivatives of

$$f(x, y) = 3x^2 + 9xy^2 + 4y^3 - 3x - 6y + 1$$

8. Find the first partial derivatives of

$$f(x, y) = \ln(x^2 + y^2 + 5)$$

9. Find the first partial derivatives of

$$f(x, y) = x^2 y^3 \sqrt{z}$$

10. Find the second partial derivatives of

$$z = \dfrac{x}{x^2 + y^2}$$

11. Find the relative extrema of

$$f(x, y) = 3x^2 + 4y^2 - 6x + 16y - 4$$

12. Find the relative extrema of

$$f(x, y) = 4xy - x^4 - y^4$$

13. Use Lagrange multipliers to find the minimum of $f(x, y) = xy$ subject to the constraint $4x - y = 16$.

14. Use Lagrange multipliers to find the minimum of $f(x, y) = x^2 - 16x + y^2 - 8y + 12$ subject to the constraint $x + y = 4$.

15. Find the least squares regression line for the points $(-3, 7)$, $(1, 5)$, $(8, -2)$, and $(4, 4)$.

16. Find the least squares regression quadratic for the points $(-5, 8)$, $(-1, 2)$, $(1, 3)$, and $(5, 5)$.

17. Evaluate $\int_0^3 \int_0^{\sqrt{x}} xy^3 \, dy \, dx$.

18. Evaluate $\int_{-1}^2 \int_0^{3y} (x^2 - 4xy) \, dx \, dy$.

19. Set up a double integral to find the area of the indicated region.

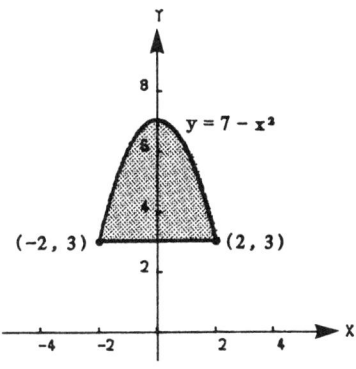

20. Set up a double integral to find the area of the indicated region.

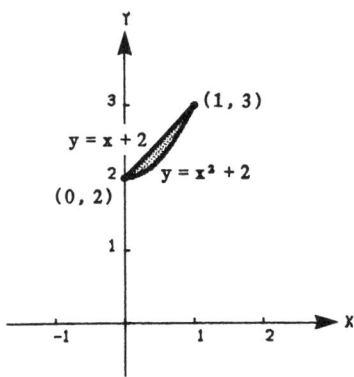

Chapter 9 Taylor Polynomials and Series

Section 9.1 Sequences

1. Write the first five terms of $\{2^n\}$.
   Solution:

   $$2, 4, 8, 16, 32, \ldots$$

3. Write the first five terms of $\{(-1/2)^n\}$.
   Solution:

   $$-\frac{1}{2}, \frac{1}{4}, -\frac{1}{8}, \frac{1}{16}, -\frac{1}{32}, \ldots$$

5. Write the first five terms of $\{3^n/n!\}$.
   Solution:

   $$3, \frac{9}{2}, \frac{27}{6}, \frac{81}{24}, \frac{243}{120}, \ldots$$

7. Write the first five terms of $\{(-1)^n/n^2\}$.
   Solution:

   $$-1, \frac{1}{4}, -\frac{1}{9}, \frac{1}{16}, -\frac{1}{25}, \ldots$$

9. Write an expression for the nth term of the sequence $1, 4, 7, 10, \ldots$.
   Solution:

   $$a_n = 3n - 2$$

11. Write an expression for the nth term of the sequence $-1, 2, 7, 14, 23, \ldots$.
    Solution:

    $$a_n = n^2 - 2$$

13. Write an expression for the nth term of the sequence $2/3, 3/4, 4/5, 5/6, \ldots$.
    Solution:

    $$a_n = \frac{n+1}{n+2}$$

15. Write an expression for the nth term of the sequence $2, -1, 1/2, -1/4, 1/8, \ldots$.
    Solution:

    $$a_n = \frac{(-1)^{n-1}}{2^{n-2}}$$

17. Write an expression for the nth term of the sequence $2, [1 + (1/2)], [1 + (1/3)], [1 + (1/4)], \ldots$.
    Solution:

    $$a_n = 1 + \frac{1}{n} = \frac{n+1}{n}$$

Section 9.1

19. Write an expression for the nth term of the sequence
1, -1, 1, -1, 1, -1, ...
Solution:
$$a_n = (-1)^{n-1}$$

21. Write an expression for the nth term of the sequence
1, 1/2, 1/6, 1/24, 1/120, ...
Solution:
$$a_n = 1/n!$$

23. Determine the convergence or divergence of $a_n = 5/n$. If the sequence converges, find its limit.
Solution: This sequence converges since
$$\lim_{n \to \infty} = \frac{5}{n} = 0$$

25. Determine the convergence or divergence of $a_n = (n + 1)/n$. If the sequence converges, find its limit.
Solution: This sequence converges since
$$\lim_{n \to \infty} = \frac{n + 1}{n} = 1$$

27. Determine the convergence or divergence of $a_n = (-1)^n [n/(n + 1)]$. If the sequence converges, find its limit.
Solution: This sequence diverges since
$$\lim_{n \to \infty} (-1)^n \frac{n}{n + 1} \quad \text{does not exist.}$$

29. Determine the convergence or divergence of $a_n = (3n^2 - n + 4)/(2n^2 + 1)$. If the sequence converges, find its limit.
Solution: This sequence converges since
$$\lim_{n \to \infty} = \frac{3n^2 - n + 4}{2n^2 + 1} = \frac{3}{2}$$

31. Determine the convergence or divergence of $a_n = (n^2 - 1)/(n + 1)$. If the sequence converges, find its limit.
Solution: This sequence diverges since
$$\lim_{n \to \infty} \frac{n^2 - 1}{n + 1} = \infty$$

33. Determine the convergence or divergence of $a_n = [1 + (-1)^n]/n$. If the sequence converges, find its limit.
Solution: This sequence converges since
$$\lim_{n \to \infty} \frac{1 + (-1)^n}{n} = 0$$

35. Determine the convergence or divergence of $a_n = 3 - [1/(2^n)]$. If the sequence converges, find its limit.
    Solution: This sequence converges since
    $$\lim_{n \to \infty} \left(3 - \frac{1}{2^n}\right) = 3$$

37. Determine the convergence or divergence of $a_n = 3^n/4^n$. If the sequence converges, find its limit.
    Solution: This sequence converges since
    $$\lim_{n \to \infty} \frac{3^n}{4^n} = \lim_{n \to \infty} \left(\frac{3}{4}\right)^n = 0$$

39. Determine the convergence or divergence of $a_n = [(n+1)!]/n!$. If the sequence converges, find its limit.
    Solution: This sequence diverges since
    $$\lim_{n \to \infty} \frac{(n+1)!}{n!} = \lim_{n \to \infty} \frac{(n+1)n!}{n!}$$
    $$= \lim_{n \to \infty} (n+1) = \infty$$

41. Consider the sequence $\{A_n\}$, whose nth term is given by $A_n = P[1 + (r/12)]^n$, where P is the principal. A_n is the amount at compound interest after n months, and r is the annual percentage rate. (a) Is $\{A_n\}$ a convergent sequence? (b) Find the first ten terms of the sequence if P = \$9,000 and r = 0.115.
    Solution:

    (a) $\{A_n\}$ is not a convergent sequence since
    $$\lim_{n \to \infty} P\left[1 + \frac{r}{12}\right]^n = \infty$$

    (b) P = 9000, r = 0.115, $A_n = 9000[1 + (0.115/12)]^n$

    9086.25, 9173.33, 9261.24, 9349.99, 9439.60, 9530.06, 9621.39, 9713.59, 9806.68, 9900.66, ...

43. The sum of the first n positive integers is given by $S_n = [n(n+1)]/2$, n = 1, 2, 3, ... (a) Compute the first five terms of this sequence and verify that each term is the correct sum, and (b) find the sum of the first fifty positive integers.
    Solution:

    (a) $S_1 = 1(1+1)/2 = 1$
        $S_2 = 2(2+1)/2 = 3 = 1 + 2$
        $S_3 = 3(3+1)/2 = 6 = 1 + 2 + 3$
        $S_4 = 4(4+1)/2 = 10 = 1 + 2 + 3 + 4$
        $S_5 = 5(5+1)/2 = 15 = 1 + 2 + 3 + 4 + 5$

    (b) $S_{50} = 50(50+1)/2 = 1275$

Section 9.1

45. A government program that currently costs taxpayers $2.5 billion per year is to be cut back by 20% per year. (a) Write an expression for the amount budgeted for this program after n years, (b) compute the budgets for the first four years, and (c) determine the convergence or divergence of the sequence of reduced budgets. If the sequence converges, find its limit.

Solution:

(a) $A_1 = 2.5 - 0.2(2.5) = 2.5(0.8)$
$A_2 = A_1 - 0.2A_1 = 0.8A_1 = 2.5(0.8)^2$
$A_3 = A_2 - 0.2A_2 = 0.8A_2 = 2.5(0.8)^3$
$\vdots$
$A_n = 2.5(0.8)^n$

(b) $A_1 = \$2$ billion
$A_2 = \$1.6$ billion
$A_3 = \$1.28$ billion
$A_4 = \$1.024$ billion

(c) This sequence converges since

$$\lim_{n \to \infty} 2.5(0.8)^n = 0$$

47. Consider an idealized population with the characteristic that each population member produces 1 offspring at the end of every time period. If each population member has a lifespan of 3 time periods and the population begins with 10 newborn members, then the following table gives the population during the first five time periods.

Age bracket	Time period				
	1	2	3	4	5
0-1	10	10	20	40	70
1-2		10	10	20	40
2-3			10	10	20
Total	10	20	40	70	130

The sequence for the total population has the property that $S_n = S_{n-1} + S_{n-2} + S_{n-3}$, $n > 3$. Find the total population during the next five time periods.

Solution:

$S_6 = 130 + 70 + 40 = 240$
$S_7 = 240 + 130 + 70 = 440$
$S_8 = 440 + 240 + 130 = 810$
$S_9 = 810 + 440 + 240 = 1490$
$S_{10} = 1490 + 810 + 440 = 2740$

Section 9.2 Series and Convergence

1. Find the first five terms of
$$\sum_{n=1}^{\infty} \frac{1}{n^2} = 1 + \frac{1}{4} + \frac{1}{9} + \frac{1}{16} + \frac{1}{25} + \ldots$$
   Solution:
   $S_1 = 1$
   $S_2 = 5/4 = 1.25$
   $S_3 = 49/36 \approx 1.361$
   $S_4 = 205/144 \approx 1.424$
   $S_5 = 5269/3600 \approx 1.464$

3. Find the first five terms of
$$\sum_{n=1}^{\infty} \frac{3}{2^{n-1}}$$
   Solution:
   $S_1 = 3$
   $S_2 = 9/2 = 4.5$
   $S_3 = 21/4 = 5.25$
   $S_4 = 45/8 = 5.625$
   $S_5 = 93/16 = 5.8125$

5. Verify that the following infinite series diverges.
$$\sum_{n=1}^{\infty} \frac{n}{n+1} = \frac{1}{2} + \frac{2}{3} + \frac{3}{4} + \frac{4}{5} + \ldots$$
   Solution: This series diverges by nth-Term Test since
$$\lim_{n \to \infty} \frac{n}{n+1} = 1 \neq 0$$

7. Verify that the following infinite series diverges.
$$\sum_{n=1}^{\infty} \frac{n^2}{n^2+1} = \frac{1}{2} + \frac{4}{5} + \frac{9}{10} + \frac{16}{17} + \ldots$$
   Solution: This series diverges by nth-Term Test since
$$\lim_{n \to \infty} \frac{n^2}{n^2+1} = 1 \neq 0$$

9. Verify that the following infinite series diverges.
$$\sum_{n=0}^{\infty} 3\left(\frac{3}{2}\right)^n = 3 + \frac{9}{2} + \frac{27}{4} + \frac{81}{8} + \ldots$$
   Solution: This series diverges by Test for Convergence of a Geometric Series since $r = 3/2 > 1$.

11. Verify that the following infinite series diverges.
$$\sum_{n=0}^{\infty} 1000(1.055)^n = 1000 + 1055 + 1113.025 + \ldots$$
    Solution: This series diverges by Test for Convergence of a Geometric Series since $r = 1.055 > 1$.

Section 9.2

13. Verify that the following geometric series converges.
$$\sum_{n=0}^{\infty} 2\left(\frac{3}{4}\right)^n = 2 + \frac{3}{2} + \frac{9}{8} + \frac{27}{32} + \frac{81}{128} + \ldots$$
Solution: This series converges by Test for Convergence of a Geometric Series since $r = 3/4 < 1$.

15. Verify that the following geometric series converges.
$$\sum_{n=0}^{\infty} (0.9)^n = 1 + 0.9 + 0.81 + 0.729 + \ldots$$
Solution: This series converges by Test for Convergence of a Geometric Series since $r = 0.9 < 1$.

17. Find the sum of
$$\sum_{n=0}^{\infty} \left(\frac{1}{2}\right)^n = 1 + \frac{1}{2} + \frac{1}{4} + \frac{1}{8} + \ldots$$
Solution: Since $a = 1$ and $r = 1/2$, we have
$$S = \frac{1}{1 - (1/2)} = 2$$

19. Find the sum of
$$\sum_{n=0}^{\infty} \left(-\frac{1}{2}\right)^n = 1 - \frac{1}{2} + \frac{1}{4} - \frac{1}{8} + \ldots$$
Solution: Since $a = 1$ and $r = -1/2$, we have
$$S = \frac{1}{1 + (1/2)} = \frac{2}{3}$$

21. Find the sum of
$$\sum_{n=0}^{\infty} 2\left(\frac{1}{\sqrt{2}}\right)^n = 2 + \sqrt{2} + 1 + \frac{1}{\sqrt{2}} + \ldots$$
Solution: Since $a = 2$ and $r = 1/\sqrt{2}$, we have
$$S = \frac{2}{1 - (1/\sqrt{2})} = \frac{2\sqrt{2}}{\sqrt{2} - 1} \left(\frac{\sqrt{2} + 1}{\sqrt{2} + 1}\right)$$
$$= 4 + 2\sqrt{2} \approx 6.828$$

23. Find the sum of
$$1 + 0.1 + 0.01 + 0.001 + \ldots = \sum_{n=0}^{\infty} (0.1)^n$$
Solution: Since $a = 1$ and $r = 0.1$, we have
$$S = \frac{1}{1 - 0.1} = \frac{1}{0.9} = \frac{10}{9}$$

25. Find the sum of
$$3 - 1 + \frac{1}{3} - \frac{1}{9} = \sum_{n=0}^{\infty} 3\left(-\frac{1}{3}\right)^n$$
Solution: Since $a = 3$ and $r = -1/3$, we have
$$S = \frac{3}{1 + (1/3)} = \frac{9}{4}$$

27. Find the sum of
$$\sum_{n=0}^{\infty} \left(\frac{1}{2^n} - \frac{1}{3^n}\right)$$
Solution:
$$\sum_{n=0}^{\infty} \left(\frac{1}{2^n} - \frac{1}{3^n}\right) = \sum_{n=0}^{\infty} \left(\frac{1}{2}\right)^n - \sum_{n=0}^{\infty} \left(\frac{1}{3}\right)^n$$
$$= \frac{1}{1 - (1/2)} - \frac{1}{1 - (1/3)}$$
$$= 2 - \frac{3}{2} = \frac{1}{2}$$

29. Find the sum of
$$\sum_{n=0}^{\infty} \left(\frac{1}{3^n} + \frac{1}{4^n}\right)$$
Solution:
$$\sum_{n=0}^{\infty} \left(\frac{1}{3^n} + \frac{1}{4^n}\right) = \sum_{n=0}^{\infty} \left(\frac{1}{3}\right)^n + \sum_{n=0}^{\infty} \left(\frac{1}{4}\right)^n$$
$$= \frac{1}{1 - (1/3)} + \frac{1}{1 - (1/4)}$$
$$= \frac{3}{2} + \frac{4}{3} = \frac{17}{6}$$

31. Determine the convergence or divergence of
$$\sum_{n=1}^{\infty} \frac{n + 10}{10n + 1}$$
Solution: This series diverges by the nth-Term Test since
$$\lim_{n \to \infty} \frac{n + 10}{10n + 1} = \frac{1}{10} \neq 0$$

33. Determine the convergence or divergence of
$$\sum_{n=1}^{\infty} \frac{n + 1}{n}$$
Solution: This series diverges by the nth-Term Test since
$$\lim_{n \to \infty} \frac{n + 1}{n} = 1 \neq 0$$

35. Determine the convergence or divergence of
$$\sum_{n=1}^{\infty} \frac{3n - 1}{2n + 1}$$
Solution: This series diverges by the nth-Term Test since
$$\lim_{n \to \infty} \frac{3n - 1}{2n + 1} = \frac{3}{2} \neq 0$$

37. Determine the convergence or divergence of
$$\sum_{n=0}^{\infty} (1.075)^n$$
Solution: This series diverges by the Test for Convergence of a Geometric Series since $r = 1.075 > 1$.

39. Determine the convergence or divergence of
$$\sum_{n=0}^{\infty} (0.075)^n$$
Solution: This series converges by the Test for Convergence of a Geometric Series since $r = 0.075 < 1$.

41. Find the sum of the geometric series given by
$$0.666\overline{6} = 0.6 + 0.06 + 0.006 + 0.0006 + \ldots$$
Solution:
$$0.666\overline{6} = \sum_{n=0}^{\infty} 0.6(0.1)^n = \frac{0.6}{1 - 0.1} = \frac{0.6}{0.9} = \frac{2}{3}$$

43. Find the sum of the geometric series given by
$$0.36\overline{36} = 0.36 + 0.0036 + 0.000036 + \ldots$$
Solution:
$$0.36\overline{36} = \sum_{n=0}^{\infty} 0.36(0.01)^n = \frac{0.36}{1 - 0.01} = \frac{0.36}{0.99} = \frac{4}{11}$$

45. A company produces a new product for which it estimates the annual sales to be 8000 units. Suppose that in any given year 10% of the units (regardless of age) will become inoperative. (a) How many units will be in use after n years? (b) Find the market stabilization level of the product.
Solution:
(a) $$\sum_{i=0}^{n-1} 8000(0.9)^i = \frac{8000[1 - (0.9)^{(n-1)+1}]}{1 - 0.9}$$
$$= 80{,}000(1 - 0.9^n)$$
(b) $$\sum_{i=0}^{\infty} 8000(0.9)^i = \frac{8000}{1 - 0.9} = 80{,}000$$

47. A ball is dropped from a height of 16 feet. Each time it drops h feet, it rebounds 0.81h feet. Find the total distance traveled by the ball.
Solution:
$D_1 = 16$
$D_2 = 0.81(16) + 0.81(16) = 32(0.81)$
$D_3 = 16(0.81)^2 + 16(0.81)^2 = 32(0.81)^2$
$\vdots$
$$D = -16 + \sum_{n=0}^{\infty} 32(0.81)^n = -16 + \frac{32}{1 - 0.81}$$
$$= 2896/19 \approx 152.42 \text{ feet}$$

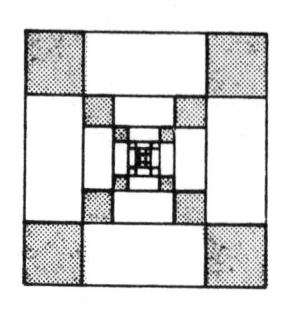

49. Find the fraction of the total area of the square that is eventually shaded if the pattern of shading shown in the figure is continued. (Note that the sides of the shaded corner squares are one-fourth those of the squares in which they are placed.)
Solution:
$$A_1 = (1/4)A$$
$$A_2 = (1/4)A_1 = (1/4)^2 A$$
$$A_3 = (1/4)A_2 = (1/4)^3 A$$
$$\vdots$$
$$A_n = (1/4)^n A$$

The shaded area is given by
$$\sum_{n=1}^{\infty} A(1/4)^n = \sum_{n=0}^{\infty} A(1/4)^n - A = \frac{A}{1 - (1/4)} - A$$
$$= \frac{4A}{3} - A = \frac{A}{3}$$
which is 1/3 the area of the square.

51. A deposit of $50 is made at the beginning of each month in an account that pays 12% interest compounded monthly. What is the balance in the account at the end of 10 years?
$$A = 50(1 + \frac{0.12}{12}) + 50(1 + \frac{0.12}{12})^2 \ldots + 50(1 + \frac{0.12}{12})^{120}$$
Solution:
$$A_{10} = 50(1 + \frac{0.12}{12}) + 50(1 + \frac{0.12}{12})^2 \ldots + 50(1 + \frac{0.12}{12})^{120}$$
$$= \sum_{n=1}^{120} 50(1 + \frac{0.12}{12})^n = -50 + \sum_{n=0}^{120} 50(1 + \frac{0.12}{12})^n$$
$$= -50 + \frac{50[1 - (1 + (0.12/12))^{121}]}{1 - (1 + (0.12/12))}$$
$$\approx \$11,616.95$$

53. Use the formula
$$A = P[(1 + \frac{r}{12})^N - 1](1 + \frac{12}{r})$$
to find the amount in an account earning 9% interest compounded monthly after deposits of $50 have been made monthly for 40 years.
Solution: Since $P = 50$, $r = 0.09$, and $N = 12t = 12(40) = 480$, the amount is
$$A = 50[(1 + \frac{0.09}{12})^{480} - 1](1 + \frac{12}{0.09})$$
$$\approx \$235,821.51$$

55. Suppose that an employer offered to pay you 1 cent the first day, and then double your wages each day thereafter. Find your total wages for working 20 days.
Solution: The daily wages would be
$D_1 = 0.01$
$D_2 = 2D_1 = 2(0.01)$
$D_3 = 2D_2 = 2^2(0.01)$
$D_4 = 2D_3 = 2^3(0.01)$
$\vdots$
$D_{20} = 2^{19}(0.01)$

Therefore, the total wages for 20 days would be
$$T = \sum_{n=1}^{20} 2^{n-1}(0.01) = \sum_{n=0}^{19} 2^n(0.01)$$
$$= \frac{0.01(1 - 2^{20})}{1 - 2} \approx \$10,485.75$$

Section 9.3 p-Series and the Ratio Test

1. Determine the convergence or divergence of the p-series
$$\sum_{n=1}^{\infty} \frac{1}{n^3}$$
Solution: This series converges since $p = 3 > 1$.

3. Determine the convergence or divergence of the p-series
$$\sum_{n=1}^{\infty} \frac{1}{\sqrt[3]{n}}$$
Solution: This series diverges since $p = 1/3 < 1$.

5. Determine the convergence or divergence of the p-series
$$\sum_{n=1}^{\infty} \frac{1}{n^{1.04}}$$
Solution: This series converges since $p = 1.04 > 1$.

7. Determine the convergence or divergence of the p-series
$$1 + \frac{1}{\sqrt{2}} + \frac{1}{\sqrt{3}} + \frac{1}{\sqrt{4}} + \ldots$$
Solution:
$$1 + \frac{1}{\sqrt{2}} + \frac{1}{\sqrt{3}} + \frac{1}{\sqrt{4}} + \ldots = \sum_{n=1}^{\infty} \frac{1}{\sqrt{n}} = \sum_{n=1}^{\infty} \frac{1}{n^{1/2}}$$

Therefore, this series diverges since $p = 1/2 < 1$.

Section 9.3

9. Determine the convergence or divergence of the p-series

$$1 + \frac{1}{2\sqrt{2}} + \frac{1}{3\sqrt{3}} + \frac{1}{4\sqrt{4}} + \cdots$$

Solution:

$$1 + \frac{1}{2\sqrt{2}} + \frac{1}{3\sqrt{3}} + \frac{1}{4\sqrt{4}} + \cdots = \sum_{n=1}^{\infty} \frac{1}{n^{3/2}}$$

Therefore, the series converges since $p = 3/2 > 1$.

11. Use the Ratio Test to determine the convergence or divergence of

$$\sum_{n=0}^{\infty} \frac{3^n}{n!}$$

Solution: Since $a_n = 3^n/n!$, we have

$$\lim_{n \to \infty} \left| \frac{a_{n+1}}{a_n} \right| = \lim_{n \to \infty} \left| \frac{3^{n+1}}{(n+1)!} \cdot \frac{n!}{3^n} \right|$$

$$= \lim_{n \to \infty} \frac{3}{n+1} = 0$$

and the series converges.

13. Use the Ratio Test to determine the convergence or divergence of

$$\sum_{n=0}^{\infty} \frac{n!}{3^n}$$

Solution: Since $a_n = n!/3^n$, we have

$$\lim_{n \to \infty} \left| \frac{a_{n+1}}{a_n} \right| = \lim_{n \to \infty} \left| \frac{(n+1)!}{3^{n+1}} \cdot \frac{3^n}{n!} \right|$$

$$= \lim_{n \to \infty} \frac{n+1}{3} = \infty$$

and the series diverges.

15. Use the Ratio Test to determine the convergence or divergence of

$$\sum_{n=1}^{\infty} \frac{n}{4^n}$$

Solution: Since $a_n = n/4^n$, we have

$$\lim_{n \to \infty} \left| \frac{a_{n+1}}{a^n} \right| = \lim_{n \to \infty} \left| \frac{n+1}{4^{n+1}} \cdot \frac{4^n}{n} \right|$$

$$= \lim_{n \to \infty} \frac{n+1}{4n} = \frac{1}{4}$$

and the series converges.

17. Use the Ratio Test to determine the convergence or divergence of
$$\sum_{n=1}^{\infty} \frac{2^n}{n^3}$$
Solution: Since $a_n = 2^n/n^3$, we have
$$\lim_{n \to \infty} \left|\frac{a_{n+1}}{a_n}\right| = \lim_{n \to \infty} \left|\frac{2^{n+1}}{(n+1)^3} \cdot \frac{n^3}{2^n}\right|$$
$$= \lim_{n \to \infty} \frac{2n^3}{n^3 + 3n^2 + 3n + 1} = 2$$

and the series diverges.

19. Use the Ratio Test to determine the convergence or divergence of
$$\sum_{n=0}^{\infty} \frac{(-1)^n 2^n}{n!}$$
Solution: Since $a_n = [(-1)^n 2^n]/n!$, we have
$$\lim_{n \to \infty} \left|\frac{a_{n+1}}{a_n}\right| = \lim_{n \to \infty} \left|\frac{(-1)^{n+1} 2^{n+1}}{(n+1)!} \cdot \frac{n!}{(-1)^n 2^n}\right|$$
$$= \lim_{n \to \infty} \left|\frac{-2}{n+1}\right| = 0$$

and the series converges.

21. Using four terms, approximate the sum of
$$\sum_{n=1}^{\infty} \frac{1}{n^5}$$
Include an estimate of the maximum error for your approximation.
Solution:
$$\sum_{n=1}^{\infty} \frac{1}{n^5} \approx 1 + \frac{1}{32} + \frac{1}{243} + \frac{1}{1024} \approx 1.036$$

The error is less than $(4^{1-5})/4 = 1/1024$, which implies that $1.036 < S < 1.037$.

23. Using ten terms, approximate the sum of
$$\sum_{n=1}^{\infty} \frac{1}{n^{3/2}}$$
Include an estimate of the maximum error for your approximation.
Solution:
$$\sum_{n=1}^{\infty} \frac{1}{n^{3/2}} \approx 1 + \frac{1}{2\sqrt{2}} + \frac{1}{3\sqrt{3}} + \ldots + \frac{1}{10\sqrt{10}} \approx 1.995$$

The error is less than $10^{1-(3/2)}/[(3/2) - 1] = 2/\sqrt{10}$ which implies that $1.995 < S < 2.628$.

25. Test for convergence or divergence using any appropriate test from this chapter.

$$\sum_{n=1}^{\infty} \frac{2n}{n+1}$$

Solution: This series diverges by the nth-Term Test since

$$\lim_{n \to \infty} \frac{2n}{n+1} = 2 \neq 0$$

27. Test for convergence or divergence using any appropriate test from this chapter.

$$\sum_{n=1}^{\infty} \frac{1}{n\sqrt{n}} = \sum_{n=1}^{\infty} \frac{1}{n^{3/2}}$$

Solution: This series converges by the p-series test since $p = 3/2 > 1$.

29. Test for convergence or divergence using any appropriate test from this chapter.

$$\sum_{n=0}^{\infty} \frac{(-1)^n 2^n}{3^n} = \sum_{n=0}^{\infty} (-\frac{2}{3})^n$$

Solution: This series converges by the Geometric Series Test since $|r| = |-2/3| = 2/3 < 1$.

31. Test for convergence or divergence using any appropriate test from this chapter.

$$\sum_{n=1}^{\infty} (\frac{1}{n^2} - \frac{1}{n^3}) = \sum_{n=1}^{\infty} \frac{1}{n^2} - \sum_{n=1}^{\infty} \frac{1}{n^3}$$

Solution: Since both series are convergent p-series, their difference is convergent.

33. Test for convergence or divergence using any appropriate test from this chapter.

$$\sum_{n=0}^{\infty} (\frac{4}{3})^n$$

Solution: This series converges by the Geometric Series Test since $r = 4/3 > 1$.

35. Test for convergence or divergence using any appropriate test from this chapter.

$$\sum_{n=1}^{\infty} \frac{n!}{3^{n-1}}$$

Solution: This series diverges by the Ratio Test since $a_n = n!/(3^{n-1})$, and

$$\lim_{n \to \infty} \left|\frac{a_{n+1}}{a_n}\right| = \lim_{n \to \infty} \left|\frac{(n+1)!}{3^n} \cdot \frac{3^{n-1}}{n!}\right|$$

$$= \lim_{n \to \infty} \frac{n+1}{3} = \infty$$

Section 9.4 Power Series and Taylor's Theorem

1. Find the radius of convergence for
$$\sum_{n=0}^{\infty} \left(\frac{x}{2}\right)^n = \sum_{n=0}^{\infty} \left(\frac{1}{2}\right)^n x^n$$
Solution:
$$R = \lim_{n \to \infty} \left|\frac{a_n}{a_{n+1}}\right| = \lim_{n \to \infty} \left|\frac{1}{2^n}\left(\frac{2^{n+1}}{1}\right)\right| = 2$$

3. Find the radius of convergence for
$$\sum_{n=1}^{\infty} \frac{(-1)^n x^n}{n}$$
Solution:
$$R = \lim_{n \to \infty} \left|\frac{a_n}{a_{n+1}}\right| = \lim_{n \to \infty} \left|\frac{(-1)^n}{n}\left(\frac{n+1}{(-1)^{n+1}}\right)\right|$$
$$= \lim_{n \to \infty} \frac{n+1}{n} = 1$$

5. Find the radius of convergence for
$$\sum_{n=0}^{\infty} \frac{x^n}{n!}$$
Solution:
$$R = \lim_{n \to \infty} \left|\frac{a_n}{a_{n+1}}\right| = \lim_{n \to \infty} \left|\frac{1}{n!}\left(\frac{(n+1)!}{1}\right)\right|$$
$$= \lim_{n \to \infty} (n+1) = \infty$$

7. Find the radius of convergence for
$$\sum_{n=0}^{\infty} n!\left(\frac{x}{2}\right)^n = \sum_{n=0}^{\infty} \frac{n!}{2^n} x^n$$
Solution:
$$R = \lim_{n \to \infty} \left|\frac{a_n}{a_{n+1}}\right| = \lim_{n \to \infty} \left|\frac{n!}{2^n} \cdot \frac{2^{n+1}}{(n+1)!}\right|$$
$$= \lim_{n \to \infty} \frac{2}{n+1} = 0$$

9. Find the radius of convergence for
$$\sum_{n=1}^{\infty} \frac{(-1)^{n+1} x^n}{4^n}$$
Solution:
$$R = \lim_{n \to \infty} \left|\frac{(-1)^{n+1}}{4^n}\left(\frac{4^{n+1}}{(-1)^{n+2}}\right)\right| = 4$$

Section 9.4

11. Find the radius of convergence for
$$\sum_{n=1}^{\infty} \frac{(-1)^{n+1}(x-5)^n}{n5^n}$$
Solution:
$$R = \lim_{n \to \infty} \left| \frac{a_n}{a_{n+1}} \right|$$
$$= \lim_{n \to \infty} \left| \frac{(-1)^{n+1}}{n5^n} \left(\frac{(n+1)5^{n+1}}{(-1)^{n+2}} \right) \right| = 5$$

13. Find the radius of convergence for
$$\sum_{n=0}^{\infty} \frac{(-1)^{n+1}(x-1)^{n+1}}{n+1}$$
Solution:
$$R = \lim_{n \to \infty} \left| \frac{a_n}{a_{n+1}} \right|$$
$$= \lim_{n \to \infty} \left| \frac{(-1)^{n+1}}{n+1} \left(\frac{n+2}{(-1)^{n+2}} \right) \right| = 1$$

15. Find the radius of convergence for
$$\sum_{n=1}^{\infty} \frac{(x-c)^{n-1}}{c^{n-1}}$$
Solution:
$$R = \lim_{n \to \infty} \left| \frac{a_n}{a_{n+1}} \right| = \lim_{n \to \infty} \left| \frac{1}{c^{n-1}} \left(\frac{c^n}{1} \right) \right| = c$$

17. Find the radius of convergence for
$$\sum_{n=1}^{\infty} \frac{n}{n+1}(-2x)^{n-1} = \sum_{n=0}^{\infty} \frac{(-2)^{n-1}n}{n+1}(x^{n-1})$$
Solution:
$$R = \lim_{n \to \infty} \left| \frac{a_n}{a_{n+1}} \right|$$
$$= \lim_{n \to \infty} \left| \frac{(-2)^{n-1}n}{n+1} \left(\frac{n+2}{(-2)^n(n+1)} \right) \right| = \frac{1}{2}$$

19. Find the radius of convergence for
$$\sum_{n=0}^{\infty} \frac{x^{2n+1}}{(2n+1)!}$$
Solution:
$$R = \lim_{n \to \infty} \left| \frac{a_n}{a_{n+1}} \right| = \lim_{n \to \infty} \left| \frac{1}{(2n-1)!} \left(\frac{(2n+1)!}{1} \right) \right|$$
$$= \lim_{n \to \infty} |(2n+1)(2n)| = \infty$$

Section 9.4 405

21. Apply Taylor's Theorem to find the power series (centered at 0) for $f(x) = e^x$.
Solution:
$$f(x) = e^x \qquad f(0) = 1$$
$$f'(x) = e^x \qquad f'(0) = 1$$
$$f''(x) = e^x \qquad f''(0) = 1$$
$$\vdots \qquad \vdots$$
$$f^{(n)}(x) = e^x \qquad f^{(n)}(0) = 1$$

The power series for f is
$$e^x = f(0) + f'(0)x + \frac{f''(0)x^2}{2!} + \ldots$$
$$= 1 + x + \frac{x^2}{2!} + \frac{x^3}{3!} + \ldots = \sum_{n=0}^{\infty} \frac{x^n}{n!}$$

and the radius of convergence is
$$R = \lim_{n \to \infty} \left| \frac{1}{n!} \cdot \frac{(n+1)!}{1} \right| = \lim_{n \to \infty} (n+1) = \infty$$

23. Apply Taylor's Theorem to find the power series (centered at 0) for $f(x) = e^{2x}$.
Solution:
$$f(x) = e^{2x} \qquad f(0) = 1$$
$$f'(x) = 2e^{2x} \qquad f'(0) = 2$$
$$f''(x) = 4e^{2x} \qquad f''(0) = 4$$
$$f'''(x) = 8e^{2x} \qquad f'''(0) = 8$$
$$\vdots$$
$$f^{(n)}(0) = 2^n$$

The power series for f is
$$e^{2x} = f(0) + f'(0)(x) + \frac{f''(0)(x)^2}{2!} + \ldots$$
$$= 1 + 2x + \frac{4x^2}{2!} + \frac{8x^3}{3!} + \ldots = \sum_{n=0}^{\infty} \frac{(2x)^n}{n!}$$

and the radius of convergence is
$$R = \lim_{n \to \infty} \left| \frac{2^n}{n!} \cdot \frac{(n+1)!}{2^{n+1}} \right| = \lim_{n \to \infty} \left(\frac{n+1}{2}\right) = \infty$$

25. Apply Taylor's Theorem to find the power series (centered at 0) for $f(x) = 1/(x+1)$.
Solution:
$$f(x) = 1/(x+1) \qquad f(0) = 1$$
$$f'(x) = -1/(x+1)^2 \qquad f'(0) = -1$$
$$f''(x) = 2/(x+1)^3 \qquad f''(0) = 2$$
$$f'''(x) = -6/(x+1)^4 \qquad f'''(0) = -6$$
$$\vdots$$
$$f^{(n)}(0) = (-1)^n n!$$

The power series for f is

$$\frac{1}{x+1} = f(0) + f'(0)x + \frac{f''(0)x^2}{2!} + \ldots$$

$$= 1 - x + \frac{2x^2}{2!} - \frac{6x^3}{3!} + \ldots + \frac{(-1)^n n! x^n}{n!} + \ldots$$

$$= 1 - x + x^2 - x^3 + \ldots = \sum_{n=0}^{\infty} (-1)^n x^n$$

and the radius of convergence is

$$R = \lim_{n \to \infty} \left| \frac{(-1)^n}{(-1)^{n+1}} \right| = \lim_{n \to \infty} |-1| = 1$$

27. Apply Taylor's Theorem to find the power series (centered at 1) for $f(x) = \sqrt{x}$.
Solution:

$$\begin{array}{ll} f(x) = \sqrt{x} & f(1) = 1 \\ f'(x) = 1/2\sqrt{x} & f'(1) = 1/2 \\ f''(x) = -1/4x\sqrt{x} & f''(1) = -1/4 \\ f'''(x) = 3/8x^2\sqrt{x} & f'''(1) = 3/8 \\ f^{(4)}(x) = -15/16x^3\sqrt{x} & f^{(4)}(1) = -15/16 \end{array}$$

Thus, the general pattern (for $n \geq 2$) is given by

$$f^{(n)}(1) = \frac{(-1)^n 1 \cdot 3 \cdot 5 \ldots (2n-3)}{2^n}$$

The power series for f is

$$\sqrt{x} = f(1) + f'(1)(x-1) + \frac{f''(1)(x-1)^2}{2!} + \ldots$$

$$= 1 + \frac{x-1}{2} - \frac{(x-1)^2}{4 \cdot 2!} + \frac{3(x-1)^3}{8 \cdot 3!} - \ldots$$

$$= 1 + \frac{1}{2}(x-1) - \sum_{n=2}^{\infty} \frac{(-1)^n 1 \cdot 3 \cdot 5 \ldots (2n-3)(x-1)^n}{2^n n!}$$

and the radius of convergence is

$$R = \lim_{n \to \infty} \left| \frac{a_n}{a_{n+1}} \right| = \lim_{n \to \infty} \left| -\frac{2(n+1)}{2n-1} \right| = 1$$

29. Apply Taylor's Theorem to find the power series (centered at 1) for $f(x) = \ln x$.
Solution:

$$\begin{array}{ll} f(x) = \ln x & f(1) = 0 \\ f'(x) = 1/x & f'(1) = 1 = 0! \\ f''(x) = -1/x^2 & f''(1) = -1 = -(1!) \\ f'''(x) = 2/x^3 & f'''(1) = 2 = 2! \\ f^{(4)}(x) = -6/x^4 & f^{(4)}(1) = -6 = -(3!) \\ \vdots & \\ & f^{(n)}(1) = (-1)^{n-1}(n-1)! \end{array}$$

Section 9.4   407

The power series for f is

$$\ln x = f(1) + f'(1)(x-1) + \frac{f''(1)(x-1)^2}{2!} + \ldots$$

$$= 0 + (x-1) - \frac{(x-1)^2}{2!} + \frac{2(x-1)^3}{3!} - \ldots$$

$$= \sum_{n=1}^{\infty} \frac{(-1)^{n-1}}{n}(x-1)^n$$

and the radius of convergence is

$$R = \lim_{n \to \infty} \left| \frac{(-1)^{n-1}}{n} \frac{(n+1)}{(-1)^n} \right| = 1$$

31. Apply Taylor's Theorem to find the binomial series (centered at 0) for $f(x) = 1/[(1+x)^2]$.
Solution:

$$f(x) = (1+x)^{-2} \qquad f(0) = 1$$
$$f'(x) = -2(1+x)^{-3} \qquad f'(0) = -2$$
$$f''(x) = 6(1+x)^{-4} \qquad f''(0) = 6$$
$$f'''(x) = -24(1+x)^{-5} \qquad f'''(0) = -24$$
$$\vdots$$
$$f^{(n)}(0) = (-1)^{(n)}(n+1)!$$

The power series for f is

$$\frac{1}{(1+x)^2} = f(0) + f'(0)x + \frac{f''(0)x^2}{2!} + \ldots$$

$$= 1 - 2x + \frac{6x^2}{2} - \frac{24x^3}{6} + \ldots$$

$$= \sum_{n=0}^{\infty} (-1)^n(n+1)x^n$$

and the radius of convergence is

$$R = \lim_{n \to \infty} \left| \frac{n+1}{n+2} \right| = 1$$

33. Apply Taylor's Theorem to find the binomial series (centered at 0) for $f(x) = 1/(\sqrt{1+x})$.
Solution:

$$f(x) = (1+x)^{-1/2} \qquad f(0) = 1$$
$$f'(x) = -(1/2)(1+x)^{-3/2} \qquad f'(0) = -1/2$$
$$f''(x) = (3/4)(1+x)^{-5/2} \qquad f''(0) = 3/4$$
$$f'''(x) = -(15/8)(1+x)^{-7/2} \qquad f'''(0) = -15/8$$

Thus, the general pattern is given by

$$f^{(n)}(0) = (-1)^n \left(\frac{1 \cdot 3 \cdot 5 \ldots (2n-1)}{2^n} \right)$$

The power series for f is

$$\frac{1}{\sqrt{1+x}} = f(0) + f'(0)x + \frac{f''(0)x^2}{2!} + \frac{f'''(x)x^3}{3!} + \ldots$$

$$= 1 - \frac{1}{2}x + \frac{1 \cdot 3 x^2}{2^2 2!} - \frac{1 \cdot 3 \cdot 5 x^3}{2^3 3!} + \ldots$$

$$= 1 + \sum_{n=1}^{\infty} \frac{(-1)^n 1 \cdot 3 \cdot 5 \ldots (2n-1)}{2^n n!} x^n$$

and the radius of convergence is

$$R = \lim_{n \to \infty} \left| \frac{1 \cdot 3 \cdot 5 \ldots (2n-1)}{2^n n!} \left(\frac{2^{n+1}(n+1)!}{1 \cdot 3 \cdot 5 \ldots (2n+1)} \right) \right|$$

$$= \lim_{n \to \infty} \frac{2(n+1)}{2n+1} = 1$$

35. Find the radius of convergence of (a) $f(x)$, (b) $f'(x)$, (c) $f''(x)$, and (d) $\int f(x)\,dx$ when

$$f(x) = \sum_{n=0}^{\infty} \left(\frac{x}{2}\right)^n$$

Solution:

(a) $f(x) = \sum_{n=0}^{\infty} \left(\frac{x}{2}\right)^n = \sum_{n=0}^{\infty} \frac{x^n}{2^n}$

$$R = \lim_{n \to \infty} \left| \frac{1}{2^n} \cdot \frac{2^{n+1}}{1} \right| = \lim_{n \to \infty} 2 = 2$$

(b) $f'(x) = \sum_{n=1}^{\infty} \frac{n x^{n-1}}{2^n}$

$$R = \lim_{n \to \infty} \left| \frac{n}{2^n} \cdot \frac{2^{n+1}}{n+1} \right| = \lim_{n \to \infty} \frac{2n}{n+1} = 2$$

(c) $f''(x) = \sum_{n=2}^{\infty} \frac{n(n-1) x^{n-2}}{2^n}$

$$R = \lim_{n \to \infty} \left| \frac{n(n-1)}{2^n} \cdot \frac{2^{n+1}}{(n+1)n} \right|$$

$$= \lim_{n \to \infty} \frac{2(n-1)}{n+1} = 2$$

(d) $\int f(x)\,dx = C + \sum_{n=0}^{\infty} \frac{x^{n+1}}{2^n (n+1)}$

$$R = \lim_{n \to \infty} \left| \frac{1}{2^n (n+1)} \cdot \frac{2^{n+1}(n+2)}{1} \right|$$

$$= \lim_{n \to \infty} \frac{2(n+2)}{n+1} = 2$$

37. Find the radius of convergence of (a) $f(x)$, (b) $f'(x)$, (c) $f''(x)$, and (d) $\int f(x)\,dx$ when

$$f(x) = \sum_{n=0}^{\infty} \frac{(x-1)^{n+1}}{n+1}$$

Solution:

(a) $\quad f(x) = \sum_{n=0}^{\infty} \dfrac{(x-1)^{n+1}}{n+1}$

$R = \lim_{n \to \infty} \left| \dfrac{1}{n+1} \cdot \dfrac{n+2}{1} \right| = \lim_{n \to \infty} \dfrac{n+2}{n+1} = 1$

(b) $\quad f'(x) = \sum_{n=0}^{\infty} \dfrac{(n+1)(x-1)^n}{n+1} = \sum_{n=0}^{\infty} (x-1)^n$

$R = \lim_{n \to \infty} 1 = 1$

(c) $\quad f''(x) = \sum_{n=1}^{\infty} n(x-1)^{n-1}$

$R = \lim_{n \to \infty} \left| \dfrac{n}{n+1} \right| = 1$

(d) $\quad \int f(x)\,dx = C + \sum_{n=0}^{\infty} \dfrac{(x-1)^{n+2}}{(n+1)(n+2)}$

$R = \lim_{n \to \infty} \left| \dfrac{1}{(n+1)(n+2)} \cdot \dfrac{(n+2)(n+3)}{1} \right|$

$= \lim_{n \to \infty} \dfrac{n+3}{n+1} = 1$

39. Use the power series for e^x to find the power series for $f(x) = e^{x^2}$.

Solution: Since the power series for e^x is

$$e^x = \sum_{n=0}^{\infty} \frac{x^n}{n!}$$

it follows that the power series for e^{x^2} is

$$e^{x^2} = \sum_{n=0}^{\infty} \frac{(x^2)^n}{n!} = \sum_{n=0}^{\infty} \frac{x^{2n}}{n!}$$

41. Differentiate the series found in Exercise 39 to find the power series for $f(x) = 2xe^{x^2}$.

Solution:

$2xe^{x^2} = \dfrac{d}{dx}[e^{x^2}] = \sum_{n=1}^{\infty} \dfrac{2nx^{2n-1}}{n!}$

$= 2 \sum_{n=1}^{\infty} \dfrac{x^{2n-1}}{(n-1)!} = 2 \sum_{n=0}^{\infty} \dfrac{x^{2n+1}}{n!}$

Section 9.4

43. Use the power series for $1/(1 + x)$ to find the power series for $f(x) = 1/(1 + x^2)$.
Solution: Since the power series for $1/(1 + x)$ is
$$f(x) = \frac{1}{1 + x} = \sum_{n=0}^{\infty} (-1)^n x^n$$
it follows that the power series for $1/(1 + x^2)$ is
$$f(x^2) = \frac{1}{1 + x^2} = \sum_{n=0}^{\infty} (-1)^n x^{2n}$$

45. Integrate the series
$$\sum_{n=0}^{\infty} (-1)^n x^{2n+1}$$
to find the power series for $f(x) = \ln(1 + x^2)$.
Solution:
$$\frac{1}{1 + x^2} = \sum_{n=0}^{\infty} (-1)^n x^{2n}$$
$$\frac{2x}{1 + x^2} = \sum_{n=0}^{\infty} (-1)^n (2x) x^{2n} = 2\sum_{n=0}^{\infty} (-1)^n x^{2n+1}$$
$$\ln(1 + x^2) = \int \frac{2x}{1 + x^2}\, dx = 2\sum_{n=0}^{\infty} \frac{(-1)^n x^{2n+2}}{2n + 2}$$
$$= \sum_{n=0}^{\infty} \frac{(-1)^n x^{2n+2}}{n + 1}$$

47. Integrate the series for $1/x$ to find the power series for $f(x) = \ln x$.
Solution:
$$\frac{1}{x} = \sum_{n=0}^{\infty} (-1)^n (x - 1)^n$$
$$\ln x = \int \frac{1}{x}\, dx = \sum_{n=0}^{\infty} \frac{(-1)^n (x - 1)^{n+1}}{n + 1}$$

49. Differentiate the series for $-1/(1 + x)$ to find the power series for $f(x) = 1/(1 + x)^2$.
Solution:
$$-\frac{1}{1 + x} = \sum_{n=0}^{\infty} (-1)^{n-1} x^n$$
$$\frac{1}{(1 + x)^2} = \frac{d}{dx}\left[-\frac{1}{1 + x}\right]$$
$$= \sum_{n=1}^{\infty} (-1)^{n-1} n x^{n-1}$$

Section 9.5 Taylor Polynomials

1. Find the Taylor polynomial (centered at 0) of degree (a) 1, (b) 2, (c) 3, and (d) 4 for $f(x) = e^x$.
   Solution:
   $$e^x = \sum_{n=0}^{\infty} \frac{1}{n!} x^n$$
   (a) $S_1(x) = 1 + x$
   (b) $S_2(x) = 1 + x + (x^2/2)$
   (c) $S_3(x) = 1 + x + (x^2/2) + (x^3/6)$
   (d) $S_4(x) = 1 + x + (x^2/2) + (x^3/6) + (x^4/24)$

3. Find the Taylor polynomial (centered at 0) of degree (a) 1, (b) 2, (c) 3, (d) 4 for $f(x) = \sqrt{x+1}$.
   Solution:
   $$\frac{1}{\sqrt{x+1}} = 1 + \frac{x}{2^1 \cdot 1!} - \frac{x^2}{2^2 \cdot 2!} + \frac{3x^3}{2^3 \cdot 3!} - \frac{3 \cdot 5 x^4}{2^4 \cdot 4!} + \ldots$$
   (a) $S_1(x) = 1 + (x/2)$
   (b) $S_2(x) = 1 + (x/2) - (x^2/8)$
   (c) $S_3(x) = 1 + (x/2) - (x^2/8) + (3x^3/48)$
   (d) $S_4(x) = 1 + (x/2) - (x^2/8) + (3x^3/48) - (15x^4/384)$

5. Find the Taylor polynomial (centered at 0) of degree (a) 2, (b) 4, (c) 6, and (d) 8 for $f(x) = 1/(1 + x^2)$.
   Solution:
   $$\frac{1}{1+x^2} = \sum_{n=0}^{\infty} (-1)^n x^{2n}$$
   (a) $S_2(x) = 1 - x^2$
   (b) $S_4(x) = 1 - x^2 + x^4$
   (c) $S_6(x) = 1 - x^2 + x^4 - x^6$
   (d) $S_8(x) = 1 - x^2 + x^4 - x^6 + x^8$

7. Complete the table using the Taylor polynomial as an approximation to $f(x) = e^{x/2}$.
   Solution:
   $S_1(x) = 1 + (x/2)$
   $S_2(x) = 1 + (x/2) + (x^2/8)$
   $S_3(x) = 1 + (x/2) + (x^2/8) + (x^3/48)$
   $S_4(x) = 1 + (x/2) + (x^2/8) + (x^3/48) + (x^4/384)$

x	0	0.25	0.50	0.75
$f(x)$	1.0000	1.1331	1.2840	1.4550
$S_1(x)$	1.0000	1.1250	1.2500	1.3750
$S_2(x)$	1.0000	1.1328	1.2813	1.4453
$S_3(x)$	1.0000	1.1331	1.2839	1.4541
$S_4(x)$	1.0000	1.1331	1.2840	1.4549

9. Use a sixth-degree Taylor polynomial centered at $c = 0$ for the function $f(x) = e^{-x}$ to approximate $f(1/2)$.
Solution:

$$S_6(x) = 1 - x + \frac{x^2}{2} - \frac{x^3}{6} + \frac{x^4}{24} - \frac{x^5}{120} + \frac{x^6}{720}$$

$$f(\frac{1}{2}) \approx 1 - \frac{1}{2} + \frac{1}{8} - \frac{1}{48} + \frac{1}{384} - \frac{1}{3840} + \frac{1}{46080}$$

$$\approx 0.607$$

11. Use a sixth-degree Taylor polynomial centered at $c = 1$ for the function $f(x) = \ln x$ to approximate $f(3/2)$.
Solution:

$$S_6(x) = (x - 1) - \frac{(x-1)^2}{2} + \ldots - \frac{(x-1)^6}{6}$$

$$f(\frac{3}{2}) \approx \frac{1}{2} - \frac{1}{8} + \frac{1}{24} - \frac{1}{64} + \frac{1}{160} - \frac{1}{384}$$

$$\approx 0.405$$

13. Use a sixth-degree Taylor polynomial centered at $c = 0$ for the function $f(x) = e^{-x^2}$ to approximate

$$\int_0^{1/2} e^{-x^2}\, dx$$

Solution:

$$S_6(x) = 1 - x^2 + \frac{x^4}{2} - \frac{x^6}{6}$$

$$\int_0^{1/2} e^{-x^2}\, dx \approx \int_0^{1/2} (1 - x^2 + \frac{x^4}{2} - \frac{x^6}{6})\, dx$$

$$= \left[x - \frac{x^3}{3} + \frac{x^5}{10} - \frac{x^7}{42} \right]_0^{1/2} \approx 0.461$$

15. Use a sixth-degree Taylor polynomial centered at $c = 0$ for the function $f(x) = 1/\sqrt{1 + x^2}$ to approximate

$$\int_0^{1/2} \frac{1}{\sqrt{1 + x^2}}\, dx$$

Solution:

$$S_6(x) = 1 - \frac{1}{2}x^2 + \frac{3}{8}x^4 - \frac{5}{16}x^6$$

$$\int_0^{1/2} \frac{1}{\sqrt{1+x^2}}\, dx \approx \int_0^{1/2} (1 - \frac{1}{2}x^2 + \frac{3}{8}x^4 - \frac{5}{16}x^6)\, dx$$

$$= \left[x - \frac{x^3}{6} + \frac{3x^5}{40} - \frac{5x^7}{112} \right]_0^{1/2}$$

$$= \frac{1}{2} - \frac{1}{48} + \frac{3}{1280} - \frac{5}{14336} \approx 0.481$$

17. Determine the degree of the Taylor polynomial centered at c = 0 to approximate $f(x) = e^x$ in the interval [-1, 1] to an accuracy of ± 0.001.
Solution: Since the (n+1)st derivative of $f(x) = e^x$ is $f^{(n+1)}(x) = e^x$, the maximum value of $|f^{(n+1)}(x)|$ in the interval [-1, 1] is $e^1 \approx 2.71828 < 3$. Therefore, the nth remainder is bounded by

$$|R_n| \leq \left|\frac{3}{(n+1)!} x^{n+1}\right|, \quad -1 \leq x \leq 1$$

$$|R_n| \leq \frac{3}{(n+1)!}(1)^{n+1}$$

When n = 5,

$$\frac{3}{(n+1)!} = 0.0041667$$

When n = 6,

$$\frac{3}{(n+1)!} = 0.000595$$

Thus, n = 6 will approximate e^x with an error less than 0.001 in [-1, 1].

19. Determine the maximum error guaranteed by Taylor's Remainder Theorem when

$$1 - x + \frac{x^2}{2!} - \frac{x^3}{3!} + \frac{x^4}{4!} - \frac{x^5}{5!}$$

is used to approximate $f(x) = e^{-x}$ in the interval [0, 1].
Solution:

$$|R_5| \leq \frac{f^{(6)}(z)}{6!} x^6 = \frac{e^{-z}}{6!} x^6$$

Since $e^{-z} \leq 1$ in the interval [0, 1], it follows that

$$R_5 \leq \frac{1}{6!} \approx 0.0083$$

● Section 9.6 Newton's Method

1. Complete one iteration of Newton's Method for $f(x) = x^2 - 3$ using the initial estimate $x_1 = 1.7$.
Solution:

$$x_2 = x_1 - \frac{f(x_1)}{f'(x_1)}$$

$$= 1.7 - \frac{(1.7)^2 - 3}{2(1.7)} \approx 1.732$$

3. Approximate the indicated zero of $f(x) = x^3 + x - 1$. Use Newton's Method, continuing the process until successive approximations differ by less than 0.001.

Section 9.6

Solution:

$$f'(x) = 3x^2 + 1$$

n	x_n	$f(x_n)$	$f'(x_n)$	$\dfrac{f(x_n)}{f'(x_n)}$	$x_n - \dfrac{f(x_n)}{f'(x_n)}$
1	0.5000	−0.3750	1.7500	−0.2143	0.7143
2	0.7143	0.0787	2.5306	0.0311	0.6832
3	0.6832	0.0021	2.4002	0.0009	0.6823

Approximation: $x \approx 0.682$

5. Approximate the indicated zero of $f(x) = 3\sqrt{x-1} - x$. Use Newton's Method, continuing the process until successive approximations differ by less than 0.001.
Solution:

$$f'(x) = \frac{3}{2\sqrt{x-1}} - 1$$

n	x_n	$f(x_n)$	$f'(x_n)$	$\dfrac{f(x_n)}{f'(x_n)}$	$x_n - \dfrac{f(x_n)}{f'(x_n)}$
1	1.2000	0.1416	2.3541	0.0602	1.1398
2	1.1398	−0.0180	3.0113	−0.0060	1.1458
3	1.1458	−0.0003	2.9283	−0.0001	1.1459

Approximation: $x \approx 1.146$

7. Approximate the indicated zeros of $f(x) = x^3 - 27x - 27$. Use Newton's Method, continuing the process until successive approximations differ by less than 0.001.
Solution:

$$f'(x) = 3x^2 - 27$$

n	x_n	$f(x_n)$	$f'(x_n)$	$\dfrac{f(x_n)}{f'(x_n)}$	$x_n - \dfrac{f(x_n)}{f'(x_n)}$
1	−5.0000	−17.0000	48.0000	−0.3542	−4.6458
2	−4.6458	−1.8371	37.7513	−0.0487	−4.5972
3	−4.5972	−0.0329	36.4019	−0.0009	−4.5963

n	x_n	$f(x_n)$	$f'(x_n)$	$\dfrac{f(x_n)}{f'(x_n)}$	$x_n - \dfrac{f(x_n)}{f'(x_n)}$
1	−1.0000	−1.0000	−24.0000	0.0417	−1.0417
2	−1.0417	−0.0053	−23.7448	0.0002	−1.0419

n	x_n	$f(x_n)$	$f'(x_n)$	$\dfrac{f(x_n)}{f'(x_n)}$	$x_n - \dfrac{f(x_n)}{f'(x_n)}$
1	6.0000	27.0000	81.0000	0.3333	5.6667
2	5.6667	1.9630	69.3333	0.0283	5.6384
3	5.6384	0.0136	68.3731	0.0002	5.6382

Approximations: $x \approx -4.596, -1.042, 5.638$

Section 9.6

9. Approximate the indicated zero of $f(x) = \ln x + x$. Use Newton's Method, continuing the process until successive approximations differ by less than 0.001.
Solution:

$$f'(x) = \frac{1}{x} + 1$$

n	x_n	$f(x_n)$	$f'(x_n)$	$\dfrac{f(x_n)}{f'(x_n)}$	$x_n - \dfrac{f(x_n)}{f'(x_n)}$
1	0.6000	0.0892	2.1667	0.4120	0.5588
2	0.5588	-0.0231	2.7895	-0.0083	0.5671
3	0.5671	-0.0002	3.7634	-0.0001	0.5672

Approximation: $x \approx 0.567$

11. Approximate the indicated zeros of $f(x) = e^{-x^2} - x^2$. Use Newton's Method, continuing the process until successive approximations differ by less than 0.001.
Solution:

$$f'(x) = -2xe^{-x^2} - 2x = -2x(e^{-x^2} + 1)$$

n	x_n	$f(x_n)$	$f'(x_n)$	$\dfrac{f(x_n)}{f'(x_n)}$	$x_n - \dfrac{f(x_n)}{f'(x_n)}$
1	0.8000	-0.1127	-2.4437	0.0461	0.7539
2	0.7539	-0.0019	-2.3619	0.0008	0.7531

Approximations: $x \approx \pm 0.753$

13. Apply Newton's Method to approximate the x-value of the point of intersection of $f(x) = 3 - x$ and $g(x) = \ln x$. Continue the process until two successive approximations differ by less than 0.001.
Solution: Let $3 - x = \ln x$ and define

$$h(x) = \ln x + x - 3$$

then $h'(x) = (1/x) + 1$.

n	x_n	$h(x_n)$	$h'(x_n)$	$\dfrac{h(x_n)}{h'(x_n)}$	$x_n - \dfrac{h(x_n)}{h'(x_n)}$
1	2.2000	-0.0115	1.4545	-0.0079	2.2079
2	2.2079	-0.0001	1.4529	-0.0000	2.2079

Approximation: $x \approx 2.208$

15. Apply Newton's Method to approximate the x-value of the point of intersection of $f(x) = 3 - x$ and $g(x) = 1/(x^2 + 1)$. Continue the process until two successive approximations differ by less than 0.001.
Solution: Let $3 - x = 1/(x^2 + 1)$ and define

$$h(x) = \frac{1}{x^2 + 1} + x - 3$$

then $h'(x) = -2x/(x^2 + 1)^2 + 1$.

n	x_n	$h(x_n)$	$h'(x_n)$	$\dfrac{h(x_n)}{h'(x_n)}$	$x_n - \dfrac{h(x_n)}{h'(x_n)}$
1	3.0000	0.1000	0.9400	0.1064	2.8936
2	2.8936	0.0003	0.9341	0.0003	2.8933

Approximation: $x \approx 2.893$

17. Apply Newton's Method to $y = 2x^3 - 6x^2 + 6x - 1$, using the initial estimate $x_1 = 1$, and explain why the method fails.
    Solution: Newton's Method fails because $f'(x_1) = 0$.

19. Apply Newton's Method to $y = -x^3 + 3x^2 - x + 1$, using the initial estimate $x_1 = 1$, and explain why the method fails.
    Solution: Newton's Method fails because

    $$\lim_{x \to \infty} x_n = \begin{cases} 1 = x_1 = x_3 = \ldots \\ 0 = x_2 = x_4 = \ldots \end{cases}$$

    Therefore the limit does not exist.

21. Use the result of Exercise 20 to approximate $\sqrt[3]{7}$ to three decimal places.
    Solution: Define $f(x) = x^3 - 7$, then $f'(x) = 3x^2$

    $$x_{i+1} = \frac{2x_i^3 + 7}{3x_i^2}$$

i	1	2	3	4
x_i	2.0000	1.9167	1.9129	1.9129

    $\sqrt[3]{7} \approx 1.913$

23. Use the result of Exercise 20 to approximate $\sqrt[5]{40}$ to three decimal places.
    Solution: Define $f(x) = x^5 - 40$, then $f'(x) = 5x^4$

    $$x_{i+1} = x_i - \frac{x_i^5 - 40}{5x_i^4} = \frac{4x_i^5 + 40}{5x_i^4}$$

i	1	2	3	4
x_i	2.0000	2.1000	2.0914	2.0913

    $\sqrt[5]{40} \approx 2.091$

25. In the Introductory Example of this section the equation $2x^3 - 7x - 1 = 0$ must be solved to find the coordinates of the point on the graph of $y = 4 - x^2$ that is closest to the point $(1, 0)$. Use Newton's Method to solve for x (accurate to three decimal places). What is the y-coordinate of the point?

Section 9.6   417

Solution: Since $f'(x) = 6x^2 - 7$, we have the following table.

n	x_n	$f(x_n)$	$f'(x_n)$	$\dfrac{f(x_n)}{f'(x_n)}$	$x_n - \dfrac{f(x_n)}{f'(x_n)}$
1	1.9000	-0.5820	14.6600	-0.0397	1.9397
2	1.9397	0.0181	15.5746	0.0012	1.9385

Approximation: $x \approx 1.939$, $y \approx 0.240$
Point nearest (1, 0): (1.939, 0.240)

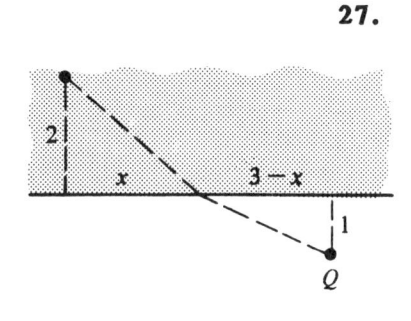

27. A man is in a boat 2 miles from the nearest point on the coast as shown in the figure. He is to go to a point Q, which is 3 miles down the coast and 1 mile inland. If he can row at 3 mi/hr and walk at 4 mi/hr, toward what point on the coast should he row in order to reach point Q in the least time?

Solution: The time is given by

$$T = \frac{\sqrt{x^2 + 4}}{3} + \frac{\sqrt{x^2 - 6x + 10}}{4}$$

To minimize the time, we set dT/dx equal to zero and solve for x. This produces the equation

$$7x^4 - 42x^3 + 43x^2 + 216x - 324 = 0$$

Let $f(x) = 7x^4 - 42x^3 + 43x^2 + 216x - 324$. Since $f(1) = -100$ and $f(2) = 56$, the solution is in the interval (1, 2).

n	x_n	$f(x_n)$	$f'(x_n)$	$\dfrac{f(x_n)}{f'(x_n)}$	$x_n - \dfrac{f(x_n)}{f'(x_n)}$
1	1.7000	19.5887	135.6240	0.1444	1.5556
2	1.5556	-1.0414	150.2782	-0.0069	1.5625
3	1.5629	-0.0092	149.5693	-0.0001	1.5626

Approximation: $x \approx 1.563$ miles

29. The concentration C of a certain chemical in the bloodstream t hours after injection into muscle tissue is given by $C = (3t^2 + t)/(50 + t^3)$. When is the concentration the greatest?

Solution: To maximize C, we set dC/dt equal to zero and solve for t. This produces

$$C' = -\frac{3t^4 - 2t^3 + 300t + 50}{(50 + 5^3)^2} = 0$$

Let $f(x) = 3t^4 + 2t^3 - 300t - 50$. Since $f(4) = -354$ and $f(5) = 575$, the solution is in the interval (4, 5).

n	x_n	$f(x_n)$	$f'(x_n)$	$\dfrac{f(x_n)}{f'(x_n)}$	$x_n - \dfrac{f(x_n)}{f'(x_n)}$
1	4.5000	12.4375	915.0000	0.0136	4.4864
2	4.4864	0.0658	904.3822	0.0001	4.4863

Approximation: $t \approx 4.486$ hours

● Review Exercises for Chapter 9

1. Find the general term of the sequence

$$\frac{1}{3}, \frac{2}{5}, \frac{3}{7}, \frac{4}{9}, \ldots$$

Solution:

$$a_n = \frac{n}{2n+1}, \quad n = 1, 2, 3, \ldots$$

OR

$$a_n = \frac{n+1}{2n+3}, \quad n = 0, 1, 2, \ldots$$

3. Find the general term of the sequence

$$\frac{1}{3}, -\frac{2}{9}, \frac{4}{27}, -\frac{8}{81}, \ldots$$

Solution:

$$a_n = (-1)^{n-1}\left(\frac{2^{n-1}}{3^n}\right), \quad n = 1, 2, 3, \ldots$$

OR

$$a_n = (-1)^n\left(\frac{2^n}{3^{n+1}}\right), \quad n = 0, 1, 2, \ldots$$

5. Determine the convergence or divergence of the sequence with the general term

$$a_n = \frac{n+1}{n^2}$$

Solution: The sequence converges since

$$\lim_{n \to \infty} \frac{n+1}{n^2} = 0$$

7. Determine the convergence or divergence of the sequence with the general term

$$a_n = \frac{n^3}{n^2+1}$$

Solution: The sequence diverges since

$$\lim_{n \to \infty} \frac{n^3}{n^2+1} = \infty$$

9. Determine the convergence or divergence of the sequence with the general term

$$a_n = 5 + \frac{1}{3^n}$$

Solution: The sequence converges since

$$\lim_{n \to \infty} \left(5 + \frac{1}{3^n}\right) = 5 + 0 = 5$$

Review Exercises for Chapter 9 419

11. Find the first five terms of the sequence of partial sums for the series
$$\sum_{n=0}^{\infty} \left(\frac{3}{2}\right)^n$$
Solution:

$$S_0 = 1$$

$$S_1 = 1 + \frac{3}{2} = \frac{5}{2} = 2.5$$

$$S_2 = 1 + \frac{3}{2} + \frac{9}{4} = \frac{19}{4} = 4.75$$

$$S_3 = 1 + \frac{3}{2} + \frac{9}{4} + \frac{27}{8} = \frac{65}{8} = 8.125$$

$$S_4 = 1 + \frac{3}{2} + \frac{9}{4} + \frac{27}{8} + \frac{81}{16} = \frac{211}{16} = 13.1875$$

13. Find the first five terms of the sequence of partial sums for the series
$$\sum_{n=1}^{\infty} \frac{(-1)^{n+1}}{(2n)!}$$
Solution:

$$S_1 = \frac{1}{2!} = \frac{1}{2} = 0.5$$

$$S_2 = \frac{1}{2!} - \frac{1}{4!} = \frac{11}{24} \approx 0.4583$$

$$S_3 = \frac{1}{2!} - \frac{1}{4!} + \frac{1}{6!} = \frac{331}{720} \approx 0.4597$$

$$S_4 = \frac{1}{2!} - \frac{1}{4!} + \frac{1}{6!} - \frac{1}{8!} = \frac{18,535}{40,320} \approx 0.4597$$

$$S_5 = \frac{1}{2!} - \frac{1}{4!} + \frac{1}{6!} - \frac{1}{8!} + \frac{1}{10!} = \frac{1,668,151}{3,628,800} \approx 0.4597$$

15. Find the sum of the series
$$\sum_{n=0}^{\infty} \left(\frac{1}{5}\right)^n$$
Solution:

$$\sum_{n=0}^{\infty} \left(\frac{1}{5}\right)^n = \frac{1}{1 - (1/5)} = \frac{5}{4}$$

17. Find the sum of the series
$$\sum_{n=0}^{\infty} \left(\frac{1}{4^n} - \frac{1}{3^n}\right)$$
Solution:
$$\sum_{n=0}^{\infty} \left(\frac{1}{4^n} - \frac{1}{3^n}\right) = \sum_{n=0}^{\infty} \left(\frac{1}{4}\right)^n - \sum_{n=0}^{\infty} \left(\frac{1}{3}\right)^n$$
$$= \frac{1}{1 - (1/4)} - \frac{1}{1 - (1/3)}$$
$$= \frac{4}{3} - \frac{3}{2} = -\frac{1}{6}$$

19. Express the repeating decimal $0.\overline{0909}$ as the ratio of two integers by finding the sum of the geometric series.
Solution:
$$0.\overline{0909} = 0.09 + 0.0009 + 0.000009 + \ldots$$
$$= 0.09 + 0.09(0.01) + 0.09(0.01)^2 + \ldots$$
$$= \sum_{n=0}^{\infty} 0.09(0.01)^n$$
$$= \frac{0.09}{1 - 0.01} = \frac{0.09}{0.99} = \frac{1}{11}$$

21. Determine the convergence or divergence of the series
$$\sum_{n=1}^{\infty} \frac{1}{n^{5/2}}$$
Solution: This series converges by the p-series test since with $p = 5/2 > 1$.

23. Determine the convergence or divergence of the series
$$\sum_{n=1}^{\infty} \frac{n^2 + 1}{n(n + 1)}$$
Solution: This series diverges by the nth-Term Test since
$$\lim_{n \to \infty} \frac{n^2 + 1}{n(n + 1)} = 1 \neq 0$$

25. Determine the convergence or divergence of the series
$$\sum_{n=1}^{\infty} \frac{2^n}{n}$$
Solution: This series diverges by the Ratio Test since
$$\lim_{n \to \infty} \left|\frac{2^{n+1}}{n + 1} \cdot \frac{n}{2^n}\right| = \lim_{n \to \infty} \frac{2n}{n + 1} = 2 > 1$$

27. Determine the convergence or divergence of the series
$$\sum_{n=1}^{\infty} \frac{n4^n}{n!}$$
Solution: This series converges by the Ratio Test since
$$\lim_{n \to \infty} \left| \frac{(n+1)4^{n+1}}{(n+1)!} \cdot \frac{n!}{n4^n} \right| = \lim_{n \to \infty} \frac{4}{n} = 0 < 1$$

29. Determine the convergence or divergence of the series
$$\sum_{n=1}^{\infty} \frac{\sqrt[3]{n^2}}{n}$$
Solution: This series diverges by the p-series test since $p = 1/3 < 1$.
$$\sum_{n=1}^{\infty} \frac{\sqrt[3]{n^2}}{n} = \sum_{n=1}^{\infty} \frac{n^{2/3}}{n} = \sum_{n=1}^{\infty} \frac{1}{n^{1/3}}$$

31. Find the radius of convergence of the power series
$$\sum_{n=0}^{\infty} \frac{(-1)^n (x-2)^n}{(n+1)^2}$$
Solution:
$$R = \lim_{n \to \infty} \left| \frac{(-1)^n}{(n+1)^2} \cdot \frac{(n+2)^2}{(-1)^{n+1}} \right|$$
$$= \lim_{n \to \infty} \left| -\frac{n^2 + 4n + 4}{n^2 + 2n + 1} \right| = 1$$

33. Find the radius of convergence of the power series
$$\sum_{n=0}^{\infty} n!(x-2)^n$$
Solution:
$$R = \lim_{n \to \infty} \left| \frac{n!}{(n+1)!} \right| = \lim_{n \to \infty} \frac{1}{n+1} = 0$$

35. Apply Taylor's Theorem to find the power series for $f(x) = e^{-2x}$ centered at $c = 0$.
Solution:
$$\begin{aligned} f(x) &= e^{-2x} & f(0) &= 1 \\ f'(x) &= -2e^{-2x} & f'(0) &= -2 \\ f''(x) &= 4e^{-2x} & f''(0) &= 4 \\ f'''(x) &= -8e^{-2x} & f'''(0) &= -8 \\ & & &\vdots \\ & & f^{(n)}(0) &= (-2)^n \end{aligned}$$

The power series for f is
$$e^{-2x} = 1 - 2x + \frac{4x^2}{2!} - \frac{8x^3}{3!} + \ldots = \sum_{n=0}^{\infty} \frac{(-2)^n x^n}{n!}$$

37. Apply Taylor's Theorem to find the power series for $f(x) = 1/x$ centered at $c = -1$.
Solution:

$$f(x) = 1/x \qquad f(-1) = -1$$
$$f'(x) = -1/x^2 \qquad f'(-1) = -1$$
$$f''(x) = 2/x^3 \qquad f''(-1) = -2$$
$$f'''(x) = -6/x^4 \qquad f'''(-1) = -6$$
$$\vdots$$
$$f^{(n)}(-1) = -(n!)$$

The power series for f is

$$\frac{1}{x} = f(-1) + f'(-1)(x+1) + \frac{f''(-1)(x+1)^2}{2!} + \ldots$$

$$= -1 - (x+1) - \frac{2(x+1)^2}{2!} - \frac{6(x+1)^3}{3!} - \ldots$$

$$= -[1 + (x+1) + (x+1)^2 + (x+1)^3 + \ldots]$$

$$= -\sum_{n=0}^{\infty} (x+1)^n$$

39. Find the series representation for

$$f(x) = \int_0^x \ln(t+1)\,dt$$

Solution:

$$\frac{1}{t+1} = 1 - t + t^2 - t^3 + t^4 - t^5 + \ldots$$

$$\ln(t+1) = \int \frac{1}{t+1}\,dt = t - \frac{t^2}{2} + \frac{t^3}{3} - \frac{t^4}{4} + \ldots$$

$$= \sum_{n=1}^{\infty} \frac{(-1)^{n+1} t^n}{n}$$

$$\int_0^x \ln(t+1)\,dt = \sum_{n=1}^{\infty} \frac{(-1)^{n+1} t^{n+1}}{n(n+1)} \bigg]_0^x$$

$$= \sum_{n=1}^{\infty} \frac{(-1)^{n+1} x^{n+1}}{n(n+1)}$$

41. Use a Taylor polynomial of sixth degree to approximate $\ln 1.75$.
Solution:

$$\ln x = \sum_{n=0}^{\infty} \frac{(-1)^{n-1}(x-1)^n}{n}$$

$$S_6(x) = (x-1) - \frac{(x-1)^2}{2} + \frac{(x-1)^3}{3} - \ldots - \frac{(x-1)^6}{6}$$

$$S_6(1.75) = 0.75 - \frac{(0.75)^2}{2} + \frac{(0.75)^3}{3} - \ldots - \frac{(0.75)^6}{6}$$

$$\approx 0.548$$

Review Exercises for Chapter 9

43. Use a Taylor polynomial of sixth degree to approximate $e^{0.6}$.
Solution:

$$e^x = \sum_{n=0}^{\infty} \frac{x^n}{n!}$$

$$S_6(x) = 1 + x + \frac{x^2}{2!} + \frac{x^3}{3!} + \frac{x^4}{4!} + \frac{x^5}{5!} + \frac{x^6}{6!}$$

$$S_6(0.6) = 1 + 0.6 + \frac{(0.6)^2}{2} + \ldots + \frac{(0.6)^5}{120} + \frac{(0.6)^6}{720}$$

$$\approx 1.822$$

45. Use a Taylor polynomial of sixth degree to approximate
$$\int_0^{0.3} \sqrt{1 + x^3} \, dx$$
Solution:

$$\sqrt{1 + x^3} = 1 + \frac{x^3}{2} - \frac{x^6}{8} + \ldots$$

$$\int_0^{0.3} \sqrt{1 + x^3} \, dx \approx \left[x + \frac{x^4}{8} - \frac{x^7}{56} \right]_0^{0.3}$$

$$= 0.3 + \frac{(0.3)^4}{8} - \frac{(0.3)^7}{56}$$

$$\approx 0.301$$

47. Sketch the graph of $f(x) = e^x$ and its fifth-degree Taylor polynomial on the same axes.
Solution:

$$f(x) = e^x = \sum_{n=0}^{\infty} \frac{x^n}{n!}$$

$$S_5(x) = 1 + x + \frac{x^2}{2} + \frac{x^3}{6} + \frac{x^4}{24} + \frac{x^5}{120}$$

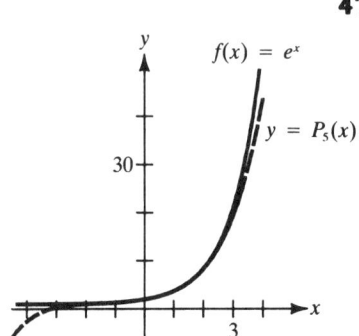

49. Use Newton's Method to approximate to three decimal places the zeros of $f(x) = x^3 - 3x - 1$.
Solution: From the graph we see that there are three zeros.

$$f'(x) = 3x^2 - 3$$

n	x_n	$f(x_n)$	$f'(x_n)$	$\dfrac{f(x_n)}{f'(x_n)}$	$x_n - \dfrac{f(x_n)}{f'(x_n)}$
1	-1.5000	0.1250	3.7500	0.0333	-1.5333
2	-1.5333	-0.0050	4.0533	-0.0012	-1.5321
3	-1.5321	-0.0000	4.0419	-0.0000	-1.5321

n	x_n	$f(x_n)$	$f'(x_n)$	$\dfrac{f(x_n)}{f'(x_n)}$	$x_n - \dfrac{f(x_n)}{f'(x_n)}$
1	-0.5000	0.3750	-2.2500	-0.1667	-0.3333
2	-0.3333	-0.0370	-2.6667	0.0139	-0.3472
3	-0.3472	-0.0002	-2.6383	0.0001	-0.3473

n	x_n	$f(x_n)$	$f'(x_n)$	$\dfrac{f(x_n)}{f'(x_n)}$	$x_n - \dfrac{f(x_n)}{f'(x_n)}$
1	2.0000	1.0000	9.0000	0.1111	1.8889
2	1.8889	0.0727	7.7037	0.0094	1.8795
3	1.8795	0.0005	7.5970	0.0001	1.8794

Approximations: $x \approx -1.532, -0.347, 1.879$

51. Use Newton's Method to approximate to three decimal places the x-value of the points of intersection of the graphs of $f(x) = x^4$ and $g(x) = x + 3$.
    Solution: From the graph we see that there are two zeros. Let $x^4 = x + 3$ and define $h(x) = x^4 - x - 3$, then $h'(x) = 4x^3 - 1$.

n	x_n	$h(x_n)$	$h'(x_n)$	$\dfrac{h(x_n)}{h'(x_n)}$	$x_n - \dfrac{h(x_n)}{h'(x_n)}$
1	-1.2000	0.2736	-7.9120	-0.0346	-1.1654
2	-1.1654	0.0101	-7.3315	-0.0014	-1.1640
3	-1.1640	0.0000	-7.3090	-0.0000	-1.1640

n	x_n	$h(x_n)$	$h'(x_n)$	$\dfrac{h(x_n)}{h'(x_n)}$	$x_n - \dfrac{h(x_n)}{h'(x_n)}$
1	1.4000	-0.5584	9.9760	-0.0560	1.4560
2	1.4560	0.0378	11.3459	0.0033	1.4526
3	1.4526	0.0001	11.2612	0.0000	1.4526

Approximations: $x \approx -1.164, 1.453$

53. Show that the Ratio Test is inconclusive for a p-series.
    Solution:
    $$\sum_{n=0}^{\infty} \frac{1}{n^p}, \qquad a_n = \frac{1}{n^p}$$
    $$\lim_{n \to \infty} \left| \frac{a_{n+1}}{a_n} \right| = \lim_{n \to \infty} \left| \frac{1/(n+1)^p}{1/n^p} \right|$$
    $$= \lim_{n \to \infty} \left| \frac{n^p}{(n+1)^p} \right| = 1$$

    When the limit is one, the Ratio Test is inconclusive.

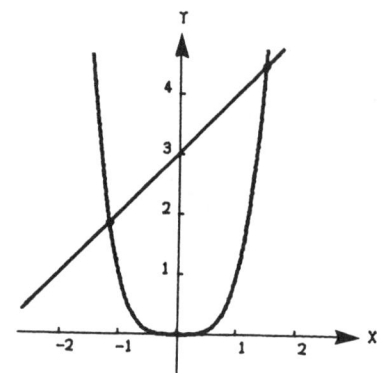

PRACTICE TEST FOR CHAPTER 9

1. Find the general term of the sequence
$$\frac{1}{2}, \frac{2}{5}, \frac{3}{10}, \frac{4}{17}, \frac{5}{26}, \ldots$$

2. Find the general term of the sequence
$$5, -7, 9, -11, 13, \ldots$$

3. Determine the convergence or divergence of the sequence whose general term is
$$a_n = \frac{n^2}{3n^2 + 4}$$

4. Determine the convergence or divergence of the sequence whose general term is
$$a_n = \frac{4n}{\sqrt{n^2 + 1}}$$

5. Find the sum of the series
$$\sum_{n=0}^{\infty} \left(\frac{1}{5^n} - \frac{1}{7^n}\right)$$

6. Determine the convergence or divergence of the series
$$\sum_{n=1}^{\infty} \frac{3^n}{n!}$$

7. Determine the convergence or divergence of the series
$$\sum_{n=1}^{\infty} \frac{1}{n\sqrt[3]{n}}$$

8. Determine the convergence or divergence of the series
$$\sum_{n=1}^{\infty} \frac{n}{2n + 3}$$

9. Determine the convergence or divergence of the series
$$\sum_{n=0}^{\infty} \frac{(-1)^n 6^n}{5^n}$$

10. Determine the convergence or divergence of the series
$$\sum_{n=1}^{\infty} \frac{\sqrt[3]{n}}{\sqrt{n}}$$

11. Determine the convergence or divergence of the series
$$\sum_{n=1}^{\infty} \frac{5^n n!}{(n + 1)!}$$

12. Determine the convergence or divergence of the series
$$\sum_{n=0}^{\infty} 4(0.27)^n$$

13. Determine the convergence or divergence of the series
$$\sum_{n=1}^{\infty} (1 + \frac{1}{3^n})$$

14. Find the radius of convergence of the power series
$$\sum_{n=0}^{\infty} \frac{(-1)^n (x-3)^n}{(n+4)^2}$$

15. Find the radius of convergence of the power series
$$\sum_{n=0}^{\infty} \frac{x^n}{(n+1)!}$$

16. Apply Taylor's Theorem to find the power series (centered at 0) for
$$f(x) = e^{-4x}$$

17. Apply Taylor's Theorem to find the power series (centered at 1) for
$$f(x) = \frac{1}{\sqrt[3]{x}}$$

18. Use the ninth-degree Taylor Polynomial for e^{x^3} to approximate the value of
$$\int_0^{0.213} e^{x^3} \, dx$$

19. Use Newton's Method to approximate the zero of the function $f(x) = x^3 + x - 3$. (Make your approximation good to three decimal places.)

20. Use Newton's Method to approximate $\sqrt[4]{10}$ to three decimal places.

Chapter 10 The Trigonometric Functions

Section 10.1 Radian Measure of Angles

1. Determine two coterminal angles (one positive and one negative) for (a) 36° and (b) −45°. Give the answers in degrees.
   Solution:
   (a) Positive: 36° + 360° = 396°
       Negative: 36° − 360° = −324°
   (b) Positive: −45° + 360° = 315°
       Negative: −45° − 360° = −405°

3. Determine two coterminal angles (one positive and one negative) for (a) 300° and (b) 740°. Give the answers in degrees.
   Solution:
   (a) Positive: 300° + 360° = 660°
       Negative: 300° − 360° = −60°
   (b) Positive: 740° − 2(360°) = 20°
       Negative: 740° − 3(360°) = −340°

5. Determine two coterminal angles (one positive and one negative) for (a) π/9 and (b) 4π/3. Give the answers in radians.
   Solution:
   (a) Positive: (π/9) + 2π = 19π/9
       Negative: (π/9) − 2π = −17π/9
   (b) Positive: (4π/3) + 2π = 10π/3
       Negative: (4π/3) − 2π = −2π/3

7. Determine two coterminal angles (one positive and one negative) for (a) −9π/4 and (b) −2π/15. Give the answers in radians.
   Solution:
   (a) Positive: (−9π/4) + 2(2π) = 7π/4
       Negative: (−9π/4) + 2π = −π/4
   (b) Positive: (−2π/15) + 2π = 28π/15
       Negative: (−2π/15) − 2π = −32π/15

9. Express 30° is radian measure as a multiple of π.
   Solution:
   $$30° \left(\frac{\pi \text{ radians}}{180°} \right) = \frac{\pi}{6} \text{ radians}$$

11. Express 315° is radian measure as a multiple of π.
    Solution:
    $$315° \left(\frac{\pi \text{ radians}}{180°} \right) = \frac{7\pi}{4} \text{ radians}$$

13. Express −20° is radian measure as a multiple of π.
    Solution:
    $$-20° \left(\frac{\pi \text{ radians}}{180°} \right) = -\frac{\pi}{9} \text{ radians}$$

15. Express $-270°$ in radian measure as a multiple of π.
Solution:
$$-270°\left(\frac{\pi \text{ radians}}{180°}\right) = -\frac{3\pi}{2} \text{ radians}$$

17. Express $3\pi/2$ in degree measure.
Solution:
$$\frac{3\pi}{2}\left(\frac{180°}{\pi}\right) = 270°$$

19. Express $-7\pi/12$ in degree measure.
Solution:
$$-\frac{7\pi}{12}\left(\frac{180°}{\pi}\right) = -105°$$

21. Express $7\pi/3$ in degree measure.
Solution:
$$\frac{7\pi}{3}\left(\frac{180°}{\pi}\right) = 420°$$

23. Express $11\pi/6$ in degree measure.
Solution:
$$\frac{11\pi}{6}\left(\frac{180°}{\pi}\right) = 330°$$

25. Solve the triangle for θ and c.
Solution: The angle θ is $\theta = 90° - 30° = 60°$. Since the vertical side is $c/2$, we can use the Pythagorean Theorem to find the length of the hypotenuse as follows
$$c^2 = (5\sqrt{3})^2 + (c/2)^2$$ 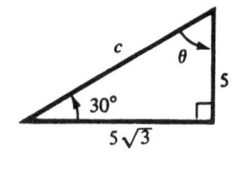
$$3c^2/4 = 75$$
$$c^2 = 100$$
$$c = 10$$

27. Solve the triangle for θ and a.
Solution: The angle θ is 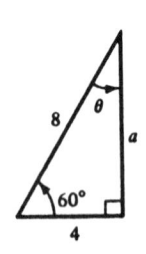
$$\theta = 90° - 60° = 30°$$
By the Pythagorean Theorem the value of a is
$$a = \sqrt{8^2 - 4^2} = \sqrt{64 - 16}$$
$$= \sqrt{48}$$
$$= 4\sqrt{3}$$

29. Solve the triangle for θ.
Solution: Since the triangle is iscoseles, we have 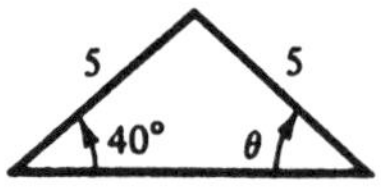
$$\theta = 40°$$

31. A person 6 feet tall standing 12 feet from a streetlight casts a shadow 8 feet long. What is the height of the streetlight?
   Solution: Using similar triangles, we have

   $$\frac{h}{20} = \frac{6}{8}$$

   which implies that

   $$h = 15 \text{ feet}$$

33. Let r represent the radius of a circle, θ the central angle (measured in radians), and s the length of the arc subtended by the angle. Use the relationship θ = s/r to complete the following table.
   Solution:

r	8 ft	15 in	85 cm	24 in	12,963/π mi
s	12 ft	24 in	200.28 cm	96 in	8642 mi
θ	1.5	1.6	3π/4	4	2π/3

35. A man bends his elbow through 75°. The distance from his elbow to the tip of his index finger is $18\frac{3}{4}$ inches.
   (a) Find the radian measure of this angle.
   (b) Find the distance the tip of the index finger moves.
   Solution:
   (a) The radian measure is

   $$75°\left(\frac{\pi \text{ radians}}{180°}\right) = \frac{5\pi}{12} \text{ radians}$$

   (b) The distance moved is

   $$s = \frac{5\pi}{12}(18.75) = 7.8125\pi \text{ in} \approx 24.54 \text{ in}$$

37. Assuming that the earth is a sphere of radius 4000 miles, what is the difference in latitude of two cities, one of which is 325 miles due north of the other? [Latitude lines on the earth run parallel to the equator and measure the angle from the equator to a point north or south of the equator.]
   Solution: Using the equation θ = s/r, we have

   $$\theta = \frac{325}{4000} = \frac{13}{160}$$

   $$= \frac{13}{160}\left(\frac{180°}{\pi}\right)$$

   $$= \left(\frac{117}{8\pi}\right)°$$

   $$\approx 4.655°$$

Section 10.2 The Trigonometric Functions

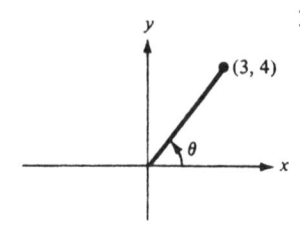

1. Determine all six trigonometric functions for θ.
   Solution: Since $x = 3$ and $y = 4$, it follows that $r = \sqrt{3^2 + 4^2} = 5$. Therefore, we have

   $\sin\theta = 4/5 \qquad \csc\theta = 5/4$
   $\cos\theta = 3/5 \qquad \sec\theta = 5/3$
   $\tan\theta = 4/3 \qquad \cot\theta = 3/4$

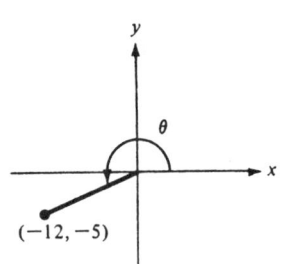

3. Determine all six trigonometric functions for θ.
   Solution: Since $x = -12$ and $y = -5$, it follows that $r = \sqrt{(-12)^2 + (-5)^2} = 13$. Therefore, we have

   $\sin\theta = -5/13 \qquad \csc\theta = -13/5$
   $\cos\theta = -12/13 \qquad \sec\theta = -13/12$
   $\tan\theta = 5/12 \qquad \cot\theta = 12/5$

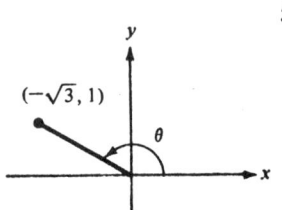

5. Determine all six trigonometric functions for θ.
   Solution: Since $x = -\sqrt{3}$ and $y = 1$, it follows that $r = \sqrt{3 + 1^2} = 2$. Therefore, we have

   $\sin\theta = 1/2 \qquad \csc\theta = 2$
   $\cos\theta = -\sqrt{3}/2 \qquad \sec\theta = -2\sqrt{3}/3$
   $\tan\theta = -\sqrt{3}/3 \qquad \cot\theta = -\sqrt{3}$

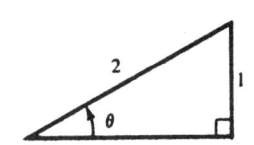

7. Find $\csc\theta$ given $\sin\theta = 1/2$.
   Solution:

   $$\csc\theta = \frac{1}{\sin\theta} = \frac{1}{1/2} = 2$$

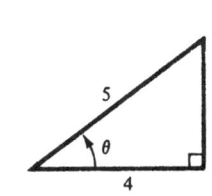

9. Find $\cot\theta$ given $\cos\theta = 4/5$.
   Solution: Since $x = 4$ and $r = 5$, the length of the opposite side is $y = \sqrt{5^2 - 4^2} = 3$. Therefore, we have

   $$\cot\theta = \frac{x}{y} = \frac{4}{3}$$

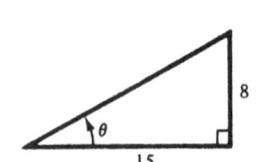

11. Find $\sec\theta$ given $\cot\theta = 15/8$.
    Solution: Since $x = 8$ and $y = 15$, the length of the hypotenuse is $r = \sqrt{15^2 + 8^2} = 17$. Therefore, we have

    $$\sec\theta = \frac{r}{x} = \frac{17}{8}$$

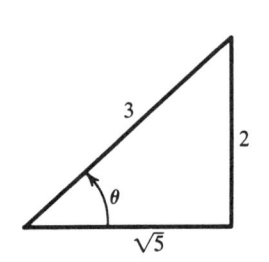

13. Sketch a right triangle corresponding to $\sin\theta = 2/3$, and find the other five trigonometric functions of θ.
    Solution: Since $y = 2$ and $r = 3$, the length of the adjacent side is $x = \sqrt{3^2 - 2^2} = \sqrt{5}$. Therefore,

    $\sin\theta = 2/3 \qquad \csc\theta = 3/2$
    $\cos\theta = \sqrt{5}/3 \qquad \sec\theta = 3\sqrt{5}/5$
    $\tan\theta = 2\sqrt{5}/5 \qquad \cot\theta = \sqrt{5}/2$

Section 10.2

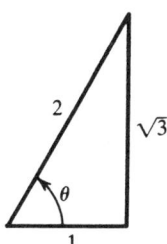

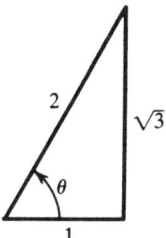

15. Sketch a right triangle corresponding to $\sec \theta = 2$, and find the other five trigonometric functions of θ.
    Solution: Since $x = 1$ and $r = 2$, the length of the opposite side is $y = \sqrt{2^2 - 1^2} = \sqrt{3}$. Therefore,

    $\sin \theta = \sqrt{3}/2$ $\csc \theta = 2\sqrt{3}/3$
    $\cos \theta = 1/2$ $\sec \theta = 2$
    $\tan \theta = \sqrt{3}$ $\cot \theta = \sqrt{3}/3$

17. Sketch a right triangle corresponding to $\tan \theta = 3$, and find the other five trigonometric functions of θ.
    Solution: Since $x = 1$ and $y = 3$, the length of the hypothenuse is $r = \sqrt{1^2 + 3^2} = \sqrt{10}$. Therefore,

    $\sin \theta = 3\sqrt{10}/10$ $\csc \theta = \sqrt{10}/3$
    $\cos \theta = \sqrt{10}/10$ $\sec \theta = \sqrt{10}$
    $\tan \theta = 3$ $\cot \theta = 1/3$

19. Given that $\sin \theta < 0$ and $\cos \theta < 0$, determine the quadrant in which θ lies.
    Solution: Since the sine is negative and the cosine is negative, θ must lie in Quadrant III.

21. Given that $\sin \theta > 0$ and $\sec \theta > 0$, determine the quadrant in which θ lies.
    Solution: Since the sine is positive and the secant is positive, θ must lie in Quadrant I.

23. Given that $\csc \theta > 0$ and $\tan \theta < 0$, determine the quadrant in which θ lies.
    Solution: Since the cosecant is positive and the tangent is negative, θ must lie in Quadrant II.

25. Evaluate the sine, cosine, and tangent of
    (a) $60°$ (b) $2\pi/3$
    Do not use a calculator.
    Solution:
    (a) $\sin 60° = \sqrt{3}/2$ (b) $\sin (2\pi/3) = \sqrt{3}/2$
         $\cos 60° = 1/2$ $\cos (2\pi/3) = -1/2$
         $\tan 60° = \sqrt{3}$ $\tan (2\pi/3) = -\sqrt{3}$

27. Evaluate the sine, cosine, and tangent of
    (a) $-\pi/6$ (b) $150°$
    Do not use a calculator.
    Solution:
    (a) $\sin (-\pi/6) = -1/2$ (b) $\sin 150° = 1/2$
         $\cos (-\pi/6) = \sqrt{3}/2$ $\cos 150° = -\sqrt{3}/2$
         $\tan (-\pi/6) = -\sqrt{3}/3$ $\tan 150° = -\sqrt{3}/3$

29. Evaluate the sine, cosine, and tangent of
    (a) $225°$ (b) $-225°$
    Do not use a calculator.
    Solution:
    (a) $\sin 225° = -\sqrt{2}/2$ (b) $\sin (-225°) = \sqrt{2}/2$
         $\cos 225° = -\sqrt{2}/2$ $\cos (-225°) = -\sqrt{2}/2$
         $\tan 225° = 1$ $\tan (-225°) = -1$

Section 10.2

31. Evaluate the sine, cosine, and tangent of
   (a) 750°   (b) 510°
   Do not use a calculator.
   Solution:
   (a) $\sin 750° = 1/2$ (b) $\sin 510° = 1/2$
        $\cos 750° = \sqrt{3}/2$ $\cos 510° = -\sqrt{3}/2$
        $\tan 750° = \sqrt{3}/3$ $\tan 510° = -\sqrt{3}/3$

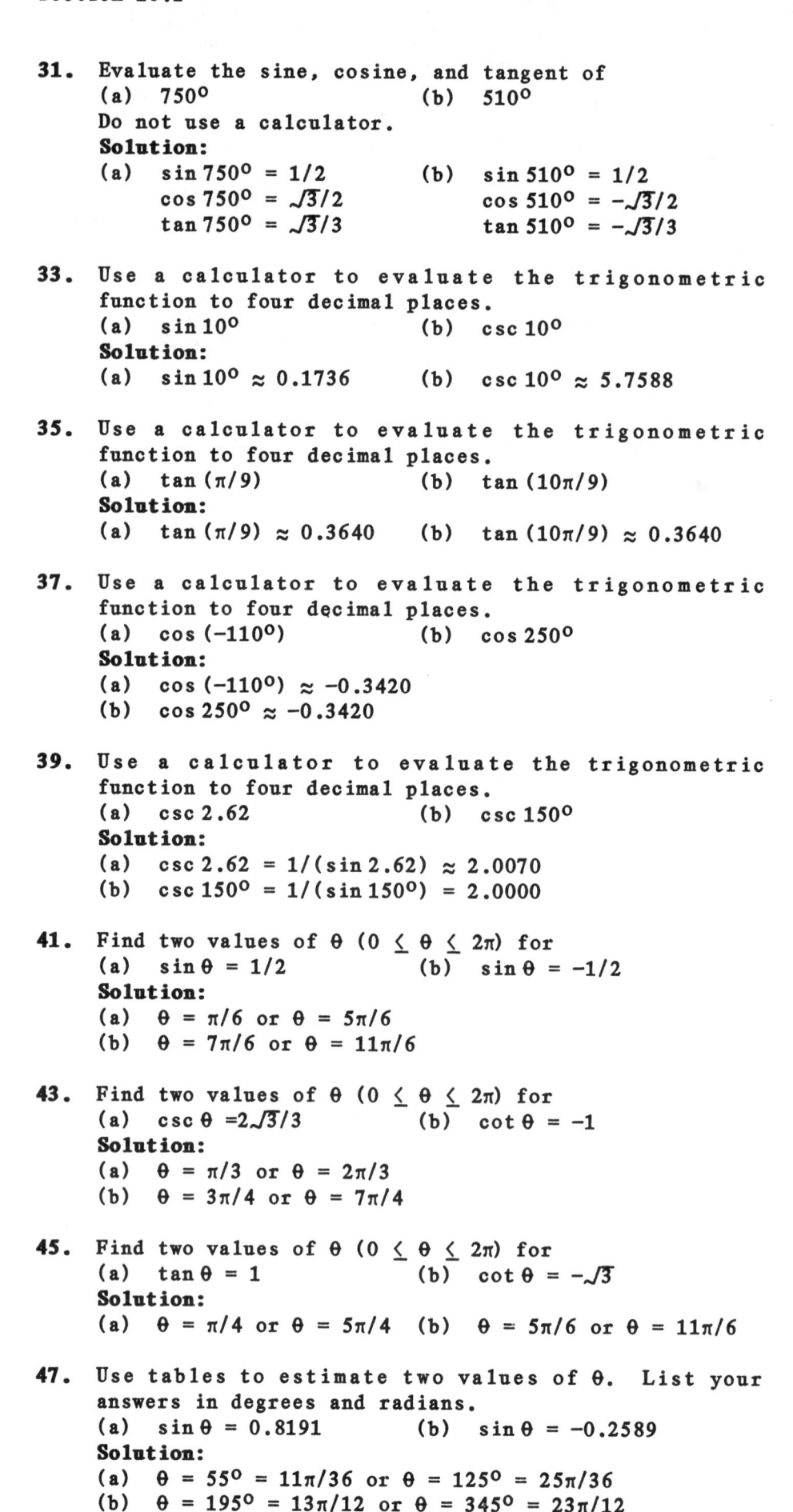

33. Use a calculator to evaluate the trigonometric function to four decimal places.
   (a) $\sin 10°$ (b) $\csc 10°$
   Solution:
   (a) $\sin 10° \approx 0.1736$ (b) $\csc 10° \approx 5.7588$

35. Use a calculator to evaluate the trigonometric function to four decimal places.
   (a) $\tan(\pi/9)$ (b) $\tan(10\pi/9)$
   Solution:
   (a) $\tan(\pi/9) \approx 0.3640$ (b) $\tan(10\pi/9) \approx 0.3640$

37. Use a calculator to evaluate the trigonometric function to four decimal places.
   (a) $\cos(-110°)$ (b) $\cos 250°$
   Solution:
   (a) $\cos(-110°) \approx -0.3420$
   (b) $\cos 250° \approx -0.3420$

39. Use a calculator to evaluate the trigonometric function to four decimal places.
   (a) $\csc 2.62$ (b) $\csc 150°$
   Solution:
   (a) $\csc 2.62 = 1/(\sin 2.62) \approx 2.0070$
   (b) $\csc 150° = 1/(\sin 150°) = 2.0000$

41. Find two values of θ ($0 \le \theta \le 2\pi$) for
   (a) $\sin \theta = 1/2$ (b) $\sin \theta = -1/2$
   Solution:
   (a) $\theta = \pi/6$ or $\theta = 5\pi/6$
   (b) $\theta = 7\pi/6$ or $\theta = 11\pi/6$

43. Find two values of θ ($0 \le \theta \le 2\pi$) for
   (a) $\csc \theta = 2\sqrt{3}/3$ (b) $\cot \theta = -1$
   Solution:
   (a) $\theta = \pi/3$ or $\theta = 2\pi/3$
   (b) $\theta = 3\pi/4$ or $\theta = 7\pi/4$

45. Find two values of θ ($0 \le \theta \le 2\pi$) for
   (a) $\tan \theta = 1$ (b) $\cot \theta = -\sqrt{3}$
   Solution:
   (a) $\theta = \pi/4$ or $\theta = 5\pi/4$ (b) $\theta = 5\pi/6$ or $\theta = 11\pi/6$

47. Use tables to estimate two values of θ. List your answers in degrees and radians.
   (a) $\sin \theta = 0.8191$ (b) $\sin \theta = -0.2589$
   Solution:
   (a) $\theta = 55° = 11\pi/36$ or $\theta = 125° = 25\pi/36$
   (b) $\theta = 195° = 13\pi/12$ or $\theta = 345° = 23\pi/12$

Section 10.2

49. Use tables to estimate two values of θ. List your answers in degrees and radians.
(a) $\cos\theta = 0.9848$ (b) $\cos\theta = -0.5890$
Solution:
(a) $\theta = 10° = \pi/18$ or $\theta = 350° = 35\pi/18$
(b) $\theta = 126° = 7\pi/10$ or $\theta = 234° = 13\pi/10$

51. Use tables to estimate two values of θ. List your answers in degrees and radians.
(a) $\tan\theta = 1.192$ (b) $\tan\theta = -8.144$
Solution:
(a) $\theta = 50° = 5\pi/18$ or $\theta = 230° = 23\pi/18$
(b) $\theta = 97° = 97\pi/180$ or $\theta = 277° = 277\pi/180$

53. Solve for θ ($0 \le \theta \le 2\pi$) in the equation

$$2\sin^2\theta = 1$$

Solution: Solving for $\sin\theta$ produces

$$\sin\theta = \pm\frac{\sqrt{2}}{2}$$
$$\theta = \frac{\pi}{4}, \frac{3\pi}{4}, \frac{5\pi}{4}, \frac{7\pi}{4}$$

55. Solve for θ ($0 \le \theta \le 2\pi$) in the equation

$$\tan^2\theta - \tan\theta = 0$$
Solution:
$$\tan\theta(\tan\theta - 1) = 0$$

$\tan\theta = 0$ or $\tan\theta = 1$

$\theta = 0, \pi, 2\pi$ $\qquad \theta = \dfrac{\pi}{4}, \dfrac{5\pi}{4}$

57. Solve for θ ($0 \le \theta \le 2\pi$) in the equation

$$\sin 2\theta + 2\sin\theta = 0$$
Solution: Using the identity $\sin 2\theta = 2\sin\theta\cos\theta$, we have

$$2\sin\theta\cos\theta + 2\sin\theta = 0$$
$$\sin\theta(\cos\theta + 1) = 0$$

$\sin\theta = 0$ or $\cos\theta = -1$

$\theta = 0, \pi, 2\pi$ $\qquad \theta = \pi$

59. Solve for θ ($0 \le \theta \le 2\pi$) in the equation

$$\sin\theta = \cos\theta$$
Solution: Dividing both sides by $\cos\theta$ produces

$$\tan\theta = 1$$
$$\theta = \frac{\pi}{4}, \frac{5\pi}{4}$$

61. Solve for θ (0 ≤ θ ≤ 2π) in the equation

$$\cos^2 \theta + \sin \theta = 1$$

Solution: Using the identity $\cos^2 \theta = 1 - \sin^2 \theta$ produces

$$1 - \sin^2 \theta + \sin \theta = 1$$
$$\sin^2 \theta - \sin \theta = 0$$
$$\sin \theta (\sin \theta - 1) = 0$$

$\sin \theta = 0$ or $\sin \theta = 1$

$\theta = 0, \pi, 2\pi$ $\theta = \dfrac{\pi}{2}$

63. Solve for y.
Solution: Since

$$\tan 30° = \frac{1}{\sqrt{3}} = \frac{y}{100}$$

it follows that

$$y = \frac{100}{\sqrt{3}} = \frac{100\sqrt{3}}{3}$$

65. Solve for x.
Solution: Since

$$\cot 60° = \frac{1}{\sqrt{3}} = \frac{x}{25}$$

it follows that

$$x = \frac{25}{\sqrt{3}} = \frac{25\sqrt{3}}{3}$$

67. Solve for r.
Solution: Since

$$\sin 40° = \frac{10}{r} \approx 0.6428$$

it follows that

$$r = \frac{10}{0.6428} \approx 15.5572$$

69. Solve for y.
Solution: Since

$$\sin 50° = \frac{y}{12} \approx 0.7660$$

it follows that

$$y = 12(0.7660) \approx 9.1925$$

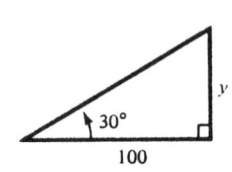

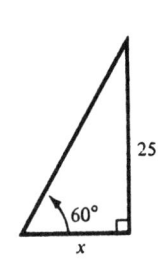

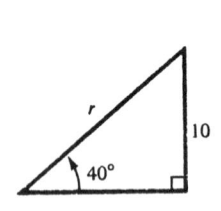

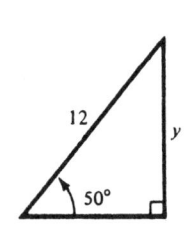

Section 10.2

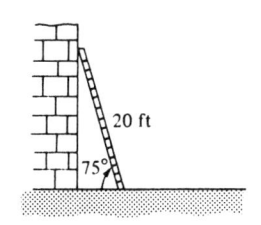

71. A 20-foot ladder leaning against the side of a house makes a 75° angle with the ground. How far up the side of the house does the ladder reach?
Solution: Let h be the height of the ladder. Then
$$\sin 75° = \frac{h}{20} \approx 0.9659$$
and
$$h = 20(0.9659) \approx 19.3185 \text{ feet}$$

73. From a 150-foot observation tower on the coast, a Coast Guard officer sights a boat in difficulty. The angle of depression of the boat is 4°. How far is the boat from the shoreline?
Solution: Let x be the distance from the shore. Then
$$\cot 4° = \frac{x}{150} \approx 14.3007$$
and
$$x = 150(14.3007) \approx 2145.1 \text{ feet}$$

75. The average daily temperature (in degrees Fahrenheit) for a certain city is given by
$$T(t) = 45 - 23 \cos\left[\frac{2\pi}{365}(t - 32)\right]$$
where t is the time in days with t = 1 corresponding to January 1. Find the average temperature on the following days.
(a) January 1   (b) July 4 (t = 185)
(c) October 18 (t = 291)
Solution:
(a) For January 1, we have t = 1 and the average temperature is
$$T(1) = 45 - 23 \cos\left[\frac{2\pi}{365}(-31)\right]$$
$$\approx 45 - 23 \cos(-0.5336)$$
$$\approx 45 - 19.8 = 25.2° \text{ Fahrenheit}$$

(b) For July 4, we have t = 185 and the average temperature is
$$T(185) = 45 - 23 \cos\left[\frac{2\pi}{365}(153)\right]$$
$$\approx 45 - 23 \cos(2.6338)$$
$$\approx 45 + 20.1 = 65.1° \text{ Fahrenheit}$$

(c) For October 18, we have t = 291 and the average temperature is
$$T(291) = 45 - 23 \cos\left[\frac{2\pi}{365}(259)\right]$$
$$\approx 45 - 23 \cos(4.4585)$$
$$\approx 45 + 5.8 = 50.8° \text{ Fahrenheit}$$

Section 10.3 Graphs of Trigonometric Functions

1. Find the period and amplitude of $y = 2 \sin 2x$.
   Solution:
   Period: $2\pi/2 = \pi$ Amplitude: 2

3. Find the period and amplitude of $y = (3/2) \cos (x/2)$.
   Solution:
   Period: $2\pi/(1/2) = 4\pi$ Amplitude: 3/2

5. Find the period and amplitude of $y = (1/2) \sin \pi x$.
   Solution:
   Period: $2\pi/\pi = 2$ Amplitude: 1/2

7. Find the period and amplitude of $y = -2 \sin x$.
   Solution:
   Period: $2\pi/1 = 2\pi$ Amplitude: 2

9. Find the period and amplitude of $y = -2 \sin 10x$.
   Solution:
   Period: $2\pi/10 = \pi/5$ Amplitude: 2

11. Find the period and amplitude of $y = (1/2) \cos (2x/3)$.
    Solution:
    Period: $2\pi/(2/3) = 3\pi$ Amplitude: 1/2

13. Find the period and amplitude of $y = 3 \sin 4\pi x$.
    Solution:
    Period: $2\pi/4\pi = 1/2$ Amplitude: 3

15. Find the period $y = 5 \tan 2x$.
    Solution:
    Period: $\pi/2$

17. Find the period $y = 3 \sec 5x$.
    Solution:
    Period: $2\pi/5$

19. Find the period $y = \cot (\pi x/3)$.
    Solution:
    Period: $\pi/(\pi/3) = 3$

21. Match $y = \sec 2x$ with the correct graph.
    Solution: The graph of this function has a period of π and matches graph (c).

23. Match $y = \cot \pi x$ with the correct graph.
    Solution: The graph of this function has a period of 1 and matches graph (f).

25. Match $y = 2 \csc (x/2)$ with the correct graph.
    Solution: The graph of this function has a period of 4π and matches graph (b).

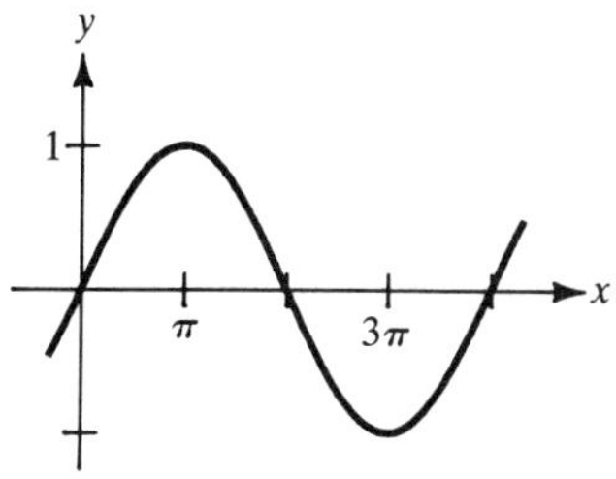

Section 10.3

27. Sketch the graph of $y = \sin(x/2)$.
Solution:
    Period: 4π
    Amplitude: 1
    x-intercepts: $(0, 0), (2\pi, 0), (4\pi, 0)$
    maximum: $(\pi, 1)$
    minimum: $(3\pi, -1)$

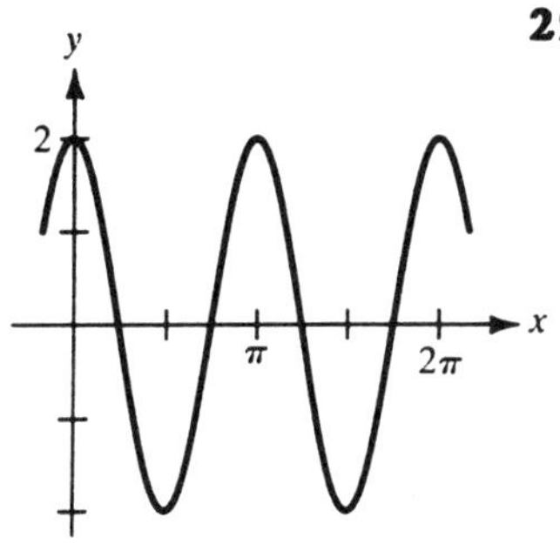

29. Sketch the graph of $y = 2\cos 2x$.
Solution:
    Period: π
    Amplitude: 2
    x-intercepts: $(\pi/4, 0), (3\pi/4, 0), (5\pi/4, 0)$
    maximum: $(0, 2), (\pi, 2)$
    minimum: $(\pi/2, -2), (3\pi/2, -2)$

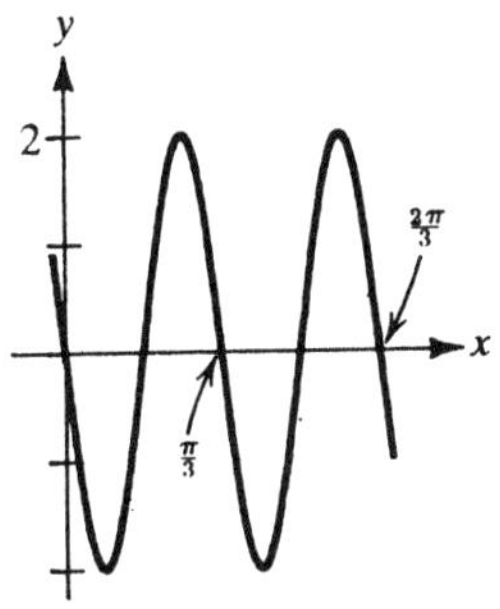

31. Sketch the graph of $y = -2\sin 6x$.
Solution:
    Period: $\pi/3$
    Amplitude: 2
    x-intercepts: $(0, 0), (\pi/6, 0), (\pi/3, 0),$
                    $(\pi/2, 0), (2\pi/3, 0)$
    maximum: $(\pi/4, 2), (7\pi/4, 2)$
    minimum: $(\pi/12, -2), (5\pi/12, -2)$

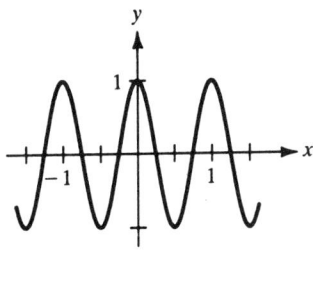

33. Sketch the graph of $y = \cos 2\pi x$.
Solution:
    Period: 1
    Amplitude: 1
    x-intercepts: $(-3/4, 0), (-1/4, 0),$
                    $(1/4, 0), (3/4, 0)$
    maximum: $(-1, 1), (0, 1), (1, 1)$
    minimum: $(-1/2, -1), (1/2, 1)$

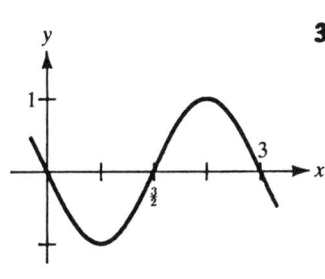

35. Sketch the graph of $y = -\sin(2\pi x/3)$.
Solution:
    Period: 3
    Amplitude: 1
    x-intercepts: $(0, 0), (3/2, 0), (3, 0)$
    maximum: $(9/4, 1)$
    minimum: $(3/4, -1)$

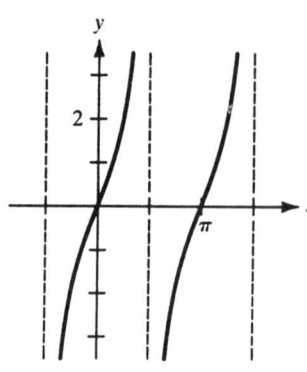

37. Sketch the graph of $y = 2 \tan x$.
Solution:
Period: π
x-intercepts: $(0, 0)$
asymptotes: $x = -\pi/2$, $x = \pi/2$

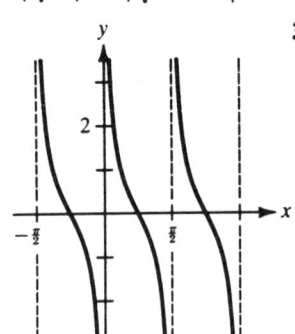

39. Sketch the graph of $y = \cot 2x$.
Solution:
Period: $\pi/2$
x-intercepts: $(-3\pi/4, 0)$, $(-\pi/4, 0)$,
$(\pi/4, 0)$, $(3\pi/4, 0)$
asymptotes: $x = -\pi/2$, $x = 0$, $x = \pi/2$

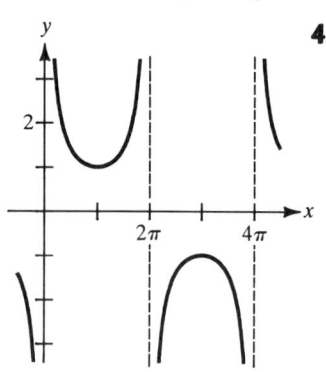

41. Sketch the graph of $y = \csc(x/2)$.
Solution:
Period: 4π
asymptotes: $x = 0$, $x = 2\pi$, $x = 4\pi$
rel. minimum: $(\pi, 1)$
rel. maximum: $(3\pi, -1)$

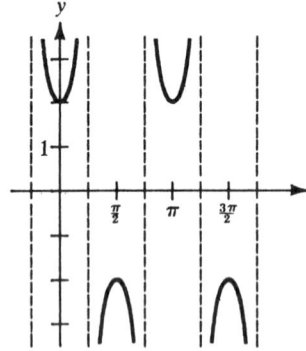

43. Sketch the graph of $y = 2 \sec 2x$.
Solution:
Period: π
asymptotes: $x = -\pi/4$, $x = \pi/4$, $x = 3\pi/4$,
$x = 5\pi/4$, $x = 7\pi/4$
rel. minimum: $(0, 2)$, $(\pi, 2)$
rel. maximum: $(\pi/2, -2)$, $(3\pi/2, -2)$

45. Sketch the graph of $y = \csc 2\pi x$.
Solution:
Period: 1
asymptotes: $x = -3/2$, $x = -1$, $x = -1/2$, $x = 0$
$x = 1/2$, $x = 1$, $x = 3/2$
rel. minimum: $(-7/4, 1)$, $(-3/4, 1)$,
$(1/4, 1)$, $(5/4, 1)$
rel. maximum: $(-5/4, -1)$, $(-1/4, -1)$,
$(3/4, -1)$, $(7/4, -1)$

Section 10.3	439

47. For a person at rest, the velocity v (in liters per second) of air flow into and out of the lungs during a respiratory cycle is

$$v(t) = 0.85 \sin \frac{\pi t}{3}$$

where t is the time in seconds. Inhalation occurs when v > 0 and exhalation occurs when v < 0. (a) Find the time for one full respiratory cycle. (b) Find the number of cycles per minute. (c) Sketch the graph of the velocity function.
Solution:
(a) Period = $2\pi/(\pi/3)$ = 6 seconds
(b) Cycles per minute = 60/6 = 10
(c) See accompanying graph.

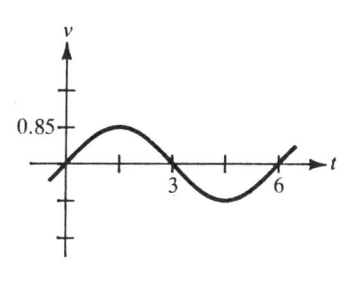

49. When tuning a piano, a technician strikes a tuning fork for the A above middle C and sets up wave motion that can be approximated by

$$y = 0.001 \sin 880\pi t$$

where t is the time in seconds. (a) What is the period p of this function? (b) What is the frequency f of this note? (f = 1/p) (c) Sketch the graph of this function. 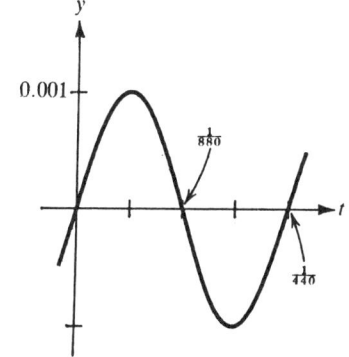
Solution:
(a) Period = $2\pi/880\pi$ = 1/440
(b) Frequency = 1/period = 440
(c) See accompanying graph.

51. Sketch the graph of the sales function

$$S(t) = 22.3 - 3.4 \cos \frac{\pi t}{6}$$

over one year where S is sales in thousands of units and t is the time in months with t = 1 corresponding to January. 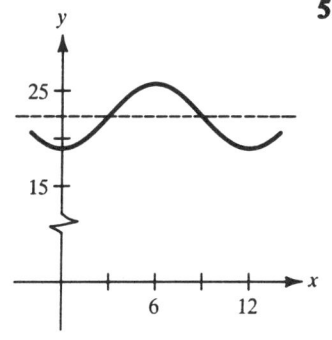
Solution: The graph has a period of 12, an amplitude of 3.4, and oscillates about the line y = 22.3 as shown in the accompanying graph.

53. Complete a table (using a calculator set in radian mode) to estimate

$$\lim_{x \to 0} \frac{1 - \cos x}{x}$$

Solution:

x	-1.0	-0.1	-0.01	-0.001	0.001	0.01	0.1	1.0
f(x)	-0.46	-0.05	-0.005	-0.0005	0.0005	0.005	0.05	0.46

From this table, we estimate that

$$\lim_{x \to 0} \frac{1 - \cos x}{x} = 0$$

55. Complete a table (using a calculator set in radian mode) to estimate

$$\lim_{x \to 0} \frac{\sin x}{5x}$$

Solution:

x	-1.0	-0.1	-0.01	-0.001	0.001	0.01	0.1	1.0
f(x)	0.17	0.2	0.20	0.200	0.200	0.20	0.2	0.17

From this table, we estimate that

$$\lim_{x \to 0} \frac{\sin x}{5x} = \frac{1}{5}$$

● **Section 10.4 Derivatives of Trigonometric Functions**

1. Find the derivative of $y = x^2 - \cos x$.
Solution:
$y' = 2x + \sin x$

3. Find the derivative of $y = (1/x) - 3 \sin x$.
Solution:

$$y' = -\frac{1}{x^2} - 3 \cos x$$

5. Find the derivative of $f(x) = 4\sqrt{x} + 3 \cos x$.
Solution:

$$f'(x) = \frac{2}{\sqrt{x}} - 3 \sin x$$

7. Find the derivative of $f(t) = t^2 \sin t$.
Solution:
$f'(t) = t^2 \cos t + 2t \sin t$

9. Find the derivative of $g(t) = (\cos t)/t$.
Solution:

$$g'(t) = \frac{t(-\sin t) - (\cos t)(1)}{t^2}$$

$$= -\frac{t \sin t + \cos t}{t^2}$$

11. Find the derivative of $y = \tan x + x^2$.
Solution:
$y' = \sec^2 x + 2x$

13. Find the derivative of $y = 5x \csc x$.
Solution:
$y' = 5x(-\csc x \cot x) + 5 \csc x$
$ = 5 \csc x - 5x \csc x \cot x$
$ = 5 \csc x(1 - x \cot x)$

Section 10.4   441

15. Find the derivative of $y = \sin 4x$.
Solution:
$$y' = 4 \cos 4x$$

17. Find the derivative of $y = \sec x^2$.
Solution:
$$y' = 2x \sec x^2 \tan x^2$$

19. Find the derivative of $y = (1/2) \csc 2x$.
Solution:
$$y' = \frac{1}{2}(2)(-\csc 2x \cot 2x)$$
$$= -\csc 2x \cot 2x$$

21. Find the derivative of $y = x \sin(1/x)$.
Solution:
$$y' = x(\frac{1}{x^2})(-\cos\frac{1}{x}) + (\sin\frac{1}{x})(1)$$
$$= \sin\frac{1}{x} - \frac{1}{x}\cos\frac{1}{x}$$

23. Find the derivative of $y = 3 \tan 4x$.
Solution:
$$y' = 12 \sec^2 4x$$

25. Find the derivative of $y = \cos^2 x$.
Solution:
$$y = (\cos x)^2$$
$$y' = 2(\cos x)(-\sin x) = -2 \cos x \sin x$$
$$= -\sin 2x$$

27. Find the derivative of $y = \cos^2 x - \sin^2 x$.
Solution:
$$y' = -2 \cos x \sin x - 2 \cos x \sin x$$
$$= -4 \cos x \sin x$$
$$= -2 \sin 2x$$

29. Find the derivative of $y = (\cos x)/(\sin x)$.
Solution:
$$y = \cot x$$
$$y' = -\csc^2 x$$

31. Find the derivative of $y = \ln |\sin x|$.
Solution:
$$y' = \frac{\cos x}{\sin x} = \cot x$$

33. Find the derivative of $y = \ln |\csc x - \cot x|$.
Solution:
$$y' = \frac{-\csc x \cot x + \csc^2 x}{\csc x - \cot x}$$
$$= \frac{\csc x(\csc x - \cot x)}{\csc x - \cot x} = \csc x$$

35. Find the derivative of $y = \tan x - x$.
Solution:
$$y' = \sec^2 x - 1 = \tan^2 x$$

37. Find the derivative of $y = \sqrt{\sin x}$.
Solution:
$$y' = \frac{1}{2}(\sin x)^{-1/2} \cos x$$
$$= \frac{\cos x}{2\sqrt{\sin x}}$$

39. Find the derivative of $y = (1/2)(x \tan x - \sec x)$.
Solution:
$$y' = \frac{1}{2}(x \sec^2 x + \tan x - \sec x \tan x)$$

41. Use implicit differentiation to find y' in the equation $\sin x + \cos 2y = 1$. Evaluate y' at $(\pi/2, \pi/4)$.
Solution:
$$\sin x + \cos 2y = 1$$
$$\cos x - 2 \sin 2y \frac{dy}{dx} = 0$$
$$\frac{dy}{dx} = \frac{\cos x}{2 \sin 2y}$$

At $(\pi/2, \pi/4)$, we have $dy/dx = 0$.

43. Verify that $y = 2 \sin x + 3 \cos x$ is a solution of the differential equation $y'' + y = 0$.
Solution: Since
$$y' = 2 \cos x - 3 \sin x$$
and
$$y'' = -2 \sin x + 3 \cos x$$
it follows that $y'' + y = 0$.

45. Verify that $y = \cos 2x + \sin 2x$ is a solution of the differential equation $y'' + 4y = 0$.
Solution: Since
$$y' = -2 \sin 2x + 2 \cos 2x$$
and
$$y'' = -4 \cos 2x - 4 \sin 2x$$
it follows that $y'' + 4y = 0$.

47. Find the slope of the tangent line to $y = \sin 3x$ at the point $(0, 0)$. Compare this to the number of complete cycles in the interval $[0, 2\pi]$.
Solution: Since $y' = 3 \cos 3x$, the slope of the tangent line at $(0, 0)$ is 3. There are 3 complete cycles of the graph in the interval $[0, 2\pi]$.

49. Find the slope of the tangent line to $y = \sin 2x$ at the point $(0, 0)$. Compare this to the number of complete cycles in the interval $[0, 2\pi]$.
Solution: Since $y' = 2\cos 2x$, the slope of the tangent line at $(0, 0)$ is 2. There are 2 complete cycles of the graph in the interval $[0, 2\pi]$.

51. Find the slope of the tangent line to $y = \sin x$ at the point $(0, 0)$. Compare this to the number of complete cycles in the interval $[0, 2\pi]$.
Solution: Since $y' = \cos x$, the slope of the tangent line at $(0, 0)$ is 1. There is 1 complete cycle of the graph in the interval $[0, 2\pi]$.

53. Find an equation of the tangent line to the graph of $f(x) = \tan x$ at the point $(-\pi/4, -1)$.
Solution: Since $y' = \sec^2 x$, the slope of the tangent line at $(-\pi/4, -1)$ is

$$m = \sec^2(-\pi/4) = 2$$

Therefore, the equation of the tangent line is

$$y + 1 = 2(x - \frac{\pi}{4})$$

$$y = 2x + (\frac{\pi}{2} - 1)$$

55. Sketch the graph of $f(x) = 2\sin x + \sin 2x$ on the interval $[0, 2\pi]$.
Solution: The first derivative is zero when

$$2\cos x + 2\cos 2x = 0$$
$$2[\cos x + 2\cos^2 x - 1] = 0$$
$$2\cos^2 x + \cos x - 1 = 0$$
$$(2\cos x - 1)(\cos x + 1) = 0$$
$$\cos x = 1/2 \quad \text{or} \quad \cos x = -1$$

Critical numbers: $x = \pi/3, 5\pi/3, \pi$
Relative maximum: $(\pi/3, 3\sqrt{3}/2)$
Relative minimum: $(5\pi/3, -3\sqrt{3}/2)$
Point of inflection: $(\pi, 0)$

57. Sketch the graph of $f(x) = x - 2\sin x$ on the interval $[0, 4\pi]$.
Solution: The first derivative is zero when

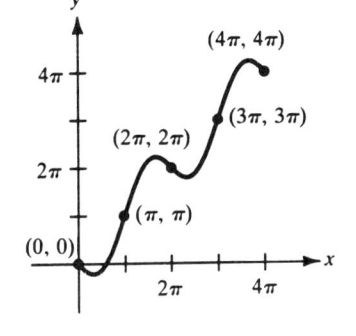

$$1 - 2\cos x = 0$$
$$\cos x = 1/2$$

Critical numbers: $x = \pi/3, 5\pi/3, 7\pi/3, 11\pi/3$
Relative minima: $(\pi/3, [\pi/3] - \sqrt{3})$
$\quad\quad\quad\quad\quad\quad\,\,(7\pi/3, [7\pi/3] - \sqrt{3})$,
Relative maxima: $(5\pi/3, [5\pi/3] + \sqrt{3})$,
$\quad\quad\quad\quad\quad\quad\,\,(11\pi/3, [11\pi/3] + \sqrt{3})$
Points of inflection: $(0, 0), (\pi, \pi), (2\pi, 2\pi),$
$\quad\quad\quad\quad\quad\quad\quad\,\,(3\pi, 3\pi), (4\pi, 4\pi)$

59. Electricity sales in the United States have had both an increasing annual sales pattern and a seasonal monthly sales pattern. For the years 1978 and 1979, the sales pattern can be approximated by the model

$$S(t) = 164.68 + 0.56t + 13.60 \cos \frac{\pi t}{3}$$

where S is the sales per month in billions of kilowatt hours and t is the time in months, with $t = 1$ corresponding to January 1978.
(a) Find the relative extrema of this function for the years 1978 and 1979.
(b) Use this model to predict the sales in August 1980. (Use $t = 31.5$.)

Solution:
(a) Since $S'(t) = 0.56 - (13.60\pi/3) \sin(\pi t/3)$, it follows that $S'(t) = 0$ when $t = 0.0376 + 3n$. Therefore, the relative minima and maxima are

Relative max: (0.0376, 178.29), (6.0376, 181.65), (12.0376, 185.01), (18.0376, 188.37)
Relative min: (3.0376, 152.79), (9.0376, 156.15), (15.0376, 159.51), (21.0376, 162.87)

(b) When $t = 31.5$, the sales are

$$S(31.5) = 164.68 + 0.56(31.5) + 13.60 \cos \frac{\pi(31.5)}{3}$$

$$= 182.32 \text{ billion kilowatt hours}$$

61. The normal average daily temperature in degrees Fahrenheit for a certain city is given by

$$T(t) = 45 - 23 \cos \frac{2\pi(t - 32)}{365}$$

where t is the time in days with $t = 1$ corresponding to January 1. Find the expected date of (a) the warmest day, and (b) the coldest day.

Solution: The derivative of T is

$$T'(t) = 23\left(\frac{2\pi}{365}\right) \sin \frac{2\pi(t - 32)}{365}$$

which is zero when

$$\sin \frac{2\pi(t - 32)}{365} = 0$$

which implies that $t = 32$ or $t = (365/2) + 32 = 214.5$.
(a) The warmest day occurs when $t = 214.5$, on August 2 and August 3.
(b) The coldest day occurs when $t = 32$ on February 1.

63. Apply Taylor's Theorem to verify the power series centered at $c = 0$ for $\sin x$, and find the radius of convergence.

$$\sin x = x - \frac{x^3}{3!} + \frac{x^5}{5!} - \frac{x^7}{7!} + \cdots + \frac{(-1)^n x^{2n+1}}{(2n + 1)!} + \cdots$$

Section 10.4

Solution:

$$f(x) = \sin x \qquad f(0) = 0$$
$$f'(x) = \cos x \qquad f'(0) = 1$$
$$f''(x) = -\sin x \qquad f''(0) = 0$$
$$f'''(x) = -\cos x \qquad f'''(0) = -1$$

Therefore the Taylor series is

$$f(x) = f(0) + f'(0)x + \frac{f''(0)x^2}{2!} + \cdots$$

$$= x - \frac{x^3}{3!} + \frac{x^5}{5!} - \frac{x^7}{7!} + \cdots + \frac{(-1)^n x^{2n+1}}{(2n+1)!} + \cdots$$

$$= \sum_{n=0}^{\infty} (-1)^n \cdot \frac{x^{2n+1}}{(2n+1)!}$$

The radius of convergence is given by

$$R = \lim_{n \to \infty} \left| \frac{1}{(2n+1)!} \cdot \frac{(2n+3)!}{1} \right| = \infty$$

65. Use the power series for $\sin x$ to find the power series of $f(x) = \sin x^2$.

Solution:

$$\sin x^2 = \sum_{n=0}^{\infty} (-1)^n \cdot \frac{(x^2)^{2n+1}}{(2n+1)!}$$

$$= \sum_{n=0}^{\infty} (-1)^n \cdot \frac{x^{4n+2}}{(2n+1)!}$$

67. Use the power series for $\cos x$ to find the power series of $h(x) = \cos 2x$.

Solution:

$$\cos 2x = \sum_{n=0}^{\infty} (-1)^n \cdot \frac{(2x)^{2n}}{(2n)!}$$

69. Differentiate the appropriate power series to verify the derivative formula

$$\frac{d}{dx}[\sin x] = \cos x$$

Solution: Since

$$\sin x = x - \frac{x^3}{3!} + \frac{x^5}{5!} - \frac{x^7}{7!} + \cdots$$

it follows that

$$\frac{d}{dx}[\sin x] = 1 - \frac{x^2}{2!} + \frac{x^4}{4!} - \frac{x^6}{6!} + \cdots$$

which is the power series for $\cos x$.

Section 10.5 Integrals of Trigonometric Functions

1. Evaluate $\int (2 \sin x + 3 \cos x)\, dx$

Solution:
$$\int (2 \sin x + 3 \cos x)\, dx = -2 \cos x + 3 \sin x + C$$

3. Evaluate $\int (1 - \csc t \cot t)\, dt$

Solution:
$$\int (1 - \csc t \cot t)\, dt = t + \csc t + C$$

5. Evaluate $\int (\sec^2 \theta - \sin \theta)\, d\theta$

Solution:
$$\int (\sec^2 \theta - \sin \theta)\, d\theta = \tan \theta + \cos \theta + C$$

7. Evaluate $\int \sin 2x\, dx$

Solution:
$$\int \sin 2x\, dx = \frac{1}{2} \int 2 \sin 2x\, dx = -\frac{1}{2} \cos 2x + C$$

9. Evaluate $\int x \cos x^2\, dx$

Solution:
$$\int x \cos x^2\, dx = \frac{1}{2} \int 2x \cos x^2\, dx = \frac{1}{2} \sin x^2 + C$$

11. Evaluate $\int \sec^2 \frac{x}{2}\, dx$

Solution:
$$\int \sec^2 \frac{x}{2}\, dx = 2 \int \frac{1}{2} \sec^2 \frac{x}{2}\, dx = 2 \tan \frac{x}{2} + C$$

13. Evaluate $\int \tan 3x\, dx$

Solution:
$$\int \tan 3x\, dx = \frac{1}{3} \int 3 \tan 3x\, dx$$
$$= -\frac{1}{3} \ln |\cos 3x| + C$$

15. Evaluate $\int \tan^4 x \sec^2 x \, dx$

Solution:
$$\int \tan^4 x \sec^2 x \, dx = \frac{\tan^5 x}{5} + C$$

17. Evaluate $\int \cot \pi x \, dx$

Solution:
$$\int \cot \pi x \, dx = \frac{1}{\pi} \int \pi \cot \pi x \, dx$$
$$= \frac{1}{\pi} \ln |\sin \pi x| + C$$

19. Evaluate $\int \csc 2x \, dx$

Solution:
$$\int \csc 2x \, dx = \frac{1}{2} \int 2 \csc 2x \, dx$$
$$= \frac{1}{2} \ln |\csc 2x - \cot 2x| + C$$

21. Evaluate $\int \frac{\sec^2 x}{\tan x} \, dx$

Solution:
$$\int \frac{\sec^2 x}{\tan x} \, dx = \ln |\tan x| + C$$

23. Evaluate $\int \frac{\sec x \tan x}{\sec x - 1} \, dx$

Solution:
$$\int \frac{\sec x \tan x}{\sec x - 1} \, dx = \ln |\sec x - 1| + C$$

25. Evaluate $\int \frac{\sin x}{1 + \cos x} \, dx$

Solution:
$$\int \frac{\sin x}{1 + \cos x} \, dx = -\ln |1 + \cos x| + C$$

27. Evaluate $\int \frac{\csc^2 x}{\cot^3 x} \, dx$

Solution:
$$\int \frac{\csc^2 x}{\cot^3 x} \, dx = -\int \cot^{-3} x (-\csc^2 x) \, dx$$
$$= -\frac{\cot^{-2} x}{-2} + C = \frac{1}{2} \tan^2 x + C$$

Section 10.5

29. Evaluate $\int e^x \cos e^x \, dx$

Solution:
$$\int e^x \cos e^x \, dx = \sin e^x + C$$

31. Evaluate $\int e^{-x} \tan e^{-x} \, dx$

Solution:
$$\int e^{-x} \tan e^{-x} \, dx = -\int (-e^{-x}) \tan e^{-x} \, dx$$
$$= \ln |\cos e^{-x}| + C$$

33. Evaluate $\int (\sin 2x + \cos 2x)^2 \, dx$

Solution:
$$\int (\sin 2x + \cos 2x)^2 \, dx$$
$$= \int (\sin^2 2x + 2 \sin 2x \cos 2x + \cos^2 2x) \, dx$$
$$= \int (1 + \sin 4x) \, dx = x - \frac{1}{4} \cos 4x + C$$

35. Evaluate $\int x \cos x \, dx$

Solution: Using integration by parts, we let $u = x$ and $dv = \cos x \, dx$. Then $du = dx$ and $v = \sin x$.
$$\int x \cos x \, dx = x \sin x - \int \sin x \, dx$$
$$= x \sin x + \cos x + C$$

37. Evaluate $\int x \sec^2 x \, dx$

Solution: Using integration by parts, we let $u = x$ and $dv = \sec^2 x \, dx$. Then $du = dx$ and $v = \tan x$.
$$\int x \sec^2 x \, dx = x \tan x - \int \tan x \, dx$$
$$= x \tan x + \ln |\cos x| + C$$

39. Evaluate $\int_0^{\pi/2} \cos \frac{2x}{3} \, dx$

Solution:
$$\int_0^{\pi/2} \cos \frac{2x}{3} \, dx = \frac{3}{2} \sin \frac{2x}{3} \Big]_0^{\pi/2} = \frac{3}{2}(\sin \frac{\pi}{3}) = \frac{3\sqrt{3}}{4}$$

Section 10.5   449

41. Evaluate $\int_{\pi/2}^{2\pi/3} \sec^2 \frac{x}{2} \, dx$

Solution:

$$\int_{\pi/2}^{2\pi/3} \sec^2 \frac{x}{2} \, dx = 2 \tan \frac{x}{2} \Big]_{\pi/2}^{2\pi/3}$$

$$= 2(\sqrt{3} - 1) \approx 1.4641$$

43. Evaluate $\int_{\pi/12}^{\pi/4} \csc 2x \cot 2x \, dx$

Solution:

$$\int_{\pi/12}^{\pi/4} \csc 2x \cot 2x \, dx = -\frac{1}{2} \csc 2x \Big]_{\pi/12}^{\pi/4}$$

$$= -\frac{1}{2} [\csc \frac{\pi}{2} - \csc \frac{\pi}{6}]$$

$$= -\frac{1}{2} [1 - 2] = \frac{1}{2}$$

45. Evaluate $\int_0^1 \sec(1-x) \tan(1-x) \, dx$

Solution:

$$\int_0^1 \sec(1-x) \tan(1-x) \, dx = -\sec(1-x) \Big]_0^1$$

$$= -[1 - \sec 1]$$
$$\approx 0.8508$$

47. Find the area of the region bounded by the graphs of $y = \cos(x/2)$ and $y = 0$, $0 \leq x \leq \pi$.

Solution:

$$\text{Area} = \int_0^\pi \cos \frac{x}{2} \, dx = 2 \sin \frac{x}{2} \Big]_0^\pi$$

$$= 2 \text{ square units}$$

49. Find the area of the region bounded by the graphs of $y = \tan x$ and $y = 0$, $0 \leq x \leq \pi/4$.

Solution:

$$\text{Area} = \int_0^{\pi/4} \tan x \, dx = -\ln |\cos x| \Big]_0^{\pi/4}$$

$$= -\ln(1/\sqrt{2})$$

$$= \ln \sqrt{2}$$

$$\approx 0.3466 \text{ square units}$$

51. Find the area of the region bounded by the graphs of $y = \sin x + \cos 2x$ and $y = 0$, $0 \le x \le \pi$.
Solution:

$$\text{Area} = \int_0^\pi (\sin x + \cos 2x)\, dx$$

$$= \left[-\cos x + \frac{1}{2} \sin 2x \right]_0^\pi$$

$$= 2 \text{ square units}$$

53. Find the volume of the solid generated by revolving the region bounded by the graphs of $y = \sec x$, $y = 0$, $x = 0$ and $x = \pi/4$ about the x-axis.
Solution:

$$V = \pi \int_0^{\pi/4} \sec^2 x\, dx = \pi \tan x \Big]_0^{\pi/4}$$

$$= \pi \text{ cubic units}$$

55. Find the general solution of the first-order linear differential equation $y' + 2y = \sin x$.
Solution: Since $P(x) = 2$ and $Q(x) = \sin x$, the integrating factor is

$$u(x) = e^{\int 2\, dx} = e^{2x}$$

and the solution is

$$y = e^{-2x} \int e^{2x} \sin x\, dx$$

Using integration by parts, we find the solution to be

$$y = e^{-2x} \left[\left(\frac{1}{5}\right) e^{2x} (2 \sin x - \cos x) + C \right]$$

$$= \frac{1}{5}(2 \sin x - \cos x) + Ce^{-2x}$$

57. Use the Taylor polynomial

$$\frac{\sin x}{x} \approx 1 - \frac{x^2}{6} + \frac{x^4}{120} - \frac{x^6}{5040}$$

to approximate the integral

$$\int_0^{\pi/2} \frac{\sin x}{x}\, dx$$

Solution:

$$\int_0^{\pi/2} \frac{\sin x}{x}\, dx \approx \left[x - \frac{x^3}{18} + \frac{x^5}{600} - \frac{x^7}{35280} \right]_0^{\pi/2}$$

$$\approx 1.3707$$

59. Approximate the integral
$$\int_0^{\pi/2} f(x)\,dx, \qquad f(x) = \begin{cases} (\sin x)/x, & x > 0 \\ 1, & x = 0 \end{cases}$$
letting n = 4 and using (a) the Trapezoidal Rule and (b) Simpson's Rule.

Solution:

(a) $\dfrac{\pi/2}{8}[f(0) + 2f(\dfrac{\pi}{8}) + 2f(\dfrac{\pi}{4}) + 2f(\dfrac{3\pi}{8}) + f(\dfrac{\pi}{2})] \approx 1.3655$

(b) $\dfrac{\pi/2}{12}[f(0) + 4f(\dfrac{\pi}{8}) + 2f(\dfrac{\pi}{4}) + 4f(\dfrac{3\pi}{8}) + f(\dfrac{\pi}{2})] \approx 1.3708$

61. The minimum stockpile level of gasoline in the United States can be approximated by the model
$$Q(t) = 217 + 13\cos\dfrac{\pi(t-3)}{6}$$
where Q is measured in millions of barrels of gasoline and t is the time in months, with t = 1 corresponding to January. Find the average minimum level given by this model during:

(a) the first quarter ($0 \le t \le 3$)
(b) the second quarter ($3 \le t \le 6$)
(c) the entire year   ($0 \le t \le 12$)

Solution:

(a) $\dfrac{1}{3}\int_0^3 [217 + 13\cos\dfrac{\pi(t-3)}{6}]\,dt$

$= \dfrac{1}{3}\left[217t + \dfrac{78}{\pi}\sin\dfrac{\pi(t-3)}{6}\right]_0^3$

$= \dfrac{1}{3}(651 + \dfrac{78}{\pi}) \approx 225.28$ million barrels

(b) $\dfrac{1}{3}\int_3^6 [217 + 13\cos\dfrac{\pi(t-3)}{6}]\,dt$

$= \dfrac{1}{3}\left[217t + \dfrac{78}{\pi}\sin\dfrac{\pi(t-3)}{6}\right]_3^6$

$= \dfrac{1}{3}(1302 + \dfrac{78}{\pi} - 651)$

≈ 225.28 million barrels

(c) $\dfrac{1}{12}\int_0^{12} [217 + 13\cos\dfrac{\pi(t-3)}{6}]\,dt$

$= \dfrac{1}{12}\left[217t + \dfrac{78}{\pi}\sin\dfrac{\pi(t-3)}{6}\right]_0^{12}$

$= \dfrac{1}{12}(2604 - \dfrac{78}{\pi} + \dfrac{78}{\pi})$

≈ 217 million barrels

63. For a person at rest, the velocity v (in liters per second) of air flow into and out of the lungs during a respiratory cycle is

$$v(t) = 0.85 \sin \frac{\pi t}{3}$$

where t is the time in seconds. Find the volume (in liters) of air inhaled during one cycle by integrating this function over the interval [0, 3].
Solution: The volume is given by

$$V = \int_0^3 0.85 \sin \frac{\pi t}{3} \, dt$$

$$= -\frac{3}{\pi}(0.85) \cos \frac{\pi t}{3} \Big]_0^3$$

$$= -\frac{2.55}{\pi}(-1 - 1) \approx 1.6234 \text{ liters}$$

65. Suppose that the temperature (in degrees Fahrenheit) is given by

$$T(t) = 72 + 12 \sin \frac{\pi(t - 8)}{12}$$

where t is the time in hours, with t = 0 representing midnight. Furthermore, suppose that it costs $0.10 to cool a particular house 1° for 1 hour.
(a) Find the cost C of cooling this house if the thermostat is set at 72° and the cost is given by

$$C = 0.1 \int_8^{20} [72 + 12 \sin \frac{\pi(t - 8)}{12} - 72] \, dt$$

(b) Find the savings in resetting the thermostat to 78° by evaluating the integral

$$C = 0.1 \int_{10}^{18} [72 + 12 \sin \frac{\pi(t - 8)}{12} - 78] \, dt$$

Solution:

(a) $$C = 0.1 \int_8^{20} 12 \sin \frac{\pi(t - 8)}{12} \, dt$$

$$= -\frac{14.4}{\pi} \cos \frac{\pi(t - 8)}{12} \Big]_8^{20}$$

$$= -\frac{14.4}{\pi}(-1 - 1) \approx \$9.17$$

(b) $$C = 0.1 \int_{10}^{18} [12 \sin \frac{\pi(t - 8)}{12} - 6] \, dt$$

$$= \left[-\frac{14.4}{\pi} \cos \frac{\pi(t - 8)}{12} - 0.6t\right]_{10}^{18} \approx \$3.14$$

Savings ≈ 9.17 − 3.14 = $6.03

Review Exercises for Chapter 10

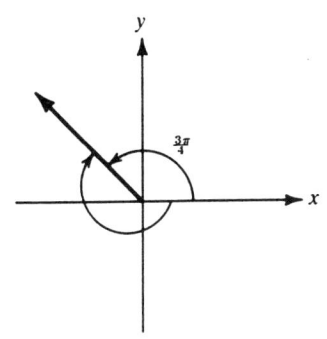

1. Sketch the angle $\theta = 11\pi/4$ in standard position and give a positive and negative coterminal angle.
Solution:
Positive coterminal angle:
$$\frac{11\pi}{4} - 2\pi = \frac{3\pi}{4}$$
Negative coterminal angle:
$$\frac{11\pi}{4} - 2(2\pi) = -\frac{5\pi}{4}$$

3. Convert the angle $\theta = 5\pi/7$ from radian to degree measure.
Solution:
$$\frac{5\pi}{7}\left(\frac{180°}{\pi}\right) = \left(\frac{900}{7}\right)° \approx 128.571°$$

5. Convert the angle $\theta = 480°$ from degree to radian measure.
Solution:
$$480°\left(\frac{\pi \text{ radians}}{180°}\right) = \frac{8\pi}{3} \text{ radians}$$

7. Find the reference angle for $\theta = -6\pi/5$.
Solution: The reference angle is $\pi/5$.

9. Find the six trigonometric functions of θ in standard position with terminal point at $(-7, 2)$.
Solution: Since $x = -7$ and $y = 2$, it follows that $r = \sqrt{(-7)^2 + 2^2} = \sqrt{53}$.

$\sin\theta = 2\sqrt{53}/53 \qquad \csc\theta = \sqrt{53}/2$
$\cos\theta = -7\sqrt{53}/53 \qquad \sec\theta = -\sqrt{53}/7$
$\tan\theta = -2/7 \qquad \cot\theta = -7/2$

11. Given $\sec\theta = 6/5$ and $\tan\theta < 0$, find the remaining five trigonometric functions.
Solution: Since the secant is positive and the tangent is negative, θ lies in Quadrant IV. Therefore, $r = 6$, $x = 5$, and $y = -\sqrt{6^2 - 5^2} = -\sqrt{11}$.

$\sin\theta = -\sqrt{11}/6 \qquad \csc\theta = -6\sqrt{11}/11$
$\cos\theta = 5/6 \qquad \sec\theta = 6/5$
$\tan\theta = -\sqrt{11}/5 \qquad \cot\theta = -5\sqrt{11}/11$

13. Find two values of θ in degrees ($0° \leq \theta \leq 360°$) and in radians ($0 \leq \theta \leq 2\pi$) such that $\cos\theta = -\sqrt{2}/2$.
Solution:
In Quadrant II: $\theta = 3\pi/4 = 135°$
In Quadrant III: $\theta = 5\pi/4 = 225°$

15. Find two values of θ in degrees (0° ≤ θ ≤ 360°) and in radians (0 ≤ θ ≤ 2π) such that csc θ = -2.
Solution:
In Quadrant III: θ = 7π/6 = 210°
In Quadrant IV: θ = 11π/6 = 330°

17. Find two values of θ in degrees (0° ≤ θ ≤ 360°) and in radians (0 ≤ θ ≤ 2π) such that sin θ = 0.8387.
Solution:
In Quadrant I: θ = 57° = 19π/60
In Quadrant II: θ = 123° = 41π/60

19. Find two values of θ in degrees (0° ≤ θ ≤ 360°) and in radians (0 ≤ θ ≤ 2π) such that sec θ = -1.0353.
Solution:
In Quadrant II: θ = 165° = 11π/12
In Quadrant III: θ = 195° = 13π/12

21. Sketch the graph of $f(x) = 3 \sin(2x/5)$.
Solution:
Period: $2\pi/(2/5) = 5\pi$
Amplitude: 3
x-intercepts: $(0, 0)$, $(5\pi/2, 0)$, $(5\pi, 0)$
Maximum: $(5\pi/4, 3)$
Minimum: $(15\pi/4, -3)$

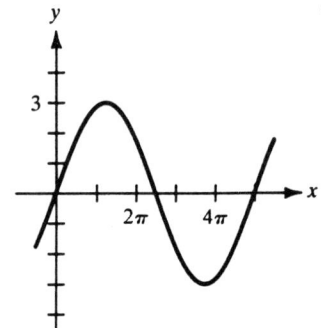

23. Sketch the graph of $f(x) = -\tan(\pi x/4)$.
Solution:
Period: $\pi/(\pi/4) = 4$
x-intercepts: $(-4, 0)$, $(0, 0)$, $(4, 0)$
Vertical asymptotes: $x = -2$, $x = 2$

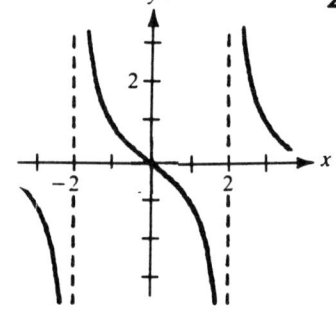

25. An observer 2.5 miles from the launch pad of a space shuttle measures the angle of elevation to the base of the vehicle to be 28° soon after liftoff. How high is the shuttle at that instant if you assume that the shuttle is still moving vertically?
Solution: If we let y be the height, then

$$\tan 28° = \frac{y}{2.5}$$

which implies that

$$y = 2.5 \tan 28° \approx 1.329 \text{ miles} \approx 7{,}018.6 \text{ feet}$$

27. Find the derivative of $y = \cos 5\pi x$.
Solution:
$y' = -5\pi \sin 5\pi x$

Review Exercises for Chapter 10

29. Find the derivative of $y = -x \tan x$.
Solution:
$$y' = -x \sec^2 x - \tan x$$

31. Find the derivative of $y = (\sin x)/x^2$.
Solution:
$$y' = \frac{x^2 \cos x - 2x \sin x}{x^4} = \frac{x \cos x - 2 \sin x}{x^3}$$

33. Find the derivative of $y = 3 \sin^2 4x + x$.
Solution:
$$y' = 6 \sin 4x (4 \cos 4x) + 1$$
$$= 24 \sin 4x \cos 4x + 1 = 12 \sin 8x + 1$$

35. Find the derivative of $y = 2 \csc^3 x$.
Solution:
$$y' = 6 \csc^2 x(-\csc x \cot x) = -6 \csc^3 x \cot x$$

37. Find the derivative of $y = e^x \tan x$.
Solution:
$$y' = e^x \sec^2 x + e^x \tan x = e^x(\sec^2 x + \tan x)$$

39. Find the derivative of $x = 2 + \sin y$.
Solution: By implicit differentiation, we have
$$1 = 0 + \cos y \left(\frac{dy}{dx}\right)$$
$$\frac{dy}{dx} = \frac{1}{\cos y} = \sec y$$

41. Find the derivative of $\cos(x + y) = x$.
Solution: By implicit differentiation, we have
$$-\sin(x + y)\left(1 + \frac{dy}{dx}\right) = 1$$
$$1 + \frac{dy}{dx} = -\csc(x + y)$$
$$\frac{dy}{dx} = -\csc(x + y) - 1$$

43. Find the second derivative of $f(x) = \cot x$.
Solution:
$$f'(x) = -\csc^2 x$$
$$f''(x) = -2\csc x(-\csc x \cot x) = 2 \csc^2 x \cot x$$

45. Find the second derivative of $h(x) = (\cos x)/x$.
Solution:
$$h'(x) = \frac{x(-\sin x) - \cos x}{x^2} = -\frac{x \sin x + \cos x}{x^2}$$
$$h''(x) = -\frac{x^2(x \cos x) - (x \sin x + \cos x)(2x)}{x^4}$$
$$= \frac{-x^2 \cos x + 2x \sin x + 2 \cos x}{x^3}$$

47. Evaluate $\int \csc 2x \cot 2x \, dx$

Solution:

$$\int \csc 2x \cot 2x \, dx = \frac{1}{2} \int 2\csc 2x \cot 2x \, dx$$

$$= -\frac{1}{2} \csc 2x + C$$

49. Evaluate $\int \tan \frac{\pi x}{4} \, dx$

Solution:

$$\int \tan \frac{\pi x}{4} \, dx = \frac{4}{\pi} \int \left(\tan \frac{\pi x}{4}\right)\left(\frac{\pi}{4}\right) dx$$

$$= -\frac{4}{\pi} \ln \left|\cos \frac{\pi x}{4}\right| + C$$

51. Evaluate $\int \tan^n x \sec^2 x \, dx$, $n \neq -1$.

Solution:

$$\int \tan^n x \sec^2 x \, dx = \frac{1}{n+1} \tan^{n+1} x + C, \; n \neq -1$$

53. Evaluate $\int (3 - 5 \sin 5\pi x) \, dx$

Solution:

$$\int (3 - 5 \sin 5\pi x) \, dx = \int 3 \, dx - \frac{1}{\pi} \int 5\pi \sin 5\pi x \, dx$$

$$= 3x + \frac{1}{\pi} \cos 5\pi x + C$$

55. Evaluate $\int x \cos x^2 \, dx$

Solution:

$$\int x \cos x^2 \, dx = \frac{1}{2} \int 2x \cos x^2 \, dx = \frac{1}{2} \sin x^2 + C$$

57. Show that $y = (a + \ln |\cos x|) \cos x + (b + x) \sin x$ is a solution of $y'' + y = \sec x$.

Solution: Since

$$y' = (b + x - \tan x) \cos x + (1 - a - \ln |\cos x|) \sin x$$
$$y'' = (2 - a - \sec^2 x - \ln |\cos x|) \cos x$$
$$\qquad - (b + x - 2 \tan x) \sin x$$

it follows that

$$y'' + y = (2 - \sec^2 x) \cos x + 2 \tan x \sin x$$
$$= 2 \cos x - \sec x + 2 \sin^2 x \sec x$$
$$= 2 \cos x - \sec x + 2 \sec x - 2 \cos x$$
$$= \sec x$$

59. Find an equation of the tangent line to the graph of $y = x \cos x$ at the point $(\pi/2, 0)$.
Solution:
$$y' = -x \sin x + \cos x$$

When $x = \pi/2$, $y' = -\pi/2$. Thus, the equation of the tangent line is
$$y = -\frac{\pi}{2}\left(x - \frac{\pi}{2}\right)$$

61. Domestic energy consumption in the United States is seasonal. Suppose the consumption is approximated by the model
$$Q(t) = 6.9 + \cos\frac{\pi(2t - 1)}{12}$$
where Q is the total consumption in quads (quadrillion BTUs) and t is the time in months, with $0 \leq t \leq 1$ corresponding to January. Find the dates for which this model predicts the greatest and the least consumption, respectively.
Solution:
$$Q'(t) = -\frac{\pi}{6}\sin\frac{\pi(2t - 1)}{12}$$

$Q'(t) = 0$ when $t = 1/2$ or $t = 13/2$. The greatest consumption occurs when $t = 1/2$ (January 15) and the least consumption occurs when $t = 13/2$ (July 15).

63. Find the area of the largest rectangle that can be inscribed between one arch of the graph of $y = \cos x$ and the x-axis. Compare the area of the rectangle with the area bounded by one arch of the cosine function and the x-axis.
Solution: The area of the rectangle is given by $A = 2x \cos x$. Since
$$\frac{dA}{dx} = -2x \sin x + 2 \cos x$$
it follows that $dA/dx = 0$ when
$$\cos x = x \sin x \quad \Longrightarrow \quad 0 = x - \cot x$$

Letting $f(x) = x - \cot x$, we can approximate the zero of f using Newton's Method as follows

n	x_n	$f(x_n)$	$f'(x_n)$	$f(x_n)/f'(x_n)$	x_{n+1}
1	1.0000	0.3579	2.4123	0.1484	0.8516
2	0.8516	-0.0240	2.7667	-0.0087	0.8603
3	0.8603	-0.0001	2.7403	-0.0000	0.8603

Thus, $x \approx 0.860$, $\cos x \approx 0.652$ and the area of the rectangle is $A = 2x \cos x \approx 1.12$. The area bounded by one arc of the cosine function is
$$A = 2\int_0^{\pi/2} \cos x \, dx = 2 \sin x \Big]_0^{\pi/2} = 2$$

PRACTICE TEST FOR CHAPTER 10

1. (a) Express $12\pi/23$ in degree measure.
   (b) Express $105°$ in radian measure.

2. Determine two coterminal angles (one positive and one negative) for the given angle.
   (a) $-220°$ Give your answers in degrees.
   (b) $7\pi/9$ Give your answers in radians.

3. Find the six trigonometric functions of the angle θ if it is in standard position and the terminal side passes through the point $(12, -5)$.

4. Solve for θ ($0 \leq \theta < 2\pi$): $\sin^2 \theta + \cos \theta = 1$.

5. Sketch the graph of the given function.
   (a) $y = 3 \sin(x/4)$
   (b) $y = \tan 2\pi x$

6. Find the derivative of $y = 3x - 3\cos x$.

7. Find the derivative of $f(x) = x^2 \tan x$.

8. Find the derivative of $g(x) = \sin^3 x$.

9. Find the derivative of $y = \dfrac{\sec x}{x^2}$.

10. Find the derivative of $y = \sin 5x \cos 5x$.

11. Find the derivative of $y = \sqrt{\csc x}$.

12. Find the derivative of $y = \ln|\sec x + \tan x|$.

13. Find the derivative of $f(x) = \cot e^{2x}$.

14. Find dy/dx: $\sin(x^2 + y) = 3x$.

15. Find dy/dx: $\tan x - \cot 3y = 4$.

16. Evaluate $\int \cos 4x \, dx$.

17. Evaluate $\int \csc^2 \dfrac{x}{8} \, dx$.

18. Evaluate $\int x \tan x^2 \, dx$.

19. Evaluate $\int \sin^5 x \cos x \, dx$.

20. Evaluate $\int \dfrac{\cos^2 x}{\sin x} \, dx$.

21. Evaluate $\int e^{\tan x} \sec^2 x \, dx$.

22. Evaluate $\int \dfrac{\sin x}{1 + \cos x} \, dx$.

23. Evaluate $\int (\sec x - \tan x)^2 \, dx$.

24. Evaluate $\int x \cos x \, dx$.

25. Evaluate $\int_0^{\pi/4} (2x - \cos x) \, dx$.

ANSWERS TO CHAPTER 0 PRACTICE TEST

1. Rational

2. (a) Satisfies
   (b) Does not satisfy
   (c) Satisfies
   (d) Satisfies

3. $x \geq 3$

4. $-1 < x < 7$

5. $\sqrt{19} > 13/3$

6. (a) $d = 10$
   (b) Midpoint = 2

7. $-11/3 \leq x \leq 3$

8. $x < -5$ or $x > 33/5$

9. $-25/2 < x < 55/2$

10. $|x - 1| \leq 4$

11. $3x^5$

12. 1

13. $2xy\sqrt[3]{4x}$

14. $\dfrac{1}{4}(x + 1)^{-1/3}(x + 7)$

15. $x < 5$

16. $(3x + 2)(x - 7)$

17. $(5x + 9)(5x - 9)$

18. $(x + 2)(x^2 - 2x + 4)$

19. $-3 \pm \sqrt{11}$

20. $-1, 2, 3$

21. $\dfrac{-3}{(x - 1)(x + 3)}$

22. $\dfrac{x + 13}{2\sqrt{x + 5}}$

23. $\dfrac{1}{\sqrt{x}(x + 2)^{3/2}}$

24. $\dfrac{3y\sqrt{y^2 + 9}}{y^2 + 9}$

25. $-\dfrac{1}{2(\sqrt{x} - \sqrt{x + 7})}$

ANSWERS TO CHAPTER 1 PRACTICE TEST

1. $d = \sqrt{82}$

2. Midpoint = $(1, 3)$

3. Collinear

4. $x = \pm 3\sqrt{5}$

5. x-intercepts: $(\pm 2, 0)$
   y-intercept: $(0, 4)$

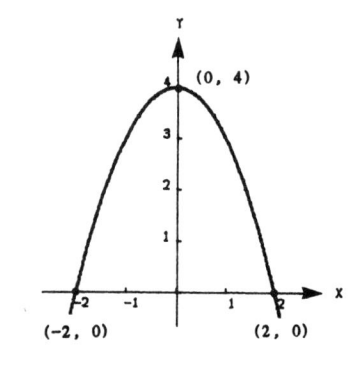

6. x-intercept: $(2, 0)$
   No y-intercept

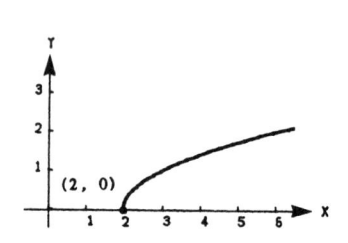

Answers to Practice Tests

7. x-intercept: (3, 0)
   y-intercept: (0, 3)

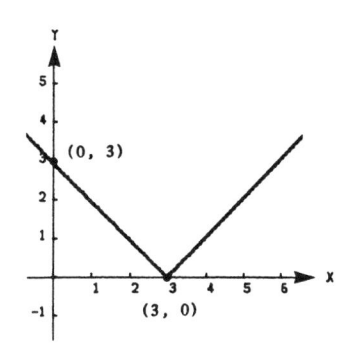

8. $(x - 4)^2 + (y + 1)^2 = 9$
   Center: (4, -1)
   Radius: 3

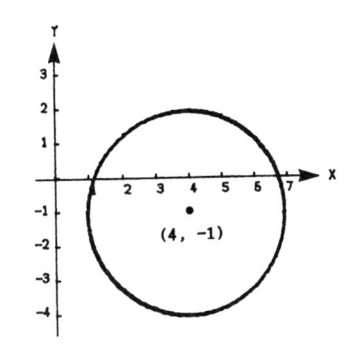

9. (0, -5) and (4, -3)

10. $6x - y - 38 = 0$

11. $2x - 3y + 1 = 0$

12. $x - 6 = 0$

13. $5x + 2y - 6 = 0$

14. (a) 4
    (b) 31
    (c) $x^2 - 10x + 20$
    (d) $x^2 + 2x(\Delta x) + (\Delta x)^2 - 5$

15. Domain: $(-\infty, 3]$
    Range: $[0, \infty)$

16. (a) $2x^2 + 1$
    (b) $4(x + 1)(x + 2)$

17. $f^{-1}(x) = \sqrt[3]{x - 6}$

18. 22

19. 12

20. Does not exist

21. $\dfrac{1}{2\sqrt{5}}$

22. 5

23. Discontinuities: $x = \pm 8$
    $x = 8$ is removable.

24. $x = 3$ is a nonremovable discontinuity.

25.

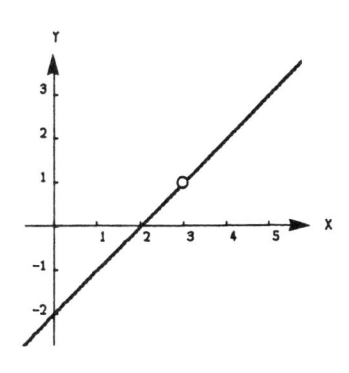

ANSWERS TO CHAPTER 2 PRACTICE TEST

1.
$$f(x + \Delta x) = 2(x + \Delta x)^2 + 3(x + \Delta x) - 5$$
$$f(x + \Delta x) - f(x) = 4x\Delta x + 2(\Delta x)^2 + 3\Delta x$$
$$\frac{f(x + \Delta x) - f(x)}{\Delta x} = 4x + 2\Delta x + 3$$
$$\lim_{\Delta x \to 0} \frac{f(x + \Delta x) - f(x)}{\Delta x} = 4x + 3$$

2.
$$f(x + \Delta x) = \frac{1}{x + \Delta x - 4}$$
$$f(x + \Delta x) - f(x) = \frac{-\Delta x}{(x + \Delta x - 4)(x - 4)}$$
$$\frac{f(x + \Delta x) - f(x)}{\Delta x} = \frac{-1}{(x + \Delta x - 4)(x - 4)}$$
$$\lim_{\Delta x \to 0} \frac{f(x + \Delta x) - f(x)}{\Delta x} = -\frac{1}{(x - 4)^2}$$

3. $x - 4y + 2 = 0$

4. $15x^2 - 12x + 15$

5. $\dfrac{4x - 2}{x^3}$

6. $\dfrac{2}{3\sqrt[3]{x}} + \dfrac{3}{5\sqrt[5]{x^2}}$

7. Average rate of change: 4
Instantaneous rates of change:
$f'(0) = 0$, $f'(2) = 12$

8. Marginal cost $= 4.31 - 0.0002x$

9. $5x^4 + 28x^3 - 39x^2 - 56x + 36$

10. $-\dfrac{x^2 + 14x + 8}{(x^2 - 8)^2}$

11. $\dfrac{3x^4 + 14x^3 - 45x^2}{(x + 5)^2}$

12. $-\dfrac{3x^2 + 4x + 1}{2\sqrt{x}(x^2 + 4x - 1)^2}$

13. $72(6x - 5)^{11}$

14. $-\dfrac{12}{\sqrt{4 - 3x}}$

15. $\dfrac{18x}{(x^2 + 1)^4}$

16. $\dfrac{10}{(x + 2)\sqrt{10x(x + 2)}}$

17. $24x - 54$

18. $-\dfrac{15}{16(3 - x)^{7/2}}$

19. $-\dfrac{x^4}{y^4}$

20. $-\dfrac{2(xy^3 + 1)}{3(x^2y^2 - 1)}$

21. $\dfrac{8\sqrt{xy + 4} + y}{10\sqrt{xy + 4} - x}$

22. $-\dfrac{8x^2}{y^2(x^3 - 4)^2}$

23. $5/12$

24. $5/4\pi$

25. $1/8\pi$

ANSWERS TO CHAPTER 3 PRACTICE TEST

1. Increasing: $(-\infty, 0), (4, \infty)$
   Decreasing: $(0, 4)$

2. Increasing: $(-\infty, 2/3)$
   Decreasing: $(2/3, 1)$

3. Relative minimum: $(2, -45)$

4. Relative minimum: $(-3, 0)$

5. Maximum: $(5, 0)$
   Minimum: $(2, -9)$

6. No inflection points

7. Points of inflection:
   $(-\frac{1}{\sqrt{3}}, \frac{1}{4}), (\frac{1}{\sqrt{3}}, \frac{1}{4})$

8. first: $10\sqrt{6}$, second: $\frac{10\sqrt{6}}{3}$

9. $x = 250$ ft, $y = 375$ ft

10. $x \approx 13{,}333$ units

11. $p = \$14{,}100$

12. -1

13. $-\infty$

14. -2

15. Intercept: $(0, 0)$
    Vertical asymptotes: $x = \pm 3$
    Horizontal asymptote: $y = 1$
    Relative maximum: $(0, 0)$
    No inflection points

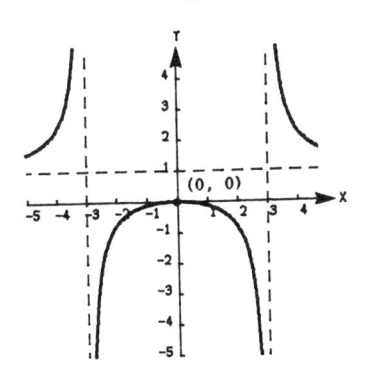

16. Intercepts: $(-2, 0), (0, 2/5)$
    Horizontal asymptote: $y = 0$
    Relative maximum: $(1, 1/2)$
    Relative minimum: $(-5, -1/10)$

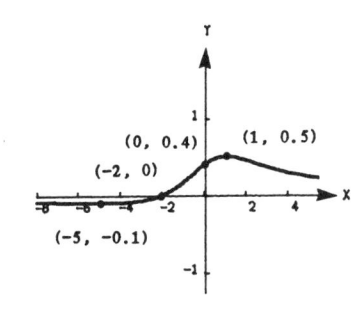

17. Intercept: $(0, -1)$
    No relative extrema
    Inflection point: $(-1, -2)$

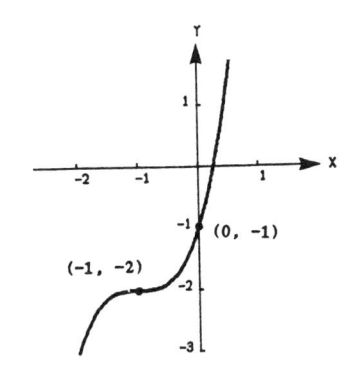

18. Intercepts: $(0, 4), (2, 0)$
    Relative minimum: $(2, 0)$
    No inflection points

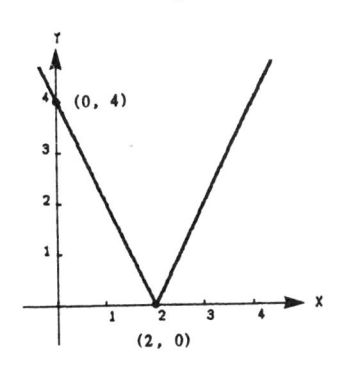

19. Intercepts: (2, 0), (0, $\sqrt[3]{4}$)
    Relative minimum: (2, 0)
    No inflection points

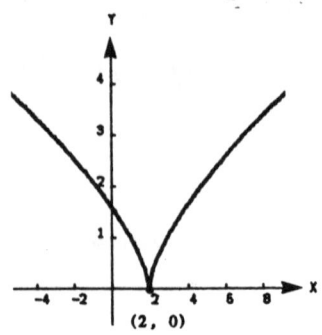

20. $\sqrt[3]{65} \approx 4.0208$

ANSWERS TO CHAPTER 4 PRACTICE TEST

1. $x^3 - 4x^2 + 5x + C$

2. $\dfrac{x^4}{4} + \dfrac{7x^3}{3} - 2x^2 - 28x + C$

3. $\dfrac{x^2}{2} - 9x - \dfrac{1}{x} + C$

4. $-\dfrac{1}{5}(1 - x^4)^{5/4} + C$

5. $\dfrac{9}{14}(7x)^{2/3} + C$

6. $-\dfrac{2}{33}(6 - 11x)^{3/2} + C$

7. $\dfrac{4}{5}x^{5/4} + \dfrac{6}{7}x^{7/6} + C$

8. $-\dfrac{1}{3x^3} + \dfrac{1}{4x^4} + C$

9. $x - x^3 + \dfrac{3}{5}x^5 - \dfrac{1}{7}x^7 + C$

10. $-\dfrac{5}{12(1 + 3x^2)^2} + C$

11. -3

12. $\dfrac{381}{7}$

13. 3

14. $A = 36$

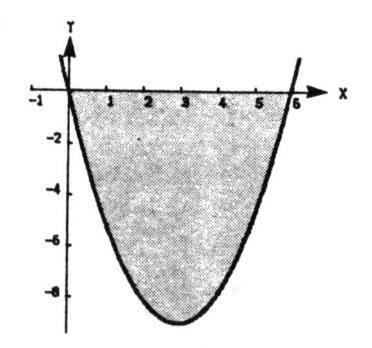

15. $A = \dfrac{1}{2}$

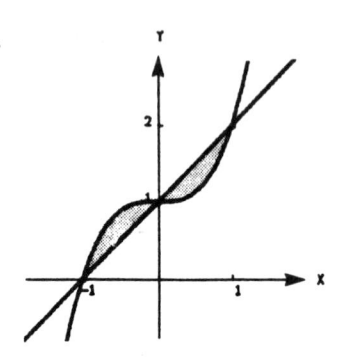

16. $A = \dfrac{2}{3}$

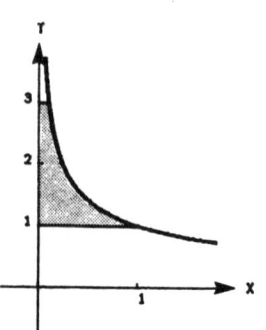

17. 1.4949

18. 0.1472

19. 3π

20. $\dfrac{5000\pi}{3}$

ANSWERS TO CHAPTER 5 PRACTICE TEST

1. (a) 81
   (b) 1/32
   (c) 1

2. (a) $x = 2$
   (b) $x = 32$
   (c) $x = 5$

3. (a) [graph of exponential growth through (0, 1)]

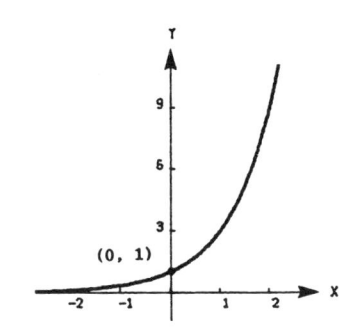

   (b) [graph of exponential decay through (0, 1)]

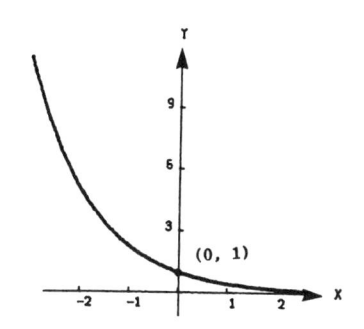

4. (a) $A \approx \$3540.28$
   (b) $A \approx \$3618.46$
   (c) $A \approx \$3626.06$

5. $6xe^{3x^2}$

6. $\dfrac{e^{\sqrt[3]{x}}}{3\sqrt[3]{x^2}}$

7. $\dfrac{e^x - e^{-x}}{2\sqrt{e^x + e^{-x}}}$

8. $x^2 e^{2x}(2x + 3)$

9. $\dfrac{xe^x - e^x - 3}{4x^2}$

10. $\left(\dfrac{1}{7}\right)e^{7x} + C$

11. $\left(\dfrac{1}{8}\right)e^{4x^2} + C$

12. $\left(\dfrac{1}{16}\right)(1 + 4e^x)^4 + C$

13. $\left(\dfrac{1}{2}\right)e^{2x} + 4e^x + 4x + C$

14. $\left(\dfrac{1}{2}\right)e^{2x} - 4x - e^{-x} + C$

15. $e^{1.6094\ldots} = 5$

16. (a) [graph of ln-type curve through (-1, 0) and (0, 0.69)]

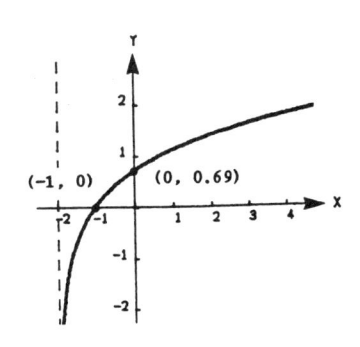

    (b) [graph of ln-type curve]

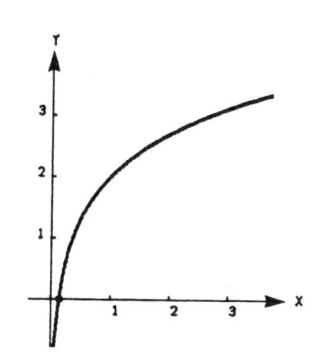

17. (a) $\ln\dfrac{3x + 1}{2x - 5}$

    (b) $\ln\dfrac{x^4}{y^3\sqrt{z}}$

18. (a) $x = e^{17}$

    (b) $x = \dfrac{\ln 2}{3\ln 5}$

19. $\dfrac{6}{6x - 7}$

20. $\dfrac{8x + 30}{x(4x + 10)}$

21. $\dfrac{1}{x(x + 3)}$

22. $x^3(1 + 4 \ln x)$

23. $\dfrac{1}{2x\sqrt{\ln x + 1}}$

24. $\ln|x + 6| + C$

25. $-(1/3) \ln|8 - x^3| + C$

26. $\dfrac{1}{3} \ln(1 + 3e^x) + C$

27. $(\ln x)^7/7 + C$

28. $\dfrac{x^2}{2} + x + 6 \ln|x - 1| + C$

29. (a) $y = 7e^{-0.7611t}$
    (b) $y = 0.1501e^{0.4970t}$

30. $t \approx 5.776$ years

● ANSWERS TO CHAPTER 6 PRACTICE TEST

1. $\dfrac{2}{5}(x + 3)^{3/2}(x - 2) + C$

2. $-\dfrac{x - 1}{(x - 2)^2} + C$

3. $\dfrac{2}{3} \ln|3\sqrt{x} + 1| + C$

4. $\dfrac{1}{4} e^{2x}(2x - 1) + C$

5. $\dfrac{x^4}{16}[4(\ln x) - 1] + C$

6. $\dfrac{2}{35}(x - 6)^{3/2}(5x^2 + 24x + 96) + C$

7. $\ln\left|\dfrac{x + 3}{x - 2}\right| + C$

8. $\ln\left|\dfrac{x^3}{(x + 4)^2}\right| + C$

9. $5 \ln|x + 2| + \dfrac{7}{x + 2} + C$

10. $-\dfrac{\sqrt{16 - x^2}}{16x} + C$

11. $x[(\ln x)^3 - 3(\ln x)^2 + 6(\ln x) - 6] + C$

12. (a) 15.567
    (b) 15.505

13. (a) 1.191
    (b) 1.196

14. Convergent, 6

15. Divergent

16. Divergent

17. $k = 1/4$

18. (a) 25/64
    (b) 63/64

19. (a) 4
    (b) $(4\sqrt{5})/5$
    (c) 4

20. (a) $6 \ln(3/2) \approx 2.433$
    (b) $\sqrt{6 - 36(\ln 2)^2} \approx 0.286$
    (c) 12/5

ANSWERS TO CHAPTER 7 PRACTICE TEST

1. $y' = 3x^2 - 4 - (C/x^2)$
   $xy' + y = x[3x^2 - 4 - (C/x^2)] + [x^3 - 4x + (C/x)]$
   $ = 3x^3 - 4x - (C/x) + x^3 - 4x + (C/x)$
   $ = 4x^3 - 8x = 4x(x^2 - 2)$

2. $y' = -5Ce^{-5x}$
   $y'' = 25Ce^{-5x}$
   $y''' = -125Ce^{-5x}$
   $y''' + 125y = -125Ce^{-5x} + 125Ce^{-5x} = 0$

3. $y^4 = 2x^2 + 8x + C$

4. $y = C(x - 1) - 4$

5. $y(\ln y - 1) = x(e^x - 1) + C$

6. $y = (1/2)e^{-2x} + Ce^{-4x}$

7. $y = -\dfrac{e^{1/x^2}}{2x^2} + Ce^{1/x^2}$

8. $y = -3x^2 + \dfrac{1}{4} + Cx^4$

9. $y = 30 - 26e^{-0.0523t}$

10. $y = 1000e^{-2.9957e^{-0.1553t}}$

ANSWERS TO CHAPTER 8 PRACTICE TEST

1. (a) $d = 14\sqrt{2}$
   (b) Midpoint = $(4, 2, -2)$

2. $(x - 1)^2 + (y + 3)^2 + z^2 = 5$

3. Center: $(2, -1, -4)$
   Radius: $\sqrt{21}$

4. (a) x-intercept: $(8, 0, 0)$
       y-intercept: $(0, 3, 0)$
       z-intercept: $(0, 0, 4)$

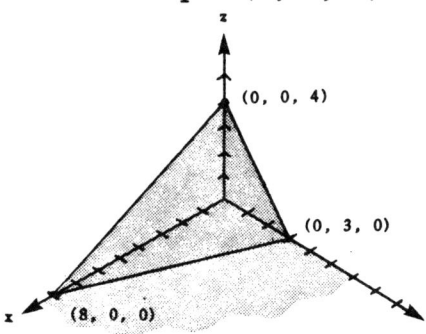

   (b) $y = 2$
       Parallel to xz-plane

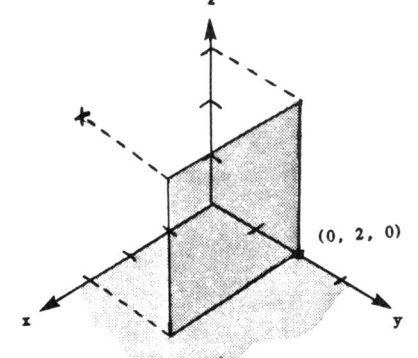

5. (a) Hyperboloid of one sheet
   (b) Elliptic paraboloid

6. (a) Domain: $x + y < 3$
   (b) Domain: all points in the xy-plane except the origin

7. $f_x(x, y) = 6x + 9y^2 - 3$
   $f_y(x, y) = 18xy + 12y^2 - 6$

8. $f_x(x, y) = 2x/(x^2 + y^2 + 5)$
   $f_y(x, y) = 2y/(x^2 + y^2 + 5)$

9. $\partial w/\partial x = 2xy^3\sqrt{z}$
   $\partial w/\partial y = 3x^2y^2\sqrt{z}$
   $\partial w/\partial z = (x^2y^3)/(2\sqrt{z})$

10. $\partial^2 w/\partial x^2 = 2x(x^2 - 3y^2)/(x^2 + y^2)^3$
    $\partial^2 w/\partial y\partial x = 2y(3x^2 - y^2)/(x^2 + y^2)^3$
    $\partial^2 w/\partial x\partial y = 2y(3x^2 - y^2)/(x^2 + y^2)^3$
    $\partial^2 w/\partial y^2 = 2x(3y^2 - x^2)/(x^2 + y^2)^3$

11. Relative minimum: (1, -2, -23)
12. Saddle point: (0, 0, 0)
    Relative maxima:
    (1, 1, 2), (-1, -1, 2)
13. $f(2, -8) = -16$
14. $f(4, 0) = -36$
15. $y = \dfrac{1}{65}(-51x + 355)$
16. $y = \dfrac{1}{6}x^2 - \dfrac{7}{26}x + \dfrac{7}{3}$
17. 81/16
18. -135/4
19. $A = \displaystyle\int_{-2}^{2}\int_{3}^{7-x^2} dy\,dx = \int_{3}^{7}\int_{-\sqrt{7-y}}^{\sqrt{7-y}} dx\,dy$
20. $A = \displaystyle\int_{0}^{1}\int_{x^2+2}^{x+2} dy\,dx = \int_{2}^{3}\int_{y-2}^{\sqrt{y-2}} dx\,dy$

● ANSWERS TO CHAPTER 9 PRACTICE TEST

1. $a_n = \dfrac{n}{n^2 + 1}$
2. $a_n = (-1)^{n-1}(2n + 3)$
3. Converges to 1/3
4. Converges to 4
5. 1/12
6. Converges by Ratio Test
7. Converges since it is a p-series with p = 4/3 > 1
8. Diverges by the nth-Term Test
9. Diverges since it is a geometric series with $|r| = |-6/5| = 6/5 > 1$
10. Diverges since it is a p-series with p = 1/6 < 1
11. Diverges by Ratio Test
12. Converges since it is a geometric series with $|r| = |0.27| = 0.27 < 1$
13. Diverges by the nth-Term Test
14. R = 1
15. $R = \lim_{n\to\infty}(n + 2) = \infty$
16. $e^{-4x} = \displaystyle\sum_{n=0}^{\infty}\dfrac{(-4x)^n}{n!}$
17. $\dfrac{1}{\sqrt[3]{x}} = 1 + \displaystyle\sum_{n=1}^{\infty}\dfrac{(-1)^n 1\cdot 4\cdot 7\cdots(3n - 2)(x - 1)^n}{3^n n!}$
18. 0.214
19. $x \approx 1.213$
20. $\sqrt[4]{10} \approx 1.778$

● ANSWERS TO CHAPTER 10 PRACTICE TEST

1. (a) 93.913°
   (b) 7π/12
2. (a) 140°, -580°
   (b) 25π/9, -11π/9

Answers to Practice Tests

3. $\sin\theta = y/r = -5/13$
   $\cos\theta = x/r = 12/13$
   $\tan\theta = y/x = -5/12$
   $\csc\theta = r/y = -13/5$
   $\sec\theta = r/x = 13/12$
   $\cot\theta = x/y = -12/5$

4. $\theta = 0, \pi/2, 3\pi/2$

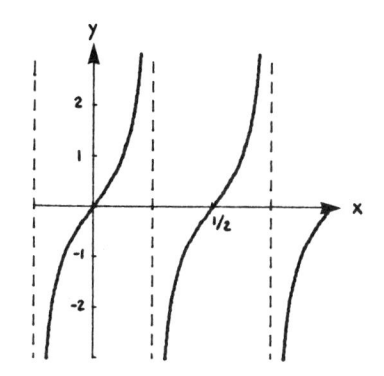

5. (a) Period: 8π
   Amplitude: 3

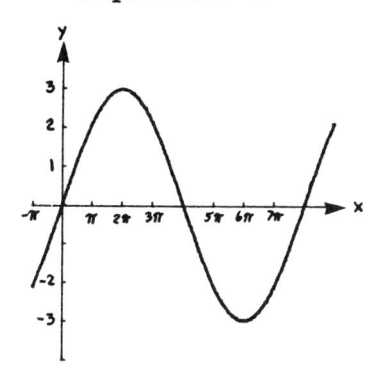

   (b) Period: $1/2$

6. $3(1 + \sin x)$

7. $x(x \sec^2 x + 2 \tan x)$

8. $3 \sin^2 x \cos x$

9. $\dfrac{\sec x(x \tan x - 2)}{x^3}$

10. $5 \cos 10$

11. $-\dfrac{1}{2}\sqrt{\csc x}\,\cot x$

12. $\sec x$

13. $-2e^{2x} \csc^2 e^{2x}$

14. $3 \sec(x^2 + y) - 2x$

15. $-\dfrac{1}{3}\sin^2 3y \sec^2 x$

16. $\dfrac{1}{4}\sin 4x + C$

17. $-8 \cot \dfrac{x}{8} + C$

18. $-\dfrac{1}{2}\ln|\cos x^2| + C$

19. $\dfrac{\sin^6 x}{6} + C$

20. $-\ln|\csc x - \cot x| + \cos x + C$

21. $e^{\tan x} + C$

22. $-\ln|1 + \cos x| + C$

23. $2 \tan x - 2 \sec x - x + C$

24. $x \sin x + \cos x + C$

25. $\dfrac{\pi^2 - 8\sqrt{2}}{16}$

1 2 3 4 5 6 7 8 9 10